HAZARDOUS WASTE SITE REMEDIATION
Source Control

Domenic Grasso
Associate Professor
Departments of Civil and Chemical Engineering
University of Connecticut
Storrs, CT

LEWIS PUBLISHERS
Boca Raton Ann Arbor London Tokyo

The material contained in this document was prepared as a course pack, under a federal grant from the United States Environmental Protection Agency (U.S. EPA), to assist university professors in preparing notes on hazardous waste site remediation. Consequently, narrative portions of the book may be somewhat incomplete.

The material in this document has not been subject to EPA policy review. The views expressed by individual authors are their own and do not necessarily reflect those of the U.S. EPA. Mention of trade names, products, or services does not convey, and should not be interpreted as conveying, official EPA approval, endorsement, or recommendation.

This document was produced through The University of Connecticut, under the direction of Professor Domenic Grasso, through EPA Cooperative Agreement #X-818009-01-0, Thomas R. De Kay, Ph.D., U.S. Environmental Protection Agency, Washington, D.C. 20460, Project Officer.

Catalog information is available from the Library of Congress.
ISBN 1-56670-056-6

Lewis Publishers is an imprint of CRC Press

Direct all inquiries to CRC Press, Inc., 2000 Corporate Blvd., N.W.,
Boca Raton, Florida 33431.

PRINTED IN THE UNITED STATES OF AMERICA
1 2 3 4 5 6 7 8 9 0
Printed on acid-free paper

Table of Contents

PREFACE

The remediation of hazardous waste sites is an area of major environmental concern. The variability in character of contaminated sites and associated safety considerations make their solution both unique and challenging. It is incumbent upon environmental engineering educators to properly prepare practitioners to solve these remediation problems. The evolution of sanitary engineering into a more holistic environmental engineering discipline has proceeded at varying rates at different universities. Consequently, the ability of existing programs to address the more unique problems associated with hazardous waste treatment may not be equally developed among various universities. In fact, some schools may not offer any appropriate course work in the area of hazardous waste treatment. Although many of the fundamental principles stressed in traditional sanitary engineering programs are applicable to the solution of more contemporary hazardous waste problems, the actual implementation of these principles is not always clear nor directly transferable.

The development of new course material is a tedious and time-consuming process. Under a grant from the U.S. Environmental Protection Agency, The University of Connecticut undertook a project involving an extensive review of published and unpublished literature to prepare a complete graduate-level course pack. The course pack focuses on a quantitative approach to source control and does not cover traditional processes (e.g., GAC, air stripping) that are most probably addressed in other courses. Hazardous waste specific processes such as soil vapor extraction, soil washing, bioremediation, and thermal processes are stressed. Although the course presumes knowledge of fundamental environmental engineering principles, a short review of basic concepts such as organic geochemistry, material balance, mass transfer, thermodynamics, and kinetics as they pertain to the remediation of hazardous waste problems precedes the discussion of specific treatment processes.

Information utilized in the preparation of this course pack was based, in large measure, on materials available through the Center for Environmental Research Information, USEPA, Cincinnati, OH. Additional information, such as pilot study results and supporting documentary videotapes were obtained from other organizations involved in hazardous waste clean-up, such as the US Army Toxic and Hazardous Materials Agency. Treatment processes covered incorporate, to the extent possible, process description and theory, design protocol and considerations, performance expectations, safety issues, and economic considerations.

The work was conducted under the overall guidance of the USEPA. At EPA, special thanks are extended to Dr. Walter Kovalick, Director of the Technology Innovation Office, for the foresight to fund the project, Dr. Thomas DeKay for guidance and administrative assistance, and Mr. Edward Barth and Mr. George Huffman for review of technology modules. The Association of Environmental Engineering Professors (AEEP) agreed to assist in identifying a board of advisors who are recognized experts in hazardous waste treatment. Board members consisted of Professor Thomas Keinath, Clemson University; Professor Richard Luthy, Carnegie-Mellon University; Professor Richard Magee, New Jersey Institute of Technology; Professor Neil Hutzler, Michigan Technological University; Professor Walter J. Weber, Jr.; The University of Michigan; and, Professor Kenneth Williamson, Oregon State University. I would like to thank the AEEP reviewers for their helpful comments. Professor Mackenzie Davis at Michigan State University, provided detailed comments that were greatly appreciated on a variety of modules. Additionally, in order to provide a practice perspective, Environ Corporation reviewed technology modules and provided copious comments. I would like to thank Mr. William Stone and all the the reviewers at Environ. Environ was also the senior contributor to the Risk Assessment Module.

Many students, both graduate and undergraduate, contributed in no small fashion to the realization of this final product. Some served as typists and proofreaders, others as draftsmen, while others were major contributors to the composition of the individual modules. I would like to thank those students who coauthored the individual modules: Messrs. Rajat Garg and Harish Pesari (Bioremediation), Ms. Heide-Marie Dudek (Chemical Destruction and Soil Washing), Mr. Daniel Servetas (Thermal Processes), and Mr. David Harvey (Critical Issues in Technology Selection). Professor George Hoag (University of Connecticut) was also very helpful in coauthoring the Soil Vapor Extraction (SVE) Module and the Critical Issues in Technology Selection Module. I am indebted to Professor David Ahlfeld (University of Connecticut) for his assistance in composing the groundwater portion of the Engineering Principles Module and Dr. Craig Myler (U.S. Military Academy) for his assistance in composing the Materials Handling Module. I am also thankful to Dr. Paul Johnson of Shell Development for allowing the use of HyperVentilate in the SVE Module. Dr. Rakesh Govind (University of Cincinnati) assisted in the Bioremediation Module. Mr. Mark Wilkins was helpful in preparation of figures and diagrams for the modules. I am especially grateful to Mr. Alfred Kovalick and Ms. Colleen Martin. They were indispensable in the preparation of the final manuscript.

An additional module on waste minimization has also been added. This module is the result of a project conducted at Clarkson University under the direction of Professors Anthony Collins and Joseph DePinto and sponsored by the New York Center for Hazardous Waste Management. In addition to a project description a didactic interactive computer program is also provided.

I would finally like to thank Professor George Hoag, Director of the Environmental Research Institute at Connecticut and the entire ERI staff for going beyond the call of duty in assisting in the completion of this project.

Domenic Grasso
Storrs, Connecticut

A – Regulations

RESOURCE CONSERVATION AND RECOVERY ACT (RCRA)

Instructional objectives:

- List goals of RCRA
- Differentiate between solid waste and hazardous waste and consider exclusions
- Discuss and determine different generator categories
- Compare differences in regulations among generators, TSDFs, and transporters
- Discuss procedure for tracking hazardous wastes from source to disposal
- Review processes involved in corrective action
- Examine purpose of RCRA Land Disposal Restrictions
- Cover enforcement actions available to EPA

1. GENERAL

In 1976, Congress enacted the Resource Conservation and Recovery Act (RCRA), an amendment to the Solid Waste Disposal Act of 1965 (SWDA), which called for regulation by the EPA of solid and hazardous waste disposal. The statute can be found at P.L. 94-580, 90 Stat. 2795 (1976), 42 USC 6901 *et seq*.. RCRA worked as an enabling statute giving general directives within the statute while assigning EPA primary regulatory authority.

The goals Congress set in RCRA include:

- Protection of human health and the environment
- Reduction of waste and conservation of energy and natural resources
- Reduction or termination of hazardous waste generation

Table 1.1 lists the subtitles and the sections of the statute under which they are covered.

Table 1.1: Subtitles of RCRA 1976

RCRA as enacted in 1976		Sections under P.L. 94-580
Subtitle A	General provisions	1001-1008
Subtitle B	Office of Solid Waste; Authorities of the administrator	2001-2006
Subtitle C	Hazardous waste management	3001-30011
Subtitle D	State or regional solid waste plans	4001-4009
Subtitle E	Duties of the secretary of commerce in resource and recovery	5001-5004
Subtitle F	Federal responsibilities	6001-6004
Subtitle G	Miscellaneous provisions	7001-7009
Subtitle H	Information	8001-8007

In 1984, a significant series of amendments to RCRA were enacted entitled Hazardous and Solid Waste Amendments (HSWA). The amendments were created in response to EPA's inability over the previous eight years to promulgate effective and satisfactory regulations. HSWA gave more specific guidelines for regulations, established a timetable for promulgation reinforced by the installation of Minimum Regulatory Controls (MRCs or "hammers"), and expanded hazardous waste regulation. Each of these measures was taken to promote the rapid promulgation of regulations. For example, specificity within the Act, to the point of resembling regulations, precludes the need for extensive research by the EPA to justify its regulations. The MRCs or "hammers", harsh penalties for violators which would be triggered in the event EPA failed to issue regulations by a certain date, are substituted for the complete lack of regulation.

The hazardous waste management program was expanded in response to increased public awareness and newly acquired information on hazardous waste health risks and environmental effects.

HSWA amended many of the 1976 provisions and added some new sections as shown in Table 1.2 below.

Table 1.2: Subtitle C of RCRA with HSWA Amendments and Additions

RCRA section number	Content of section	HSWA section number	Amendments
3001	Identification and listing of hazardous waste	221	Small quantity generator waste
		222	Listing and delisting of hazardous waste
		223	Clarification of household waste exclusion
3002	Standards applicable to generators of hazardous waste	224	Waste minimization
3003	Standards applicable to hazardous waste transporters		
3004	Standards applicable to hazardous waste treatment, storage, and disposal facilities	201	Landfill disposal of hazardous waste
		202	Minimum technological requirements
		203	Groundwater monitoring
		204	Burning and blending of hazardous wastes
		205	Direct action (financial responsibility)
		206	Continuing releases at permitted facilities (corrective action)
		207	Corrective action beyond facility boundaries; Underground tanks
		208	Financial responsibility for corrective action
		209	Mixing wastes and other special wastes
3005	Permits for treatment, storage, and disposal of hazardous waste	211	Authority for permit to construct hazardous waste treatment, storage, and disposal facility
		212	Permit life
		213	Interim Status
		214	New and innovative treatment technologies
		215	Existing (interim status) surface impoundments
		224	Waste minimization

...continued

Table 1.2: Subtitle C of RCRA with HSWA Amendments and Additions (continued)

RCRA section number	Content of section	HSWA section number	Amendments
3006	Authorized state hazardous waste programs	225	Basis of authorization
		226	Availability of information
		227	Interim authorization of state programs
		228	Application of amendments to authorized states
3007	Inspections	229	Federal facilities
		230	State-operated facilities
		231	Mandatory inspections
3008	Federal enforcement	232	Federal enforcement (knowing endangerment)
		233	Interim status corrective action orders
3009	Retention of state authority		
3010	Effective date	234	Effective date of regulations
3011	Authorization of assistance to States		
	New sections in Subtitle C		
3014	Restrictions on recycled oil	241	Management of used oil
		242	Recovery and recycling of used oil
3015	Expansion during interim status		
3016	Inventory of Federal agency hazardous waste facilities		
3017	Export of hazardous waste		
3018	Domestic sewage		
3019	Exposure information and health assessments		

2. RCRA SUBTITLE C PROGRAM

2.1. General

Subtitle C of RCRA is divided into six major components:

- Requirements of identifying and listing hazardous wastes
- A system for tracking hazardous waste from "cradle to grave"
- Standards for generators and transporters of hazardous waste and for hazardous waste treatment, storage and disposal facilities
- A permitting program for TSDFs
- Authorization for states to administer the hazardous waste program
- Provisions concerning enforcement

2.2. Identification of a Hazardous Waste

2.2.1. General

Hazardous wastes are a subset of solid wastes, both of which have precise regulatory definitions under RCRA. 40 CFR Part 261 provides the regulatory definitions for solid wastes and hazardous wastes.

2.2.2. Definition of a Solid Waste

In order for a waste to be considered a hazardous waste it must first be classified as a solid waste. A solid waste is defined by the Act as the following:

> Any garbage, refuse, sludge from a waste treatment plant, water supply treatment plant or air pollution control facility and other discarded material including solid, liquid, semisolid or contained gaseous material resulting from industrial, commercial, mining and agricultural operations and from community activities §6903(27)

The regulatory definition is found at 40 CFR 261.2(a)(1). A solid waste is defined by the regulations as any "discarded material" not expressly excluded by the regulations. A material is considered "discarded" if it is "abandoned", "recycled", or "considered inherently waste-like."

Specific wastes are excluded from the definition of solid waste:

- Domestic sewage
- Wastes that are discharged to a publicly-owned sewage treatment works
- Discharges of industrial wastewater subject to regulation under the Clean Water Act
- Certain secondary materials that are reclaimed or reused

Congress excluded the following wastes:

- Oil and gas production wastes
- Mining wastes
- Wastes from the combustion of coal or other fossil fuels
- Cement kiln dust waste

If any of these wastes contains compounds that are not inherently present in the wastes, they could then be considered a hazardous waste.

Also excluded are:

- Household waste
- Solid waste returned to the soils as fertilizers
- Specified chromium-containing wastes (40 CFR Part 261.4(b)(6)
- Discarded arsenical-treated wood
- Certain residues of hazardous waste in empty containers

2.2.3. Definition of a Hazardous Waste

A hazardous waste is defined in the statutes as:

"a solid waste, or combination of solid wastes, which because of its quantity, concentration, or physical, chemical, of infectious characteristics may:

a) "cause, or significantly contribute to, an increase in mortality or an increase in serious irreversible, or incapacitating reversible, illness;" or

b) "pose a substantial present or potential hazard to human health or the environment when improperly treated, stored, transported, or disposed of, or otherwise managed."

A material may be classified as a hazardous waste if it meets the criteria of any of the following four categories:

- Listed wastes
- Characteristic wastes
- Mixed wastes
- "Derived from" wastes

2.2.3.1. Listed Wastes

Listed wastes are wastes specifically regulated by the EPA and appear on one of four lists found in 40 CFR 261.3. The four lists fall into three categories of wastes.

Nonspecific source wastes (40 CFR 261.31) – generic wastes, commonly produced by manufacturing and industrial processes

Specific source wastes (40 CFR 261.32) – wastes from specifically identified industries such as wood preserving and petroleum refining

Commercial chemical products (40 CFR 261.33(e) and (f)) – wastes that include specific commercial chemical products or manufacturing chemical intermediates such as creosote and kepone

Appendix VIII to 40 CFR Part 261 lists the hazardous constituents EPA checks for when deciding whether to list a hazardous waste under RCRA. The presence of an Appendix VIII chemical in a waste does not insure that the waste will be listed. Each waste requires a separate regulatory action and is determined based on eleven factors such as concentration, persistence, and degradation of the constituent in the waste.

There are several other criteria by which wastes are determined to be hazardous. The waste may be listed if it:

- Exhibits one of the four characteristics of a hazardous waste: ignitability, corrosivity, toxicity, and reactivity (see section 2.2.3.2 for further discussion)
- Proves to be acutely hazardous based on health assessments
- Belongs to a type or class of waste, which, according to the Administrator, is typically or frequently hazardous under the definition of hazardous waste found in Section 1004(5) of the Act

2.2.3.2. Characteristic Wastes

Wastes may be regulated for exhibiting one of the four characteristics of a hazardous waste as determined by the EPA.

Ignitability – A solid waste that exhibits any of the following properties is considered a hazardous waste due to ignitability:

- A liquid, except aqueous solutions containing less than 24% alcohol, that has a flash point less than 60°C
- A nonliquid capable, under normal conditions, of spontaneous and sustained combustion
- An ignitable compressed gas under Department of Transportation (DOT) regulations
- An oxidizer under DOT regulations

Corrosivity – A solid waste which exhibits any of the following properties is considered a hazardous waste due to corrosivity:

- An aqueous material with pH less than or equal to two or greater than or equal to 12.5
- A liquid than corrodes steel at a rate greater than 1/4 inch per year at a temperature of 55°C

Reactivity – A solid waste exhibits the characteristic of reactivity if a representative sample of the waste has any of the following properties:

- It is normally unstable and readily undergoes violent change without detonation
- It reacts violently with water
- It forms potentially explosive mixtures with water
- When mixed with water, it generates toxic gases, vapors, or fumes in a quantity sufficient to present a danger to human health or the environment
- It is a cyanide or sulfide bearing waste, which, when exposed to pH conditions between 2 and 12.5, can generate toxic gases, vapors or fumes in a quantity sufficient to present a danger to human health or the environment
- It is capable of detonation or explosive reaction if subjected to a strong initiating source or if heated under confinement
- It is readily capable of detonation or explosive decomposition or reaction at standard temperature and pressure
- It is a forbidden explosive as defined in 49 CFR 173.51, a Class A explosive as defined in 49 CFR 173.53, or a Class B explosive as defined in 49 CFR 173.88

Toxicity – A solid waste exhibits the characteristic of toxicity if, using approved methods (Toxicity Characteristic Leaching Procedure (TCLP)), the extract from the waste contains any of the contaminants listed in a concentration equal to or greater than the values listed Table 2.1.

Table 2.1: Toxicity Characteristic Constituents and Regulatory Levels for the TCLP Test *

Constituent	Regulatory Level (mg/L)
arsenic	5.0
barium	100.0
benzene	0.5
cadmium	1.0
carbon tetrachloride	0.5
chlordane	0.03
chlorobenzene	100.0
chloroform	6.0
chromium	5.0
o-cresol **	200.0
m-cresol **	200.0
p-cresol **	200.0
cresol **	200.0
2,4-D	10.0
1,4-dichlorobenzene	7.5
1,2-dichloroethane	0.5
1,1-dichloroethylene	0.7
2,4-dinitrotoluene	0.13
endrin	0.02
heptachlor (and its hydroxide)	0.008
hexachlorobenzene	0.14
hexachloroethane	3.0
lead	5.0
lindane	0.4
mercury	0.2
methoxychlor	10.0
methyl ethyl ketone	200.0
nitrobenzene	2.0
pentachlorophenol	100.0
pyridine	5.0
selenium	1.0
silver	5.0
tetrachloroethylene	0.7
toxaphene	0.5
trichloroethylene	0.5
2,4,5-trichlorophenol	400.0
2,4,6-trichlorophenol	2.0
2,4,5-TP (silvex)	1.0
vinyl chloride	0.2

* Procedure and Levels promulgated March 29, 1990

** If o-, m-, p-cresol concentrations cannot be differentiated, the total cresol concentration is used

2.2.3.3. Mixed wastes

The entire volume of any waste mixture containing a listed hazardous waste, regardless of concentration, is considered a hazardous waste with the following exceptions:

- The listed waste in the mixture was listed solely because it exhibits a hazardous waste characteristic and the mixture no longer exhibits that characteristic
- A wastewater discharge subject to regulation by the Clean Water Act and the hazardous waste mixed with the wastewater is one of the following:
  - Mixtures of nonhazardous wastes and characteristic hazardous wastes if the mixture no longer exhibits any of the characteristics
  - Concentrations of carcinogenic and noncarcinogenic spent solvents listed in 40 CFR Part 261.31 in the wastewater, provided the concentration does not exceed 1 ppm and 25 ppm respectively
  - *De minimis* losses of discarded commercial chemical products or intermediaries used as raw materials in manufacturing or produced as by-products
  - Laboratory wastewater containing small amounts of listed toxic wastes
  - Heat exchanger bundle cleaning sludge (petroleum refining industry)

2.2.3.4. "Derived From" Wastes

EPA's "derived from" rule regulates hazardous waste, which is defined as "any solid waste generated from the treatment, storage, or disposal of a hazardous waste; including any sludge, spill residue, ash, emission control dust or leachate (but not including precipitation run-off)" 40 CFR 261.4 (c)(2)(i), with specified exclusions.

2.2.4. Hazardous Waste Delisting

Delisting of a hazardous waste falls under the category of General Rule making, which is discussed in 40 CFR 260.2 (Subpart C). Rule making petitions allow the public and regulated community to modify or revoke regulations including:

- Adding a testing or analytical method for identifying a hazardous waste
- Excluding (delisting) a waste produced at a particular facility

Delisting a hazardous waste for a particular facility requires a review of:

- The criteria for which the waste was originally listed
- Any other factors or constituents that the Administrator has reasonable basis to believe may cause the waste to be hazardous

The petitioner bares the responsibility for proving the waste is not hazardous based on the constituents for which it was originally listed; EPA must determine if the waste is hazardous for other reasons. EPA must have reasonable basis for requesting additional information or denying a petition. A Notice and Comment period is also required prior to granting or denying a petition on the basis of additional factor or constituents.

Delisting under RCRA requires specific procedures for submitting and evaluating petitions, analyzing data, and proper sampling protocols.

Delisting decisions incorporate data quantitatively and qualitatively. To facilitate evaluation, EPA employs an analytical approach using a groundwater transport model—the Vertical and Horizontal

Spread model (VHS)—that can be used to reasonably predict a worst-case contamination of groundwater.

The VHS model attempts to simulate the dispersion of toxicants in the vertical and horizontal direction in an aquifer. The model incorporates several reasonable worst-case assumptions by EPA definition. The overall approach is used to predict reasonable worst-case contaminant levels in the groundwater of receptor wells while estimating the ability of an aquifer to dilute the toxicant from a specific volume of waste. The concentration of listed metal constituents derived from the extraction of a given waste material is then compared to health-based standards for these constituents.

2.3. Generator Status

2.3.1. General

A generator is an entity whose act or process produces a hazardous waste, or whose act first causes a hazardous waste to become subject to regulation. Generator regulations are set forth in 40 CFR Part 262.

There are three categories of generators:

- A "large quantity generator" (LQG) produces more than 1000 kg of hazardous waste or 1 kg of acutely hazardous waste in a calendar month
- A "small quantity generator" (SQG) produces between 100 and 1000 kg of hazardous wastes in a calendar month
- A "conditionally exempt small quantity generator" produces no more than 100 kg of hazardous waste in a calendar month

"Acute" hazardous wastes are identified in sections 261.31, 261.32 and 261.33(e) of the regulations. If a generator produces more than 1 kg of acute hazardous waste in a calendar month, or more than 100 kg of residue, contaminated soil, waste or other debris resulting from a spill of an acute hazardous waste, that waste is subject to full regulation 261(f).

Generator status is determined on a month to month a basis. For example, a generator that produces 90 kg of hazardous waste one month is conditionally exempt, but if 1200 kg of hazardous waste is produced the next month, the generator is subject to the full LQG regulations.

2.3.2. General Regulatory Requirements

The regulatory requirements for hazardous waste generators include:

- Obtaining an EPA ID number
- Observing proper handling of hazardous waste before transport
- Following accumulation and storage requirements
- Manifesting of hazardous waste
- Following record-keeping and reporting procedures

2.3.3. Large and Small Quantity Generator Requirements

2.3.3.1. EPA Identification Number

Large and small quantity generators cannot treat, store, dispose of, transport, or offer for transportation, hazardous waste without having received an EPA Identification (ID) number. Nor can they offer the hazardous waste to transporters or to treatment, storage, or disposal facilities that do not have an EPA ID number (40 CFR Part 262.12).

2.3.3.2. Handling Before Transportation

Pre-transport regulations apply only when generators ship waste off-site. Requirements include:

- Proper packaging to prevent leakage of hazardous waste during both normal and potentially dangerous transport conditions
- Labeling, marking and placarding of packaged wastes to identify the characteristics and dangers associated with transportation of these wastes

2.3.3.3. Accumulation of Waste

These regulations (40 CFR Part 262.3) cover the packaging, labeling, marking, and placarding of hazardous waste after it has been generated. For SQGs, hazardous waste may be accumulated on-site for 180 days or less as long as less than 6,000 kg is accumulated on-site; however, longer periods, up to 270 days, are allowed if shipment is over 200 miles to TSDF (40 CFR 262.34 (e)). LQGs may accumulate hazardous waste for 90 days.

Generators must:

- Properly store waste
- Have an emergency plan
- Have personnel training

2.3.3.4. Manifesting

The Manifest system is the realized form of "cradle-to-grave" hazardous waste management. The manifest, or Uniform Hazardous Waste Manifest (UHWM), is a waste shipment invoice that accompanies the waste from point of origin to point of final disposal. Generators are required, with some exceptions, to use manifests when transporting hazardous waste off-site for treatment, storage, or disposal (40 CFR Part 262.20). RCRA/HSWA manifests must contain:

- Name and EPA identification number of the generator, the transporter(s), and the facility where the waste is to be treated, stored, or disposed
- Department of Transportation (DOT) description of the waste being transported
- Amount of waste being transported
- Address of the treatment, storage, or disposal facility where the waste is being sent
- Certification that the generator has a waste reduction program
- Certification that the treatment, storage or disposal method chosen is the most practicable method available

Within 30 days of the arrival of the waste with its manifest, the TSDF must send a copy of the manifest with all the appropriate signatures and information to the original generator. LQGs must issue an investigation and notify the EPA, called "exception reporting", if a copy of the manifest fails to arrive.

Compliance with the manifest tracking system requires:

- Transporters and TSDFs should not accept waste not accompanied by a manifest
- Manifests must be signed upon acceptance and a copy must be retained by the generator, transporter, or TSDF
- Copies of manifests must be retained by the generator, transporter, or TSDF for three years

2.3.3.5. Record-Keeping and Reporting

Small Quantity Generators (SQG) are required to:

- Retain copies of records, manifests, and reports for three years
- Retain records of test results and waste analysis, or any other records made in accordance with the hazardous waste determination 40 CFR part 262.40

Large Quantity Generators (LQG) are required to:

- Provide biennial reports which identify the generator, facilities that managed the generators waste, and waste transporters; describe the waste sent off-site; and discuss the measures taken toward reducing the volume and toxicity of wastes generated and the results of those efforts
- Provide exception reporting
- Retain three years of reports, manifests, and test records

2.3.5. Conditionally Exempt Generators

Conditionally exempt generators are required to:

- Identify the waste to determine whether it is a hazardous waste
- Not accumulate more than 1,000 kg of hazardous waste at any time
- Treat or dispose of waste on site, or ensure waste is sent to a:
  - — Permitted or interim status TSDF
  - — Permitted municipal or industrial solid waste facility
  - — Recycling facility

2.4. Transporter Standards

Transporters are regulated both by the EPA and the U.S. Department of Transportation (DOT) through the Hazardous Materials Transportation Act (P.L. 93-633, 49 U.S.C. 1801 et seq.). DOT regulations are set forth in volume 49 of the CFR and cover preparation for shipment, regulating container manufacturing, and transportation, including highway, railway, water, and air. The procedures are specific to the type of waste and the mode of transportation. As waste characterization, packaging, and labeling are essential to compliance with DOT regulations, the responsibilities of the generator and transporter overlap.

EPA transporter regulations, set forth in 40 CFR Part 263, deal primarily with manifest tracking and discharge cleanup. The transporter is responsible for following the manifest tracking system by not accepting waste without a manifest, insuring the manifest is signed by the accepting TSDF, and retaining a copy of the manifest for three years. Transporters are expected to be prepared to handle an accidental discharge to reduce or prevent harm to human health or the environment and will accept limited responsibility for the clean-up of contamination.

2.5. Treatment, Storage, and Disposal Facilities (TSDF)

2.5.1. General

TSDF regulations are set in 40 CFR Parts 264 and 265. Subtitle C requires all TSDFs handling hazardous waste to obtain an operating permit and abide by the TDSF regulations and performance standards. The definition of a TSDF follows (260.10):

> Treatment – Any method, technique, or process, including neutralization, designed to change the physical, chemical, or biological character or composition of any hazardous waste so as to neutralize it, or render it non-hazardous or less hazardous, or to recover it, make it safer to transport, store, or dispose of, or amenable for recovery, storage or volume reduction.
>
> Storage – The holding of hazardous waste for a temporary period, at the end of which the hazardous waste is treated, disposed of, or stored elsewhere.
>
> Disposal – The discharge, deposit, injection, dumping, spilling, leaking, or placing of any solid waste or hazardous waste into or on any land or water so that any constituent thereof may enter the environment or be emitted into the air or discharge into any waters, including ground waters.

2.5.2. Regulatory Requirements

Facilities submit applications for either a Part A or Part B permit according to their circumstances. Part A permits are grandfather permits for facilities existing on November 19, 1980, or for facilities that treat wastes that have been regulated since November 19, 1980. A Part A permit grants these existing facilities interim status, allowing them to continue operating until a final permit can be issued. Interim status standards are presented in 40 CFR Part 265.

Part B is a final permit that replaces a Part A or is a permit for a new facility. No specific form exists for a Part B application. Rather, Part B application requirements are the final regulations for TSDFs set forth in Part 264 and are specific to the type of TSDF seeking a permit. The applying TSDF is required to provide general information as well as information specific to the facility location and type. All Part A permits will eventually be replaced by Part B permits.

Interim status standards cover primarily administrative concerns. The administrative and non-technical requirements apply to both interim and permitted facilities; technical requirements are applied to finally permitted facilities and are specific to the type of facility.

2.5.2.1. Administrative and Non-Technical Requirements

Administrative and non-technical requirements are presented in Subparts A through E of 40 CFR Parts 264 and 265. Subpart A defines who is subject to the regulations. Subpart B specifies facility standards. Subparts C and D set forth the requirements for the preparedness and prevention, and the contingency plan, and emergency procedures. Subpart E sets forth the manifest system, record-keeping, and reporting requirements.

General administrative requirements, which must be met by all TSDFs requiring a RCRA permit, are listed in the following:

- All facilities must obtain an EPA ID number
- A general waste analysis must be conducted before a waste is treated, stored, or disposed of at a TSDF
- Security systems must be installed to prevent unauthorized entrance into the active portions of the TSDF
- Operators must schedule regular, preventative inspections as appropriate to the specific facility and within the minimum limit set by the regulations
- Personnel must be trained in their areas and in emergency procedures
- Special procedures and precautions should be used when handling ignitable, reactive, or incompatible wastes
- Locations standards must be met regarding salt domes and underground mines for all TSDFs and floodplains an seismic restrictions for finally permitted facilities
- Facilities must take measures to prevent or prepare for the event of an emergency, including a set contingency plan and emergency procedures
- Facilities must comply with manifest, record-keeping, and reporting requirements
- Closure and Post-closure requirements, including closure and post-closure plans (groundwater monitoring, facility security, and general maintenance) must be prepared and approved
- Proof of adequate financial resources to compensate for third-party liability and post-closure costs is required, including third-party liability insurance for sudden and non-sudden occurrences

2.5.2.2. Technical Requirements

Technical requirements relate to the specific type of TSDF being regulated. Regulations are established for interim status and finally permitted facilities according to the following divisions:

- Containers
- Hazardous waste tanks
- Surface impoundments
- Landfills
- Waste piles
- Land treatment units
- Incinerators
- Other thermal treatment units
- Chemical, physical, and biological treatment units
- Underground injection wells

Technical standards are discussed in 40 CFR Parts 264 and 265 Subparts F through BB.

2.6. The Corrective Action Process

2.6.1. General

In HSWA, Congress enables the EPA to address releases of hazardous waste through corrective action. Corrective action requirements are imposed through permits or enforcement orders. Permits issued to RCRA facilities must, at a minimum, contain schedules of compliance to address releases and include provisions for financial assurance to cover the cost of implementing the corrective measures. EPA has estimated that corrective action at RCRA facilities could encompass about 2,000 TSDFs, take until the year 2025 to complete, and be as large and costly as the current Superfund program.

2.6.2. Definitions

Solid Waste Management Unit (SWMU) – waste management units from which hazardous wastes or constituents may migrate, even if the unit was not intended for the management of hazardous waste. Additionally, any areas that become contaminated as a result of routine and systematic releases from process/industrial discharge are SWMUs.

Regulated Units – a subset of SWMUs, a regulated unit is any surface impoundment, waste pile, land treatment unit, or landfill that received waste after July 26, 1982.

Hazardous constituents – Substances which are listed in 40 CFR Part 261, Appendix VIII.

2.6.3. Scope of Corrective Action

The scope of the corrective action required may vary between regulated units at permitted facilities and other solid waste management units at permitted or interim status facilities. Releases to ground water from regulated units are set forth under 40 CFR Part 264, Subpart F. Corrective action for releases of hazardous wastes or constituents from any SWMU at a TSDF that is seeking or subject to a RCRA permit is set forth in sections 3004(u) and (v) of RCRA (40 CFR 264.101). Section 3004(v) authorizes the EPA to impose corrective action requirements for releases which migrated beyond facility boundaries. Section 3008(h) authorized the EPA to require corrective action whenever there is or has been a release of hazardous waste or constituents from an interim status RCRA facility.

Section 3004(u), added under HSWA, allows the EPA to require corrective action for all releases of hazardous waste constituents from "waste management units" regardless of when the waste was placed in the units. This includes releases resulting from placement of waste in a waste management unit by a previous owner or operator of the facility. It was a significant expansion of the EPA's authority to require corrective actions at all solid waste management units at facilities seeking a RCRA treatment, storage, or disposal permit, as distinguished from limiting the remedial measures to hazardous waste management units only.

2.6.4. Correction Action Components

2.6.4.1. General

The corrective action process has four main components, each of which comprises a number of steps. The decision as to which steps must be observed is made by the EPA on a facility-by-facility basis.

2.6.4.2. RCRA Facility Assessment (RFA)

The RCRA facility assessment, often performed by the EPA, is a preliminary investigation. The objectives of the RFA are:

- Identification of SWMUs and collection of existing information on contaminant releases
- Identification of releases or suspected releases needing further investigation.

The RFA includes a review of pertinent existing information at the facility. If necessary, a visual site inspection and sampling visit are performed.

Based upon the finding of the RFA, one or more of the following will occur:

- No further action
- A RCRA Facility Investigation (RFI), comparable to RI under Superfund, by the facility owner or operator is required where the information collected indicates a release(s) or suspected release(s) warrants further investigation
- Interim corrective measures by the owner or operator are required where the regulatory agency believes that expedited action should be taken to protect human health and the environment
- In cases where problems associated with permitted releases are found, the regulatory agency will refer such releases to the appropriate permitting authorities.

2.6.4.3. RCRA Facility Investigation (RFI)

The RCRA Facility Investigation may be required either under a permit schedule of compliance or an enforcement order. Based on the results of the RFI, the regulatory agency will outline, in detail, specific conditions in the permits or enforcement orders, often performed by the responsible party. The RFI may range from an examination of a small specific activity to a complex multimedia study. If the investigation verifies a release, further characterization of the release is required.

RFI includes interpretation by the regulatory agency of release characterization data against established health and environmental criteria to determine whether a Corrective Measure Study (CMS) is necessary. Identifying and implementing interim corrective measures may also be conducted during the RFI.

2.6.4.4. Corrective Measures Study (CMS)

If the potential need for corrective measures is determined during the RFI process, the owner or operator is then responsible for performing a Corrective Measures Study (CMS), analogous to FS under Superfund. The owner or operator will identify and recommend specific measures that will correct the release. Information generated during the RFI will be used to aid in the selection and implementation of these measures. Criteria used to assess CMS and recommended corrective action schemes are proposed in the Federal Register.

2.6.4.5. Corrective Measures Implementation (CMI)

CMI includes designing, construction, operating, maintaining, and monitoring selected corrective measures. If the remedy is not properly implemented, the EPA will direct the facility to take additional action. Financial assurance must be demonstrated prior to implementation.

RCRA authorizes the EPA to impose, upon the owner/operator of a waste disposal facility seeking a full RCRA permit, corrective action obligations for any release of hazardous waste into the environment, regardless of the time at which waste was placed in the facility and regardless of whether the current owner/operator was responsible for the release. For interim status facilities, RCRA allows EPA to order corrective actions for any release into the environment. "Corrective action" requires a cleanup to satisfactory standards according to the EPA.

An owner/operator of a closed facility may be subject to EPA authority to clean up these facilities even though that owner/ operator was not the party responsible for the activity that originally subjected the property to regulation.

Section 3004(v) allows EPA to require owners and operators of permitted hazardous waste facilities to clean up contamination that has migrated beyond the property boundary. Owners/ operators of any disposal facility may be required, pursuant to section 3013 of RCRA, to monitor, analyze, test, and report to the EPA on the nature and extent of hazards that the EPA suspects.

2.7. Enforcement

2.7.1. General

There are two essential aspects of the enforcement program (this section from k-1):

- Compliance monitoring
- Enforcement actions

2.7.2. Compliance Monitoring

The two primary methods of collecting compliance monitoring data are:

- Inspection by State or EPA officials
  - Annually at all Federal or State-operated facilities
  - At least once every two years at each TSDF
- Six types of inspections:
  - Compliance Evaluation Inspection (CEI)
  - Case Development Inspection (CDI)
  - Comprehensive Ground-Water Monitoring Evaluation (CME)
  - Compliance Sampling Inspection (CSI)
  - Operations and Maintenance Inspection (O&M)
  - Laboratory Audit
- Examinations of the reports that each handler is required to submit

2.7.3. Enforcement Actions

The EPA, or authorized state, has the following enforcement options:

- Administrative Actions
- Civil Actions
- Criminal Actions

2.7.3.1. Administrative Actions

An administrative action is a non-judicial enforcement action taken by EPA or a State under its own authority. There are two types of administrative actions:

- Informal Actions – Any communication from an agency that notifies the handler of a problem. This includes notices of violation (NOV) and notices of deficiency (NOD)
- Administrative Orders – Administrative orders impose enforceable legal duties
  - Compliance Orders—Section 3008 of RCRA allows the EPA to issue orders which require any person who is not in compliance to take steps toward becoming so. The order can contain a penalty of up to $25,000/day for each day of non-compliance. The facility's permit or interim status may be suspended or revoked
  - Corrective Action Orders—Section 3008(h) allows the EPA to issue orders which require corrective action at an interim status facilities when there is evidence of a release of a hazardous waste into the environment. Past problems at RCRA facilities may be cleaned up using this mechanism.. Orders can suspend interim status and impose penalties of up to $25,000/day for each day of non-compliance
  - Section 3013 Orders—used when EPA finds a substantial hazard to human health and the environment
  - Section 7003 Orders— used when the handling of a nonhazardous or hazardous waste creates an imminent and substantial endangerment to health or the environment.

2.7.3.2. Civil Actions

A civil action is defined as a formal lawsuit against a person who has either failed to comply with some statutory or regulatory requirement or administrative order, or has contributed to a release of hazardous waste or constituents. There are four types of civil actions:

- Compliance action
- Corrective action
- Monitoring and analysis
- Imminent hazard

2.7.3.3. Criminal Actions

Seven acts identified in Section 3008 of RCRA are subject to criminal action and carry criminal penalties. The first six acts carry a penalty of up to $50,000 per day or from two to five years in jail. These acts are knowingly:

- Transporting waste to a non-permitted facility
- Treating, storing, or disposing of waste without a permit or in violation of a material condition of a permit or interim status standard
- Omitting important information from, or making a false statement in, a label, manifest, report, permit, or compliance document
- Generating, storing, treating, or disposing of waste without compling with RCRA's record-keeping and reporting requirements
- Transporting waste without a manifest
- Exporting a waste without the consent of the receiving country

The seventh criminal act, which carries a possible penalty of up to $250,000 or 15 years in prison for an individual or a $1,000,000 fine for a corporation, is knowingly:

- Transporting, treating, storing, disposing, or exporting of any hazardous waste in such a way that another person is placed in imminent danger of death or serious bodily injury.

3. SUBTITLE I: UNDERGROUND STORAGE TANKS

3.1. General

Regulation of underground storage tanks (USTs) was added under HSWA in Subtitle I. Presently, USTs are considered to be one of the major sources of groundwater contamination in the country.

Title VI of the 1984 Hazardous and Solid Waste Amendments to Subtitle I of RCRA added new provisions to Subtitle I regarding the regulation of USTs: 42 USC 6901 and RCRA 9001. Unlike the main body of RCRA, these regulations apply to USTs used for storage of useful materials in addition to those used for wastes.

3.2. Definition of an UST

RCRA defines a UST as "any one or combination of tanks (including underground pipes connected thereto) that is used to contain an accumulation of regulated substances, and the volume of which.... is ten percent or more beneath the surface of the ground." RCRA §9001(1).

Excluded from the definition of UST, and therefore from the regulations, are home heating oil tanks, septic tanks, otherwise regulated pipelines, USTs with less than an 1,100 gallon capacity used for noncommercial purposes, and tanks with fuel oil used for heating commercial and industrial establishments RCRA. §9001(1)(A).

3.3. UST Requirements

The EPA codified UST regulations in 40 CFR 280 that mandate leak detection systems, release reporting, record-keeping requirements, performance standards, and closure requirements. The regulations distinguish between "new" and "existing" tanks. Existing tanks are those that were installed prior to December 22, 1988. Depending on the age and condition of the existing tank, some requirements, such as leak detection systems and cathodic protection, must be met prior to 1998. All existing tanks will be required to meet the new performance and safety standards by December, 1998.

In brief, the new performance standards require all tanks and piping to have the following:

- Corrosion protection
- Leak detection systems
- Spill/overfill prevention systems

Additionally:

- All new tanks must meet proper installation requirements
- Owners and operators are required to comply with reporting and record-keeping requirements

- Owners must report releases, take immediate action to prevent further release, and identify and mitigate any further hazards. (Part 280.61). If a release occurs, further actions are prescribed by the regulations, including site characterization, free product removal, investigations for soil and groundwater cleanup, and a corrective action plan
- Owners must prove they have sufficient funds to cleanup any leaks and to compensate third parties for damage suffered as a result of the leak. For most owners, this means they must have insurance coverage for at least $1 million per occurrence (Part 280.93).

4. RCRA LAND DISPOSAL RESTRICTIONS

4.1. General

Land Disposal Restrictions, set forth in 40 CFR Part 266, were created in HSWA to redirect disposal options away from land disposal.

Land disposal restrictions insisted on improved treatment technologies for wastes, waste minimization, recycling, and complete prohibition of the disposal of wastes and minimized disposal of all others.

4.2. Types of Wastes

In HSWA, Congress divided, listed, and characterized wastes into five groups and set schedules for EPA to develop treatment standards for these groups.

- Solvents and Dioxins
- Liquid and solid "California List" Wastes
- "First-, Second-, and Third-Third" Wastes
- Newly Listed Wastes

Land disposal includes placement in any of the following:

- Landfill
- Surface impoundment
- Waste pile
- Injection well
- Land treatment facility
- Underground mine or cave
- Salt dome or salt bed formation
- Concrete vault or bunker

The RCRA Land Disposal Restrictions was promulgated in three stages: first-third, second-third, and third-third. The third-third rule effectively forbids the land disposal of any hazardous waste for which a treatment standard has not been issued.

The most significant elements of the third-third rule are:

- Treatment standards are proposed for all characteristic hazardous wastes
- Treatment standards are proposed for multi-source leachates
- Treatment standards under the "first-third" or "second-third" rules that were specified as "no land disposal" are replaced by either concentration-based standards or required treatment methods for non-wastewaters only
- Procedures are established for the handling of lab packs that contain a large number of wastes
- An interpretation is provided for when dilution is an acceptable practice
- An explanation is provided of how the California list restrictions will be phased out once specific treatment standards are finalized for all hazardous wastes destined for land disposal

Treatment standards may either be concentration-based treatment standards or be based on specified treatment methods. Concentration limits are based upon EPA's evaluation of potentially applicable technologies and the selection of one or more "best demonstrated available technology" (BDAT). The concentration of hazardous constituents in any treatment residues cannot be higher than those that would be obtained using BDAT. Treatment standards based on specific technologies are used when it is impossible to establish concentration limits, i.e. when no acceptable analytical method exists for a hazardous waste constituent.

5. RCRA/CERCLA RELATIONSHIPS

5.1. General

The Comprehensive Environmental Response, Compensation and Liability Act of 1980 (CERCLA) and the Superfund Amendments and Reauthorization Act (SARA) were enacted to deal with the remediation of existing hazardous waste sites. Superfund was designed to facilitate the remediation of the nation's worst abandoned hazardous waste sites. RCRA and CERCLA regulations overlap in a number of areas including the disposal of CERCLA wastes, corrective action, and imminent hazards.

5.2. Disposal of CERCLA Wastes

When wastes from a Superfund site are taken off-site for treatment, storage, or disposal, the facility to which they are brought must have a RCRA permit or be operating under interim status. In addition, any individual moving waste off a Superfund site must adhere to RCRA's generator and transporter requirements.

5.3. Corrective Action

Corrective action under RCRA assumes some of the tactics and responsibility of CERCLA. RCRA, which primarily deals with prevention of releases and management of current waste, regulates past releases at TSDFs and employs the CERCLA legal strategy, which assigns liability to the present owner/operator regardless of the actual responsible party.

5.4. Imminent Hazard

Both CERCLA and RCRA contain provisions that allow the EPA to require persons contributing to an imminent hazard to take the necessary actions to clean it up. RCRA's imminent hazard section (Section 7003) provides regulations for both nonhazardous and hazardous solid waste releases whereas CERCLA (Section 106) regulates the actual or threatened release of a hazardous substance.

As the current owner becomes responsible for previous releases, 7003 of RCRA creates an implied duty to investigate if the purchaser wishes to avoid cleanup liability. Thereafter, it imposes a duty to take reasonable action to abate the imminent and substantial hazard.

6. RCRA'S RELATIONSHIP TO OTHER ENVIRONMENTAL ACTS

6.1. OSHA

Regulation of asbestos at the federal level consists primarily of regulations adopted pursuant to OSHA, Toxic Substance Control Act and Clean Air Act. The regulatory agencies primarily responsible for asbestos regulation are OSHA and the EPA.

6.2. Toxic Substances Control Act (TSCA)

The Toxic Substances Control Act was promulgated in 1976 and provides a regulatory framework for comprehensively dealing with risks posed by the manufacture and use of chemical substances. TSCA was enacted in large part due to the discovery of widespread contamination by polychlorinated biphenyls (PCBs) and the EPA's lack of regulatory tools to control PCB material. Regulations of PCBs are set forth under TSCA, thus they are not regulated under RCRA (unless they are mixed with a RCRA waste).

6.2. Clean Air Act (CAA)

The Clean Air Act consists of:

Title I - Attainment and Maintenance of Air Quality Standards

Title II - Motor Vehicles and Fuels

Title III - Toxic Air Pollution

Hazardous materials treatment technologies will be most affected by Title III. Title II requires maximum achievable control technology (MACT), or best control technology on plants that are major sources of air toxins. A list of 189 toxic air pollutants has been established, defining major sources as those with the potential to emit at least 10 tons per year of any one of the listed pollutants or 25 tons of any of the listed pollutants in combination.

The EPA is required within 10 years to issue MACT standards for all categories of major sources. MACT standards are to require the maximum reductions in toxic emissions that EPA determines are achievable, taking cost and other factors into account. For new sources, MACT must be at least as stringent as the most stringent emissions level achieved in practice by a similar source.

6.3. FIFRA

FIFRA is the Federal Insecticide, Fungicide and Rodenticide Act. Its purpose is to regulate the sale, use, and disposal of "pesticides", where "pesticide" is defined as:

- Any substance or mixture of substances intended for preventing, destroying, repelling, or mitigating any pest
- Any substance or mixture of substances intended for use as a plant regulator, defoliant, or desiccant.

The term "pest" means:

- Any insect, rodent, menatode, fungus, or weed
- Any other form of terrestrial or aquatic plant or animal life or virus, bacteria, or other microorganism (except viruses, bacteria, or other microorganisms on or in living man or other living animal) which EPA declares to be a pest under section 25(c)(1)

The regulations implementing FIFRA are at 40 CFR Parts 152 to 180. In general, EPA has authority to register pesticides, establish use standards, issue experimental use permits, and enforce violations of the act.

Pesticides may also be regulated under RCRA and accompanying regulations, 40 CFR Parts 124, 260-270. If pesticides are "disposed of" as wastes rather than used as a product, they may be considered a hazardous waste and subject to stringent handling requirements. Section 40 CFR § 261.7 sets forth what determines an empty pesticide container.

As long as farmers dispose of the pesticide "on their own farm in a manner consistent with the disposal instruction on the pesticide label," they can be largely exempt from hazardous waste regulations §262.70.

B-Regulations-CERCLA

COMPREHENSIVE ENVIRONMENTAL RESPONSE, COMPENSATION, AND LIABILITY ACT (CERCLA)

1. INTRODUCTION

Instructional objectives:

- Summarize CERCLA objectives
- Explain how CERCLA interfaces with SARA and RCRA
- Discuss Superfund uses and financing
- Discuss aspects of liability and financial responsibility
- Provide a general overview of the CERCLA system

1.1. Enactment of CERCLA

Comprehensive Environmental Response, Compensation, and Liability Act (CERCLA) (P.L. 96-510; 42 USCA § 9601 - 9657) was passed by Congress in 1980.

The main objectives of CERCLA are to:

- Develop a comprehensive program to prioritize and clean-up the worst existing hazardous waste sites
- Establish a fund for clean-up
- Authorize the federal government to impose cost and liability on responsible parties
- Authorize EPA to take any emergency and long term (more than six months) response action.
- Provide a post closure liability fund to interact with RCRA to ensure adequate financial coverage for the ultimate closure of present and future hazardous waste management operations
- Advance scientific and technological capabilities in all aspects of hazardous waste management, treatment, and disposal

1.2. Funding

Funding for site clean-up under CERCLA and meeting the objectives listed above primarily comes from two sources:

- Hazardous Substances Release Trust Fund (Superfund)
    - — Aided by the Superfund Amendments and Reauthorization Act of 1984 (SARA)
    - — Amounts to approximately $8.5 billion (increased to this amount under SARA)
    - — Used when potentially responsible parties (PRPs) do not respond to EPA requests for clean-up
    - — Recovered costs from PRPs after remediation, if possible

- Post Closure Liability Trust Fund
  - Amounts to approximately $250 million
  - Used for present and future remedial site cleanups
  - Provides a financial guarantee against all possible post closure remedial costs for up to 30 years after site closure

Money for these funds comes from:

- Taxes on petroleum and other chemicals
- An environmental tax on corporations
- Costs recovered from responsible parties
- Punitive damages and penalties imposed under CERCLA
- Earned interests
- $ 1.25 billion from general appropriations

1.3. Relationship with RCRA and SARA

1.3.1. RCRA

RCRA directs current hazardous waste management, focusing on prevention of spills (although RCRA does address releases); CERCLA addresses years of hazardous waste mismanagement. CERCLA was developed to address some of the weaknesses of both the RCRA regulatory program and the tort liability process, such as:

- Cleanup requirements of sites that were closed before November 19, 1980 (i.e. when RCRA became final)
- Definition of responsible parties and their duties
- Establishment of a mechanism to fix costs
- Implementation of a clear liability standard

CERCLA remediation activities require compliance with RCRA regulations. For example, removing hazardous waste from a Superfund site requires a RCRA permit. The Superfund agency or contractor becomes a RCRA "generator" and must subsequently comply with all RCRA requirements. Superfund wastes that are disposed or destroyed on-site may still require RCRA closure and post-closure compliance. Thus, both CERCLA and RCRA often depend on each other and often have complimentary roles.

1.3.2. SARA

The Superfund Amendments and Reauthorization Act of 1986 (SARA) was signed on October 17, 1986, as an amendment to CERCLA. SARA amended CERCLA in three ways:

- It increased the size of the Superfund by providing an additional $8.5 billion (as compared to $1.6 billion initially)
- It made the remedial standards more stringent, consequently increasing the cost and time of Superfund site clean-ups
- It emphasized remedial actions that permanently and significantly reduce the volume, toxicity, or mobility of hazardous wastes

2. SOME IMPORTANT TERMS

The following are some important terms pertaining to CERCLA:

Applicable or Relevant and Appropriate Requirement (ARAR) - CERCLA § 122(d). A federal or state standard that is more stringent than CERCLA standards and is legally applicable to the substance or is relevant and appropriate under the circumstances on which the minimum clean-up level is set.

Hazard Ranking System (HRS) - CERCLA § 105(c)

- — Determines the priority at which the given site will be placed on the National Priority List (NPL).
- — Ranks facilities in terms of potential threat based upon hazardous substance, route of release, characteristics and amount of substance released, and likely targets.
- — Assesses the probability of contamination from three sources/routes: ground water; surface water, or air.
- — Scores range from 0-100 and a score of at least 28.5 is required for a site to be placed on the NPL.
- — Places a strong bias toward human health effects.
- — Places least priority on threats that are hazards only to the environment.

National Contingency Plan (NCP) - CERCLA § 105, 40 CFR Part 300.

- — Sets forth the procedures and standards for responding to releases of contaminants and hazardous substances.
- — Provides basic guidelines for the federal response actions.
- — Sets forth the HRS.

National Priority List (NPL) - CERCLA § 105.

- — Lists all the sites throughout the country requiring a clean-up under CERCLA
- — Orders sites undertaken for clean-up according to priority given by the HRS score
- — Is revised three times a year

Potentially Responsible Parties (PRP) - CERCLA § 107(a).

- — Parties or individuals identified by EPA as potentially liable under CERCLA for the expenses incurred during clean up.
- — Includes generators, transporters, parties responsible for storing up hazardous substances, as well as present or former owners/operators of certain facilities or real estate property where hazardous wastes have been treated, stored, or disposed.

Remedial Action - CERCLA § 101(24).

- — Actions taken by EPA to clean up a hazardous waste site.
- — Taken instead of, or in addition to, removal action.
- — May include on-site measures like dredging, excavation, recycling, transport, storage, destruction, or secure disposition of hazardous substances.

Removal Action - CERCLA § 101(23).

- — Action taken to respond promptly to an urgent need.
- — May be in response to a threat of release.
- — May require alternative water supplies, temporary evacuation, limiting access, etc.
- — May be followed by a remedial action depending on whether the treatment is permanent or not.

Response Action - CERCLA § 101(25). Action taken to remedy, remove, or clean-up the site. Response actions also include enforcement related activities.

Remedial Investigation / Feasibility Study (RI/FS) - 40 CFR § 300.68.

- Extensive technical study undertaken by the government or the PRP's to evaluate the scope of contamination (RI) and best remedy for the contamination (FS)
- Alternatives evaluated by FS must be consistent with the NCP and the best alternative evaluated by FS may then be implemented at a Superfund site
- An RI/FS may include a variety of activities such as monitoring, sampling, screening the best alternative, etc.

Record of Decision (ROD) - CERCLA § 113(k).

- Is a written record of the chosen remedy for the cleanup action.
- Normally follows an RI/FS, and addresses any comments that may have been offered by the PRP or other interested parties (such as the general public, State, or cooperating federal agency).
- Lists the reasons and explanation for selecting the particular remedy and is part of the administrative record.
- Signed by the Regional administrator as a formal "binding" decision
- Serves as the basis for recovery of costs from PRPs

Release - CERCLA § 101(22).

- Defined by EPA as any discharge into the environment, in the form of spilling, leaking, pumping, pouring, emitting, emptying, discharging, injecting, escaping, leaching, dumping, or discarding of barrels, containers and any closed receptacles containing hazardous substance, pollutant or contaminant
- Some releases are excluded by EPA (e.g. releases solely exposing workers in a work place, engine exhaust emissions etc.)

Hazardous Substances - CERCLA § 101(14). Any substance, mixture, solution, element, or compound that, when discharged into the environment, may present substantial danger to public health/welfare or the environment. List of lists including:

- Any substance designated as a hazardous pollutant under § 311 of the Clean Water Act
- Any toxic pollutant listed under §307(a) of the Clean Water Act
- Any hazardous substance listed under RCRA or having hazardous characteristics as defined by RCRA
- Any hazardous chemical substance against which EPA has taken action under § 7 of the TSCA
- Any substance designated as hazardous under §102 of CERCLA
- Exclusions include petroleum, natural gas, synthetic gas, and any waste exempt from regulation under RCRA by Congress

Pollutant/ Chemical - CERCLA § 101(33). Any element, substance, compound, or mixture, including disease causing agents, which, after exposure, ingestion, inhalation or assimilation into an organism (directly or through the food chain), will or may cause death, disease, behavioral abnormalities, cancer, genetic mutation, physiological malfunctions, or physical deformations in such organisms or their offsprings.

Environment - 50, Federal Register, 13462. All navigable waters and waters of the contiguous zone, and any other surface water, ground water, drinking water supply, landsurface or subsurface strata, or ambient air within the United States or under the jurisdiction of the United States. A hazardous substance may not be released into the environment even if it remains entirely on plant or facility grounds.

Facility - CERCLA § 101(9). Any building, structure, installation, equipment, pipe or pipeline, well, pit, pond, lagoon, impoundment, ditch, landfill, storage container, motor vehicle, rolling stock, aircraft or any site or area where a hazardous substance has been deposited, stored disposed of, placed or otherwise come to be located is termed as "facility". The term however excludes consumer product in consumer use or any vessel.

3. REMEDIAL ACTION UNDER CERCLA

The complete structure of Remedial Action under CERCLA can be broadly classified in six sections that follow in accordance with the National Contingency Plan (NCP).

1) Reporting and identification
2) Selection of response action
3) Preliminary assessment / Site investigation (PA/SI)
4) Remedial investigation / Feasibility study (RI/FS)
5) Remedial design / Remedial action (RD/RA)
6) Operation and Maintenance / Post closure monitoring

An outline of the above processes is given in a sequential manner in the form of a flow chart in Figure 3.1.

3.1. Reporting and Identification

Two kinds of environmental damage or a potential for environmental damage reported under CERCLA are:

- Spills and hazardous substance releases
- Facility notification

Spills and Releases – In case of a spill or release, for off-shore and on-shore facilities:

- Person in charge must immediately notify the National Response Center (1-800-924-8802) of any release of reportable quantity

  "Reportable quantity" varies with the characteristics of the substance released and varies from 1 pound in some cases and ranges up to 5000 pounds in others. EPA's regulations in 40 CFR §302.4 set forth the list of 717 hazardous substances and the reportable quantities for each of them. For non-listed hazardous wastes exhibiting characteristics of ignitability, corrosivity, or reactivity, EPA has set a reportable quantity of 100 pounds (40 CFR §302.5). If a reportable quantity has not been defined for a particular substance, then it is assumed to be one pound (CERCLA §102(b)). The period over which the quantity release should be assessed has been set as 24 hours (40 CFR §302.6). Some releases are excluded from the reporting requirements. A list of such spills can be obtained from CERCLA §103. Some examples of the excluded spills and releases are Federally permitted releases (CERCLA §101(10)), consumer products in consumer use, work place exposure and engine exhaust emissions.

- Responsibility begins when the release is noticed

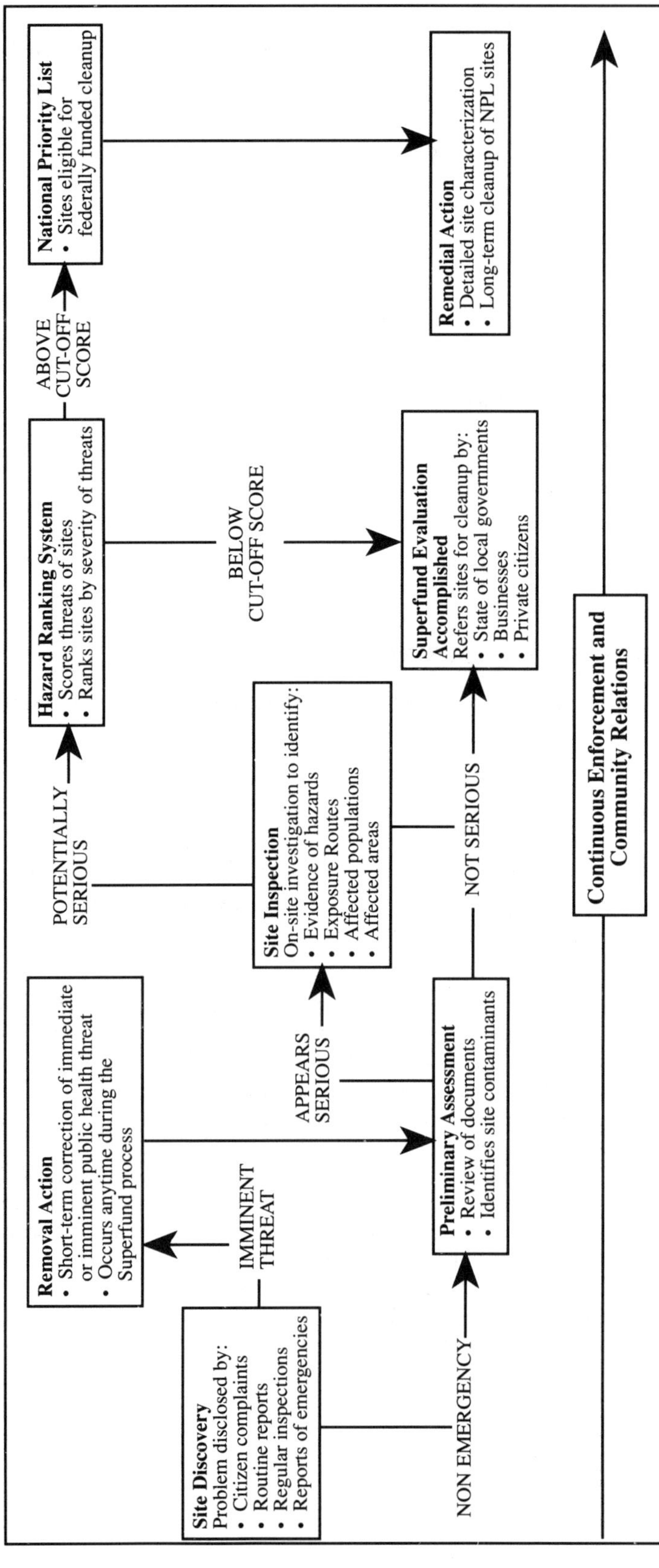

FIGURE 3.1
Superfund Cleanup Process
Regulations CERCLA

Facility Notification – Under the requirements of CERCLA §103(c), EPA must be notified of a facility where hazardous substances are or have been treated, stored, or disposed of. Facilities covered under RCRA notification requirements are exempted from such procedures under CERCLA. EPA's facility notification form identifies the facility, the amount and type of hazardous waste to be found there and any known, suspected, or likely releases of such waste from the facility (CERCLA §103(c)). Facility notification applies only to hazardous wastes as defined by RCRA §300.1. Facility notification is not necessary for all hazardous substances under CERCLA.

Persons responsible for a facility notification are:

- Present owner/operator of such facility.
- Those who owned/operated the facility at the time of disposal.
- Those who accepted hazardous waste for transport and selected such a facility for treatment, storage or disposal.

Persons who do not fulfill their obligations under Facility Notification can be imposed with a financial penalty, imprisonment, or a combination of the two.

Some facilities are exempted for such notification requirements. A complete list of exemptions can be obtained from 46 Federal Register, pg. 22,144, April 15, 1981. Some examples of excluded facilities are facilities receiving only household wastes, consumer products for consumer use and applications of registered pesticides.

Other Reporting Requirements – When a release occurs, local publication of the release information with reasonable notice is required in order to inform potential injured parties. Details of this requirement can be found in CERCLA §111(g).

Title III of SARA (the Emergency Plan and Community Right To Know Act of 1986) establishes detailed provisions for emergency planning and for reporting of inventories and spills of hazardous substances and extremely hazardous substances 42 USC §11001 to 11050.

All notification requirements are reported to:

- the National Response Center (NRC)
- the Appropriate state agency
- Coast guard headquarters, (if at an off-shore or shore facility)

Upon notification of a release or threat of release, the NRC notifies the on-scene coordinator (OSC). The OSC in turn notifies the governor of the affected state and begins the response procedure.

3.2. Preliminary Assessment/ Site investigation

A preliminary assessment follows immediately after identification of the potential hazard. This study is often a quick analysis based on readily available information such as:

- Site management practices
- Information from generators
- Photographs, literature, and personnel interviews
- Site visits, if additional information is needed

The study determines:

- Risk presented by a potentially hazardous waste site
- Sources of contamination that provide a qualitative characterization of the transport mechanisms by which the contaminants are migrating
- Consequences likely to be associated with the contaminants and their effect on human health and environment.

A preliminary assessment determines the urgency of the situation and identifies any non-federal parties willing and able to undertake proper response. If it is determined that no remedy is required at the present time, the site is monitored.

Site investigations provide more detailed information that is quantified by the Hazardous Ranking System (HRS). HRS assesses the relative threat posed by a release or potential release based on data gathered relatively quickly and inexpensively, conserving Superfund resources. Each site is assessed according to criteria and assigned a numerical score between 1-100. Sites must score a minimum of 28.5 to be placed on the National Priority List (NPL). These sites are then eligible for remediation under CERCLA. This site investigation may involve limited sampling to better assess the potential for release or threat of release. If the preliminary assessment and site investigation indicate that the site poses a significant threat to human health and environment, EPA selects its response action for the remediation. The selection of the response action depends on the urgency of the situation as described in the next section.

3.3. Selection of Response Action

Whenever EPA determines that there may be an imminent and substantial endangerment to the public health or welfare or the environment due to an actual release or a threat of release of a hazardous substance, pollutant, or contaminant, CERCLA authorizes EPA to take a response action. These response actions must be consistent with the NCP.

There are three types of responses that may be taken under CERCLA:

- Removal Action
- Remedial Action
- Enforcement Action

Removal and remedial actions are the clean-up actions taken under CERCLA and have been defined earlier in section 2. Enforcement actions, either administrative or judicial, are always initiated at the time a site is discovered. All three actions may be taken in combination with each other or separately at any CERCLA site. As illustrated in the flow chart, the choice of the response action is based on the nature and urgency of the problem.

EPA is authorized to use the $8.5 billion fund for these response actions. This fund is recouped later in a cost recovery action against PRP's under §107. EPA can use this fund subject to certain requirements for uses as mentioned in CERCLA §111. Examples of the allowed uses of the fund are government response costs, natural resource damages, and abatement action costs.

Instead of actually conducting a response action and bringing a cost recovery action, EPA may, in certain circumstances, have an action brought under §106 to require PRP's to undertake the response action themselves (CERCLA §104(a) and § 106).

Exclusions: — Unless EPA determines a release or threat of release constitutes a public health or environmental emergency and that no other authorized and capable person will respond, EPA, under CERCLA, will not respond to the following releases:

- Releases from an unaltered naturally occurring substance or a naturally occurring substance altered solely by naturally occurring processes from a location where it is naturally found (e.g. radon gas).
- Releases from products that are part of the structure of, and result in exposure within, residential buildings or businesses or community structures (e.g. asbestos).
- Releases into public or private drinking water supplies due to deterioration of the system through ordinary use.

Selection of the response action marks the end of the investigation phase. If the response action selected is a remedial action, an HRS scoring is required in order to determine the appropriate placement of the site on the NPL.

3.4. Remedial investigation/ Feasibility study

An RI/FS is a detailed investigation of the Superfund site for characterizing the nature and extent of risks posed by uncontrolled hazardous waste sites and evaluating the potential remedy options. RI/FS is a dynamic and flexible process and is tailored to the needs of the site; it is not a rigid step-by-step approach that must be conducted identically at each and every site. The objective of RI/FS process is not the unobtainable goal of removing 'all' uncertainty associated with the site, but rather to gather information sufficient to support an informed risk management decision as to an appropriate remedy for the given site.

The RI serves as the mechanism for collecting data to characterize site conditions, determine the nature of waste, assess risk to human health and environment, and conduct treatability testing as necessary to evaluate the potential performance and cost of the treatment technologies that are being considered. The latter also supports the designs of selected remedies.

The FS involves the development, screening and detailed evaluation of alternative remedial actions. The data collected in the RI influences the development of the remedial alternatives in FS, which in turn affects the data needs and scope of treatability studies and additional field investigations (see Figure 3.1).

The various steps involved in the RI/FS study are described below.

3.4.1 Scoping

On the basis of the data collected in the Preliminary Assessment/ Site Investigation phase, site management planning is undertaken to preliminarily identify likely remedial action objectives and whether interim actions may be necessary or appropriate, and to establish whether the site may be remedied best as one or several separate operable units.

Typical scoping activities include:

- Initiating the identification and discussion of potential ARARs with the support agency.
- Determining the types of future decisions and identifying information needed to support those decisions.
- Assembling a technical advisory committee to assist in the activities, to serve as a review board for important ascertainable goals, and to monitor progress as required during the study.

- Preparing the work plan, the sampling and analysis plan (SAP) (consisting of quality assurance project plan (QAPP) and the field sampling plan (FSP)), the health and safety plan, and the community relations plan.

3.4.2 Site Characterization

During the characterization, field sampling and lab analysis are initiated. Data quality objectives are revised as appropriate, based on an improved understanding of the site, to facilitate a more efficient and accurate characterization of the site, resulting in reductions in time and cost. A preliminary site characterization summary is prepared to determine the feasibility of potential technologies and assist both the lead and support agencies with the initial identification of ARARs.

A baseline risk assessment is developed to identify the existing or potential risks that may be posed to human health and the environment by the site, providing valuable input to the development and evaluation of the alternatives during the FS.

3.4.3 Development and Screening of Alternatives

The development of the alternatives usually begins during or soon after scoping when likely response scenarios may be first identified. Development of the alternatives requires:

- Identifying remedial action objectives.
- Identifying potential treatment, resource recovery, and containment technologies that will satisfy these objectives.
- Screening the technologies based on their effectiveness, implementability, and cost.
- Assembling technologies and their associated containment or disposal requirements into alternatives for the containment media of the site or for the operable unit.

Alternatives for the specific media and site areas either can be conducted separately or confined to comprehensive alternatives for the entire site. The approach is flexible to allow alternatives to be combined at various points in the process.

A practicable range of treatment alternatives should be developed. The upper bound of the range should be an alternative that would eliminate, to the extent possible, the need for long term management (including monitoring) of the site. Other alternatives should be developed that rely on various ranges of treatment as the principal elements, but require some long term management of portions of the site that do not constitute 'principal threats'. In addition, one or more containment option(s) involving little or no treatment should be developed and a no-action alternative should be developed.

Once these potential alternatives are developed, they are screened out to reduce the number of alternatives to be analyzed in detail in order to minimize the resources dedicated to evaluating options that are less promising. The screening process involves evaluating alternatives with respect to their cost, implementability, effectiveness, and environmental and institutional considerations. It is usually conducted on a general basis and with limited effort (relative to the detailed analysis) because the information necessary to fully evaluate the alternatives may not be complete at this point in the process.

3.4.4 Treatability Investigations

If the available site and treatment data are insufficient to evaluate available alternatives, a treatability study may be necessary to evaluate particular technologies on site-specific wastes. Generally, treatability tests involve bench scale testing to gather information to assess feasibility of a technology. In certain cases, a pilot-scale study may be required.

3.4.5 Detailed Analysis

Once sufficient data is available, alternatives are evaluated in detail with respect to the guidelines that EPA has developed to address the statutory requirements and performances of CERCLA:

- Compliance with ARARs
- Reduction of toxicity, mobility, or volume
- Short-term effectiveness
- Long-term effectiveness and permanence
- Implementability
- Cost
- Community acceptance
- State acceptance
- Overall protection of public health and the environment

The alternatives are analyzed individually against each criterion and then compared against one another to determine their respective strengths and weaknesses and identify the key tradeoffs that must be balanced for that site. The results of this detailed analysis are summarized and presented to the decision-maker so that an appropriate remedy consistent with CERCLA can be selected.

3.5. Remedial Design Implementation

The selected alternative developed after a detailed analysis in the RI/FS study is recorded in the Record of Decision. The ROD lists the advantages and reasons of why the remedy selected in the RI/FS was selected over other alternatives. Before a treatment plan is finally chosen and considered for the design implementation, it is presented to the public for a Notice and Comment period. If there are any critical comments, suggestions, or modifications and EPA determines that they should be incorporated or modified in the final alternative, the treatment plan is modified and the final plan is presented in the ROD and for design implementation.

At this point, EPA invites PRPs to do the remedial design and implementation. This is the last chance for the PRPs to enter the remedial process. If the PRPs agree to conduct the remediation they carry out the selected design process; if, however, the PRPs do not agree to conduct the remediation, EPA uses Superfund for the remedial design and implementation and seeks the expenses later from the PRPs (see figure 3.1).

Once it is decided who should carry out the design procedure, the remedial design is initiated and technical drawings, specifications, cost estimates beyond the feasibility study and other supporting documents are furnished. Once the design details are established, EPA determines whether the design meets the applicable or relevant and appropriate requirements (ARARs). If it is determined by EPA that ARARs are not satisfied, then the remedy is redesigned (see Figure 3.1). Some remedial actions are exempted from meeting ARARs (CERCLA § 121(d)(4)). Some examples of ARAR-exempted remedies are: when the remedial action is only part of the total remedial action, which will attain such ARARs; or when compliance with the standard will present a greater risk to health and environment and compliance is technically impractical.

Once all the ARARs have been met, bids for the contract are invited and the contractor with the best suitable offer is given the contract to implement the design. Construction at the site, the last step in the remedial design/implementation process, then begins.

3.6. Post Closure Monitoring

The standards established for land disposal closures under RCRA also apply under Superfund. Once a total remedial action is complete, post remedy requirements may be applicable. Post remedy requirements are dictated by the amount of hazardous substance remaining at the site after closure. If the hazardous substances remain at the site, post-closure groundwater monitoring is required. Only in those instances of a "clean closure" (where no hazardous substances remain at the site and no ground water contamination is present) is it possible to avoid post closure groundwater monitoring. RCRA post closure requirements specify a minimum of thirty years of groundwater monitoring. This time limit may be extended at the discretion of EPA.

4. LIABILITY UNDER CERCLA

4.1. Nature of Liability

4.1.1. Potentially Responsible Parties

The problem of identifying the owner of a hazardous waste site can be difficult and sometimes impossible. CERCLA, § 107(a), defines liable parties as:

- Any current owners and or operators of a facility or vessel
- Any owner or operator at the time of release or disposal
- Any person, who by contract, agreement, or otherwise, arranged for disposal or treatment of a hazardous substance at a facility or for transport of a hazardous substance to a facility from which a release has occurred
- Any person who transported a hazardous substance to a facility

Some specific examples include:

- Current owner
- Directors
- State governments
- Beneficiary of trust
- Lessee
- Transporters
- Persons who arranged for disposal
- Lender
- Past owner
- Joint venturers
- Operators
- Estate
- Successor corporations
- Departments within the federal government
- Officers
- Shareholders
- Local governments
- Trustee
- Lessor
- Engineers
- Designers

Some explanations are noted below:

lender - may be liable if they obtained the property through foreclosure

short term past owner - "conduit owner" is liable since a limited-time consideration in ownership is not enforceable

operators - may be liable if they oversee daily operations

state government - may be liable if they inherit property via escheat or are proven negligent in response action

lessor or property owner - may be liable even though he is not involved with daily operations

successor corporation - may be liable if they take on all responsibilities of the corporation they buy

trustee - may be liable if he is involved with daily operations, but not if he divorces himself from that responsibility early-on

estate - may be liable, but a statute of limitations may be enforced

manufactures - may be liable if the continued sale of their product is knowingly being discarded illegally

persons who arranged for disposal - may be liable if the defendant owned or possessed hazardous substance, arranged for transport, a release occurs, and the hazardous substance caused the cost of clean-up

This liability issue is the impetus for due diligence studies—a reasonable and prudent investigation for releases on a site—prior to property transfers.

4.1.2. Strict Liability

Liability, not explicitly defined in CERCLA, is strict, meaning that no exceptions are made to any particular party whether it be a private citizen or a government agency. The U.S. Code (42 U.S.C. § 9601) states:

- Liability is interpreted strictly without regard to a particular situation or person
- Liability extends to every PRP of a facility or vessel regardless of their degree of ownership or responsibility
- CERCLA is immediately enforceable for releases occurring before the enactment of the law
- Liability is enforceable for willful misconduct and non-cooperation in engagement of response actions, contractual or not

CERCLA, § 107, expands and qualifies responsibility assessing liability for response costs, natural resource damages, health assessments, and other indirect costs. "Indirect costs" may include almost any cost, even secretarial or public information costs, warranted by a particular situation.

4.2. Extent of Liability

According to CERCLA, a PRP or several PRPs "inherit" liability, but the extent to which liability is enforceable is limited depending on the specifics of the case. The share of a person's interest in a facility or vessel must be disclosed, and the following may influence that share:

- Degree of each party's involvement
- Relative amounts of wastes discharged
- Degree of toxicity
- Degree to which each party's contribution can be distinguished
- Degree of care exercised
- Degree of cooperation shown before and after response

4.2.1. Joint and Several Liability

Degrees of liability are not explicitly defined in CERCLA. The Board of Arbitrators (see sub-heading "K - Cost Recovery Actions") may decide the degree of liability imposed on a PRP. Joint and Several liability addresses situations in which there are several parties responsible for the release or the threat of release, such as a landfill used by several manufacturers. In such cases, any one party may be held liable for the entire cost of clean-up, regardless of the party's level of contribution. This prevents waste of Superfund resources on extensive litigation required to obtain compensation from all the involved parties.

Under this heading, there are three theories of liability:

Consent of Action - the defendants acted jointly with each other and caused a release.

Alternative Liability - several defendants, not acting in consent, caused a release as a result of their actions compiled together.

Enterprise Liability - the industry as a whole, as a result of standard practices, caused a release where no single firm could be held liable (market share liability).

The Superfund Amendments and Reauthorization Act (SARA) addresses the topic of liability more thoroughly.

4.2.2. Indemnification

A plea of indemnification (innocence) by a PRP may be ruled in court as an exemption of some or all penalties, liabilities, or portions of such penalties or liabilities as defined by CERCLA.

> An example of indemnification may be the "innocent land owner" who, after proper inspection procedures, obtains real property (real estate) which contains a hazardous substance, undisclosed to him. If substantiated, the present owner may be relieved of liability that may direct itself to the past owner.

Additionally, CERCLA provides several "escape hatches" for PRPs under § 107. These defenses are explained under sub-section "Liability Defenses."

4.3. Financial Responsibility & Insurance Requirements

In addition to the stipulations imposed on PRPs in § 107, PRPs may be liable for the following:

- Costs incurred by Federal or State government for response or remedial costs
- Any additional costs incurred by private or third party involvement in the response action
- Damages for remediation or replacement of natural resources
- Costs of certain related health studies on employees
- Damages due to negligence of abnormally dangerous activities based on torts of trespass or nuisance with regard to a hazardous substance

Insurance requirements are related to two separate types of operations: stationary operations and mobile operations.

4.3.1. Stationary Operations

Facilities (on-shore and off-shore)—Facilities are classified by independent insurance companies. This classification is based on "Superfund" payment experience, court settlements and civil judgements. Further requirements for certification may be found in 40 U.S.C. § 20.

4.3.2. Mobile Operations

Vessels (ships)—Detailed fines and insurance requirements are listed in § 108 of CERCLA. Generally, ship transport companies must keep the minimum requirements for insurance on their largest vessel in the fleet, but are not required to have insurance coverage for each individual vessel. The size of vessel, type, and quantity of hazardous substances transported are also considered.

In addition, vessels may be denied clearance to dock or leave port (if destination is another U.S. port) for lack of required insurance documents:

- Insurance
- Surety bonds
- Guarantee
- Self-insurer qualifications

Additional information on shipping requirements (i.e. packaging, container regulations, etc.) is found in 49 U.S.C. § 178 and § 179 and 46 U.S.C. § 151.

Transporters (rolling stock)—Detailed fines and insurance requirements are defined in the Motor Carrier Act (PL 96-296, 1980) and Amendments. Generally transporters must carry proof of the same insurance requirements stated under vessels in order to transport bulk or non-bulk hazardous substances intrastate or interstate. Additional figures are listed in 49 U.S.C. § 386 and § 387. The transportation of hazardous substances is also regulated by the Hazardous Materials Transportation Act (HMTA), which lists all transportable hazardous substances.

4.4. Liens

All costs and damages for which persons are liable constitute a lien on real property (i.e., facilities, on-shore and off-shore, vessels, rolling stock) by the U.S. government for payment of response actions under § 107(a) of CERCLA.

Moreover, 42 U.S.C. § 9607 specifically explains the following:

- The lien is for the benefit of the U.S. and is used to recover response costs only
- The lien is imposed on real property at the beginning of the response action or upon receipt of a certified letter from the U.S. government (EPA) to the facility owner . This states that a release has occurred, names the PRPs, and advises them of their rights under CERCLA
- The lien has priority over previous liens filed against the property
- The lien, in the case of a vessel, constitutes a maritime lien imposed by the U.S. government.

4.6. Post Closure Liability

The Post Closure Liability Trust Fund, created by CERCLA § 232, allocates funds toward maintenance of sites after remediation has ended. To make use of this fund, the owner's facility must be:

- In compliance with closure regulation under the Solid Waste Disposal Act; (SWDA, 1965)
- Closed for a minimum of five years
- Continually monitored

After all these conditions are met, the government may accept the transfer of liability under Post Closure. If the government determines for any reason that the owner has not complied with the statements above, liability may be transferred back to the owner. Amendments for insurance requirements and liability are contained in SARA.

4.7. Liability Defenses

CERCLA § 221 encourages Congress to study common law defenses that were later reported under SARA. Defense falls into two categories: expressed defenses and affirmative defenses.

4.7.1. Expressed Defenses

Defenses, explicitly outlined in SARA (§ 107), for which a PRP is not held liable are termed expressed defenses.

A PRP can claim no liability if the act is:

- An act of God
- An act of war
- An act of omission of a third party (non-contractual)
- Any combination of the above
- In accordance with the National Contingency Plan (NCP)
- A contractual transfer of title in which the current owner has no evidence of the presence of hazardous waste substances (innocent land owner defense) if:
  - — The current owner made an inquiry in accordance with standard practices for site assessment
  - — The current owner, in the eyes of the court, has no specialized knowledge that would have led him to believe that the presence of a hazardous substance existed on the site
  - — The previous owner did not disclose to the current owner that a release had actually occurred

4.7.2. Affirmative Defenses

Defenses not explicitly outlined in SARA, for which a PRP bears the burden of proof, are termed affirmative defenses.

A PRP, upon burden of proof, can claim no liability if:

- A release or threat of a release did not occur

- Response activities and costs are not consistent with the NCP regarding the extent of remediation or removal performed.

4.7.3. Statute of Limitations

The time limitations for cost recovery actions defined in §112 of SARA are summarized below:

- Cost recovery for response actions must commence:
  - No later than three years after the removal action has been completed
  - No later than six years from the initial response action construction
- Cost recovery for damages to natural resources must commence:
  - Within three years after the completion of the remedial action for any facility listed on the NPL
  - Within three years after the date of discovered loss in connection with a release or the date of publishing of natural resource damage estimates by the federal government
- Cost recovery for other than the above:
  - An action for additional cost recovery must be brought within three years from the judgement date of the initial cost recovery action
  - An action for contribution in regard to response costs or natural resource damages must be brought within three years of the date of judgement
- An action to recover payment for a response action contractor under an indemnification agreement must be commenced within three years after payment
- An action based on subrogation for payment of a claim, must be brought within three years after payment of the claim (Subrogation is the substitution of one claimant for another so that the new claimant succeeds to the former's rights)

In all the cases including minors, the date of the minor's eighteenth birthday or the date when a legal guardian is appointed, will be "the running date" when a minor may be considered liable for anything under CERCLA.

4.8. Parties Rendering Assistance

Parties Rendering Assistance (PRAs) shall not be liable for response costs or natural resource damages under § 107 of SARA if they are in accordance with the NCP (revised under § 105 of SARA). PRAs, if they are asked by the government to carry out emergency response actions, shall not incur liability, assuming they are in no way negligent or guilty of misconduct. Costs incurred by PRAs are recoverable from the Superfund, in most circumstances. State and local governments may be considered PRAs (see sub-section "Cost Recovery Actions").

4.9. Bankruptcy

CERCLA does not specifically address bankruptcy. However, bankruptcy has been used as a defense mechanism in federal and civil courts.

Based on legislative history, the following cannot exclude a PRP from liability:

- The PRP abandoned the site and it was known that a hazardous substance was released
- The PRP files for bankruptcy protection (automatic stay)

- The PRP's estate is depleted via a federal lien, which has priority over other liens
- The process of escheat is exercised

The U.S. Bankruptcy Code is often subject to review in regards to the specific cases listed above.

4.10. Settlement and the Superfund

4.10.1. Superfund Usage

CERCLA § 221 enacted the Hazardous Waste Response Trust Fund or "Superfund" in order to finance clean-ups of abandoned sites for which no PRP is immediately recognized.

CERCLA § 111 explains the uses of the fund:

- Replacement costs of long term and short term loss of natural resources
- Replacement costs of resources incurred by the state or federal government
- Costs of start-ups for response actions and remediation
- Costs of epidemiologic studies for personal injury (see sub-section "Risk Assessment")
- Costs of start-ups for maintenance of hazardous substance strike-forces, emergency forces and task forces under the NCP
- Costs of employee health and safety programs structured under OSHA and the EPA

If a response action is required by the EPA, the following parties may carry it out:

- EPA
- State government
- PRP
- A third party

The EPA gives PRPs the option of carrying out response or remedial actions themselves. If the PRPs refuses to perform these actions, they may bear the total cost of action as performed by the government or a third party.

4.10.2. Settlement Procedures

The EPA may use its authority to expedite remedial actions and minimize litigation by using the board of Arbitrators (created by CERCLA § 112) to help with:

Non-Binding Allocations of Responsibility - where a percentage share is determined (by the Board) for response costs by PRPs

Mixed Funding Clean-ups - where partial percentage payments by any combination of PRPs, state, local and federal governments is planned

De Minimis Settlements - where action is taken by the Board to rid small claims of several PRPs to simplify the settlement process

Covenants Not to Sue - where the board initiates an agreement between the EPA and PRPs that restricts PRPs from being sued twice

Contribution Protection - where action is taken by the Board which safeguards "settling" PRPs from litigation of non-settling PRPs

Response Action Contractors (RACs) - who bring claims against the fund for work not under contract (however they are liable for negligence and misconduct)

PRPs may be subject to consent decrees, administrative orders, or declaratory judgements brought by the EPA acting for the federal government. To this end, the EPA has the right to request and exchange information from any person or facility via:

Questionnaire that inquires as to which chemicals are found and what waste practices are used at the facility

General Notice Letter informs PRPs of liability for response costs (PRPs must notify other PRPs involved)

Special Notice Letter for RI/FS - RD/RA, which triggers a 60 day moratorium after implementation of a good fourth offer begins

4.11. Cost Recovery Actions

As stated under § 112 of CERCLA, cost recovery actions secure response costs from state or local governments, PRPs, or private parties.

4.11.1. Recoverable Costs

Since response costs are not explicitly mentioned in CERCLA, the following has been deemed recoverable through legislative history in conjunction with § 111 (uses of the Fund):

- Site investigation costs
- Sampling and analysis costs
- Hazardous substance removal costs
- Certain medical testing costs
- Costs of relocation of area residents
- Site security costs
- Costs to insure adequate water supply
- Temporary drum storage costs
- Future response costs consistent with NCP
- Certain litigation costs
- Board of Arbitrators costs

4.11.2. Non-Recoverable Costs

Non-recoverable costs expressed in CERCLA § 110 , § 111 and § 107 include the following:

- Costs of business losses due to closure
- Certain medical testing costs
- Certain litigation costs
- Costs due to loss of natural resources claimed by non-governmental parties

- Certain personal injury costs
- Death

4.11.3. Government Cost Recovery Action

The federal government, in order to establish liability, must prove the following:

- Each site is a CERCLA site
- A release has occurred or is occurring
- Response costs have resulted from the release
- Defendant is a PRP under CERCLA
- Costs recovered must be consistent with the NCP

4.11.4. Private Party Cost Recovery Action

Private parties bear the burden of proof and, in order to recover costs, must:

- Have incurred some response costs
- Demonstrate only a casual contact with a release or causation
- Act consistently with the NCP

4.12. Responsibilities of State and Federal Governments

CERCLA was enacted to cover situations neglected under RCRA on a federal level (i.e., abandoned sites), but states have also enacted individual clean-up efforts.

4.12.1. State Enforcement

State government may:

- Impose regulations that are stricter than federal regulations
- Enact their own "Superfund" programs and trust funds
- Fund all or portions of remedial clean-ups
- Enter into contracts with the federal government, PRPs, and contractors (RACs)

4.12.2. Federal Enforcement

Federal government may:

- Force states or contractors to start response action that incurs costs recoverable at a later date
- Force evacuation of surrounding area
- Force the state to comply with EPA procedures

Generally, the federal government, the EPA, and the Army Corps of Engineers control response and remedial actions that are supported by individual states

C - Critical Issues in Technology Selection

CRITICAL ISSUES IN TECHNOLOGY SELECTION

1. GOALS OF TECHNOLOGY SELECTION

The goal of technology selection is to choose the most cost effective and reliable treatment technology to achieve a desired clean-up objective in a reasonable amount of time. In selecting an appropriate remedial scheme the decision-maker must combine technical judgment, cost, liability, and occasionally statutory criteria to arrive at a decision [1].

The following module will attempt to highlight some of the important issues that should be considered when a remediation plan is being devised. First, regulatory issues will be considered. Second, engineering considerations in technology selection will be discussed.

2. REGULATORY ISSUES IN TECHNOLOGY SELECTION

CERCLA and RCRA are the two most important federal regulations that are of concern in hazardous waste site remediation. Some states also have regulations with regard to site remediation. Both federal and state requirements must be complied with during remediation. Results from a University of North Dakota study (Table 2.1) indicate that compliance with government regulations is an important consideration in selecting a hazardous waste treatment process.

Table 2.1[2]
Selection Criteria for Hazardous Waste Treatment Technology

Criteria	Weighting
Government regulations	0.4
Economics	0.3
Public relations	0.2
Process technology	0.1
Total	1.0

In this section RCRA and CERCLA programs will be considered separately. A more detailed discussion of RCRA and CERCLA issues is provided in the Regulatory Modules.

2.1 Superfund Requirements (CERCLA)

2.1.1 Hazardous Ranking System (HRS)

The basic goals of an environmental cleanup, under either state or federal programs, is the prevention of adverse impacts on human health and the environment. To determine whether or not a site qualifies for Superfund assistance, the site is evaluated using the Hazardous Ranking System (HRS). When using the HRS, the EPA considers the following criteria:

- The potential to harm humans or the environment via groundwater, surface water, or air (S_{gw}).

- The potential for harm if the substance can explode or is flammable (S_{fe}).
- The potential for harm from direct contact with hazardous substance at the site (S_{dc}).

A numerical value for each of these routes of exposure is determined considering several factors. Table 2.2 [2] lists factors that should be considered in ranking contaminant migration.

Table 2.2
Rating Factors for Hazard Ranking System (Contaminant Migration)[2]

Category	Groundwater Route	Surface Water Route	Air Route
Route Characteristics	Depth to aquifer in concern	Facility slope and intervening terrain	
	Net Precipitation	1-year 24-hour rainfall	
	Permeability of unsaturated zone	Distance to nearest surface water	
	Physical state	Physical state	
Waste Characteristics	Toxicity	Toxicity	Reactivity, toxicity and quantity
	Quantity	Quantity	
	Mobility	Mobility	
Media Use	Distance to nearest well and population served	Distance to sensitive environment	Land Use
			Population within a 4 mile radius
		Population served/distance to water intake downstream	Distance to sensitive environment

Through use of the HRS, the EPA develops the National Priority List (NPL). Using the NPL, an assessment is made regarding which waste sites are the largest threat to human health and the environment. The NPL is used to determine where remedial actions are to be undertaken first.

2.1.2. Remedial Action Process

If a site is placed on the NPL, it then is eligible for federal assistance in the clean-up proceedings. Under CERCLA, there is a systematic approach, contained in the National Contingency Plan (NCP), which outlines the criteria that must be considered in developing a remedial plan. The first requirement under CERCLA is a site characterization called a remedial investigation (RI) and a remedial alternative investigation termed a feasibility study (FS). Usually these two studies are completed at the same time and are referred to as an RI/FS study. From the outcome of the RI/FS study, and after public comment, a remedy is selected and is written into the Record of Decision (ROD).

In arriving at a remedial decision there are three categories of criteria that must be considered as per the NCP [3]:

- <u>Threshold Criteria</u>
  1. Overall protection of human health and the environment
  2. Compliance with applicable or relevant and appropriate requirements (ARARs)

- Primary Balancing Criteria
    3. Long-term effectiveness and permanence
    4. Reduction of toxicity, mobility, or volume through treatment
    5. Short-term effectiveness
    6. Implementability
    7. Cost
- Modifying Criteria
    8. State acceptance
    9. Community acceptance

2.1.3. Applicable or Relevant and Appropriate Requirements (ARARs)

2.1.3.1. Volume and Toxicity of Waste

When selecting a technology for remediation CERCLA § 9621 (b) (1) states that "remedial actions, in which treatment permanently and significantly reduces the volume, toxicity, or mobility of the hazardous substances, pollutants, and contaminants, is a principal element (in technology selection) and is preferred over remedial actions not involving such treatment." [4] Containment is not preferred because it is not viewed by the EPA as being a permanent remedy and long-term uncertainties associated with containment are high. EPA does not encourage a continuing liability that may pose another clean-up problem down the road when containment eventually fails.

2.1.3.2. On-Site vs. Off-Site Treatment and Disposal

Off-site disposal is not favored under CERCLA. Off-site transportation and disposal of untreated waste is the "least favored" alternative where "practicable" treatment alternatives exist. In order to encourage on-site treatment, Congress in CERCLA § 9261 (e) (1) does not require any federal, state, or local permits "for the portion of any removal or remediation action conducted entirely on-site" [4]. Normally under RCRA, any treatment of hazardous waste requires a permit under section 3005 before such treatment can begin. CERCLA § 9261 (e) (1) significantly reduces the delays that could occur due to permitting. Typically, the RCRA permit process for a treatment and disposal facility (TSDF) takes 1-to-2 years. [5]

2.2. RCRA Requirements

2.2.1. RCRA vs. CERCLA

CERCLA and RCRA are sometimes referred to "as being like flip sides of a coin." [6] While Superfund was created to address the problems of past waste disposal methods, RCRA was enacted to ensure that past methods of disposal are corrected. RCRA's major purpose is to provide a "cradle-to-grave" framework for tracking the generation, transportation, treatment, and disposal of hazardous waste. Although in SWDA § 3004 (v), EPA is authorized to "require that *corrective action* be taken beyond the facility boundary were appropriate to protect human health and the environment" (emphasis added) [7]. EPA has interpreted the term "corrective action" to include a "full range of possible actions, from studies and interim measures to full cleanups." [8] By this authorization, Congress makes it possible for EPA to issue administrative orders or bring court action to require cleanup at RCRA facilities operating under interim status.

Overall, CERCLA is considerably broader than RCRA in addressing remedial actions at waste sites. The RCRA provisions only apply to RCRA-related facilities while CERCLA can be applied anywhere there is a "release or threat of a release". Another difference is that under RCRA, unlike CERCLA, there is no method for the federal government to financially assist in the cleanup cost; all of the cleanup work is completed by the regulated community (no government intervention).

2.2.2. RCRA Corrective Actions

Under RCRA and CERCLA, while the nomenclature in the remedial action process are different, the procedures are similar. As with CERCLA, the first step in a RCRA corrective action plan is an evaluation to determine if an emergency removal is warranted. If no emergency removal is required then the EPA will conduct a RCRA Facility Assessment (RFA), which is similar to the RI/FS study in the CERCLA procedure. After this study is completed, the owner of the facility determines the extent of the problem and proposes a solution. If the plan is approved by the EPA and the State, it is then implemented. The only major difference between the CERCLA and RCRA procedures is that RCRA corrective actions do not have to be ranked using the Hazardous Ranking System [8]. Under RCRA, it is EPA's aim to "allow considerable procedural flexibility as well as flexibility in selecting appropriate remedies." [6]

2.3. State Requirements

Under CERCLA and RCRA, states are allowed to develop their own treatment requirements separate from the federal government. The state plans must be approved by the EPA. For EPA approval, the requirement of the State plan must be more stringent than the federal requirements. For the most part, State plans comply with federal regulations and go no further. In some cases where a state has a large number of waste sites, such as New Jersey, extensive state requirements have been developed involving risk assessments and computer modeling [9]. At least 41 states and 1 territory have established funding mechanisms to deal with clean-up cost at hazardous waste sites that do not otherwise qualify under CERCLA [6].

2.4. Cleanup Standards

The age old question of "How Clean is Clean?" is one that scientists, engineers, and legislators have debated and will continue to debate for the foreseeable future.

2.4.1. End-Point Criteria

There are many approaches used to identify end-point criteria. Below is a list of some possible approaches with consideration of issues related to each suggestion:

- Background or Non-Detectable Levels
  - Policy based on avoidance of issue. Good only for regulators.
  - Only based on analytical chemistry measuring precision.
  - Public alarm is fed by lack of achieving goals.
- Zero Risk no acceptable level
  - Is cost prohibitive.
  - Is it the best way to spend societies limited resources?
  - Perhaps only good for food

- Site-By-Site Risk Assessment
  - Based on pathways of exposure
  - Is time consuming and is sometimes not appropriate for small sites.
  - Is the residual contamination safe?
- Technology Based Clean-up
  - Results in most costly destruction technology being required at most sites
  - Would ignore worker exposure and ambient risk with excavation and incineration
  - Would reduce technology innovation
- Cost-Based Clean-up Standards
  - What risk at what cost?
  - Are small reductions in concentration worthwhile?
- Drinking Water Standards (MCL's)
  - Must consider aquifer use. If an aquifer can yield >150 gal/day, then it may be appropriate.
  - Not appropriate if the aquifer has a low yield, or if it is located in a remote area.
  - Lack of scientific certainty of chemical risk and exposure assumptions result in high uncertainty and are overprotective of risk.

2.4.2. EPA's Protection Standards

Remediation activities must be prioritized with human-health risk given top priority. A second priority, when dealing with groundwater, is to restore "reasonably expected sources of drinking water and groundwater hydrologically connected to surface waters" [10] whenever possible. It is a statutory mandate that cost be considered—"the total short-term and long-term cost of such actions (remediation), including the cost of operation and maintenance for the entire period during which the activity will be required." [4] Considering the high cost of groundwater clean-up, EPA states that "a framework should be established that ensures that environmental and public health benefit of each dollar spent is maximized." [10] Thus, in setting cleanup goals, EPA advises that the following factors be considered:

- Land-Use Characteristics
- Resource Value
- Resource Vulnerabitliy
- Social Value of Resource
- Economic Value of Resource

In the past, it has been EPA's policy, when carrying out clean-up activities, to use the Maximum Contaminant Levels (MCLs) under the Safe Drinking Water Act as "reference points" for treatment goals for groundwater that is being or could be a potential source of drinking water. Water Quality Standards under the Clean Water Act are used as reference points when the groundwater is hydrologically connected to surface water ecological systems. When an MCL is not available, EPA Health Advisory numbers or other approved health-based levels are recommended as points of reference from the health literature. In some cases, the EPA has used maximum contaminant level goals (MCLG's) under the Safe Drinking Water Act [10].

In general, EPA has set clean-up goals to achieve a total lifetime cancer risk level in the range of 10^{-4} to 10^{-6} (see Risk Assessment module). More stringent measures have been selected based on such factors as the cumulative effects of multiple contaminants, exposure from other pathways, and unusually sensitive populations.

Less stringent measures may be used only where authorized by law, based on such factors as lack of evidence of future use, cost, adverse environmental effects of remediation, and lack of technology [10]. If full restoration for portions of contaminant plume cannot be practicably carried out with current available technology, or do not warrant full restoration based on site specific factors, such as existing water quality or remote site location, then treatment may not be necessary [11]. Containment or "natural restoration" (do-nothing) alternatives may replace full restoration as a remediation goal.

3. ENGINEERING ISSUES IN TECHNOLOGY SELECTION

Waste and site characteristics are the two main issues that engineers must consider when selecting a remedial technology. Both of these issues fall under the site assessment category and would be completed during an RI/FS or an RFA. Shown in Figure 3.1 is a flow chart of the steps in selecting a remedial technology. The ideal procedure for technology design would include a dynamic iterative interaction of the following steps:

- Evaluate nature and 3-dimensional extent of contamination (phase distribution)
- Conduct a detailed review of the literature (USEPA maintains a variety of data bases with technology performance information)
- Identify clean-up goals and acceptable risks
- Estimate field parameters incorporating spatial variability
- Conduct a treatability study (bench-scale) to evaluate selected alternative(s)
- From treatability study, develop descriptive model and quantify thermodynamic and kinetic parameters
- Validate model on pilot scale for selected alternative(s)
- Conduct model sensitivity analysis
- Select and implement optimum design
- Monitor performance of full-scale remedy (evaluate efficacy of design)
- Refine remedy as needed to reflect full-scale (field) conditions

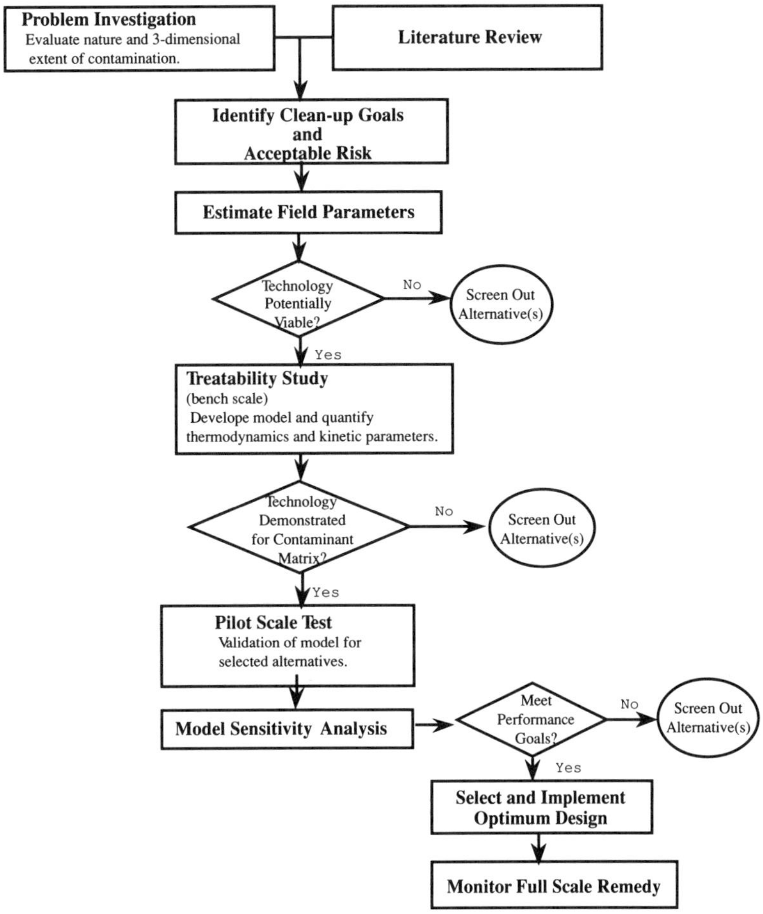

FIGURE 3.1
Technology Selection Process
Technology Selection

In the real world, as is the case with ideal procedures, it is sometimes improbable to complete. Usually due to political and economic pressures, the decision maker, in order to expedite the remediation process, must assume some risk and make decisions based on incomplete data.

3.1. Hazardous Waste Characteristics

3.1.1. Regulatory Characteristics

Congress has termed "hazardous waste" in SWDA § 1004 (5) as a "solid waste, or combination of solid waste, which, because of its quantity, concentration, or physical, chemical, or infectious characteristics, may:

1. Cause, or significantly contribute to an increase in mortality or an increase in serious irreversible, or incapacitating reversible, illness; or

2. Pose a substantial present or potential hazard to human health or the environment when improperly disposed of, or otherwise managed." [7]

Congress further mandated EPA to define which waste they would manage under Subtitle C of RCRA. EPA developed two ways in which hazardous waste could be identified as a listed waste or a characteristic waste. The four characteristics that classify a waste as hazardous are:

- Ignitability
- Corrosivity
- Reactivity
- Extraction Procedure (EP) Toxicity

If a waste exhibits any of these four characteristics it is termed hazardous and are regulated.

3.1.2. Chemical Characteristics

Table 3.1 shows some soil-based waste characteristics that are necessary to define the fate and transport of the contamination.

Table 3.1
Soil Based Waste Characteristics [12]

Chemical Class Acid Base Polar neutral Nonpolar neutral Inorganic	**Chemical Properties** Molecular weight Melting point Specific gravity Structure Water solubility
Chemical Reactivity Oxidation/Reduction Hydrolysis Precipitation Polymerization	**Soil Degradation Parameters** Half-life ($t_{1/2}$) Rate Constant (first order) Relative biodegradation
Soil Sorption Parameters Freundlich sorption constants Linear sorption based on organic carbon content (K_{OC}) Octanol-water partition coefficient (K_{OW})	**Soil Volatilization Parameters** Vapor pressure Henry's law constant Water solubility
	Soil Contamination Extent Mass released Concentration in soil Depth of contamination Date of contamination

A substance can usually exist in three phases: solid; liquid; and vapor. Depending on the different waste and environmental conditions, the contamination will usually partition into one or more phases (see Figure 3.2). It is typically not possible to obtain all of the parameters necessary to fully describe the fate and transport of the contamination. Because of economic restrictions, it is necessary to limit the characteristics to a few significant factors. The following sections, 3.1.2.1. through 3.1.2.4., are summarized from reference 13.

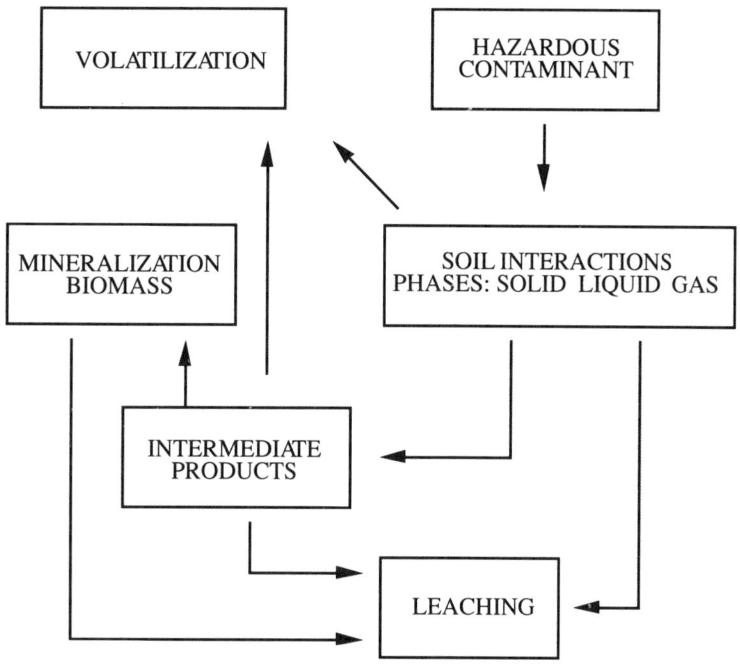

FIGURE 3.2
Fate of Hazardous Contaminants in Soils
Technology Selection

3.1.2.1. Octanol/Water Partition Coefficient

Definition — The octanol/water partition coefficient (K_{OW}) is defined as the ratio of a chemical's concentration in the octanol phase to its concentration in the aqueous phase of a two-phase octanol/water system.

$$K_{OW} = \frac{\text{Concentration in octanol phase}}{\text{Concentration in aqueous phase}}$$

Environmental Significance — The K_{OW} is a key parameter in describing the fate of an organic chemicals in environmental systems. It has been found to be related to the water solubility, soil/sediment adsorption coefficient, and the bioconcentration factors for aquatic species. The physical meaning of K_{OW} is the tendency of a chemical to partition itself between an organic phase (e.g. a fish, a soil) and an aqueous phase. Chemicals that have a low K_{OW} value (<10) may be considered relatively hydrophilic; they tend to have a high water solubility, small soil/sediment adsorption coefficients, and small bioconcentration factors for aquatic life. Conversely, a chemical with a large K_{OW} ($>10^4$)is considered hydrophobic and tends to accumulate at organic surfaces such as soil and aquatic species.

With only the K_{OW} value, an estimate can be made of three other properties of the contaminant as given in Lyman, Reehl, and Rosenblatt. With these estimated parameters, an evaluation of the partitioning of the chemical between the water/solid interface can be made. Based on these estimates, various treatment technology can be considered.

3.1.2.2. Vapor Pressure

Definition — The pressure exerted by a vapor that is in equilibrium with a liquid.

Environmental Significance — A contaminant with a high vapor pressure (>1mm Hg) is described as volatile, and therefore likely to partition itself to a greater extent into the air than a non-volatile contaminant (low vapor pressure) would. Knowledge of the volatilization rates are necessary to determine the mass of chemical that is transferred the air from the water. However, vapor pressure is not the only factor controlling volatilization. Volatilization is also dependent on the contaminants solubility, molecular weight, and physical site conditions such as the nature of the air-water interface (mass transfer coefficient). But it is possible with the evaluation of the contaminant vapor pressure to estimate the likelihood of volatilization.

3.1.2.3. Dissociation Constant

Definition — The acid dissociation constant (K_a) is the equilibrium constant of the following reaction (assuming a unit activity):

$$HA + H_2O \longleftrightarrow H_3O^+ + A$$

$$K_a = \frac{[H^+][A^-]}{[HA]}$$

Environmental Significance — By knowing the K_a value and the pH of the solution, it is possible to estimate the relative proportions of chemical species in solution. If the speciation of the chemical is known, it is possible to estimate other properties such as amount of adsorption and complexation with other ligands that occur in the solution.

3.1.2.4. Biodegradation Rate

Definition — Biodegradation rate is a time measurement of the amount of compound (usually organic) that is converted from one form to another by a bio-mass acting on the compound.

Environmental Significance — Biodegradation plays an essential role in explaining the loss (i.e. breakdown) of organic compounds in the soil. The eventual mineralization of organic compounds—i.e. their conversion to inorganic substances—can be attributed almost entirely to biodegradation. For the three other characteristics listed above there exist procedures for estimating them from handbook methods. With biodegradation, no such method yet exists. "The reason being is that the variables that control the rates are not well understood, as they have not been examined across different chemical classes."[13] The biodegradation rate must be experimentally measured for each contaminant and site condition.

The four characteristics listed above are the most important factors in controlling the fate of the contamination in the environment. Three of four of these characteristics can be estimated from handbook procedures giving the decision maker a quick way to initially evaluate different possible chemical treatment techniques. After an initial selection has been made using these roughly estimated characteristics, further study can be completed to verify the preliminary selection.

3.1.3. Characterization Parameters for Selected Treatment Techniques

In this section the characterization factors that should be considered, to verify applicability of a technology selection, are given for some typical treatment technologies. All of these tables were adapted from reference 12. A more complete analysis of this topic will be given in latter modules.

3.1.3.1. Bioremediation

TABLE 3.2
Characterization Parameters for Biological Treatment Techniques [12]

Treatment technique	Parameter	Purpose and comments
General	**Physical:**	
	Moisture content	To determine the treatability of the material and the treatment process of choice
	Field capacity	
	pH	
	Temperature	
	Oxygen availability	
	Chemical:	
	Total Organic carbon (TOC)	To determine the treatability of the material and the treatment process of choice
	Redox potential	
	Carbon:nitrogen:phosphorus ratio	To determine mineral nutrient requirements.
	Biological:	
	Soil incubation test	To determine biodegradation potentials and to quantify biodegradation rates of contaminants
	Electrolytic respirometer test	To measure oxygen uptake and biodegradation.
	Culture studies	To determine the indigenous microflora or specifically adapted microflora to be used in the inoculum during enrichment procedures.
	Bacterial enumeration test (e.g., spread-plate techniques)	To determine the bacterial population density in the inoculum
	Microbial toxicity / growth inhibition tests	To determine biological activity in the laboratory.
Microbial degradation - Aerobic - Anaerobic	**Physical:**	
	Permeability of soil	To determine ability to deliver nutrients or oxygen to matrix and to allow movement of microbes.
	Chemical / Biological:	
	Contaminant concentration and toxicity	To determine viability of microbial population in the contaminant zone
	Existence of primary substrates	To determine feasibility of co-metabolism

3.1.3.2. Solidification/Stabilization

TABLE 3.3
Characterization parameters for immobilization techniques[12]

Treatment technique	Parameter	Purpose and comments
General	**Physical:**	
	Description of materials	To determine waste handling methods (e.g., crusher shredder, removal equipment).
	Particle size analysis	To determine surface area available for binder contact and leaching.
	Moisture content	To determine amount of water to add/ remove in S/S mixing process.
	Density testing	To evaluate changes in density between untreated and treated waste.
	Strength testing	
	- Unconfined compressive strength	To evaluate changes in response to overburden stress between untreated and treated waste (e.g., material response to stress from cap).
	- Flexural strength	To evaluate material's ability to withstand loads over large area.
	- Cone index	To evaluate material's stability and bearing capacity.
	Durability testing	To evaluate durability of treated wastes (freeze-thaw and wet-dry durability.)
	Chemical:	
	pH	To evaluate changes in leaching as a function of pH.
	Interfering compounds	To evaluate viability of S/S process. (Interfering compounds are those that impede fixation reactions, cause adverse chemical reactions, generate excessive heat; interfering compounds varying with type of S/S).
	Indicator compounds	To evaluate performance of S/S (i.e. leaching)
	Leach testing	To evaluate performance of S/S
	Heat of hydration	To measure temperature changes during mixing.
Vitrification	**Physical:**	
	Depth of contamination and water table	Technology is applied to unsaturated soils
	Soil permeability	Dewatering of saturated soils may be possible. Technology is applied in unsaturated soils.
	Metal content of waste material and placement of metals within the waste	Greater than 5 to 15% by weight or significant amounts of metal interferes with process.
	Combustible liquid/solid content of waste	
	Rubble content of waste	
	Void volumes	Greater than 10 to 20% by weight interferes with process. Large, individual voids impede process.
In-situ	Presence of subsurface barriers(e.g., drums, large objects, debris, geologic formations)	To assess the feasibility of adequately delivering and mixing the S/S agents.
	Depth to first confining layer	To determine required depth of treatment.

3.1.3.3. Thermal Treatment

TABLE 3.4
Characterization parameters for thermal treatment techniques[12]

Treatment technique	Parameter	Purpose and comments
General	**Physical:**	
	Moisture content	Affects heating value and material handling
	Ash content	To determine the amount of ash that must be disposed or treated further.
	Ash fusion temperature	High temperature can cause slagging problems with inorganic salts having low melting points.
	Chemical:	
	Volatile organics	Allows determination of principal organic hazardous constituents (POHCs)
	POHCs	Allows determination of destruction removal efficiency (DRE).
	Total Chlorine, fluorine	To determine air pollution control devices for control of acid gases.
	Total sulfur, total nitrogen	Emissions of SO_2 and NO_2 are regulated; to determine air pollution devices.
	Phosphorus	Organic phosphorus compounds may contribute to refractory attack and slagging problems.
	Polychlorinated biphenyls (PCBs), dioxins (if suspected)	99.9999% DRE required for PCBs; safety considerations; incineration is required if greater than 500 ppm. PCBs present.
	Metals	Volatile metals (Hg, Pb, Cd, Zn, Ag, Sn) may require flu-gas treatment; other metals may concentrate in ash. Trivalent chromium may be oxidized to hexavalent chromium, which is more toxic. Presence of inorganic alkali salts, especially potassium and sodium sulfate, can cause slagging.
Rotary Kiln	**Physical:**	
	Particle-size distribution	Fine particle size results in high particulate loading in rotary kiln. Large particle size may present feeding problems.
Fluidized-bed	**Physical:**	
	Ash fusion temperature	For materials with a melting point less than 1600°F, particles melt and become sticky at high temperatures, which causes defluidization of the bed.
	Ash content	Ash content greater than 65% can foul the bed.
	Bulk density	As density increases, particle size must be decreased for sufficient heat transfer.

3.1.3.4. Physical/Chemical Treatment Processes

Table 3.5 lists characteristics that must be considered for physical/chemical treatment techniques. Included under the heading of physical/chemical techniques are chemical destruction, soil vapor extraction, and soil washing.

TABLE 3.5
Characterization Parameters for
Physical/Chemical Treatment Techniques[12]

Treatment technique	Parameter	Purpose and comments
General	**Physical:**	
	Type, size of debris	To determine need for pretreatment.
	Dioxins/furans, radionuclides, asbestos	To determine special waste-handling procedures.
Extraction - Aqueous - Solvent - Critical fluid - Air/steam	**Physical:**	
	Particle-size distribution	To determine volume reduction potential, pretreatment needs, solid/liquid separability.
	Clay content	To determine sorption characteristics of soil.
	Moisture content	To determine conductivity of air through soil.
	Chemical:	
	Organics	To determine concentration of target or interfering constituents, pretreatment needs, extraction medium.
	Metals (total) Metals (leachable)	To determine mobility of target constituents, postreatment needs.
	Contaminant characteristics: - vapor pressure - solubility - Henry's Law constant - Partition coefficient - Boiling point - Specific gravity	To aid in selection of extraction medium
	Total organic carbon (TOC), humic acid	To determine presence of organic matter, sorption characteristics of soil.
	Cation exchange capacity (CEC)	To determine sorption characteristics of soil.
	pH	To determine pretreatment needs, extraction medium
	Cyanides, sulfides, fluorides	To determine potential for generating toxic fumes at low pH.
Chemical dehalogenation	**Physical:**	
	Moisture content	To determine reagent requirements
	Chemical:	
	Aromatic halides	To determine concentration of target constituents, reagent requirements
	Metals	To determine concentration of other alkaline-reactive constituents, reagent requirements
	pH	To determine reagent requirements
Oxidation/ Reduction	**Physical:**	
	Total suspended solids	To determine the need for slurrying to aid mixing
	Chemical:	
	Chemical oxygen demand (COD)	To determine the presence of oxidizable organic matter, reagent requirements.
	Metals (Cr(III), Hg, Pb, Ag)	To determine the presence of constituents that could be oxidized to more toxic or mobile forms.
	pH	To determine potential chemical interferences.

TABLE 3.5 (continued)
Characterization Parameters for
***In Situ* Physical/Chemical Treatment Techniques**

Treatment technique	Parameter	Purpose and comments
(In Situ) **Vapor extraction** Vacuum extraction Steam-enhanced Hot-air-enhanced	**Physical:** Vapor pressure of contaminants Soil permeability, porosity, particle-size distribution Depth of contamination and water table	 To estimate ease of volatilization. To determine if the soil matrix will allow adequate fluid movements To determine relative distances; technology applicable in vadose zone
Soil flushing Steam/hot water Surfactant solvent	**Physical:** Presence of subsurface barriers (e.g., drums, large objects, debris, geologic formations) Hydraulic conductivity Moisture content (for vadose zone) Soil/water partition coefficient or Octanol/water partition coefficient CEC Alkalinity of soil **Chemical:** Major cation/anion present in soil	 To assess the feasibility of adequately delivering the flushing solution. To assess permeability of the soils. To calculate pore volume to determine rate of treatment. To assess removal efficiency and to correlate between field and theoretical calculations. To evaluate potential for contaminant flushing To estimate the likelihood of precipitation. To estimate the likelihood of precipitation; to estimate potential for plugging of pore volumes.

3.2. Physical Site Characteristics

3.2.1 Characterization Factors

The following is a brief outline of the factors to bear in mind when characterizing a site:

- Historical Data - Past land use and type and extent of contamination.
  - Maps
  - Land Deeds (land records)
  - Action or permits issued for or on the site
  - Industrial history of the site
- Soil Parameters
  - Hydraulic Conductivity
  - Porosity
  - Bulk Density
  - Organic Carbon Content
- Groundwater Flow
  - Direction
  - Magnitude

- Contamination Extent
  - Aerial Extent
  - Concentration Level

Table 3.6 lists some typical actions taken in a site characteristic survey [14].

Table 3.6
Typical Actions for Site Characterization[14]

Conduct geophysical survey
Ground Penetration Radar (Soil Density)
Resistivity Survey (Terrain Conductivity)
Electro Magnetic (Presence of buried metal i.e. storage drums)
Conduct soil-gas survey
Define geology by coring and split spoon sampling
Install depth-specific well cluster to evaluate hydrology
Conduct geohydraulic tests
Pump Test
Tracer Study
Perform soil analysis test
Total Organic Carbon
Grain Size Analysis
Clay content
Sample and analysis for 129+ pollutants initially
(only selected contaminants in subsequent samples)

3.2.2. Site Parameter Estimation

Over the past two decades, scientists and engineers have mainly focused on discovering and optimizing various treatment processes, such as bioremediation and SVE, in a controlled laboratory setting. Little regard has been given to the engineering considerations necessary to make these *in situ* process applicable to real-world contaminant situations. While the tables in the previous sections provide a starting point for evaluating different treatment technologies, more work is required.

Downey and Elliott give several examples in which *in situ* technologies (biodegradation, soil washing, and soil vapor extraction) have failed to meet lab performance estimates because of site specific field conditions [15]. These cases demonstrate that, because of the highly varied site and waste conditions that exist, field-scale pilot testing is a necessity. From these *in situ* test results a sensitivity analysis should be made on relevant parameters such as hydraulic conductivity, and organic carbon content. The sensitivity analysis will indicate those parameters most critically affect the site assessment and which factors should be carefully determined.

3.2.3. Diffusion Limitation

An important parameter to estimate in a pump-and-treat clean-up scheme is the sorption characteristics of the contamination to the soil. If sorption thermodynamics and/or kinetics are significant, the ability of the treatment to adequately remove the contaminant mass at the site may be compromised. In this case, contaminant removal is said to be diffusion limited.

Diffusion limitations occur when groundwater is removed "faster than the contaminants can desorb from the soil or diffuse from regions of low hydraulic conductivity to regions of higher hydraulic conductivity." [16] Data from 19 on-going and completed pump and treat remediation plans were analyzed by Haley et. al. and it was determined that sorption was important in virtually all cases [16]. If sorption properties of the contaminant are not considered during site characterization, gross underestimations of contaminant mass and remediation time are likely. For example [16]:

> At a site in South Carolina, sorption was not considered in estimating the total contaminant mass in the aquifer. After three years of pumping, the total mass actually removed was determined to be 148,000 pounds; however the groundwater concentrations comparisons (using improperly analysis data) indicated only 23,000 pounds had been removed. This difference can be largely explained due to the desorbing of contaminants from the solid to the liquid phase.

Soil cores from monitoring and extraction wells taken at depth can be analyzed to assess contaminant partitioning between solid and water phases. Organic carbon is usually identified as the most useful soil sorption parameter. By measuring the contaminant concentrations and hydraulic conductivity in these cores, a basis is provided for estimating clean-up times.

3.3. Monitoring and End-Point Validation

An issue that must be addressed in the early stages of any remedial decision is "how will the end-point of the remediation process be validated?" In other words, "when is the remediation complete?" The usual case is once the cleanup limit has been determined, the phases in which the contamination is deemed to have the largest health risk (i.e. greatest mobility) is monitored. The most important factor in end-point validation is the design of the monitoring network.

3.3.1 Monitoring Network Design

Two important issues that must be considered in monitoring well network design are [17]:

- Monitoring well size (volume of sample withdrawn)
- Monitoring well location that reflects
  - Character of contamination
  - Complexity of groundwater flow

In measuring hydraulic head, the screen length of the piezometer typically is not considered. This can cause erroneous results if the piezometer is also used as a monitoring well. Hydraulic head is not changed by the volume of water in which it is measured. If a well measures a small (point sample) volume or a large (nonpoint sample) volume no differences should be observed. For chemical concentrations, this is not the case; mixing (dilution) within larger nonpoint sampling devices will cause large plum size estimate errors to occur. Thus, all wells that are to be used as closure monitoring points must be designed as point samples. A point sample well should be designed with a small interior volume (small interior diameter), thereby limiting dilution.

A properly designed monitoring system adequately should define the contaminant plumes, width, length, and depth. General guidelines given in Domenico and Schwartz state that one line of wells down the middle of the plume will define the plumes length and depth. While a line of wells installed normal to the plumes migratory path will define the plumes width and depth. The questions of where the "middle" and "normal" positions of the plume are equally difficult to answer. An issue that must be stressed is that the design of the monitoring system should consider the overall complexity of the sites flow. Therefore sufficient hydrologic, geologic, and chemical data must be collected to adequately characterize the site so an intelligent determination of end-point monitoring stations can be made.

3.3.2. Monitoring Termination

Once contamination levels are reduced to the cleanup level in the pre-determined phase, remediation can be terminated. Although it is not typically advisable to stop monitoring at this point, take, for instance, the following example [18].

> A pump-and-treat remediation was undertaken at an IBM Dayton hazardous waste site in New Jersey. Groundwater at the site was estimated to be contaminated with 400 gal of volatile organic contaminants (mostly TCA and PCE). The maximum concentration of the contamination was 9590 ppb for TCA and 6132 ppb for PCE. Pumping was commenced in 1978 at an average rate of 300 gpm. Pumping and monitoring continued until 1984 when it was stop once VOC contamination was lowered to 100 ppb. However, after remedial operations stopped in 1984, it was discovered in 1988 that PCE concentrations rose to 12,558 ppb.

There could be a number of explanations for the possible rise in PCE concentration: desorption from soil into water, a continuing source due to presence of a NAPL or from continued waste dumping at the site, or contaminated groundwater from another site could be entering this site. In order to protect both human health and the environment, and to limit future liability against legal claims, continued monitoring is judicious. (It was later determined that the resulting contamination at this site was due to the presence of a DNAPL [16].)

4. SELECTION OF REMEDIAL TECHNOLOGY

There are a wealth of issues that must be considered in selecting a proper remedial technical action plan. No one paper could cover all of the possible circumstances that will arise when dealing with hazardous waste. This module has tried to highlight some of the most important issues:

- Regulatory Issues
- Hazardous Waste Characteristics
- Site Characteristics

If the decision maker has a sufficient understanding of a majority of the issues in each category then an adequate determination can be made on a remedial plan. But keep in mind that remediation of waste sites is not an exact science. A past EPA Superfund head described the problems associated with hazardous waste site remediation as "the 'granddaddy' of engineering judgment calls." He goes on further to say that "there may be several remedies that will do the job, depending on one's point of view of factors such as permanence, cost effectiveness, and public reaction." [6] There are no right answers, only bold hypothesis, when it comes pollution remedies.

REFERENCES

1. Sarno, D. J. Making Cleanup Decisions at Hazardous Waste Sites: The Clean Sites Approach, *Journal of the Air and Waste Management Association.* **41**: 9. 1991, pp. 1172-1175.

2. Wentz, C. A. *Hazardous Waste Management.* (McGraw Hill, New York, 1989).

3. Anderson, R. Soil Remediation Techniques at Uncontrolled Hazardous Waste Sites - Critical Review, *Journal of the Air and Waste Management Association.* **40**: 9. 1990, pp. 1232-1234.

4. U.S. Congress Comprehensive Environmental Response Compensation and Liability Act, *Selected Environmental Law Statutes.* (West Publishing Co., St. Paul, Minn, 1980).

5. Arbuckle, J. G. and e. al. *Environmental Law Handbook.* 10. (Governmental Institutes, Inc., Rockville, MD, 1989).

6. Porter, J. W. Hazardous Waste Cleanup Programs: A Critical Review, *Chemical Engineering Progress.* **96**: 4. 1989, pp. 16-25.

7. U.S. Congress Solid Waste Disposal Act, *Selected Environmental Statutes.* (West Publishing Co., St. Paul, Minn., 1976)

8. U.S. EPA *RCRA Orientation Manual.* EPA/530-SW-90-036. (Office of Solid Waste, 1990).

9. Nadim, F. *State Hazardous Waste Cleanup Standards, Regulatory Policy, and Cleanup Levels.* (Environmental Research Institute/The University of Connecticut, April 1992, 1992).

10. U. S. EPA *Protecting the Nation's Ground Water EPA's Strategy for the 1990s.* (EPA Ground-Water Task Force, May 8, 1991, 1991).

11. Longest, H. Groundwater Remediation, *Environmental Science and Technology.* **25**: 5. 1991, pp. 810-810.

12. Sims, R. C. Soil Remediation Techniques at Uncontrolled Hazardous Waste Sites - A Critical Review, *Journal of the Air and Waste Management Association.* **40**: 5. 1990, pp. 704-732.

13. Lyman, W. J., W. F. Reehl and D. H. Rosenblatt. *Handbook of Chemical Property Estimation Methods.* (McGraw-Hill, Washington DC, 1990).

14. Korte, N., P. Kearl and D. Smuin. State-of-the-Art Approach to Hazardous Waste Site Characterizations, *Environmental Management.* **13**: 6. 1989, pp. 677-684.

15. Downey, D. C. and M. G. Elliot. Performance of Selected In Situ Soil Decontamination Technology: An Air Force Perspective, *Environmental Progress.* **9**: 3. 1990, pp. 169-173.

16. Haley, J. L., B. Hanson, C. Enfield and J. Glass. Evaluating the Effectiveness of Ground Water Extraction Systems, *Ground Water Monitor Review*. **11**: 1. 1991, pp. 119-124.

17. Domenico, P. A. and F. W. Schwartz. *Physical and Chemical Hydrogeology*. (John Wilet & Sons, New York, 1990).

18. Travis, C. C. and C. B. Doty. Can contaminated aquifers at Superfund sites be remediated?, *Environmental Science and Technology*. **24**: 10. 1990, pp. 1464-1466.

D - Materials Handling

MATERIALS HANDLING

Instructional objectives:

- Describe the phases of excavation
- List the elements to be considered when creating an excavation plan
- Discuss equipment used for excavation

1. INTRODUCTION

1.1. General Information

A hazardous waste site clean-up operation involves many intricate tasks prior to, during, and after treatment of the contaminants present. One area which is often not fully planned for, nor adequately executed, is materials handling. The reason for this is most likely the perception that materials handling (e.g. - drilling, excavating, transporting, etc.) has been practiced for many years and is considered well established. Unfortunately, the additional constraints of hazardous waste site operations can make materials handling a challenging task.

The objectives of materials handling on the hazardous waste site are rather straight forward:

- To remove the contaminated soil and debris, transport it to treatment or containment facilities, prepare it for treatment, and replace the soil in an environmentally sound, safe, and economical fashion

Failure to meet any of the above objectives will result in a materials handling plan that is inadequate. The results of an inadequate materials handling plan are often increased cost, increased probability that treatment goals will not be met, and treatment equipment failure.

1.2. Hazardous Waste Site Materials Handling

Materials handling is concerned primarily with the solid phase, most often soil, at a hazardous waste site. This does not mean that the gas and liquid phases are not important on a site, only that the problems associated with them, as far as handling is concerned, are mostly small compared with the solid phase. *In situ* methods of remediation are not included in this discussion, as they have limited contaminated materials handling requirements. The principle focus will be on a hazardous waste site from which soil is excavated and either transported off-site for treatment or treated on-site and returned to the excavation.

Once excavation begins, contamination control becomes critical if spread of contamination is to be avoided. Figure 1.1 describes some typical contingencies for which planning should be made in advance.

Solid Phase
Non-Homogeneities
"Buried Treasure"
Stock Piling
Excavation Safety
Accountability
Cost Increases
Vector Spread

Liquid Phase
Surface Run-Off
Surface Run-On
Rainwater Accumulation
Groundwater
Vector Spread

Airborne Phase
Vapor Release
Contaminated Dust
Treatment Discharge

FIGURE 1.1
Contamination Contingencies
Materials Handling

1.2.1. The Airborne Phase

Anyone who has seen a construction site would probably agree that, during dry weather, dust control can be a substantial concern. On a hazardous waste site, both vapor and suspended solids can form an airborne phase that may require control. Fortunately, controls are relatively simple in most cases.

The key to controlling contamination spread by dust is through control of generation. During excavation and processing, points of motion (e.g., vehicular traffic, along conveyors, etc.) are where control efforts should be concentrated. Some effective control measures include:

- Covering of stockpiles
- Vehicle motion restrictions (i.e. - speed limits, turning areas, etc.)
- Use of fine water sprays

Respirators and protective clothing afford workers protection against inhalation of the dust and skin/eye contact with hazardous chemicals absorbed on the dust particles.

As with dust, vapor hazards generally are worst during excavation. As soil is disturbed, volatile and semi-volatile components transfer to the air. If sufficient concentrations are present, a true hazard requiring controls can exist. The concentration limits for worker protection, as well as for the battery limits of the site, should be established prior to start of clean-up operations. In addition to hazardous concentration limits, a means to monitor contaminants should be established prior to operations. A suitable monitoring system will insure operator and community safety as well as allow flexibility in the use of controls. Simple vapor hazard control includes:

- Adjustment of the rate of excavation

- Alternating zone excavation between heavy and light contamination
- Use of water sprays.

Under extreme conditions, airborne phase control may require the use of an enclosure. This may range from a windscreen up to a double walled structure under negative pressure. An example of the latter is shown in Figure 1.2 (Sprung Instant Structures, San Francisco, CA). Factors affecting the use of such a structure include the level of risk associated with the waste, proximity to populations, and unusual climactic conditions. An example of the use of enclosures is the removal of war gases from underground storage tanks.

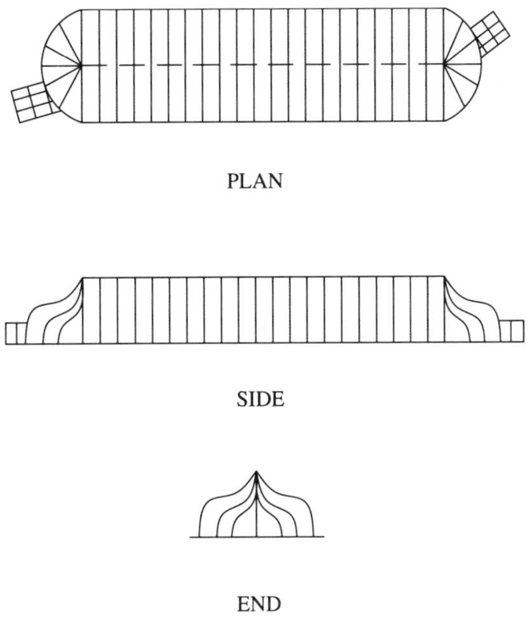

FIGURE 1.2
FULL ENCLOSURE OF AN EXCAVATION
Materials Handling

1.2.2. The Liquid Phase

The liquid phase is often not considered until it becomes a major problem. It typically stems from two sources: rain water or groundwater. As contamination is uncovered by excavation, contamination of rainwater can become a serious concern. Rainwater is commonly treated by activated carbon, but since cost of treatment is high, every effort should be made to reduce the amount that needs treatment.

Preventative measures include:

- Run-on protection in the form of dikes or ditches in conjunction with direct accumulation avoidance using covers
- Evaporation systems that reduce the volume of water to be treated (see Figure 1.3)

- Covering
- Drainage control

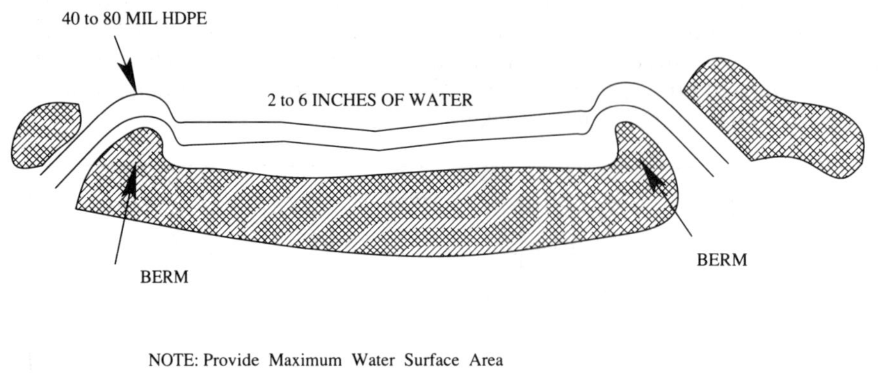

FIGURE 1.3
Simple Water Evaporation System
Materials Handling

If groundwater itself is present above the base of the excavation, consideration should be made as to the benefits of controlling the groundwater in the excavation. Again, evaporation may provide a simple system for handling groundwater.

Ecological impacts may be of concern if contaminant levels are sufficient in standing water so as to pose a threat to animals drinking the water or to animals higher on the food chain. Fencing is usually sufficient. Again, if contamination is serious enough, enclosure may be necessary.

In all the above conditions, site climatology should be considered. Scheduling during optimum time period in terms of climate may provide the lowest cost solution to the problems of airborne or liquid phase contamination.

1.2.3. The Solid Phase

The remainder of this discussion will focus on the solid phase, planning, excavation, and closure. Handling of the solid phase on a hazardous waste site can constitute on the order of 25 percent of the cost of remediation. If site analysis is included as a handling cost, this percentage increases substantially.

The contingencies in Figure 1.1 are not always simple to overcome, and should always be planned for. They constitute the reason that materials handling on a hazardous waste site is not often simple or cheap. Nonhomogeneities top the list in Figure 1.2 because it is rare that a hazardous waste site contains a single soil type, single contaminant, and even distribution of contaminants. The "buried treasure" refers to drums, explosives, equipment, and other unknowns which often surface during excavation. Stock piling of contaminated soil is often necessary to allow a continuous operation of expensive treatment systems. Excavation safety is imperative, especially when unknowns like radioactivity, flammable liquids, explosive mixtures, etc. are being excavated. Accountability requires analytical measures of the contaminants (not always a simple chemical test), treatment to standards, and decontamination. To accomplish this, handling systems must be designed to allow various processes to be performed. And the costs have to be estimated and controlled.

2. EXCAVATION/TRANSPORTATION/MIXING

The successful removal and transportation of the hazardous waste off-site requires prior planning. The aforementioned site and waste characteristics are considered and a materials handling plan is made. One should avoid the temptation to simply "dig and haul," as complications that could have been handled simply with proper prior planning will likely arise.

2.1. Detailed Excavation Plan

As part of treatment operations, a detailed excavation plan should be prepared. This plan should include:

- Limits of the excavation
- Stockpiling locations for uncontaminated and contaminated soils
- Excavation access
- Maintenance and supply of excavation equipment (i.e. fuel for excavation equipment)
- Excavation methods
- Contingency plans
- Control of excavation operations.

The excavation plan should take into account the rates of excavation and the rates of processing through treatment or transportation. Finally, the detailed excavation plan should include measures to account for the effects of wind and water.

2.2. Site Layout

The first consideration in materials handling plan or estimates must be the layout of the site. The site layout incorporates the site and waste characteristics into an operating strategy. This will include how much soil must be removed, over how large an area, weather patterns and effect of weather in operations, design, and other factors such as site access and contingencies. It is assumed by this point that sufficient information on site characteristics exists to formulate a workable plan or estimate.

2.2.1. Estimates of Excavation

Prior to digging, an estimate of excavation is a scientific guess at how much soil must be removed, how wet it will be, and what deviation from a homogeneous solid might be expected. Generally, the more data available, the better the guess. Of all information estimated at a hazardous waste site, the amount of soil for treatment is likely to have the most impact on clean-up costs. It is likely that soil treatment will be on a unit weight or unit volume of soil treated basis for time and cost estimates. This will certainly be true for a containment (hazardous waste landfill).

2.2.1.1. Soil for Treatment/Containment

The amount of soil to be treated is not normally a well established quantity as it most frequently is determined based on a treatment standard. As mentioned previously, the cost of treatment is based on the amount of soil and acceptably minimizing this amount reduces cost. From a theoretical standpoint, the amount of soil contaminated is fixed and should present no problems in terms of estimating the amount or excavating the soil. In practice, large deviations of the amount of soil estimated versus the treated are often found. One source of differences is that sampling sometimes does not fully determine the extent of contamination. Another potentially serious deviation occurs due to excavation techniques. Consider the estimate of contaminated soil and the actual amount of soil excavated in Example 2.1.

Example 2.1. A hazardous waste site is determined to require excavation of contaminated soil for treatment. The soil is contaminated only at the surface to a depth of approximately 6 inches. The site itself is rather large, approximately 1000 ft x 2000 ft. The estimate of soil for treatment is:

$$\mathbf{1000\ ft \times 2000\ ft \times 0.5\ ft \times 1yd^3/27\ ft^3 = 37{,}000\ yd^3}$$

The excavation technique used takes soil to 0.75 ft depth due to surface conditions. The resultant actual soil volume is:

$$\mathbf{1000\ ft \times 2000\ ft \times 0.75\ ft \times 1yd^3/27\ ft^3 = 56{,}000\ yd^3}$$

A difference of 50% from the estimate. The point here is to consider a reasonable excavation and treatment, due only to techniques of removal.

2.2.1.2. Limits of the Excavation

In addition to being unable to surgically remove contaminated soil, there is often a substantial amount of uncontaminated soil that must be removed in order to expose the contaminated soil. This soil may be an uncontaminated soil cap, which must be removed, or an uncontaminated soil excavated to allow proper sloping of excavation. This uncontaminated soil must be excavated, stored (normally on the site), and replaced. Care should be exercised to not contaminate this soil, causing the amount of soil needing treatment to increase. Once again, accurate estimates of site conditions are vital to make accurate excavation estimates. Figure 2.1 shows an example excavation of 1000 yd^3 of contaminated soil. Uncontaminated soil, which must be excavated, amounts to almost 1,505 as much as the contaminated soil. Failure to plan for this soil will complicate site access and control, may raise overall costs, and will result in unnecessary contamination of uncontaminated soil.

2.2.2. Plan for Contingency

Numerous contingencies are possible during excavation and transportation. They range from relatively simple occurrences, such as equipment failure, all the way to catastrophic events such as fires and explosions. The contingency plan should concentrate on those events that are correctable through the plan and leave other events to planning documents, such as the safety plan. Contingencies which occur with some frequency on hazardous waste sites are:

- Increased volume of contaminated soil
- Buried objects
- Contaminants other than those expected
- Higher concentrations of contaminants

While detailed plans aren't expected for contingencies, they should provide a direction with which to take action without substantial delays to the project. In addition, actions to take for unplanned contingencies should be described.

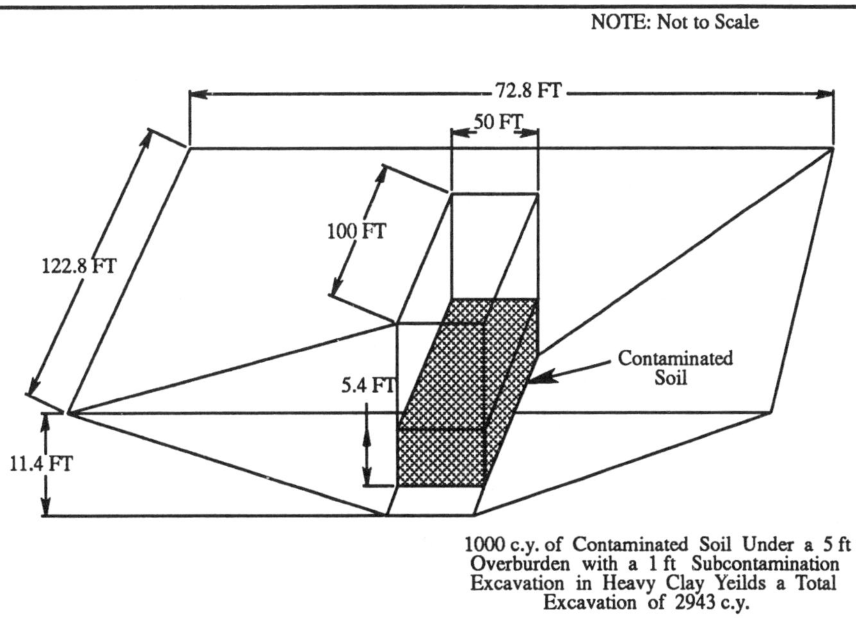

FIGURE 2.1
Example Excavation of 1000 c.y. of Soil
Materials Handling

2.2.3. Wind

The effects of wind were described previously, particularly concerning the airborne phase (see section 1.2.1) During site layout, wind direction plays a key role. To minimize the detrimental effects of wind, site layout should allow for minimum effects downwind. Figure 2.2 shows how a site should be positioned for wind effects. Note that wind directions change frequently and it is impractical to change direction of the site. For this reason, the predominant wind direction for the season in which most operations will occur should be used. Wind roses are often available from nearby airports. An often overlooked method of wind control is fencing.

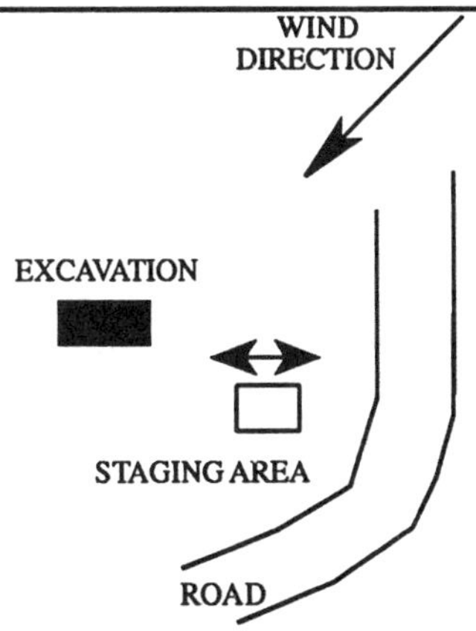

FIGURE 2.2
Proper Site Layout Considering Predominant Wind Direction
Materials Handling

2.2.4. Rain

The effects of rain on site layout are not easily accounted for. Topography of the site may not allow for rain controls to be established by layout. If possible, the excavation should be isolated using drainage ditches, etc. Careful planning by seasonal weather predictions may also reduce the problems associated with rain. It is often better to temporarily cover an excavation and reduce the amount of contaminated water due to rain accumulation than to attempt to continue operations.

2.2.5. Site Access Control

The reason for site access control, outside of safety concerns, is to minimize the spread of contamination. The rule of thumb is, unless access is required for operations to continue, don't allow it. Unfortunately, usually before remediation activity begins, access through a site is not always controlled rigidly and contamination spread invariably occurs. This spread means more equipment to be decontaminated and increased risk to personnel.

2.3. Excavation Techniques

As mentioned in the introduction, excavation techniques are often considered standard, and, to a large degree, the methods used on hazardous waste sites can be considered standard techniques with constraints. Excavation proficiency comes from years of excavation experience. A brief introduction to excavation techniques will be given here to familiarize the student with terms and references.

2.3.1. Standard Techniques

While the variation of existing equipment for excavation is fairly large, most can be categorized effectively in one of five categories of a combination of the categories. The types of excavation equipment, as well as other equipment such as dump trucks, belt conveyors, etc., have been described by Nichols in terms of their operation and use. The five categories of excavation equipment considered here are:

- Hoes
- Tractor loaders
- Clamshells
- Dipper shovels
- Draglines

This list is not exhaustive but makes up the majority of equipment in use on hazardous waste sites today.

2.3.1.1. Back Hoe Shovels

The most frequently found excavation equipment on a hazardous waste site is the back-hoe shovel. A representative back-hoe shovel is shown in Figure 2.3.

The back-hoe shovel:

- Is capable of digging below grade quickly and efficiently
- Has a depth of excavation limited by the size of the boom
  - Newer models of back-hoes are capable of depths exceeding 40 feet.
- Is good for digging trenches and skimming surface

As the majority of hazardous waste sites where excavation is required have contamination concentrated within 40 feet of the surface, the back-hoe stands out as a versatile piece of equipment capable of meeting most of the site needs. A front-end-loader with back-hoe attachment may be capable of performing all materials excavating and transferring tasks on a site. An example of a back-hoe shovel specification is shown in Figure 2.4.

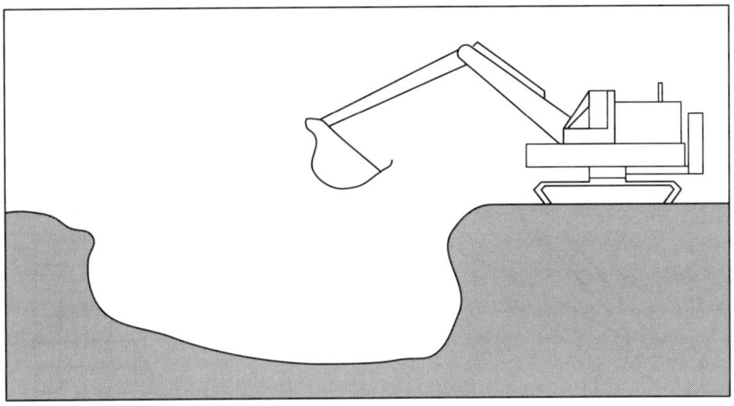

FIGURE 2.3
Example Back-Hoe Shovel
Materials Handling

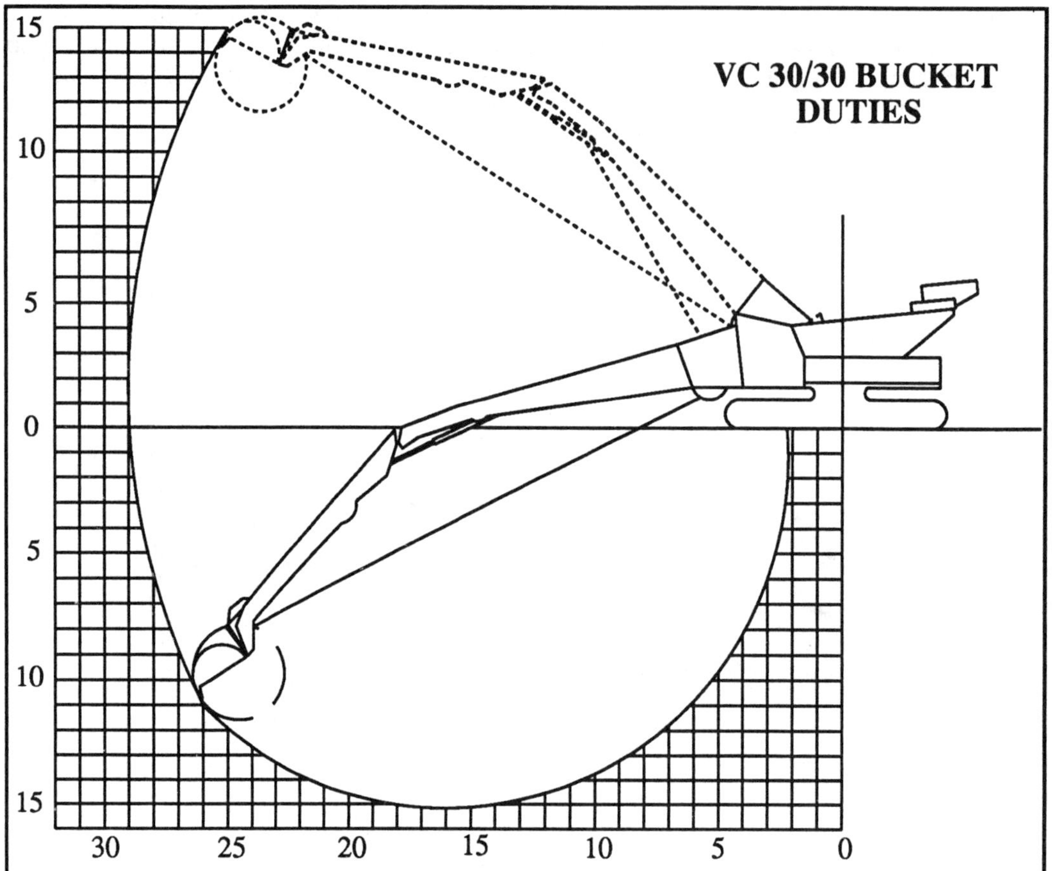

BUCKET SIZE (EXCAVATING)	1 cu. yd.
MAX LOAD AT 80% CAPACITY	3900 lbs
MACHINE WEIGHT	55.16 tons
TRAVEL SPEED	2 mph

FIGURE 2.4
Working Range and Technical Specifications of the Priestman VC 30/30 Back Hoe Shovel
Priestman, LTD, Hedon Road, Hull, North Humberside, England
Materials Handling

2.3.1.2. Tractor Loaders

Tractor loaders are versatile machines capable of moving soil from piles either to receiving processing equipment or to transportation equipment. General site work, such as simple grading, cutting banks, etc., can be performed with a tractor loader. As mentioned previously, a combination front-end-loader/back-hoe is often used on hazardous waste sites to achieve all of the excavation tasks. A front-end-loader is shown in Figure 2.5. Higher capacity loaders are normally mounted on tracked chassis.

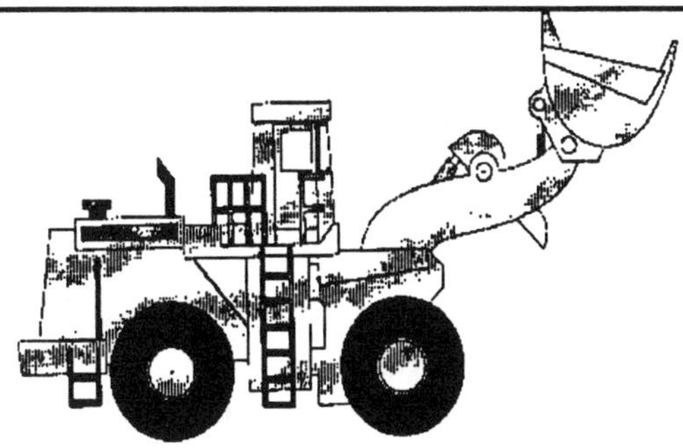

FIGURE 2.5
Front End Loader
Materials Handling

2.3.1.3. Dipper Shovels, Clamshells, and Draglines

The last three categories of excavation equipment are shovels, clamshells and draglines. Each has its advantages and disadvantages. Their use on hazardous waste sites would be determined by the site conditions.

The dipper shovel:

- Is generally restricted to operation at or just below grade
- Is capable of digging under harsh site and soil conditions, such as in heavy clays, soils containing large rocks, heavy gravel, etc.
- Is fast as it is normally equipped with a bottom opening bucket.

The clamshell:

- Is used when excavation must be made to large depths.
- Has a depth limit equal to the cable length, which can extend to over 100 feet
- Is generally slow and not suited to heavy or packed soils.

The clamshell bucket is dropped into the excavation on a cable. .

The dragline:

- Is used for surface excavation, especially in wet soils
- Is generally slow and not suited to heavy or packed soils

The bucket is "thrown" from the rig on a cable then dragged across the surface, digging as it is retrieved.

While not in frequent use at hazardous waste sites, the above rigs should not be automatically discounted. Depending on the site conditions, their particular strength or application may make them the equipment of choice. Figure 2.6 illustrates the above three pieces of equipment.

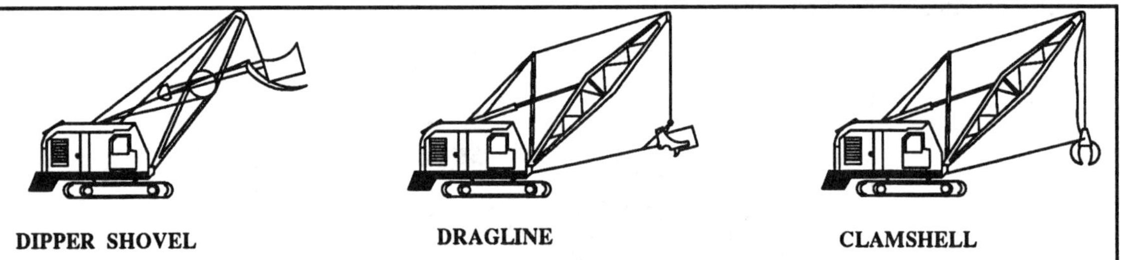

FIGURE 2.6
Dipper Shovel, Clamshell, and Dragline
Materials Handling

2.3.1.4. Nonstandard Equipment

The clean-up of hazardous waste sites is unique in many cases when it comes to materials handling equipment. The industry has responded and continues to respond to the need for solutions to equipment requirements. In some instances, modifications to existing equipment are made. In other cases, new equipment items have been designed and manufactured.

2.3.2. Contamination

Excavation of contaminated materials places workers at potential health and safety risks. Contamination may pose a danger as it reacts with the equipment being used at the site. Working with contaminated materials has resulted in the adaptation of equipment to avoid injury to personnel and dangerous interaction between equipment and contaminant. For example: rigs, which allow compressed air to be carried on board for extended periods, permit operators to continue working in level A or level B protection; cabs have been specially designed for explosion and heat protection; spark arrestors and heat shields on equipment engines help reduce risk of initiating fire or explosion; and specially designed buckets or bucket teeth can reduce spark production inside an excavation. If highly corrosive atmospheres are expected, frequent decontamination may be warranted.

2.3.3. Buried Objects

Buried objects present a difficult situation during excavation. Landfills for example may contain any number of buried items. The materials handler is faced with making decisions on how to approach, remove, and handle these items. The most frequently uncovered objects are drums and cylinders. These items may contain pressurized gases; their condition is likely to be poor due to being buried; and handling may trigger a catastrophic event. Various tools have been adapted to handle drums and cylinders at hazardous waste sites. Drum grapplers are available for attachment to excavators to allow drums to be picked up from a distance and manipulated, normally into and overpack container. Sparkless spikes are available for puncturing drums or cylinders. These are normally attached to excavation equipment to allow remote puncturing of drums or cylinders. Special equipment for safely depressurizing cylinders and repressurizing cylinders and repressurizing the contents in new cylinders is also available. In all cases, safety concerns should take precedence over operations.

2.4 Mixing

Mixing is essential as a fundamental operation for all chemical processes.

Liquid blending depends on flow current.

Due to multiple phases, soil/sludge mixing cannot employ flow currents. It is difficult to reach homogeneity.

For mixing pastes, the material must be brought to an agitator or the agitator must visit all parts of the mixture. Impeller mixers cannot be used for viscosities greater than 2KP. Heavy plastic masses and viscous material cannot flow to the suction side of the impeller.

There are various types of mixers:

- Change-can
- Kneaders
- Masticators
- Muller
- Pugmills

Kneaders are good for deformable or plastic solid. Kneaders work by first squashing a mass flat, then folding the mass over itself and squashing it again.

Pugmills are commonly used at hazardous waste sites. Pugmills consist of blades/knives set in a helical pattern on a horizontal shaft which turns in a open trough. There are two different types of pugmills: the single-shaft mills which operate in a closed chamber and the double-shaft mills which operate in an open trough. Pugmills can be operated under a vacuum.

2.4.1 Criteria for Mixer Effectiveness

- Time required
- Power
- Product characteristics

Spot samples are taken to determine the homogeneity of the mixture, using a mixing index (I_p) for calculation.

$$I_p = \frac{\sigma_o}{S} = \sqrt{\frac{(N-1)\,\mu\,(1-\mu)}{\sum x_i^2 - \bar{x}\sum x_i}} = \frac{\text{standard deviation at } t = 0}{\text{standard deviation of mixed samples}}$$

where:

$\sigma_o = \sqrt{\mu\,(1-\mu)}$

N = number of samples

μ = overall fraction of tracer in mix

I_p is unity initially and increases to ∞ as t --> ∞.

See Figure 2.7 for a representative graph of the relationship of mixing time to mixing index.

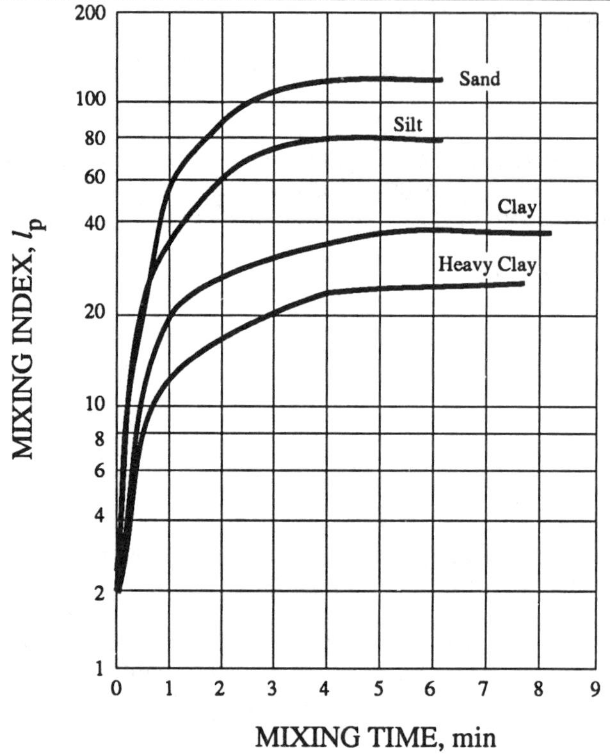

FIGURE 2.7
Mixing Index of Various Soils
Materials Handling

3. MITIGATION

Materials handlers must be concerned with the spread of the contaminants they are handling. Vapor hazards and effects of water have already been discussed. The most significant cause of contaminant spread comes from the handling of the contaminated soil. Heavy equipment tends to "track" the contaminated soil wherever the equipment travels. To a lesser extent, personnel working in the site also cause spread of contamination. The materials handler is a key player in insuring that contaminants are contained.

Operations not already discussed in terms of contaminant spread are site control and decontamination. Site control limits spread of contamination by limiting movement of contaminants and contaminated equipment. Decontamination removes the contaminant from equipment and personnel. Site control is inexpensive as it is a management technique. Decontamination on the other hand is very expensive in terms of time and resources. Maximizing site control works to minimize the amount of decontamination required.

——3.1. Site Control

Site control is a matter of common sense applied to an operation. Operating areas are established wherein contamination is expected to exist. All activities inside this area or "contamination zone" are considered contaminated. Any time any equipment, supplies, or personnel pass from one zone to another, contamination control must be considered. In planning the site layout, this requirement should be carefully considered. Equipment should be provided with enough room to operate and be serviced (fuel, oil, etc.) and stored without having to pass from the contaminated zone to the uncontaminated zone. Likewise, operation in the uncontaminated area should be planned to allow performance without having to enter the contaminated area. If transfer must be made from the contaminated zone to the uncontaminated zone, decontamination must be performed.

——3.2. Decontamination

The bulk of decontamination efforts will be performed in conjunction with materials handling or processing equipment. Decontamination may require extensive preparations and is normally time and resource intensive. Efforts should be made to minimize decontamination requirements. This is accomplished by properly planning the site layout and thinking through operations. It is a far too common occurrence to see decontamination planning independent of the site plan. If decontamination planning occurs as an afterthought, the risk of increased cost and substantial impact to operations may be great. Plan for decontamination within the framework of the operation.

An instructional videotape prepared by the U.S. Army Toxic and Hazardous Materials Agency describes decontamination for PCB contaminated equipment. The same principles and requirements apply to most contaminants at hazardous waste sites. Pay particular attention to the use and control of water as well as sampling for completeness. The level of worker protection will be dictated by the contaminant(s) present. Also note the use of excavation equipment in the video.

(show videotape)

4. COSTS

When removal of the soil is required, costs can be very high and their estimates should be carefully made. Alternatives for treatment are compared using many factors such as availability, applicability, effectiveness, safety, and cost. The cost factor will include the cost of materials handling. In order to compare alternatives on an equal basis, these costs should be estimated using the same methods. Methods of treatment that do not require excavation and removal of soil are exempt from more than simple excavation costs. Other costs which occur due to excavation are decontamination costs, site control measures costs, special protective measures, and others. We will concentrate on the cost of digging but realize that the other costs can be a substantial portion of the materials handling costs.

——4.1. The Cost of Digging

The major cost of materials handling on a hazardous waste site comes from the excavation of the soil. The materials handler must balance the cost of increased rate of excavation with the cost of stockpiling and subsequent movement. This becomes a function of the size of the excavation, the layout of the site, and the treatment process handling capability. Ideally, the rate of excavation should match the processing rate at the site. This condition is met when the contaminated soil is being transported off site in bulk. Treatment systems on site are rarely able to process materials at the rate at which they can be excavated. Therefore, selection of equipment requires more thought.

Using data from the 1991 edition of Means Heavy Construction Cost Data, Figure 4.1 shows the cost of different excavation methods versus their output capacity. It would appear from this figure that larger capacity equipment would always be preferred. This would be the case when, as mentioned before, the processing rate equaled the excavation rate. Compare now Figure 4.2 (also from Means Heavy Construction Cost Data, 1991), which provides the cost per hour of the various pieces of equipment. It is apparent that a selection based on capacity has a substantial cost advantage. For this reason, the processing rate should be matched as closely as possible with the excavation rate. It may be possible to provide an optimum cost by stockpiling contaminated soil. This point is demonstrated in Example 4.1.

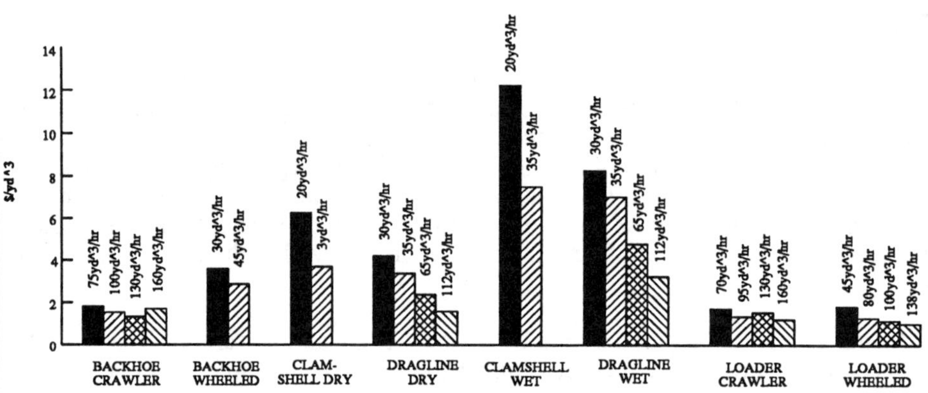

FIGURE 4.1
Cost vs. Total Hourly Capacity
Materials Handling

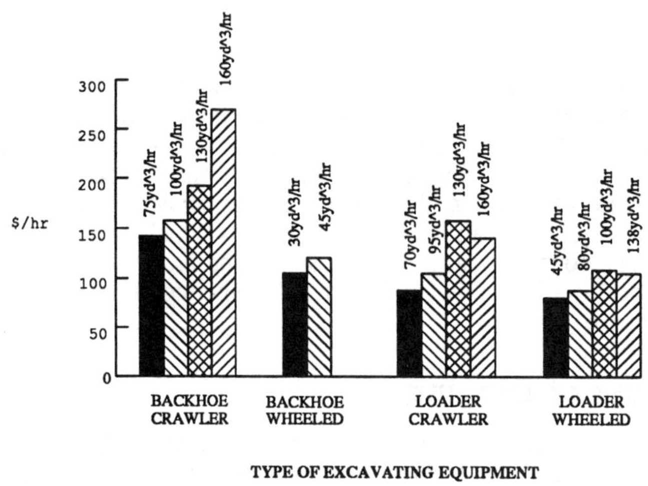

FIGURE 4.2
Cost vs. Hourly Capacity
Materials Handling

Example 4.1. Situation: A hazardous waste site requires on-site incineration of 10,000 yd^3 of contaminated soil. The excavation plan calls for excavation of 20,000 yd^3 of uncontaminated soil in addition to the contaminated soil. The excavation is not expected to exceed 14 feet depth. Sufficient site area is available for stockpiling and airborne contaminant risk is low. Select excavation equipment. The processing rate of the incinerator is 45 yd^3/hr.

Option 1: Select 3/4 yd^3 bucket wheeled back-hoe to match processing rate of 45 yd^3/hr and excavate directly to processor.

Excavation cost = \$2.81/yd^3 x 30,000 yd^3 = \$ 84,300

Option 2: Select 3 yd^3 capacity back-hoe and operate at processing rate of 45 yd^3/hr and excavate directly to processor.

Excavation cost = \$268.801hr x 1 hr/45 yd^3 x 30,000 yd3 = \$ 179,200

Option 3: Select 3 yd^3 capacity back-hoe and operate at processing rate of 160 yd^3~r, excavating to stockpile. Use 3/4 yd^3 capacity front end loader to transfer from stockpile to processor at 30 yd^3/hr

Excavation cost = \$1.68/yd^3 x 30,000 yd^3 + \$78.30/hr x 1 hr/45 yd^3 x 10,000 yd^3 = \$ 67,800

Note that only excavation costs were considered in Example 4.1. Replacement of the soil into the excavation must normally be accomplished as well. Additionally, costs for water removal, water treatment and control, dust control, site preparation, decontamination, and site exit also fall under materials handling.

4.2. Other Materials Handling Costs

Costs other than the actual digging and replacement of soil must be accounted for on a site-by-site basis. As a general rule, as protection levels increase, costs for materials handling increase. As concentration requirements for decontamination levels decrease, costs increase. Requirements for airborne and liquid contamination controls will increase handling costs not only for the controls themselves but also for the impact on operations they cause. All of the previously mentioned characteristics of a site should be taken into account when determining an estimate of handling costs.

REFERENCES

1. Terzaghi, K. and R. B. Peck. *Soil Mechanics in Engineering Practice.* 2nd ed. John Wiley and Sons, New York, New York, 1967.

2. Nichols, H. L., Jr. *How to Operate Excavation Equipment.* North Castle Books, Greenwich, Connecticut, 1954.

3. *Means Heavy Construction Cost Data.* 1991 ed. Means, Kingston, Massachusetts, 1991.

4. U.S. Army Corps of Engineers. *Laboratory Soils Testing.* Department of the Army Headquarters. Engineer Manual EM1110-2-1906. Office of the Chief of Engineers, Washington, D.C., 1986.

5. *Chemical, Biological, and Radiological (CBR) Decontamination.* Department of the Army. Technical Manual TM3-220. HQDA, Washington, D.C., 1967.

6. *Manufacturers Literature on Excavation Equipment.* R-B (Lincoln) plc, Beevor Street, Lincoln, LN6 7DJ, England,

7. *Manufacturers Literature on Enclosures.* Sprung Instant Structures, 330 Townsend Street, Suite 216, San Francisco, California,

E - Engineering Principles

ENGINEERING PRINCIPLES

1. MASS BALANCE KINETICS AND EQUILIBRIUM

Mass balance equations govern the design or analysis of environmental engineering systems. Mass balance equations are comprised of the following components:

- Transport
  - Diffusive
  - Dispersive
  - Advective
- Reactions (e.g. sorption, hydrolysis, redox)
  - Kinetics
  - Thermodynamics (equilibrium)

In general:

Accumulation or Depletion = In - Out ± Rxn

Or for incompressible fluids:

$$\frac{\partial(nc)}{\partial t} + \nabla\{cq - nD\nabla c\} \pm s = 0$$

where:

n	= porosity
D	= diffusivity
c	= concentration
s	= source/sink
q	=specific discharge

Both kinetics and thermodynamics must be considered in the source/sink as well as the transport terms. Figure 1.1 provides a representation of diffusion limitation.

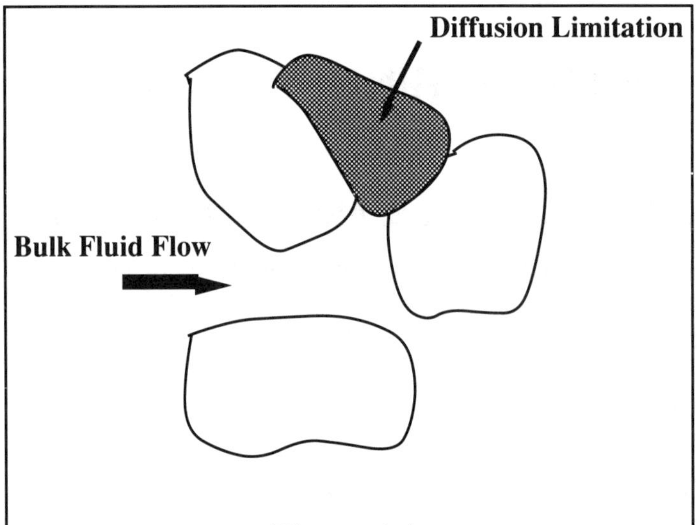

Figure 1.1
Diffusion Limitation
Engineering Principles

Mass transfer or chemical kinetics limitations may govern the reaction rather than equilibrium
An operational definition for equilibrium may be used.

Consider:

$$A \underset{k_2}{\overset{k_1}{\leftrightarrow}} B$$

$$\frac{d[A]}{dt} = k_2[B] - k_1[A]$$

At equilibrium:

$$\frac{k_1}{k_2} = \frac{[B]}{[A]} = K$$

See figure 1.2 for graph of the relationship between species concentration and time of reaction.

$$A = A_\infty + (A_o - A_\infty)\, e^{-\lambda t}$$

$$\lambda = k_1 + k_2$$

$$A_\infty = \frac{k_2}{k_1 + k_2}(A_o + B_o)$$

$$t_c = \frac{1}{\lambda} \quad \text{Characteristic time}$$

$$t >> t_c \rightarrow \text{Effective equilibrium can be assumed}$$

where:

A_o, B_o = initial concentration
A_∞ = equilibrium concentration

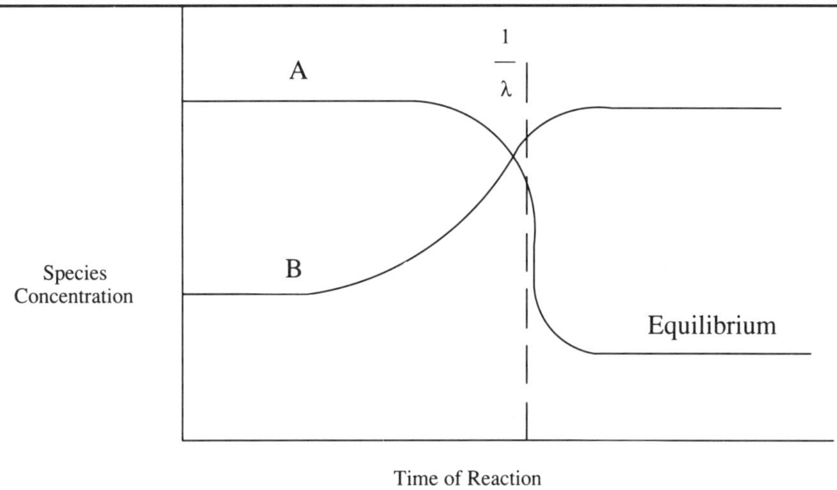

FIGURE 1.2
Relationship of Species Concentration to Time of Reaction
Engineering Principles

1.1 Partitioning (equilibrium)

Phases to consider:

- Solid
- Liquid
- Gas

Figure 1.3 depicts the vadose and saturated zone partitioning scenarios.

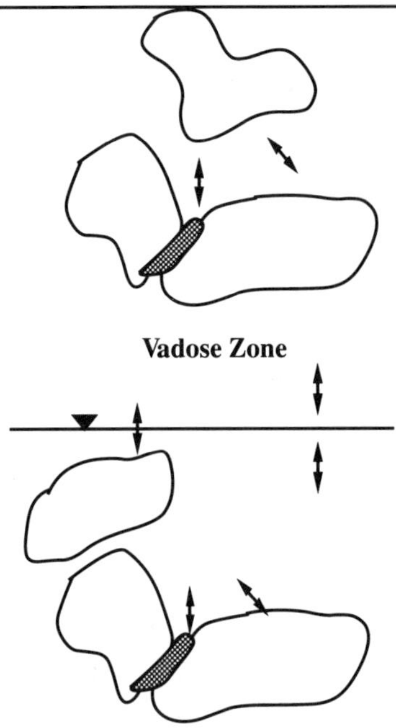

FIGURE 1.3
Partitioning Scenarios Within the Vadose and Saturated Zones
Engineering Principals

1.1.1 Gas/Liquid

Raoult's Law:

$$pA = P_A X_A \qquad X_A \to 1$$

Henry's Law:

$$pA = K_H A \text{ or } pA = K_H X_A \qquad X_A \to 0$$

1.1.2 Liquid/Solid

Isotherm:
Function of solubility:

Linear = $q = KC$

— However, the linear equation has only a limited range. Figure 1.4 shows the range in which a linear isotherm can be used.

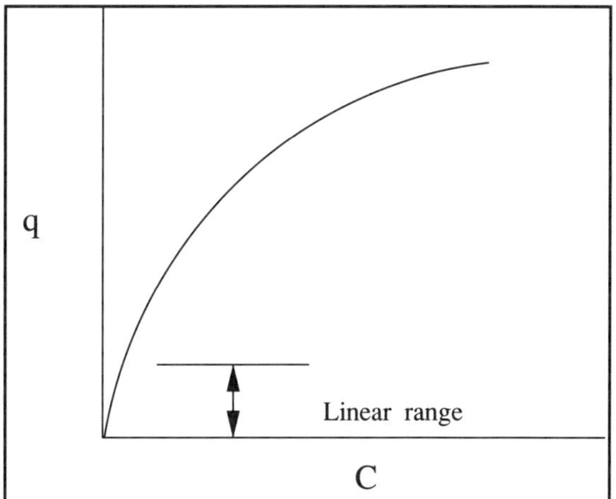

FIGURE 1.4
Isotherm
Engineering Principles

Freundlich = $q = KC^{1/n}$

— The Freundlich equation is mathematically more complex, but more versatile

One of the most common techniques used to predict the fate of organics in groundwater systems is the Octanol/water partition coefficient.

1.1.2.1 Octanol/Water Partition Coefficient

$$K_{ow} = \frac{\text{Concentration in octanol phase}}{\text{Concentration in water phase}}$$

$$\text{Range} \Rightarrow 10^{-3} \text{ to } 10^{-7}$$

Compounds tend to partition between organic phase (e.g. soil, organic matter) and aqueous phase.

If K_{OW} is low (<10), then the compound is hydrophilic. If K_{OW} is high (>10^4), then the compound can be described as hydrophobic.

There are three methods for arriving at a value for K_{OW}:

- Look it up in reference books or research articles
- Measure it experimentally
- Estimate it using correlations (the following are various estimation techniques for determining K_{OW}).

1.1.2.1.1 Fragment Constant Method

The fragment constant method is very powerful and is used for estimates in approximately 84 percent of compounds.

$$\log K_{ow} \text{ (new chemical)} = \log K_{ow} \text{ (similar chemical)} \pm \text{fragments(f)} \pm \text{Factor(F)}$$

Example:
An estimate of Kow for R-Br can be derived if Kow for R-Cl is known.

$$\log K_{ow}(\text{R-Br}) = \log K_{ow}(\text{R-Cl}) - f_{Cl} + f_{Br}$$

Factors account for structural changes as well.

1.1.2.1.2 Activity Coefficient Method

$$K_{ow} = 0.151\frac{\gamma_c^w}{\gamma_o^w}$$

γ can be estimated.

$$aA + bB = (AB) \text{ sol'n}$$

$$G = \sum_i n_i \mu_i$$

$$G_{mix} = a(\mu_A^o + RT\ln\{A\}) + b(\mu_B^o + RT\ln\{B\})$$

$$\Delta G_m = G_{mix} - G_A - G_B$$

$$\Delta G_m = G_{mix} - a\mu_A^o - b\mu_B^o$$

Activity of pure liquid = 1

$$\Delta G_m = RT[a \ln\{A\} + b \ln\{B\}]$$

$$\Delta G_m = RT[a \ln X_A + b \ln X_B] \quad \}G^I \text{ Ideal}$$
$$+ RT[a \ln \gamma_A + b \ln \gamma_B] \quad \}G^E \text{ Excess}$$

$$\frac{\partial G^E}{\partial a} = RT \ln \gamma_A \quad \text{or} \quad \frac{\partial G^E}{\partial b} = RT \ln \gamma_B$$

Assuming:

$$G^E \alpha\ a + b \text{ and } G^E = 0$$

when:

$$X_A \text{ or } X_B = 0$$

$$G^E = (a + b)\,\phi\, X_A X_B$$

Combining:

$$\ln \gamma_A = \frac{\phi_A}{RT} X_B^2$$

$$\ln \gamma_B = \frac{\phi_B}{RT} X_A^2$$

ϕ can be estimated from binary vapor-liquid data.

A shortcoming of this method is its failure to account for S^E.

1.1.2.1.3 Universal Quasi Chemical Functional Group Activity Coefficient (UNIFAC)

UNIFAC is the most common estimation method. UNIFAC has a strong theoretical base and is a computer based technique.

1.1.2.1.4 Solubility

Given S
Typically: $K_{OW}\ \alpha\ S$

Eighteen correlation equations are presented in Lyman and applied to various chemical classes. Some general statements can be made about typical relationships to solubility, as:

- Size of the molecule increases; solubility decreases
- Polarity of the molecule increases; solubility increases
- Double/triple bonds increase; solubility decreases
- Branching increases; solubility decreases
- Aliphatic > Aromatic > Cycloaliphatic (order of decreasing solubility)
- Halogenation increases; solubility also increases

1.1.2.2 K_{OW} and K_{OC}

Once K_{OW} is quantified, K_{OC} can be calculated using linear partitioning relationships.

$$K_{oc} = \frac{\text{Mass adsorbed/mass organic carbon} = q}{\text{Solution concentration} \qquad = c}$$

$$1 < K_{oc} < 10^7$$

Freundlich does not reduce to linear form at low cone; however Langmuir and BET do. Can be used for a variety of soils based only on percent of organic carbon (oc)

$$K_{oc} = \frac{K}{\%oc} 100$$

Typically:

$$\log K_{oc} = a \log\left(S \text{ or } K_{ow}\right) + b$$

(See Lyman)

Factors to consider in partitioning:

- Hysterisis
- Equilibrium versus kinetics
- Particle size distributions
  - — Fine silt and clay adsorb most
- Ionic strength
- D.O.M. — Cosolvent-competitive adsorbtion
- Character of soil organic carbon
- Temperature — sorption is exothermic
- pH —pK_a neutral species adsorb more
- colloid concentration — facilitated transport
- Solids effect — true or artifact

1.1.3 Interfacial Mass Transfer

Equilibrium is achieved when:

$$\mu_A^{solid} = \mu_A^{gas}$$

for:

$$X_A \rightarrow 0$$

Henry's Law

$$P_A = K_H X_A$$

Can calculate from ΔG.

Also, a Henry's Law-type of partitioning coefficient is often used.

$$H = \frac{VP_A}{S_A}$$

Although vapor pressure (VP) is often used to classify volatility, H is better,for:

$$X_A \rightarrow 1$$

Raoult's Law:

$$P_A = VP_A X_A$$

Used for NAPLs.

Fick's Law:

$$\frac{\partial c}{\partial t} = -D\frac{\partial c}{\partial x}$$

$$\frac{\partial c}{\partial t} = \frac{-D}{\delta}\Delta c = K(\Delta c)$$

Systems may have diffusive limitations; Therefore, equilibrium may not be a good assumption. Lyman --> methods for estimating D. Wilke/Chang is the most popular for aquatic systems.

$$D_d = \tau D + au^b$$ (Harleman and Rumer)

where:

D_d = dispersivity

τ = tortuosity

1.1.4 Precipitation/Dissolution (solubility)

Example:

$$MOH \leftrightarrow M + OH$$

$$K_{sp} = [M][OH]$$

Must include all aqueous and surface complexes.

1.1.5 Solid/gas

Appropriate isotherms.

1.1.6 Liquid/Liquid

Example:

NAPLs → Solubility(function of hydrophobicity)

2. CONTROL TECHNOLOGIES

The goal of control technologies is to confine or contain existing region of contamination, preventing any further spread of contamination.

Passive control measures include:

- Contain leachate/groundwater plume
- Divert flow
  - — Surface controls
  - — Often associated with closure procedures

Control measures may actively deal with the contamination, such as pump-and-treat remediation.

2.1 Impermeable barriers

Caveat: will typically increase upgraded head

Hydrodynamics of subsurface barrier - Imagine the barrier placed in a natural system; inside the walls the flow is zero and the head is flat. If left like this, the surrounded section will fill with water or leak if outward gradient. These problems can be solved by the placement of a cap to stop infiltration or the installation of pumps to reduce water levels and perhaps operate a water flushing system. This will keep a positive inward gradient.

2.2 Slurry walls

Slurry walls were developed in the mid-1940s for dam projects. A trench is excavate: a trench under a water/bentonite slurry and backfilled with PC or soil bentonite mix.

Excavation under slurry:

- Stabilizes walls during excavation
- Drives bentonite into soil, thus, forming a filter cake

Figure 2.1 shows the construction of a bentonite slurry wall.

Applications of a slurry wall include:

- Contain contaminated groundwater
- Divert contaminated groundwater from drinking water intake
- Divert uncontaminated groundwater flow
- Barrier for groundwater treatment system

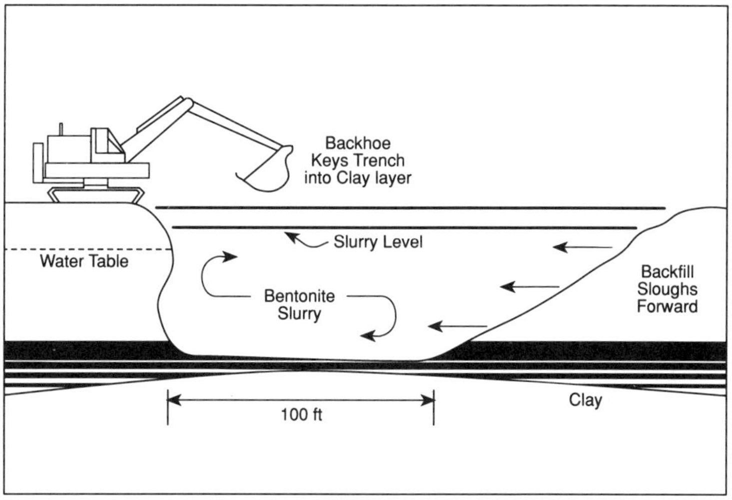

FIGURE 2.1
Construction of a Bentonite Slurry Wall
Engineering Principles

The performance of a slurry wall depends on the filter cake and the backfill permeability.

$$K = \frac{t_b}{\frac{t_b}{K_b} + \frac{2t_c}{K_c}}$$

where:

K	= wall permeability
K_c	= cake permeability
K_b	= backfill permeability
t_b	= backfill thickness
t_c	= cake thickness

For constant t_b, as K_b decreases K_c becomes insignificant, but as K_b increases K_c becomes important. Limiting $K \sim 10^{-6}$ cm/s.

Figure 2.2 shows the relationship between the wall permeability of filter cake and the backfill.

Wall thickness should be a minimum of two feet (typically thickness ranges from 2-5 feet) and the walls should extend down to impermeable layer(boundary).

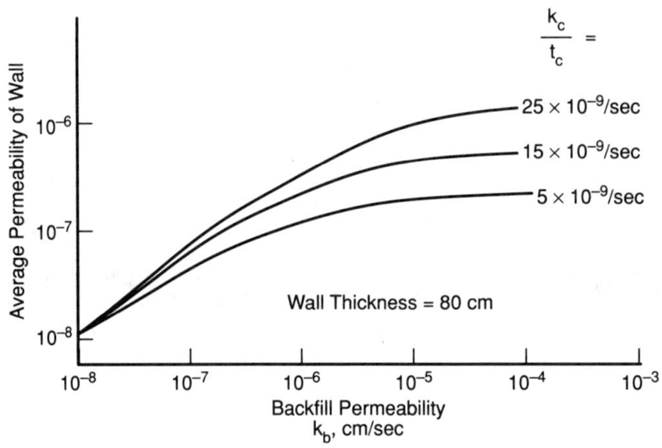

FIGURE 2.2
Theoretical Relationship Between Wall Permeability of Filter Cake and Backfill
Engineering Principles

Backfill characteristics:

- Backfill permeability decreases with fines (passing No. 200 sieve)
- Plastic fines result in permeability up to two orders of magnitude lower than non-plastic fines
- A minimum bentonite content of one percent with at least twenty percent fines is recommended (possibly up to forty to fifty percent)
- The bentonite should be completely hydrated and well-mixed before being placed in the trench
- Compatibility with leachate is very important—alcohols, HCl (75%), or NaOH (75%) are definitely not compatible (see Table 2.1)
- Life expectancy of a backfill is approximately decades

Table 2.1:
Permeability Increase Due to Leaching with Various Pollutants

Pollutant	Filter Cake	SB backfill (silty or clayey sand) 30 to 40% fines
CA^{++} or Mg^{++} at 1,000 PPM	N	N
CA^{++} or Mg^{++} at 10,000 PPM	M	M
NH_4NO_3 at 10,000 PPM	M	M
HCL (1%)	N	N
H_2SO_4 (1%)	N	N
HCL (5%)	M/H[1]	M/H[1]
NaOH (1%)	M	M
CaOH (1%)	M	M
NaOH (5%)	M	M/H[1]
Sea water	N/M	N/M
Brine (SG = 1.2)	M	M
Acid Mine Drainage $FeSO_4$ (PH-3)	N	N
Lignin (in Ca^{++} solution)	N	N
Alchol	H (failure)	M/H

N - No significant effect; permeability increase by about a factor of 2 or less at steady state

M - Moderate effect; permeability increase by factor of 2 to 5 at steady state

H - Permeability increase by factor of 5 to 10

[1] Significant dissolution likely

2.3 Grout Curtains

Grout curtains are fixed underground barriers formed by injecting grout. There are two types of grout curtains (see Table 2.2).

- Particulate (suspension) grout
    - — e.g. Portland cement
- Chemical grout
    - — e.g. Sodium silicate

Table 2.2: Types of Grout

	Significant Characteristics	Cost relative to Portland Cement
Portland Cement or Particulate Grouts	• Appropriate for higher permeability (larger grained) soils • Least expensive of all grouts when used properly • Most widely used in grouting across the U.S. (90% of all grouting)	1.0
Chemical Grouts		
Sodium silicate	• Most widely used chemical grout • At concentrations of 10-70% gives viscosity of 1.5 - 50 cP • Resistant to deterioration by freezing or thawing • Can reduce permeabilities in sands from 10^{-2} to 10^{-8} cm/s • Can be used in soils with up to 20% silt and clay at relatively low injection rates • Portland cement can be used to enhance water cutoff	2.0-5.0
Phenoplasts	• Rarely used due to high cost • Should be used with caution in areas exposed to drinking water supplies • Low viscosity • Can shrink (with impaired integrity) if excess (chemically unbound) water remains after setting; unconfined compression strength of 50-200 psi in stabilized soils	N.A.
Lignosulfate derivatives	• Rarely used due to high toxicity • Lignin can cause skin problems and hexavalent chromium is highly toxic; both are contained in these materials • Cannot be used in conjunction with Portland Cement: pH conflict • Ease of handling • Lose of integrity over time in moist soils • Initial soil strengths of 50-200 psi	1.65
Aminoplasts: e.g., urea-formaldehyde	• Rarely used due to high cost • Will gel with an acid or neutral salt • Gel time control is good	N.A.
Acrylamid grouts	• Rarely used due to toxicity • Should be used with great caution near to drinking water supplies • Readily soluble in water • Manufacter in USA prohibited; available from Japan as AV-100 • Can be used in finer soils than most grouts because low viscosities are possible (1 cP) • Excellent gel time control due to constant viscosity from time of catalysis to set/gel time • Unconfined compressive strengths of 50-200/psi in stabilized soils • Gels are permanent below the water table or in the soils approaching 100% humidity • Are vulnerable to freeze-thaw and wet-dry cycles, particularly where dry periods predominate and will fail mechanically • Due to ease of handling (low viscosity), enables more efficient installation and is often cost-competative with other grouts	4.0-10.0

Table 2.3: Grout Properties

Grout material	Catalyst material	Unconfined compressive strength (PSI) of grouted soil	Viscosity (Centipoise)	Setting time Minutes	Toxicant*	Pollutant**
Silicate base						
Low concentration	Bicarbinate	10-50	1.5	0.1-300	No	No
Low concentration	Halliburton Co. material	10-50	1.5	5-300	No	No
Low to high concentration	Siroc-Diamond Shamrock Chemical Co.	10-500	4-40	5-300	No	No
Low to high concentration	Chloride-Joosten Process	10-1000	30-50	0	No	No
Low to high concentration	Ethyl-acetate Soletanche & Halliburton	10-500	4-40	5-300	No	No
Low to high concentration	Rhone-Progil 600	—	—	—	—	—
Low to high concentration	Celoc-3 H. Baker Co.	10-500	4-25	2-200	No	No
Low to high concentration	Celoc-3X	10-250	4-25	0.3-120	No	No
Lignin base						
Blox-All	Halliburton Co. material	5-90	8-15	3-90	Yes	Yes
TDM	Cementation Co. material	50-500	2-4	5-120		
Terra-Firma	Intrusion Co. material	10-50	2-5	10-300	Yes	Yes
Lignosol	Lignosol Co. material	10-50	50	10-1000	Yes	Yes
Formal dehyde base						
Urea-Formaldehyde	Halliburton Co. material	Over 1000	10	4-60	Yes	Yes
Urea-Formaldehyde	American Cyanamide Co. material	Over 500	13	1-60	Yes	Yes
Resorcinol Formaldehyde	Cementation Co. material	Over 500	3.5	—	Yes	Yes
Tannin-Para-Formaldehyde	Borden Co. MQ-8					
Geoseal MQ-4 & MQ-5	Borden Co. material					
Unsaturated fatty acid base						
Polythixon FRD	Cementation Co. material	Over 500	10-80	25-360	No	No

* A material which must be handled using saftey precautions and/or protective clothing

** Pollutant to fresh water supplies contracted

Source: Hallburton Services, 1976

Suspension grouts are non-newtonian fluids comprised of H_2O and, primarily, Portland cement, bentonite or a combination of the two. Suspension grouts are used to seal voids in materials with high permeabilities. However, if grout is injected into too fine a medium, the result could be filtering (seeTable 2.3)

For Portland cement based grouting:

- An H_2O/Portland cement ratio of 0.6 is most effective
- The smallest voids effectively grouted with Portland cement are approximately three times cement grain size

For bentonite grouting, sand with permeability ranging from 10^{-3} to 10^{-4} cm/s can be sealed. Figure 2.3 shows the relative size of different types of grouts compared to soil grain sizes.

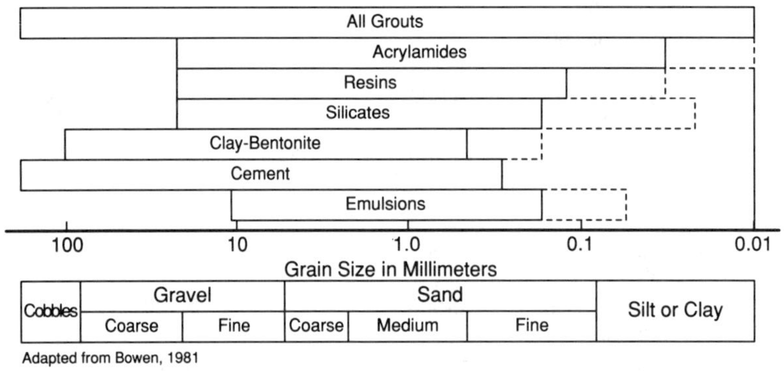

FIGURE 2.3
Applicability of Different Classes of Grout Based on Soil Grain Size
Engineering Principles

Suspension grouts are often used as pre-grout for chemical grout. In the U.S. suspension grouting is most common and accounts for fifty percent of the grouting in Europe.

Injection rates must be calculated for effective consolidation:

- Slow injection rates could result in premature consolidation
- Rapid injection rates could cause fracturing in soil formation

Figure 2.4 provides a typical example of a grout curtain.

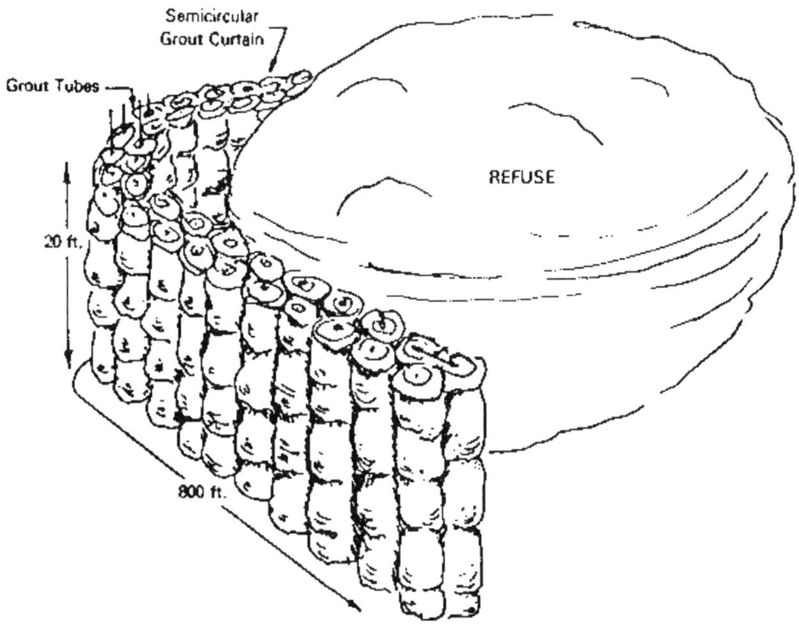

FIGURE 2.4
Semicircular Grout Curtain Around a Landfill
Engineering Principles

2.4 Sheet Pile Cut Off Walls

Sheet pile cut off walls are installed by first driving a web section of sheet piling into the ground. Each section is then interlocked at the edges and joint connections are filled with fines.

The maximum effective depth has been approximated at fifty feet; however, there have been reports of use to 98 feet.

Steel is the most common material for construction, wood and concrete are also used, but for steel corrosion could be a problem. Also, steel piles are not good for rocky soils.

3. GROUNDWATER CONTAMINATION

Contamination can be present in as many as four phases:

- Pure phase
- Soil
- Air
- Water

Many remediation methods are based on removing one of these phases from the subsurface, and, if needed, facilitating a phase transfer from an immobile phase to a more mobile one.

In groundwater remediation two elements of cleanup must be considered:

- Source control—subject of this course
- Plume control

3.1 Plume Control

A plume is a dilute contaminant spread over a large area that must be pulled back, contained, or removed. Several methods may be used to achieve remediation:

- Pump and treat
  - — Requires water removal, treatment, and disposal
  - — May be able to employ bio substrate

- Interception
  - — Place extraction wells in path of approaching plume
- Dilution
  - — Inject clean water to dilute contaminant to level of water quality standards
  - — Drains, a perforated pipe which acts lie a line of wells, may be used in the place of wells for water removal
  - — A large plume will require many wells
  - — Design requires calculation of number of wells needed and appropriate pump rates; location and pump rates determine flow and transport behavior

Figure 3.1 shows the placement of pump and treat wells, and interception wells to control plume advancement.

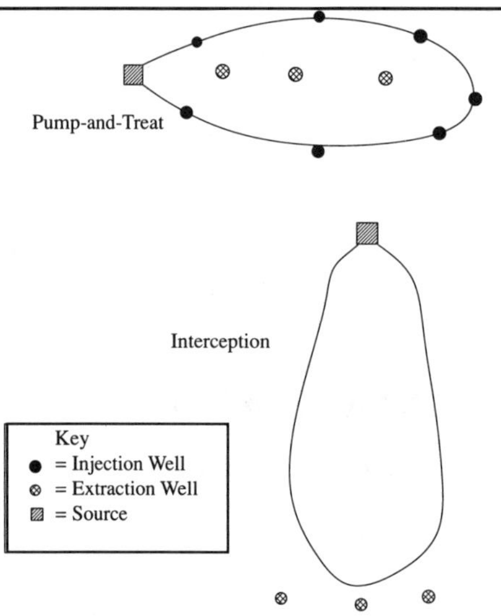

FIGURE 3.1
Plume Control Remediation
Engineering Principles

3.2 Other Transport Mechanisms

Remediation attempts may be severely impacted by

- Adsorption to soil
  - Contaminants adsorbing to soil would affect methods like pump-and-treat or water flushing, making them ineffective if the contaminant remained adsorbed to the soil after treatment.
- Decay (biological)
  - Naturally occurring bacteria, which consume or transform the contaminants, may be enhanced by the addition of any growth limiting substances (e.g. oxygen)
  - Genetically engineered bacteria may also be introduced to remediate a specific contaminant
  - Generally, this method is less successful for larger plumes as it is difficult to introduce oxygen over the whole area

3.3 Predicting Remediation

For systems that depend on drawing water into a well, it is essential to be able to predict what water will enter the well.

The cone of depression is the depression in the water table caused by pumping.

Drawdown s(r,t) is predicted by Theis:

$$s = \frac{Q}{4\pi T} W(u)$$

where:

$$W(u) \quad = \text{well function}$$

$$u = \frac{r^2 S}{4Tt}$$

Jacob-Cooper approximation for large time is:

$$s = \frac{2.3\ Q}{4\pi T} \log\left(\frac{4Tt}{r^2 S}\right)$$

Theis can be used to predict capture only approximately since it assumes a flat initial water table.

In plan view, Theis would predict flow radial to well. Drawdown will continue to expand indefinitely as there is no boundary.

Figure 3.2 depicts what would happen in the event of an existing regional groundwater flow gradient under the influence of pumping. A capture zone that would extend upgradient would be formed.

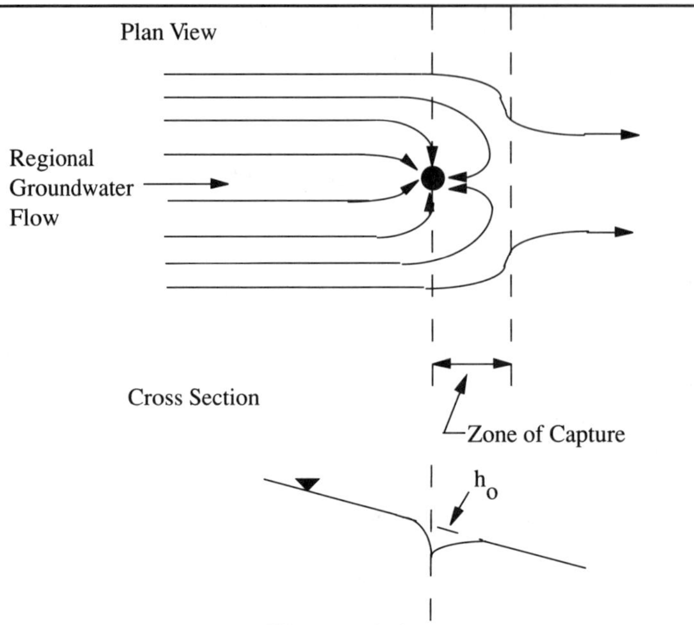

Figure 3.2
Existing Regional Flow Showing Zone of Capture
Engineering Principles

3.3.1 Capture Zone Type Curves

Capture zone type curves are used for simple design of capture wells by interception remediation.

Assume:

- K is homogeneous and isotropic
- Aquifer has constant saturated thickness of B
- Natural gradient has constant direction and magnitude
- No aquifer boundaries exist nearby (streams, no-flow boundaries, etc.)

Define U = Darcy Velocity = constant (since K and gradient are constant)

Use of type curves:

1) Draw plume on overlay at same scale as type curves
2) Align overlay so that regional flow lines coincide and the plume is centered around x-axis
3) Select the curve that contains the plume—pick off the Q/BU value
4) Compute Q = B*U*value
5) Compute drawdown that would result using Theis equation. If too large, increase number of wells and go to step 2 (make certain to account for superposition)

Jaundel and Tsang state that optimal well distance given by:

Two wells: $$\frac{Q}{\pi BU}$$

Three wells: $$\frac{1.26\ Q}{BU}$$

Four wells: $$\frac{1.2\ Q}{BU}$$

Example:

Regional gradient = 0.01

K = 75 m/day => U = 0.75 m/day

B = 30 m

Assume we pick Q/BU = 800 m
=> Q = B*U*800 = 30 m * 0.75 m/day * 800 m = 18000 m^3/day
from each well.

Jacob-Cooper approximation to Theis says:

$$s = \frac{2.3\ Q}{4\pi T} \log \frac{2.25\ Tt}{r^2 S}$$

where:

Q	= 18000 m^3/day
T	= KB = 75 * 30 = 2250 m^2/day
r	= radius just outside well, say 0.2 m
t	= long time, say 365 days
S	= small for rapid response = 1 x 10^{-4}

$$s = \frac{2.3\ (18000)}{4\pi\ (2250)} \log \frac{2.25\ (2250)\ 365}{(0.2)^2(1\mathrm{x}10^{-4})} = 17.1$$

Spacing for two wells is:

$$\frac{Q}{BU} = \frac{18000}{\pi\ (30)(0.75)} = 225\ \mathrm{m}$$

The effect of the second well on the first:

$$s = \frac{2.3\ (18000)}{4\pi\ (2250)} \log \frac{2.25\ (2250)\ 365}{(225)^2(1x10^{-4})} = 8.14\ m$$

Total drawdown ≈ 17.1 + 8.1 = 25.2 m. This is close to full thickness of the aquifer - better try two pumping wells instead of one.

For many sites, simple methods cannot be used because:

- K heterogeneous
- Regional flow is complex
- Vertical flow is significant
- Major hydrologic barriers are present
- Simple interception is not sufficient

In such cases it is necessary to use simulated models

3.4 Interacting with Modelers

Modelers need:

- Spatially distributed estimates of the boundary conditions (b.c.)
- Boundary conditions around the domain

The advantages of numerical modeling include:

- More precise representation of real system can be approached
- Model can confirm or refute hypothesis of system behavior
- Model can be used for prediction
- All data can be combined with known physical laws—tests consistency

The use of modeling is limited by:

- Large time required to create
- Special skills requirements
- Large data requirements

Some sites do not require modeling because questions are not sufficiently complicated.

Transport modeling is based on conservation of contaminant mass. For a single cell, the mass in a cell can:

- Move out of the cell by advection/dispersion
- Adsorb to soil (it leaves the water so this is mass loss)
- Decay or transform
- Leave through a pump

Transport equations take into account all these factors with varying degrees of sophistication.

The following flow equations are written as an algebraic balance equation. If cells are made very small the differential form can be used:

$$\frac{df}{dx} = \lim_{h \to 0} \frac{f(x + h) - f(x)}{h} = \frac{\Delta f}{h}$$

A differential form of flow equations is:

$$\frac{\partial}{\partial x} K_{xx} \frac{\partial h}{\partial x} + \frac{\partial}{\partial y} K_{yy} \frac{\partial h}{\partial y} + \frac{\partial}{\partial z} K_{zz} \frac{\partial h}{\partial z} - q = S \frac{\partial h}{\partial t}$$

This equation is three-dimensional and transient.

The differential form is preferred because it is:

- More compact
- Easy to manipulate
- Historically preferred

The classic approach to flow:

- State differential equation
- Discretize—define cells and nodes
  — Write algebraic approximations of the partical differential equation (PDE) at each node to yield system of equations

For numerical analysis, the smaller the cell size is, the better the approximation will be.

Model accuracy is affected by:

- Use of the wrong model (e.g. using a two-dimensional areal for a three-dimensional system)
- Use of wrong parameters—field values are only estimates
- Poor solution of balance equations—numerical implementation

A good model has small error at each level.

The purpose of site assessment is to develop a hypothesis of the behavior of groundwater flow and contaminant transport on a site which is consistent with the available data and the laws of physics. The assessment should be sufficiently detailed to address questions of importance at the site, such as “where is the contamination?”, “how much is there?”, “how fast is it moving”, and “what approaches to remediation are likely to be effective?”

Data elements to be examined:

I) Source information
  A) From soils and groundwater levels
    i) Compare soil and water concentrations with solubility levels
  B) Site history/anecdotal information
    i) employee interviews, aerial photos, chemical purchase records

II) Geology
  A) Hydraulic conductivity, porosity
  B) Presence of strata
    i) Construct cross sections
  C) Presence of fractured rock
    i) determine fracture density—equivalent porous media?
  D) Presence of confining layers or lenses

III) Groundwater flow
  A) Site hydrology—infiltration, surface waters
    i) Topography maps for regional flow features
  B) Direction of flow
    i) Collect comparable water levels
    ii) Use contouring or triangle analysis
    iii) Examine vertical head differences
  C) Rate of flow
    i) Use Darcy's Law as first approximation
    ii) Use numerical models for more sophistication
    iii) Consider anisotropy

IV) Chemical data
  A) Horizontal distribution of each constituent
    i) Use various graphical techniques
  B) Vertical distribution of each constituent
  C) Temporal variation of concentration
    i) Consider sampling error and natural variation
  D) Transport characteristics of specific compounds (density, solubility, adsorption capabilities

4. AQUEOUS PHASE TREATMENT

Once the contaminated medium has been pumped, the pumped water is treated.

4.1 Air stripping

Figure 4.1 is a schematic of an air stripping tower.

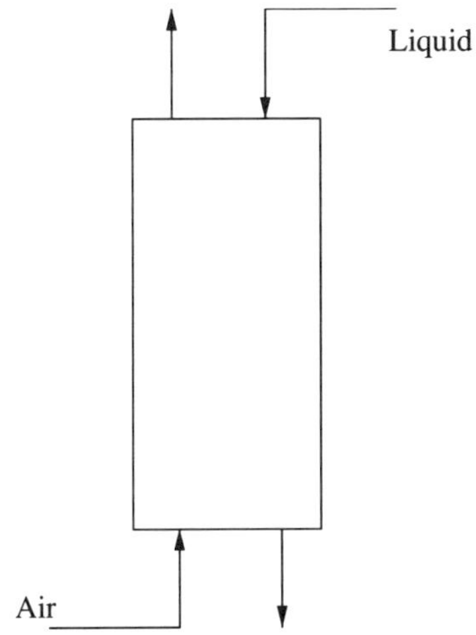

FIGURE 4.1
Air Stripping Schematic
Engineering Principles

Characteristics:

- $H > 0.003$ atm(m^3)/mol
- Low Fe2+ and TSS
- Performance > 99%
- Off-gas -- discharge/carbon/ incineration/biological treatment
- Tower: diameter, 1-12 ft; height, 5-50 ft
- Liquid loading (L): 5-30 gpm/ft^2; G/L 10:1 - 300:1
- Gas loading rate (G)
- Packing size: 0.5 - 3.5 in.

$$\left[\frac{G}{L}\right]_{min} = \frac{C_i - C_{out}}{H\,C_{in}}$$ from mass balance

Actual:

$$\left[\frac{G}{L}\right] = (2 \text{ to } 5) * \left[\frac{G}{L}\right]_{min}$$

Packing Height

$$z = \frac{L}{K_L a} \frac{R}{(R - 1)} \ln \left[\frac{\left(\frac{C_{in}}{C_{out}} \right)(R - 1) + 1}{R} \right] = (NTU)(HTU)$$

where:

L = liquid loading rate ($m^3/m^2/s$)

$K_L a$ = (1/s) Onda's Correlation

$$R = \frac{G}{L} H$$

Air velocity is typically limited to 60% of flooding.

$$\text{Pressure Drop} = 50 - 200 \frac{N}{m^2 - m} = 0.002 - .01 \frac{psi}{ft}$$

Blower Power

$$\text{Blower Brake Power (kW)} = \frac{M_{gas} RT}{283 E} \left[\left[\frac{P_{out}}{P_{in}} \right]^{-283} - 1 \right]$$

where:

M_{gas} = G * cross-sectioned area * air density (1:25 Kg/m^3)

P_{in}, P_{out} = inlet, outlet absolute pressures

E = blower efficiency

4.2 Activated Carbon Adsorption

- Solvent motivated
- Sorbent motivated

$$A = 500 - 2000 \ m^2/g$$

Where A is defined as the specific area of the activated carbon.

- PAC passes number 200 sieve
- GAC diameter > 0.1mm

Wide range of applications.

Applicable isotherms are the Freundlich and Langmuir isotherms.

Freundlich:

$$q = K_F C^{1/n}$$

Langmuir:

$$q = \frac{bQc}{1 + bc}$$

Adsorption is kinetically limited by the following:

- Film transport
- Pore/surface diffusion
- Adsorption --> Fast so therefore it is not usually the limiting factor.

Isotherms --> first cut

See Tables 4.1 and 4.2 for carbon applicability and characteristics.

Should use:

- HSDM - Homogeneous Surface Diffusion Model
- BDST - Bed Depth/ Service Time
- RSSCT - Rapid Small Scale Column Test

Table 4.1: Properties of Several Commercially Available Carbons[a]

Physical properties	ICI America Hydodarco 3000	Calgon Filtrasorb 300 (8 x 30)	Westvaco Nuchar WV-L (8 x 30)	Witco 517 (12 x 30)
Surface area, m^2/gm (BET)	600-650	950-1050	1000	1050
Apparent density, gm/cc	0.43	0.48	0.48	0.48
Density, backwashed and drained, lb/cu ft	22	26	26	30
Real density, gm/cc	2.0	2.1	2.1	2.1
Particle density, gm/cc	1.4-1.5	1.3-1.4	1.4	0.92
Effective size, mm	0.8-0.9	0.8-0.9	0.85-1.05	0.89
Uniformity coefficient	1.7	1.9 or less	1.8 or less	1.44
Pore volume, cc/gm	0.95	0.85	0.85	0.60
Mean particle diameter, mm	1.6	1.5-1.7	1.5-1.7	1.2

Table 4.1: Properties of Several Commercially Available Carbons[a] (continued)

Physical properties	ICI America Hydodarco 3000	Calgon Filtrasorb 300 (8 x 30)	Westvaco Nuchar WV-L (8 x 30)	Witco 517 (12 x 30)
Specifications				
Sieve size (U.S. std. series)				
Larger than No. 8 (max. %)	8	8	8	c
Larger than No. 12 (max. %)	c	c	c	5
Smaller than No. 30 (max. %)	5	5	5	5
Smaller than No. 40 (max. %)	c	c	c	c
Iodine No.	650	900	950	1000
Abrasion No. minimum	b	70	70	85
Ash (%)	b	8	7.5	0.5
Moisture as packed (max. %)	b	2	2	1

a Other sizes of carbon are available on request from the manufacturers

b No available data from the manufacturer

c Not applicable to this size carbon

Source: ADL, 1976

Table 4.2: Activated Carbon

Compound	Removal rating	Compound	Removal rating
Priority solvents			
acetone	P	methyl ethyl ketone	G
butanol	G	methyl isobutyl ketone	F
carbon disulfide	G	methylene chloride	P
carbon tetrachloride	G	nitrobenzene	E
chlorobenzene	E	pyridine	E
cresole	E	tetrachloroethylene	G
cyclohexanone	G	toluene	G
1,2-dichlorobenzene	E	1,1,1-trichloroethane	G
ethyl acetate	F	trichloroethane	F
ethyl ether	P	trichloromonofluoromethane	G
ethyl benzene	P	1,1,2-trichloro- 1,2,2-trifluoroethane	G
methanol	P	xylene	E

Table 4.2: Activated Carbon (continued)

Compound	Removal rating	Compound	Removal rating
Other solvents			
acetonitrile	F	2-ethoxy ethanol	P
aniline	E	ethyl carbonate	G
benzene	F	ethylene dichloride	G
bis(chloromethyl)ether	G	ethylidene dichloride	F
bromoform	E	furan	F
chloroform	P	furfural	G
cyclohexane	F	hexachloroethane	E
m-dichlorobenzene	E	2-nitropropane	F
dichlorodifluoromethane	F	2-picoline	E
1,1-dichloroethylene	P	propylene glycol	F
1,2-dichloroethylene	F	1,1,1,2-tetrachloroethane	E
1,2-dichloropropane	G	1,1,2,2-tetrachloroethane	E
1,3-dichloropropane	G	tetrahydrofuran	F
1,4-dichloro-2-butane	F	1,1,2-trichloroethane	G
1,4-dioxane	P		
Ignitables			
acrolein	P	ethylmethacrylate	P
acrylic acid	F	formaldehyde	P
allyl alchol	P	glycidylaldehyde	F
allyl chloride	P	methacrylontrile	P
2,2-bioxirane	F	2-methylaziridine	P
chloracetaldehyde	F	methyl bromide	P
chloromethyl methyl ether	F	1-methyl butadiene	P
chloroprene	F	methyl chloride	P
cumene	E	methyl chlorocarbonate	P
dimetylamine	P	methyl iscyante	P
dipropylamine	F	methyl methacrylate	P
epichlorohydrin	P	oxirane	G
ethanol	P	paraldehyde	F
ethyl acrylate	F	propylamine	P
ethylenediamine	P	thiomethanol	P
ethylencmine	P		

E = Excellent removal — Adsorbs at levels 100 mg/g carbon at C_f = 10 mg/L
Adsorbs at levels 100 mg/g carbon at C_f 1.0 mg/L
G = Good removal — Adsorbs at levels 100 mg/g carbon at C_f = 10 mg/L
Adsorbs at levels 100 mg/g carbon at C_f 1.0 mg/L
F = Fair removal — Adsorbs at levels 100 mg/g carbon at C_f = 10 mg/L
Adsorbs at levels 100 mg/g carbon at C_f 1.0 mg/L
P = Poor removal — Essentially no removal
C_f = Final concentration of pollutant at equilibrium

4.3 Biological

Biological remediation can be simply conceptualized in the following "equation":

Waste + Microorganisms + Nutrients + Oxidixed (H) Acceptor
Environmental Conditions
Stabilized Waste + Microorganisms + Reduced (H) Acceptor

4.3.1 H-Acceptor Systems

	Oxidized	**ORP***	**Reduced**
Aerobic	O_2	high	H_2O
Anoxic	NO_x		N_2,N_2O, NO
Anaerobic	SO_x		H_2S, S
Fementation	oxidized organic intermediates (glucose)		reduced organics (alcohols)
Anaerobic	CO_2, acetate, formate	low	CH_4

*Oxidation/reduction potential

4.3.2 Major Nutrients

Nutrients that are essential to the maintenance of a community of microorganisms include carbon (C), hydrogen (H), nitrogen (N), oxygen (O), and phosphorous (P).

4.3.3 Above Ground Processes

Traditional above ground processes that can be employed are:

- Activated sludge
- Fixed film processes (e.g. RBC, TF)
- Anaerobic digestion

Soil/slurry reactors have also been used, but may have mass transfer limitations. They may also require a supply of nutrients and perhaps a primary substrate if co-metabolism is to be employed.

The primary substrate should:

- Include appropriate enzyme system
- Be inexpensive
- Have low yield

4.4. Solid/Liquid Separation

Traditional processes for solid/liquid separation include:

- Sedimentation
- Flotation
- Centrifugation
- Filtration

The addition of coagulants should be avoided if possible, especially if used prior to incineration.

REFERENCES

1. Ehrenfeld, J. *Evaluation of Remedial Action Unit Operations at Hazardous Waste Disposal Sites*. Noyes Publications, Park Ridge, New Jersey, 1984.

2. McCabe, W.L. *Unit Operations of Chemical Engineering*. Ed. McGraw-Hill, New York, New York, 1985.

3. Morel, F. *Principles of Aquatic Chemistry*. Ed. John Wiley & Sons, Inc., New York, New York, 1983.

F- Soil Vapor Extraction

SOIL VAPOR EXTRACTION

Instructional Objectives:

- Understand the physical properties governing soil vapor extraction
- Describe gas and contaminant transport in the unsaturated (vadose) zone
- Determine when soil venting is feasible
- Understand and be able to utilize analytical solutions for design
- Develop a design approach for an extraction system
- Describe other technologies used sequentially or simultaneously
- Understand and accommodate non-ideal SVE conditions

1. GENERAL DESCRIPTION

1.1. Introduction

Soil Vapor Extraction (SVE) removes volatile and some semi-volatile contaminants from the vadose zone, through the application of a vacuum, using a system of well vents placed in the unsaturated zone. These systems are designed to remediate residual contaminants in unsaturated soils and can be combined with groundwater pumping wells to remediate soil previously beneath the water table [1]. The gas removed from the soil may be then discharged directly to the air or, more likely, to an off-gas treatment process.

Several U.S. patents may be applicable to the technology [2].

Key factors involved in SVE design include, but are not limited to [3,]:

- Contaminant distribution at site
- Hydrogeology
- Contaminant properties

Vapor extraction wells and injection wells are installed at appropriate locations at the contaminated site. Vacuum pumps or blowers induce a pneumatic pressure gradient in the subsurface, resulting in an air flow field about an extraction well. Ambient air enters the subsurface either passively, through leaky boundaries, or actively, by injection, if necessary.

Gas-phase contaminants are removed via advective air flow entering extraction wells. Removal of gas-phase contaminants results in pore-scale concentration gradients. This induces volatilization and contaminant diffusive transport into the advective air flow. Higher vapor pressure components are recovered first. The residual soil concentration gradually becomes richer in less volatile compounds. Consequently, as a result of compositional changes, the vapor concentrations of the contaminants and mass removal rates decrease with time [3]. Water may be withdrawn from the subsurface in separate or combined extraction wells. Water entrained in the contaminated vapor is removed by a vapor/liquid separator [4]. Typical SVE processes are illustrated in Figures 1.1, 1.3

(both adapted from reference [2]), and 1.2 (adapted from reference [5]). See reference [3] for a review of SVE state-of-the-art.

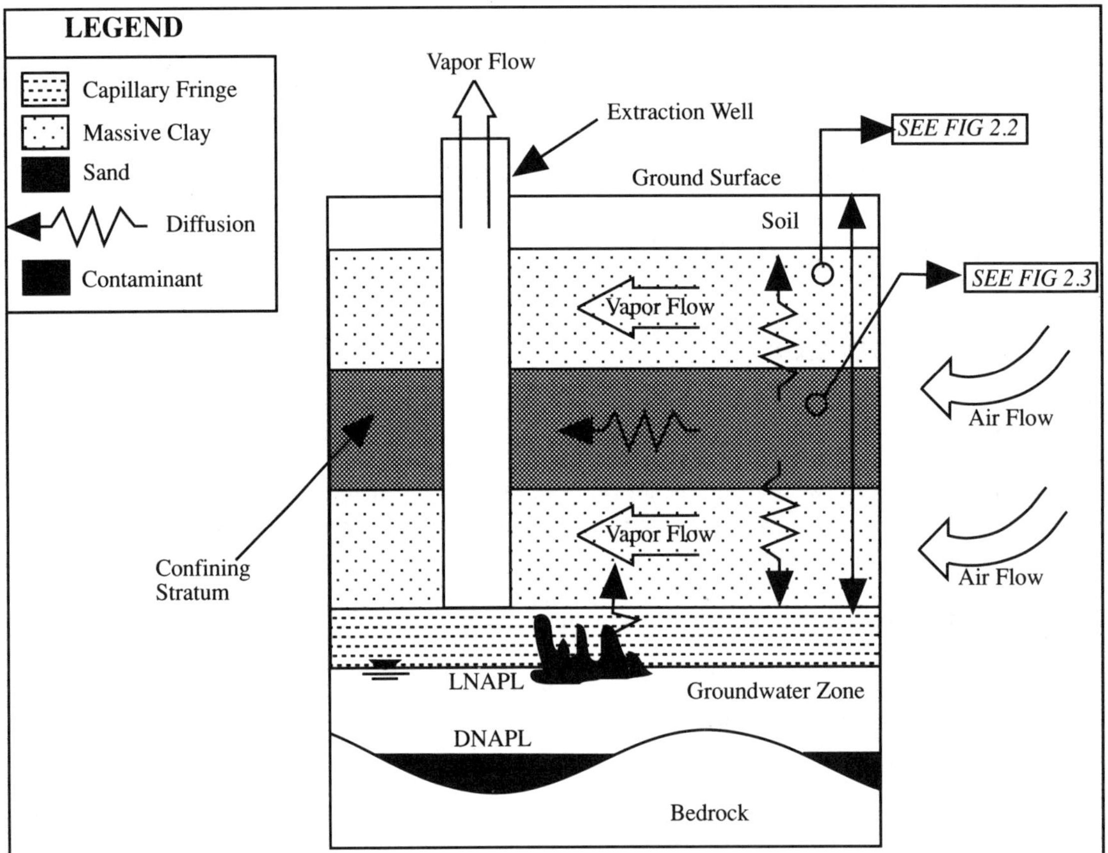

FIGURE 1.1
SVE Mechanisms
Soil Vapor Extraction

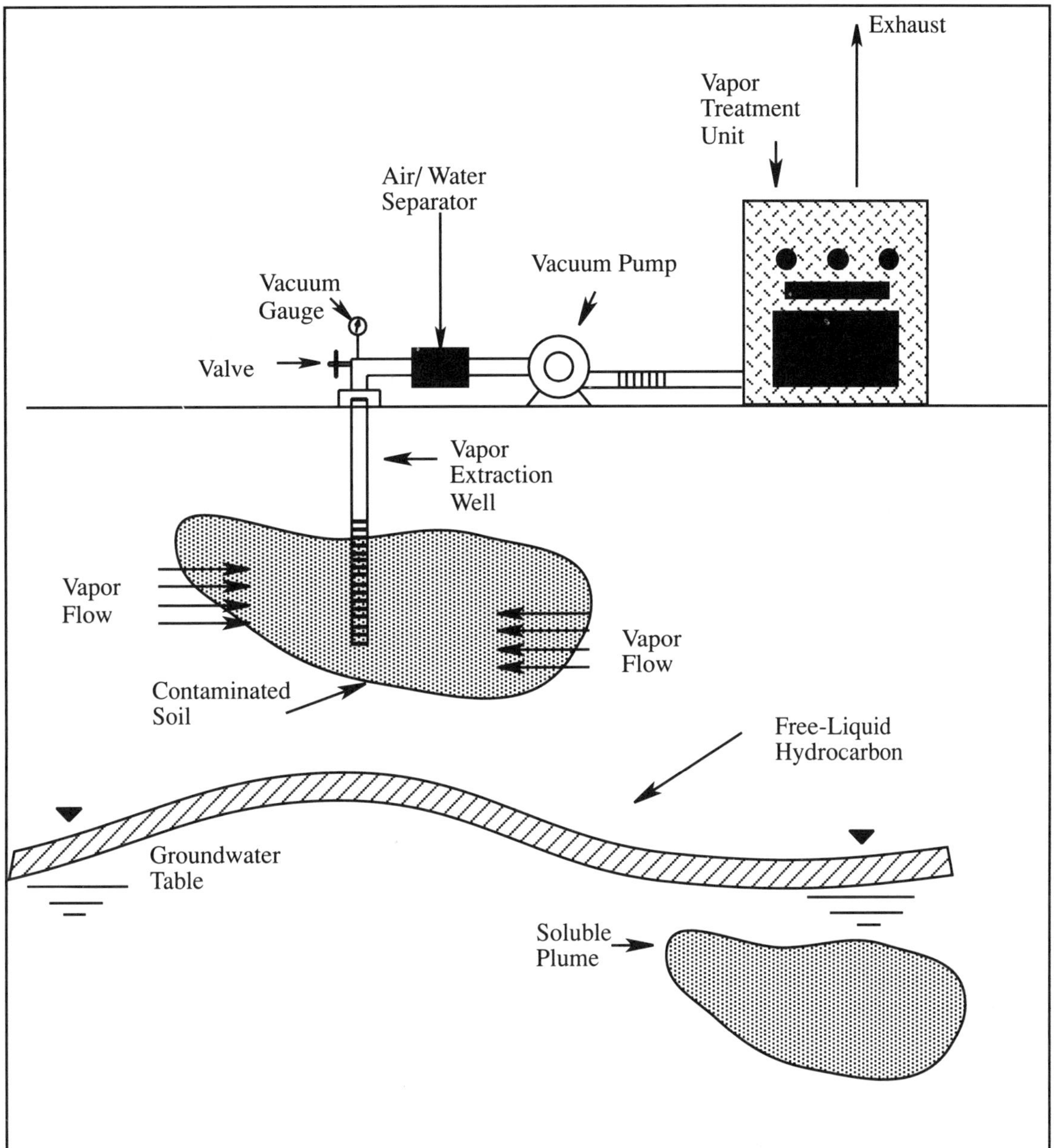

FIGURE 1.2
Pilot *In Situ* Soil Venting System
Soil Vapor Extraction

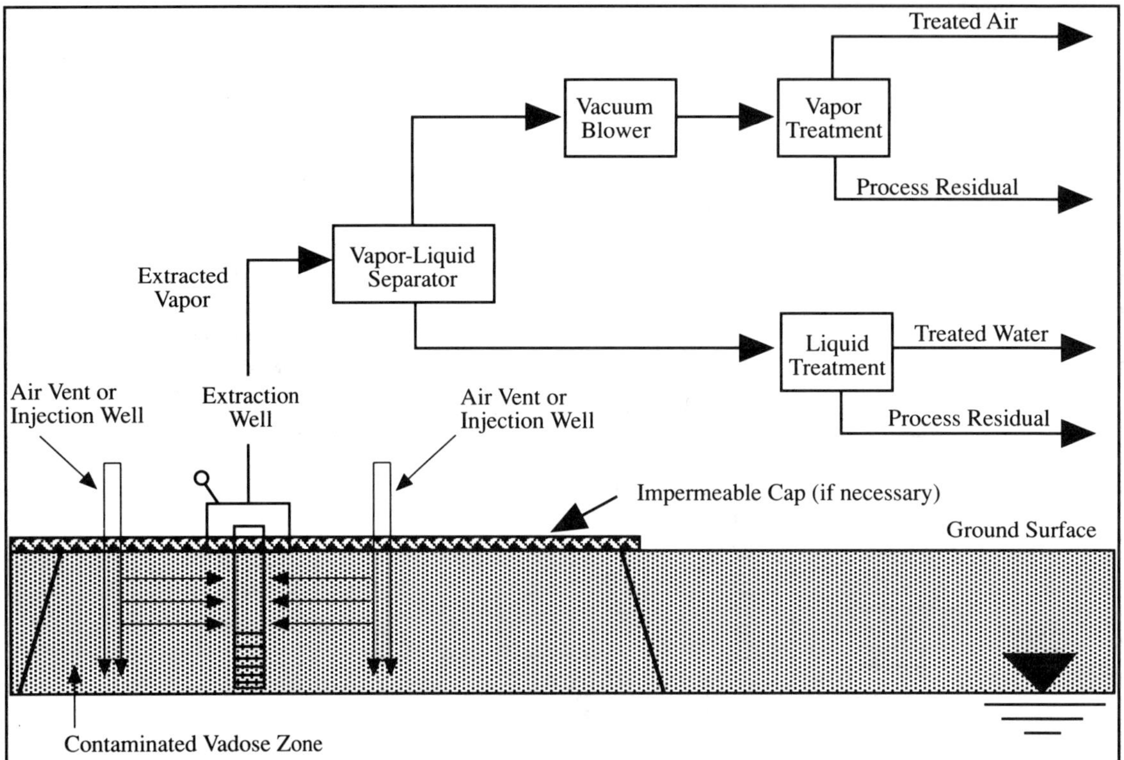

FIGURE 1.3
Generic Soil Vapor Extraction System
Soil Vapor Extraction

1.2. Compounds Removed by SVE

With the SVE process, it is feasible to treat contaminants with vapor pressures greater than 0.5 mm Hg [2]. This includes volatile organic compounds (VOCs) and some semi-volatile compounds, such as diesel fuel, kerosene, and heavy naphthas. [6]

Volatiles

- Toluene
- Ethylbenzene
- Chloroform
- Tetrachloroethylene (PCE)
- Dichloroethylene (DCE)
- Cyclohexane
- Methyl isobutyl ketone (MIBK)
- Acetone
- Tetrahydrofuran
- Carbon tetrachloride
- Tichloroethane (TCE)
- Benzene
- Xylenes
- Hexane
- Methylene chloride
- Trichloroethylene(TCE)
- Ethyl acetate
- Methyl ethyl ketone (MEK)
- Methanol
- Pyridine
- Dimethylfuran
- Clorobenzene

Semivolatiles

- Dichlorobenzene (DCB)
- Chlorobenzene
- Trichloropropane

Hydrocarbons

- Jet fuel
- Diesel
- Kerosene
- Gasoline
- Stoddard solvent
- Heavy naphthas

Table 1.1. Selected Gasoline Components and Their Chemical Properties [5]

Compound	M_w* (g/mole)	T_b (1 atm)** (C)	P_v^o 20 C† (atm)	C_{est}†† (mg/L)
n-pentane	72.2	36	0.57	1700
n-hexane	86.2	69	0.16	560
trichloroethane	133.4	75	0.132	720
benzene	78.1	80	0.10	320
cyclohexane	84.2	81	0.10	340
trichloroethylene	131.5	87	0.026	140
n-heptane	100.2	98	0.046	190
toluene	92.1	111	0.029	110
tetrachloroethylene	166	121	0.018	130
n-octane	114.2	126	0.014	65
chlorobenzene	113	132	0.012	55
p-xylene	106.2	138	0.0086	37
ethylbenzene	106.2	138	0.0092	40
m-xylene	106.2	139	0.0080	35
o-xylene	106.2	144	0.0066	29
styrene	104.1	145	0.0066	28
n-nonane	128.3	151	0.0042	22.0
n-propylbenzene	120.2	159	0.0033	16
1,2,4 trimethylbenzene	120.2	169	0.0019	9.3

Table 1.1. Selected Gasoline Components and Their Chemical Properties [5] (continued)

Compound	M_w* (g/mole)	T_b (1 atm)** (C)	P_v^o 20 C† (atm)	C_{est}†† (mg/L)
n-decane	142.3	173	0.0013	7.6
DBCP	263	196	0.0011	11
n-undecane	156.3	196	0.0006	3.8
n-dodecane	170.3	216	0.00015	1.1
naphtalene	128.2	218	0.00014	0.73
tetraethyllead	323	dec.@200C	0.0002	2.6
gasoline[1]	95		0.34	1300
weathered gasoline[2]	111		0.049	220

1 Corresponds to "fresh gasoline" defined in reference 1
2 Corresponds to "weathered" gasoline defined in reference 1
* M_w - molecular weight
** T_b (1 atm) - compound boiling point at 1 atm absolute pressure
† C_{est} - equilibrium vapor concentration-see equation 52
†† P_v^o (20 C) - vapor pressure measured at 20 C

2. FUNDAMENTALS

Figure 2.1 is a representation of soil, liquid, and gas thermodynamic and kinetic relationships relevant to the SVE process.

2.1. Introduction

Contaminants in the soil may be classified as:

NAPLs - nonaqueous phase liquids [2]

Equilibrium between vapor and liquid phases is governed by Raoult's Law.

DNAPLs - dense nonaqueous phase liquids. Includes chlorinated hydrocarbons (typically denser than water)

LNAPLs -light nonaqueous phase liquids. Includes hydrocarbons, alcohols, ketones, etc. [2]

These contaminants can be:

- Sorbed to soil

  Equilibrium between vapor and adsorbed contaminant may be described by an appropriate partitioning/isotherm equation [2]. Generally, sorption is only a water solute/soil interaction [7].

- Vaporized in pore gas phase [6]
- Dissolved in the aqueous phase

  Equilibrium between vapor and liquid phases is governed by Henry's Law for dilute solutions. Co-solvents may impact applicability of Henry's Law

- Free-floating contaminant

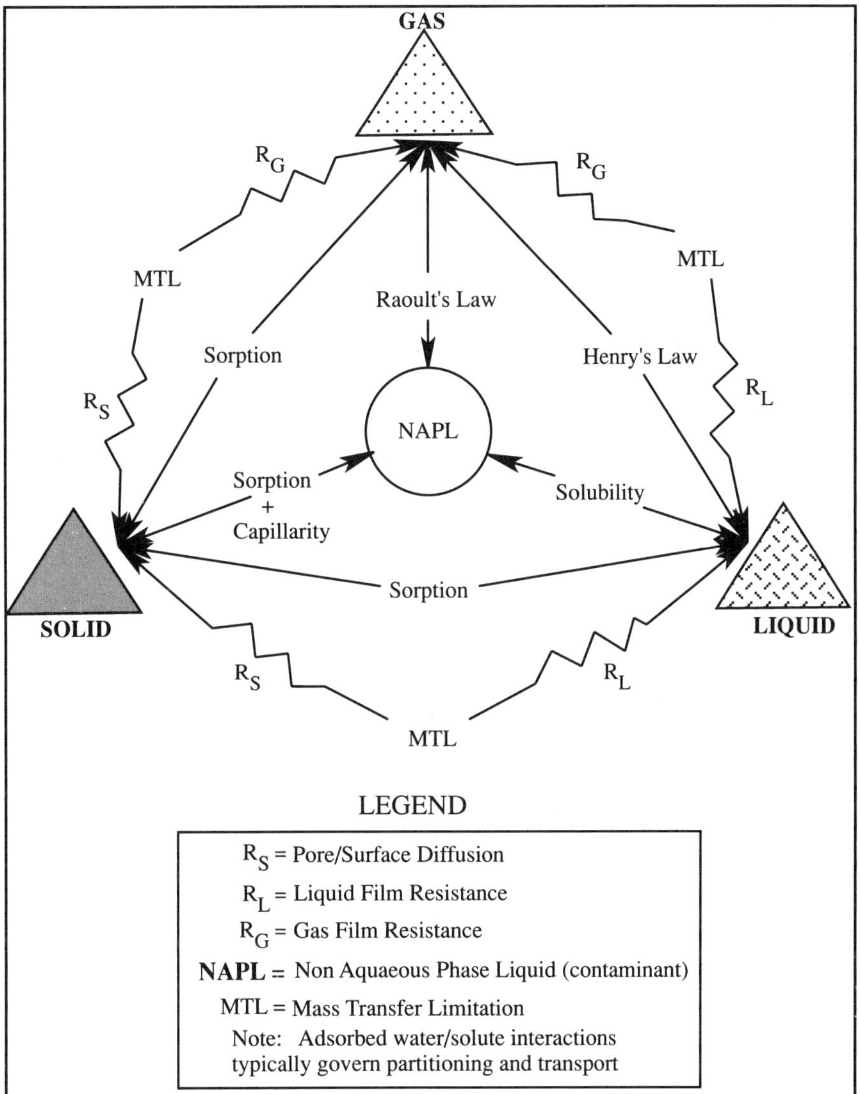

FIGURE 2.1
Generic Thermodynamic Triangle and Kinetic Circle
Soil Vapor Extraction

The soil properties, which impact SVE design, are:

- Relative intrinsic air permeability of the porous media
  - — Heterogeneities
  - — Porosity
  - — Stratigraphy
  - — Moisture content

    Since water molecules compete for soil adsorption sites and the water within pores blocks the air flow through the soil, permeability is reduced when moisture content is higher [2]. Sorption is also reduced.

- Temperature of the porous media
- Organic carbon content of the soil matrix

  Partitioning of contaminants is dependent on f_{OC}.

Advection is a process by which a contaminant is transported with the bulk flow of the medium in which it is present. See Figure 2.2 (adapted from reference [2]).

Diffusion is a process by which a contaminant moves through a medium as a result of entropic phenomena arising from concentration gradients. See Figure 2.3 (adapted from reference [2]).

The SVE process is an advective process in the sense that the gas phase containing the contaminants flows through the unsaturated zone. The gas phase favors more previous channels within soil strata. Contaminants in lower permeability zones will not be removed by advection, but must first be transported by diffusion to the higher permeability soil strata. Thus, in many cases, SVE processes may be limited by diffusion [2]. It is estimated that 40-60% of the contaminant is removed by advective transport and the remaining amount is diffusion limited [8]. This range varies significantly and is highly site specific.

2.2. Chemical Fate and Transport

2.2.1. Introduction

This section presents fundamental principles involving contaminant transport between phases. Figure 2.1 schematically illustrates the relationships between the partitioning and transport of the contaminant between the solid, liquid, and gas phases.

Properties of chemicals that will affect the contaminant transport include:

- Volatility
- Relative polarity
- Affinity for soil organic matter or organic contaminant
- Density and viscosity affect organic liquid transport and subsequent location of contaminants in the subsurface
  - LNAPLs are vertically transported through soil until they reach the capillary fringe where they tend to form pools
  - DNAPLs are vertically transported through both soil and water table until they reach an impermeable layer [2]
- Diffusivity; both in air and water

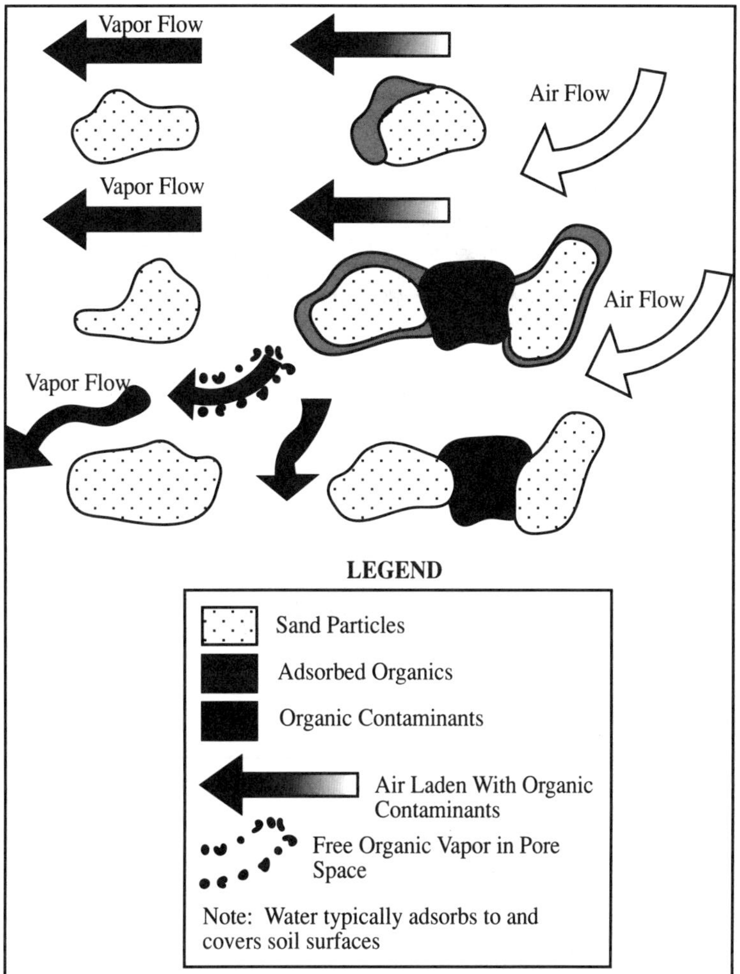

FIGURE 2.2
Advection Process
Soil Vapor Extraction

2.2.2. Contaminant/Vapor Thermodynamics

Gas and liquid phase equilibrium can be described for dilute aqueous solutions by Henry's Law and for concentrated NAPL solutions by Raoult's Law.

2.2.2.1. Raoult's Law

For concentrated solutions, Raoult's Law relates the equilibrium partial pressure of a component to its respective liquid phase mole fraction and the pure constituent vapor pressure.

Raoult's Law:

- Is applicable only when the constituent has a mole fraction greater than 0.9
- Typically applies to the equilibrium between NAPLs and the vapor phase

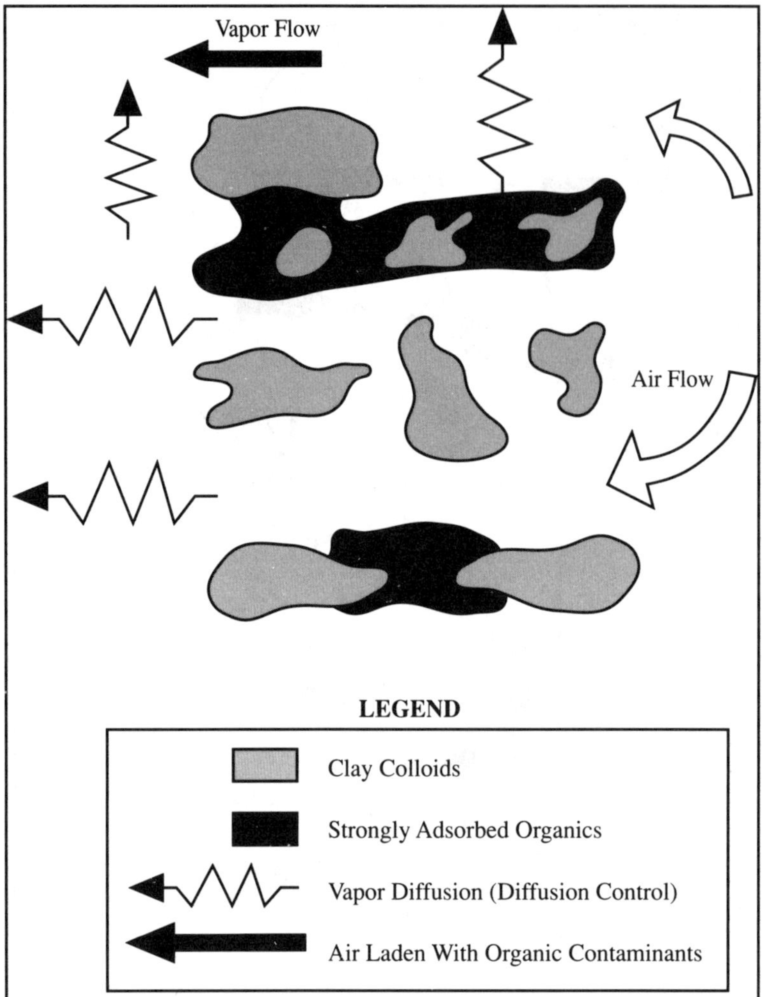

FIGURE 2.3
Diffusion Process
Soil Vapor Extraction

Raoult's Law may be written as:

$$P_i = x_i \, P_i^o \tag{1}$$

where:

P_i = the partial pressure of the constituent

x_i = the mole fraction of the constituent

P_i^o = the vapor pressure of the pure constituent, f(T)

2.2.2.2. Henry's Law

For dilute solutions, Henry's law relates the equilibrium partial pressure of a constituent to the mole fraction of the constituent in the aqueous phase. Henry's law is applicable only for dilute concentrations of the constituent, typically when $x_i < 0.15$. Henry's law is expressed as:

$$P_i = H_i \, x_i \qquad (2)$$

where:

P_i = the partial pressure of the constituent
x_i = the mole fraction of the constituent in the aqueous phase
H_i = Henry's constant for the constituent, f(T)

2.2.2.3. Effect of Temperature on Vapor Pressure

The effect of temperature on vapor pressure may be described by the Clausius-Clayperon equation, which modifies Raoult's Law. The Van't Hoff equation may be used to modify Henry's Law for temperature effects, however, this is rarely done in SVE design.

Kinetic energy of molecular translation is directly proportional to the absolute temperature and:

- Produces the forces influencing the vaporization of a liquid
- Increases the rate of vaporization and the vapor pressure, when increased

An increase in temperature will increase the rate of vaporization and the vapor pressure. The relationship between temperature and vapor pressure can be described by the Clausius-Clapeyron equation (a modification of the Van't Hoff equation):

$$\ln\left(\frac{P}{P^0}\right) = \frac{\lambda}{R}\left(\frac{1}{T^0} - \frac{1}{T}\right) \qquad (3)$$

where:

P = vapor pressure at T
P^o = vapor pressure at T^o
R = ideal gas constant
λ = molar heat of vaporization

2.2.2.4. Effects of Solubility

Constituents that are nonpolar are more likely to exist in the gas phase than polar constituents are; polar constituents exhibit attractive forces with water, making vaporization less favorable. Polar constituents typically have higher solubilities.

An increase in the liquid phase concentration of a constituent will always increase the equilibrium partial pressure.

As temperature increases, gas solubility typically decreases; consequently, steam stripping may increase removal efficiencies.

2.2.3. Contaminant/Soil Equilibrium Relationships

2.2.3.1. General

The fundamental principles of contaminant equilibrium between soil and air are basically the same as those between soil and water. However, soil and water systems must consider hydrophobic or hydrophlic effects on the sorption process.

Vapor phase diffusion is several orders of magnitude greater than liquid phase diffusion. Vapor phase diffusion typically dictates the rate of gas-phase hydrocarbon transport in the unsaturated zone when no vacuum is applied [9].

Although developed primarily to describe transport within a homogeneous phase, Fick's first law has also served as a foundation for describing interphase transport.

$$J_v = -D_v \frac{\partial C_v}{\partial x} \tag{4}$$

where:

J_v = contaminant vapor flux (mass/area-time, $M/(L^2T)$)
C_v = vapor concentration in soil (mass/volume, M/L^3)
x = distance in the direction of diffusion (length, L)
D_v = effective diffusion coefficient in soil matrix (area/time, L^2/T)

D_v has been described by the Millington and Quirk Relationship [10]:

$$D_v = \frac{\theta_a^{\frac{10}{3}}}{\varepsilon^2} D_v^{air} \tag{5}$$

where:

θ_a = air content of soil, Volume-1

ε = soil porosity

D_v^{air} = diffusion coefficient of chemical in free air (area/time, L^2/T)

An alternative way of presenting Fick's Law is to account for both soil porosity and tortuosity. Total soil porosity (ε) is subdivided into the area occupied by the air (θ_a) or effective porosity, the area occupied by water(θ_w), and that area occupied by an immiscible fluid(θ_i). Fick's law may then be written [9]:

$$F(k,T) = -\ \alpha(\varepsilon t - \theta w - \theta i) D_v^{air}\left(\frac{dc}{dx}\ (k,T)\right) \qquad (6)$$

where:

T	=	temperature (^oC)
α	=	soil tortuosity
k	=	constituent index for a multicomponant solvent

2.2.3.2. Sorption Processes

Sorption equilibrium may be described by a variety of adsorption isotherms; the most common of which are the linear adsorption isotherm and the Freundlich isotherm (for more information see [11]).

2.2.3.2.1. Linear Isotherm

Linear Isotherms are:

- Appropriate for sorption relationships in which the energetics of sorption are uniform with increasing concentration
- Appropriate when sorbent loading is low ("Henry's region" sorption)
- Appropriate for monolayer coverage
- Most commonly used at very low solute concentrations
- Used for solids exhibiting low sorption potential
- Often the result of the association of neutral, relatively nonpolar, organic molecules with soils

The linear isotherm may be written as:

$$q_e = K_D\, C_e \qquad (7)$$

where:

q_e	=	solute sorbed per unit of sorbing phase (mass/mass)
K_D	=	distribution coefficient/partition coefficient (volume/mass)
C_e	=	equilibrium concentration (mass/volume)

Advantages:

- Offers treatment for low organic soils
- Linear approximation reduces mathematical modeling complexity

- Often adequately describes sorption
- Works well for most neutral relatively, nonpolar organic molecules

Limitations:

- May not be valid over concentration ranges beyond those used for calibration
- Inappropriate for high organic content soil sorption

2.2.3.2.2. Freundlich Isotherm

The Freundlich isotherm:

- Is the most widely used nonlinear sorption model
- Is valid over a wide range of concentrations
- Describes multilayer coverage
- Was originally an empirically based model, but can be thermodynamically derived

The Freundlich isotherm is a semi-empirical model and may be written as:

$$q_e = K_F\, C_e^{1/n} \tag{8}$$

where:

K_F = sorption capacity constant
$1/n$ = sorption intensity constant

The constants may be obtained from nonlinear curve fitting.

Advantages:

- No limiting capacity
- Appropriate for sorption on heterogeneous surfaces
- Energy of sorption is a function of surface coverage
- Rapid initial uptake

Limitation:

- Nonlinear; mathematically more complex

2.2.3.3. Sorption Processes Fate and Transport

Equilibrium partition coefficients are developed on a local or microscopic scale. They are useful in terms of modeling the fate and transport of a contaminant when they are applied to the macroscopic scale.

The ability of a partition coefficient to predict the behavior of a contaminant in a subsurface system depends on:

- The relative importance of sorption processes in the context of other transformation and transport processes operative in subsurface environments

- The ability to describe the impact of microscopic processes on overall fate and transport at the macroscopic scale
- The ability to determine the level of complexity required to accurately describe overall fate and transport

The relative importance of the microscale phenomena on macroscopic contaminant transport phenomena may be estimated by:

- Conducting controlled investigations to determine rate and equilibrium parameters for microscale phenomena
- Incorporating these parameters into the reaction term in an appropriate predictive model and performing field-scale sensitivity analyses to determine the relative impact of sorption processes vs the advection and dispersion processes

In general, when the relative humidity of the soil is less than 60% (arid soils), adsorption processes are significant. When the relative humidity is greater than 60%, adsorption processes are typically negligible [8]

When sorption processes are not important, the transport model is based on mass conservation incorporating advection and dispersion.

Phenomena resulting from co-solvent effects of complex waste mixtures may cause significant deviations from equilibrium relationships.

2.2.4. Mass Transfer Considerations

2.2.4.1. General

Although local equilibrium is often assumed in the modeling of SVE processes, mass transfer limitations may occur. Presently, most researchers involved in the development of predictive SVE models have not included mass transfer resistances due to experimental and mathematical complexity. A framework for inclusion of the considerations is presented in this section.

2.2.4.2. Effect of Mass Transfer Resistances

There are three primary resistances to the transfer of a contaminant from the liquid phase to the gas phase.

They include:

- Diffusion from the bulk liquid to the liquid film
- Resistance from the liquid film to the gas film
- Resistance from the gas film to the bulk gas. This is presented graphically in Figure 2.4

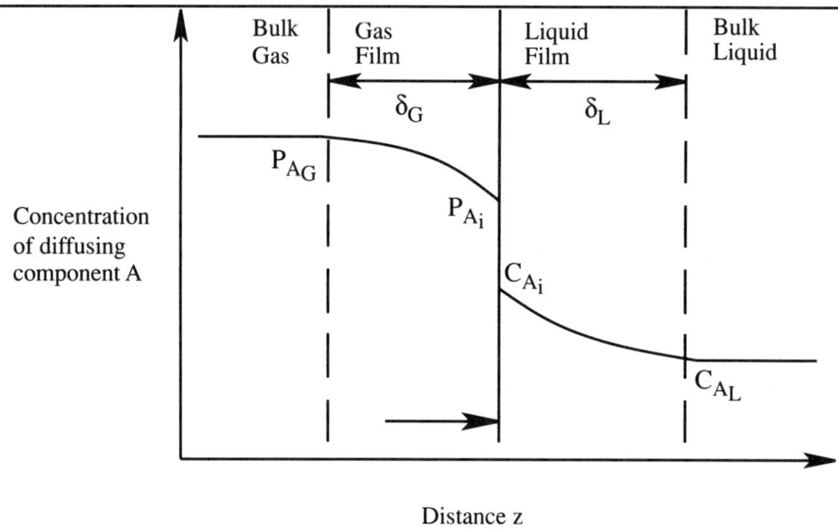

P_{A_G} = partial pressure of component A, in bulk gas phase
P_{A_i} = partial pressure of component A, at the interface
C_{A_i} = concentration of component A, at the interface
C_{A_L} = concentration of component A, in the bulk liquid phase
H = Henry's constant
δ_G = thickness of gas film
δ_L = thickness of liquid film

FIGURE 2.4
Concentration Gradients Between Two Contacting Phases
Soil Vapor Extraction

Because the SVE process exploits advection in the gas phase, gas film resistance typically has been demonstrated not to limit the rate of mass transfer. Mass transfer will be limited by either the liquid film resistance or by diffusion from the bulk liquid to the liquid film.

2.2.4.2.1. Two Film Theory

The Two Film theory may be used when the assumption of local equilibrium does not hold. This may occur when the velocity of the vapor is too fast for equilibrium to be attained. In this case, the limiting resistance to mass transfer is in the liquid film phase.

The overall mass transfer coefficient, K_L, is a function of the liquid film resistance, r_l, and the gas film resistance, r_g,. The overall resistance :

$$\frac{1}{K_L} = R = r_l + r_g \tag{9}$$

For constituents with:

- Low vapor pressures, the rate of mass transfer is controlled by the gas film resistance, r_g.

  Examples of constituents with low vapor pressures include naphthalene, dodecane, and tetramethylbenzene.

- High vapor pressures, the rate of mass transfer in controlled by the liquid film resistance, r_l

  Examples of constituents with higher vapor pressures include propane, isobutane, and isopentane.

Constituents with low vapor pressures are not candidates for the SVE process alone; therefore the liquid film resistance typically limits the rate of mass transfer in high moisture scenarios.

2.2.4.2.2. Fick's Second Law

Fick's Second Law describes diffusion from the bulk phase to the interface. This equation can be used when advective flow is minimal.

$$\frac{\partial C}{\partial t} = D\,\frac{\partial^2 C}{\partial x^2} \tag{10}$$

where:

D = the diffusion coefficient
x = Distance from gas/liquid interface
C = mole fraction of the constituent in the fluid

Equation (10) can be integrated as follows:

$$\frac{C - C^*}{C^o - C^*} = \frac{2}{\sqrt{\pi}} \int_0^{\frac{x}{\sqrt{4Dt}}} e^{-n^2}\,dn = \operatorname{erf}\left(\frac{x}{\sqrt{4Dt}}\right) \tag{11}$$

where:

C = mole fraction at time t
C_o = initial mole fraction
C_* = saturation mole fraction

Diffusion from a low permeability lens may be described by equation (11).

Figure 2.5 illustrates site specific examples of when this phenomena may occur.

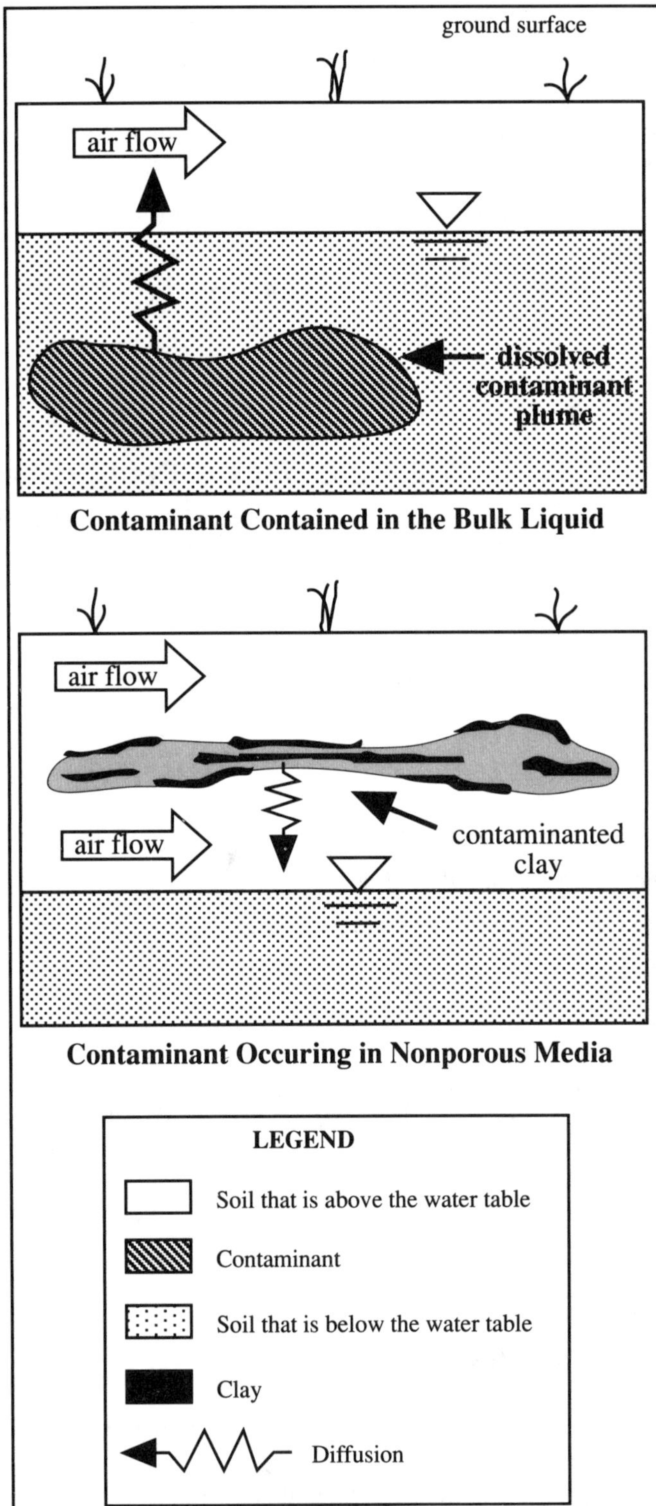

FIGURE 2.5
Examples of Diffusion Limited Stratigraphy
Soil Vapor Extraction

2.2.5. Kinetic Considerations

When mass transfer is limiting, the rate of transfer of a constituent from the aqueous phase to the vapor phase may be described by:

$$J = K_L \ (C_l - C^*) \tag{12}$$

where:

- J = molar flux rate (mol/m^2*hr)
- K_L = overall mass transfer coefficient (m/hr)
- C_l = molar concentration of constituent in the aqueous phase(mol/m^3)
- C^* = molar concentration of the constituent in the liquid film, assumed to be in equilibrium with the gas phase concentration, C_g

$$H = \frac{C_g}{C^*} \tag{13}$$

Henry's constant can be substituted into equation (12) and subsequently rearranged to yielo.

$$-\frac{dC_l}{dt} = K_L \frac{\left(C_l - \frac{C_g}{H}\right)}{L} \tag{14}$$

where:

L = depth in meters (unit volume per interfacial area)

Upon integration, equation (14) becomes:

$$C_l = \frac{C_g}{H} + (C^\circ{}_l - \frac{C_g}{H}) \ \exp(-K_L \ \frac{t}{L}) \tag{15}$$

When the gas phase concentration is significantly smaller than the liquid phase resistance, equation (15) becomes:

$$C_l = C^\circ{}_l \exp\left(\frac{-K_L t}{L}\right) \tag{16}$$

These equations are pertinent to volatilization of contaminants from high moisture soils where liquid film resistance is limiting.

3. APPLICATIONS

3.1. General Applicability

SVE can be employed:

- When soil contains contaminants with vapor pressures greater than 0.5 mm Hg
- Where soil excavation is impossible (e.g. the presence of buildings on site)
- When large volumes of contaminated soil need treatment
- When site stratigraphy has relative intrinsic air permeabilities of 10^{-2} to 10^{-5} cm/sec. (*soil-air permeability not soil-water permeability*)
- When biological degradation of the contaminant may need enhancement [2]

3.2. Advantages & Limitations of SVE

3.2.1. Advantages

There are many advantages to the SVE process:

- Requires few mechanical parts
- Has low operating and maintenance costs
- Employs conventional equipment, labor, and materials
- Can be installed rapidly
- Shows immediate results
- Does not require reagents
- Provides permanent remediation
- Is not limited to depth to groundwater.

  Successful applications have been demonstrated by Terra Vac at sites with groundwater as deep as 300ft and as shallow as 3 in [6].
- Requires no excavation
- Is not disruptive (no interruption of business operation or residential living habits)
- May be used to remediate soil underneath buildings or other structures
- May enhance biodegradation of compounds (bioventing)

3.2.2. Limitations

SVE's applicability and usefulness are limited because:

- Contaminants with low vapor pressures have poor removal efficiencies

- Additional treatment technologies may be needed (e.g.pump and treat)
- Vapors are impeded by fine-grained ("tight") soils
- Resulting vapors become an air quality concern and usually require treatment
- Soils must be permeable and fairly homogeneous for it to be most effective. Impermeable soil lenses within more permeable soils could adversely affect the recovery of volatile soil contaminants
- Toxicity of contaminant is not altered
- Cleanup times may be unpredictable due to complexity of site strata

4. PROCESS DESCRIPTION

4.1. Introduction

The SVE system, typically consists of the following components:

- Extraction System
  - One or more vapor extraction wells
  - One or more air inlet or injection wells (optional)
  - Vapor liquid separator (optional, but necessary when thermal gas treatment units are used. They also serve to protect blowers)
  - Barrier wells (optional)
  - Monitoring wells
  - Vacuum pumps and/or blowers
  - An impermeable cap (optional) [3]
- Off-gas treatment (per ARARs)
- Liquid treatment (per ARARs) [2]

4.2. Extraction System

4.2.1. Wells

4.2.1.1. Vapor extraction wells

Vapor extraction wells usually consist of a partially screened pipe placed in permeable packing, typically coarse sand or gravel [2]. The well casing is selected to be unaffected by the contaminants of concern [2]. Vents are typically cased 5 to 10 feet below the ground surface to prevent short-curcuiting of air near the vent [3].

Vapor extraction wells:

- Penetrate the contaminated soil zone but do not intercept the groundwater table
- Are spaced according to mathematical models of air flow, which are based on pilot air permeability testing results
- May be vertical wells or horizontal screens

Horizontal screens are appropriate:
— For shallow water tables
— Where groundwater removal is to be minimized
— When contamination is confined to near-surface soils

- Are used to induce a vacuum at the well head
  — The vacuum is directly related to the radius of influence.

A vapor extraction well's radius of influence is affected by:

- Vacuum applied
- Presence of air vents or air injection wells
- Stratigraphy
- Soil properties (porosity and relative permeability)

  Reported radii of influence values for permeable soils are 30-120 feet [2].

- Soil surface boundary conditions
  — Loam is less porous and acts as a confining layer, increasing the radii of influence

4.2.1.2. Monitoring Wells

Monitoring wells are placed around the extraction wells to monitor the interstitial air pressure and control the pressure gradient by allowing air to bleed into the system.

4.2.1.3. Injection and Inlet Wells

These wells are typically constructed in a similar manner as the vapor extraction wells [2].

Injection wells may be used:

- To inject air by means of blowers
- In closed-loop systems [2,4]
- To control air flow [8]

Inlet wells are:

- Passive and allow air to be drawn into the ground [2]
- Most useful when long trenches are utilized [8]

For better control over air flow, injection wells are typically preferred over inlet wells [8]. Injection and inlet wells may also be used as barriers to prevent the migration of contaminants from off-site. Injection and inlet wells may also cause anomalies in subsurface air flow fields, resulting in possible short-circuiting and dead zones. Decreases in extraction efficiency may subsequently result.

4.2.2. Pumps and Blowers

Vacuum pumps and blowers are used to:

- Reduce gas pressure in the extraction wells and induce subsurface air flow to the wells [2]
- Transport the contaminated vapor through a treatment system.[2]

Note: Water Table mounding occurs in the vicinity of an extraction well (see Figure 4.1)

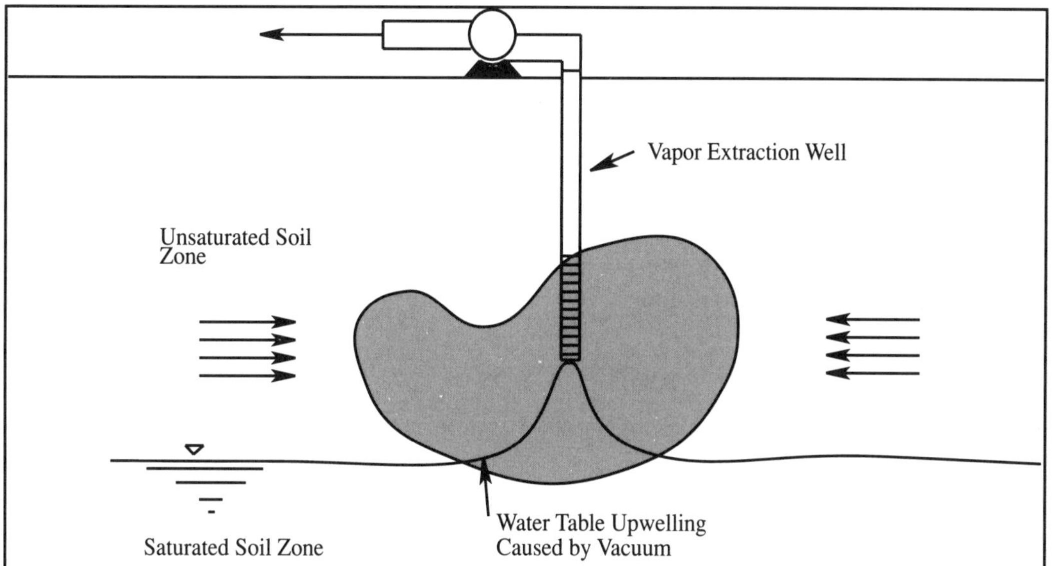

FIGURE 4.1
Water Table Rise Caused by the Applied Vacuum
Soil Vapor Extraction

4.2.2.1. Pumps

4.2.2.1.1. Liquid Ring Pump

Liquid Ring pumps:

- Have vacuums ranging from 0 to 29 inches of Mercury (760-25 Torr)
- Handle flows of 10 to 100 cfm (depending on soil strata)
- Are used often for remediation and may also be used for air permeability testing
- Allow a high vacuum to be reached without wearing out the rotor blades by taking advantage of the compressibility of water.

  Since high vacuums normally wear down the rotors and the insides of the pump casing, water is used inside the pump to reduce friction.
- Commonly used for larger jobs requiring high vacuums
- Employs mechanical seal and O ring gaskets for zero leakage

4.2.2.1.2. Rotor Lobe

- Similar to liquid ring pump
- Used for air permeability testing where high vacuums are not required.

4.2.2.1.3. Rotary Vane

- Used for testing air permeability of soil strata and low volume moderate vacuum applications
- Maximum 27" Hg
- No oil needed—low maintenance (when carbon blades used)

4.2.2.2. Regenerative Blower

- Produces low to moderate vacuums and high vapor flow rates
- Can be used for larger jobs
- Maximum pressure 114" water

4.2.3. Impermeable Caps

Impermeable caps are optional components of SVE processes.

Purposes of an impermeable cap include [2]:

- Minimizing infiltration of water from the surface
- Increasing the system's radius of influence by preventing short circuiting
- Controlling the horizontal movement of inlet air, which can by-pass contaminants.

Impermeable caps may consist of [2]:

- Plastic membranes
- Buildings
- Parking lots
- Soil layers of low permeability

4.2.4. Vapor/Liquid Separators

Vapor/Liquid separators protect blowers and pumps and may increase the efficiency of the vapor treatment systems. The entrained groundwater and condensate may require treatment and disposal as a hazardous waste depending on the type and concentration of the contaminant [2]. Vapor/liquid separators are necessary when thermal processes are used for off-gas treatment, because liquid vapor can adversely impact catalytic oxidation units and/or reduce the capacity of activated carbon.

4.3. Off-Gas Treatment

4.3.1. General

Off-gas treatment (required under ARARs) is used to remove the contaminant from the vapor stream before it is discharged to the atmosphere.

A variety of methods may be used to treat the off-gas:

- Thermal Destruction Units
  - Vapor combustion

  - Catalytic oxidation unit
- Carbon adsorption unit
- Internal combustion engines (common in California)
- Vapor condensation
- Wet scrubber
- Bioreactors

More than one off-gas treatment method may be necessary if the concentration of the contaminant varies by an order of magnitude across the site or over time.

4.3.2. Thermal Destruction Units

4.3.2.1. General

Contaminants in vapor may be destroyed by thermal incineration or catalytic oxidation. These processes are effective for a wide range of compounds. The primary advantage of thermal destruction is that the contaminant is chemically altered so that it is no longer toxic.

Vapor combustion units are typically used for contaminant concentrations greater than 12,000 ppm and catalytic oxidation is typically used when contaminant concentrations are less than 12,000 ppm. Vapor/liquid separators must be used prior to thermal destruction units.

4.3.2.2. Vapor Combustion Units

Vapor combustion processes allow for:

- > 95% efficiency
- Contaminant concentrations >12,000 ppm
  - If concentrations are < 12,000 ppm, fuel must be added to aid in combustion. [4]
- Energy recovery from high BTU content volatile contaminants
- Halogenated solvents, which produce acid gases, require the combustion unit off-gas to have a scrubber and may preclude use of these units based upon economic considerations

<u>Advantages:</u>

- No upper limit on contaminant concentration
- High compound destruction efficiencies
- Application to a wide variety of compounds
- Operation is simple

<u>Limitations:</u>

- Cost will be high if inlet concentration is less than the 10,000 ppm, requiring supplemental fuel
- May require follow-on treatment systems (e.g. carbon or catalytic oxidation)

4.3.2.3. Catalytic Oxidation Units

Catalytic oxidation is effective on hydrocarbon vapors at concentrations less than 12,000 ppm. Recently developed catalysts permit efficient destruction of halogenated compounds as well [2].

The catalytic oxidation system follows the steps below:

- The vapor is preheated to approximately 430°F in a heat exchanger
- The preheated gas enters the burner where it is heated to approximately 800°F.
- The heated gas passes through a catalyst bed where it is oxidized

  The catalyst accelerates the rate of oxidation by adsorbing the oxygen and the contaminant on the catalyst surface where they react to form carbon dioxide, water, and hydrochloric gas. The catalyst enables the oxidation reaction to occur at much lower temperatures than required by a conventional thermal incinerator.

The catalytic oxidation process:

- Is > 95% efficient
- Is effective when vapor concentrations are <12,000 ppm. Combustion of concentrations > 3,000 ppm are self sustaining; concentrations < 3,000 ppm requires fuel to aid in combustion [4].

Advantages:

- Continuous operation because it does not require the intermittent downtime needed by carbon adsorption systems
- VOC's are destroyed on-site rather than transferred into carbon and hauled off-site for thermal destruction

Limitations:

- Chlorinated hydrocarbons generate a significant amount of hydrochloric acid gas, which may require removal by a scrubber, thus increasing the cost of treatment
- High concentrations of hydrochloric acid gas may damage the catalyst, increasing the rate of catalyst replacement and the cost of treatment.

  Thermal units, which contain a metal oxide catalyst that HCl will not damage, are now being investigated. This would accommodate the treatment of chlorinated hydrocarbons.

- High water content without the aid of a liquid/vapor separator may produce droplets on the catalyst, thereby reducing the heat and oxidation efficiency

4.3.3. Carbon Adsorption Units

Carbon adsorption is the most commonly employed vapor treatment process and is adaptable to a wide range of contaminant concentrations and flow rates [2].

Advantages:

- Effective for most vapor streams
  - Some organics have poor sorption characteristics
- Most commonly used when vapor concentrations are <1,000 ppm

- High removal efficiencies

Limitations:

- Carbon must be regenerated or disposed of off-site
- The process is expensive
  — The cost doubles due to disposal of carbon
- It is a transfer process not a destructive process
- Efficiency is degraded by presence of moisture in gas phase
- Competitive adsorption effects

4.3.4. Vapor Condensation Units

Vapor Condensation Units may be used when the concentration of the contaminant in the vapor stream is high and flow rate is low.

- Condensation is usually accomplished by refrigeration
- Effectiveness is determined by the vapor pressure and temperature characteristics of the contaminant present [2]
- Condensation of the contaminant is rarely complete, requiring an additional method of treatment
- The process is expensive

4.3.5. Bioreactors

Utilizing Bioreactors for the treatment of SVE off-gas:

- Is being investigated for the treatment of halogenated and nonhalogenated hydrocarbons
- May produce nontoxic by-products
- Would require carbon substrate for an energy source, nutrients, and water

4.3.6. Diffuser Stacks or Dispersion

Diffuser stacks are a nontreatment option. For some cases, the extracted vapor can be released into the atmosphere without exceeding the air emission standards. Air transport models are used to assess ambient air concentrations when this option is proposed.

4.4. Concurrent Process Descriptions

4.4.1. General

Air sparging and bioventing are unique processes that compliment the SVE process. Air sparging aerates the saturated zone; bioventing aerates the contaminated soil allowing biodegradation.

4.4.2. Air Sparging

Air Sparging introduces air into the saturated zone, which aerates the water and stripps VOC from this zone. Air is dispersed through the saturated zone in response to an induced pressure gradient. The pressure gradient is generated by center wells and surrounding wells.

Center wells - are equipped with an air pump or blower that pumps air into the saturated zone

Surrounding wells - are equipped with vacuum pumps that remove the contaminated air from the unsaturated zone.

Air moves horizontally through the saturated zone as a result of the pressure gradient.

Aspects of air sparging to consider:

- Process is energy intensive
  - Pumping air into the saturated zone requires more energy than pumping air into the unsaturated zone
- Limited to soils that do not contain a high fraction of fines
- Care must be taken not to spread contamination
- Care must be taken not to create dead zones

  Dead Zones are areas where the induced pressure gradient is zero and may be adjusted or eliminated by varying the pumping rates.
- Process has limited field scale data
- Mathematical description of process is insufficiently developed

4.4.3. Bioventing

Bioventing is the term used for increased biological activity, which results in the degradation of the contaminant. Indigenous bacteria utilize contaminants as an energy source and produce carbon dioxide and other by-products. A new innovative technology idea is to use SVE for Bioventing.

Some aspects of bioventing include:

- Bioventing exploits natural processes
- During the SVE process, biological activity may account for 20% of the remediation [12]
  - The Air Force has demonstrated that up to 90% of jet fuel can be biologically destroyed in a bioventing system
- SVE is poor for semi-volatile fuel oil. However, > 90% of the diesel fuel may be responsive to bioremediation.[12]
- Air sparging may enhance biological activity
- The availability of biological nutrients and oxygen limit biological activity
- Oxygen is usually rate limiting.

  It typically takes 3 lbs of oxygen to remediate 1 lb of hydrocarbon
- Inhibitors include:
  - Concentrations of the contaminant that are not biologically assimilable or damage the microbial population.
  - Oxygen availability (hydrogeology impacts oxygen transport)
  - Metal contaminants
- Bioventing depends on the chemical and the soil type

It is difficult to quantify the effects of increased biodegradation. One indirect method involves monitoring CO_2/O_2 ratio in an extraction well at the edge of the contaminated zone. If the ratio increases as the vapors pass through the contaminated soil, then it is likely that bioventing is occurring [5].

SVE may also introduce nitrogen into the soil and produce an anaerobic environment. Some work has been performed on anaerobic degradation, however, this is an innovative technology with little supporting data.

> PCE will degrade anaerobically, not aerobically. However, a product of PCE anaerobic degradation is vinyl chloride, a carcinogen. Vinyl chloride will degrade aerobically, thus sequential anaerobic, aerobic degradation may be necessary for some contaminants.

4.4.4. Contaminated Water Treatment

Recovered condensate requires treatment in accordance with the State and National Pollution Discharge Elimination System (SPDES/NPDES) permit levels. Treatment must occur prior to surface water discharge or in accordance with pretreatment requirements, which occur prior to discharge to a publicly owned treatment works [2]. Contaminated water can usually be treated with carbon adsorption or air stripping.

5. FEASIBILITY

5.1. Introduction

In general, the applicability of SVE processes is based on:

- Soil and site characteristics
- Contaminants location and properties

5.2. Prescreening

Several reports and electronic data bases that may be consulted for planning and conducting SVE treatability studies, as well as for prescreening SVE for use at a specific site [2], exist.

Time since release, site conditions, soil properties, and contaminant properties will dictate SVE feasibility (see Figure 5.1 adapted from reference [13]).

5.2.1. Data Collection

The following information is required to help determine the feasibility of SVE as a treatment alternative [3].

5.2.1.1. Contaminant Information

Information that is needed to characterize contaminants includes [2]:

- The composition and concentration of contaminants in:
  - — Soil gas
  - — Soil
  - — Groundwater

- NAPL

- Diffusivity
- Henry's constant
- Vapor pressures
- Solubility
- Density
- Viscosity
- Location

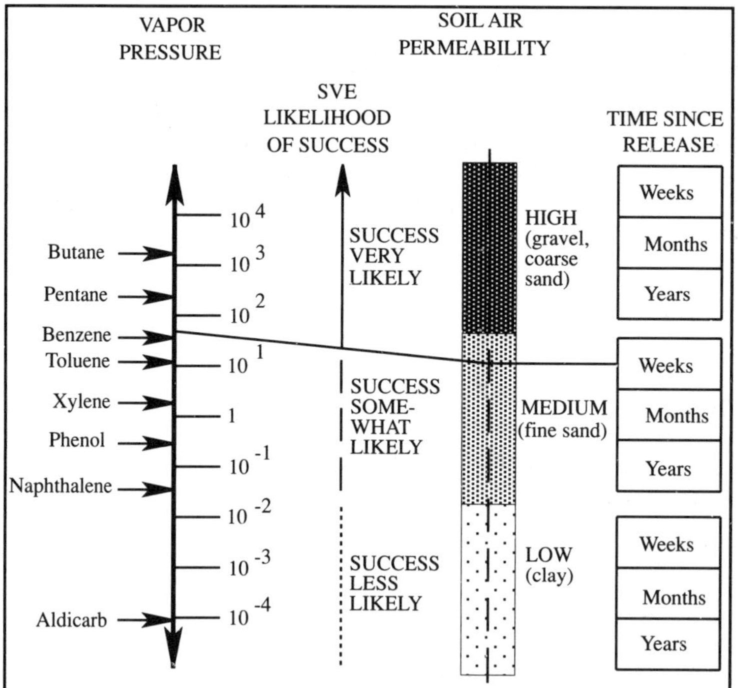

FIGURE 5.1
SVE Applicability Nomograph
Soil Vapor Extraction

5.2.1.2. Soil Gas

Analysis of contaminants in the soil gas can provide critical data regarding contaminants and their mobility and distribution at the site. Soil gas samples are taken to identify potential areas of contamination.

5.2.1.3. Soil Boring

Soil borings may be performed in those areas identified by the soil gas results to delineate the amount, location and extent of contamination.

5.2.1.4. Geological Structures

Geological structures can be identified by cone penetrometer measurements. Cone penetrometer measurements provide greater detail and supplement soil boring measurements at a reasonable cost [2].

5.2.1.5. Mapping

Once the extent of contamination and the geological structures have been identified, soil mapping should be performed to show the contaminant concentrations in relation to geological structures [2].

5.2.1.6. Site Characteristics [3]

The soil characteristics that affect the feasibility of SVE are:

- Air permeability
- Soil moisture content or degree of saturation
- Clay content
- Humic or organic carbon content
- Temperature profiles
- Site heterogeneities
- Soil heterogeneities

5.2.1.6.1. Air Permeability

The air permeability of the contaminated soils controls the rate at which air can be drawn through the soil by the applied vacuum. The lower the permeability, the more power required for a given flow rate through the soil. Intrinsic air permeabilities may be measured only within an order of magnitude. The higher the permeability of the soil, the higher feasibility of applying an SVE technology.

5.2.1.6.2. Soil Moisture Content or Degree of Saturation

Drier soils have greater availability of pore area, which permits higher air flow rates—a desirable quality for most SVE applications. However, extremely dry soils may strongly adsorb VOCs, thus, reducing contaminant removal. These VOCs are more easily desorbed when water competes with them for adsorption sites. This phenomena may limit the effectiveness of SVE in southwestern states [2].

5.2.1.6.3. Clay and Humic Content

Soils with high clay or humic content generally provide high sorption potential for VOCs, reducing feasibility. The success of SVE in clayey or silty soils depends on the presence of more permeable strata, which would permit air flow close to the tighter formations [2].

5.2.1.7. Site Conditions

5.2.1.7.1. Depth to Groundwater

When contamination extends to the groundwater that lies at depths greater than 40 ft, SVE systems may be one of the few ways to remove contaminants from the soil [2].

5.2.1.7.2. Water Infiltration

Water infiltration may affect the SVE process by:

- Reducing the rate of contaminant removal by taking up available pore space
- Increasing the amount of water entrained in the vapor, increasing residual treatment costs
- Requiring placement of an impermeable cap

5.2.1.7.3. Soil Heterogeneities

Uncertainty in the location of heterogeneities increases the difficulty in placing vapor extraction and inlet wells [2]. SVE may be favorable for a horizontally-stratified soil because the relatively impervious layers will limit the rate of vertical inflow of air from the surface and tend to extend the applied vacuum's influence from the point of extraction [3]. Heterogeneities in contaminant composition may also cause problems when considering SVE. These heterogeneities may reduce permeability throughout the layers.

5.3. Treatability Flowchart

A typical SVE treatability flowchart is illustrated in Figure 5.2 (adapted from reference [2]).

6. DESIGN

6.1. Introduction

The most important factors which impact SVE design are:

- Contaminant volatility
- Intrinsic air permeability
- Heterogeneities in the site
- Placement and number of wells

If the intrinsic air permeability data is off slightly, it can affect design results by several orders of magnitude. When a homogeneous model is used to describe a heterogeneous site, the results may be off by several orders of magnitude. According to the above factors, the number and placement of wells is a judgement call on the part of the designer and the client in terms of time and money.

The key to any successful design is the validity of the input parameters. The effects of subsurface heterogeneities are never fully accounted for in models, thus experience with SVE process design is a factor.

The design process is illustrated in Figure 6.1 (adapted from reference [5]).

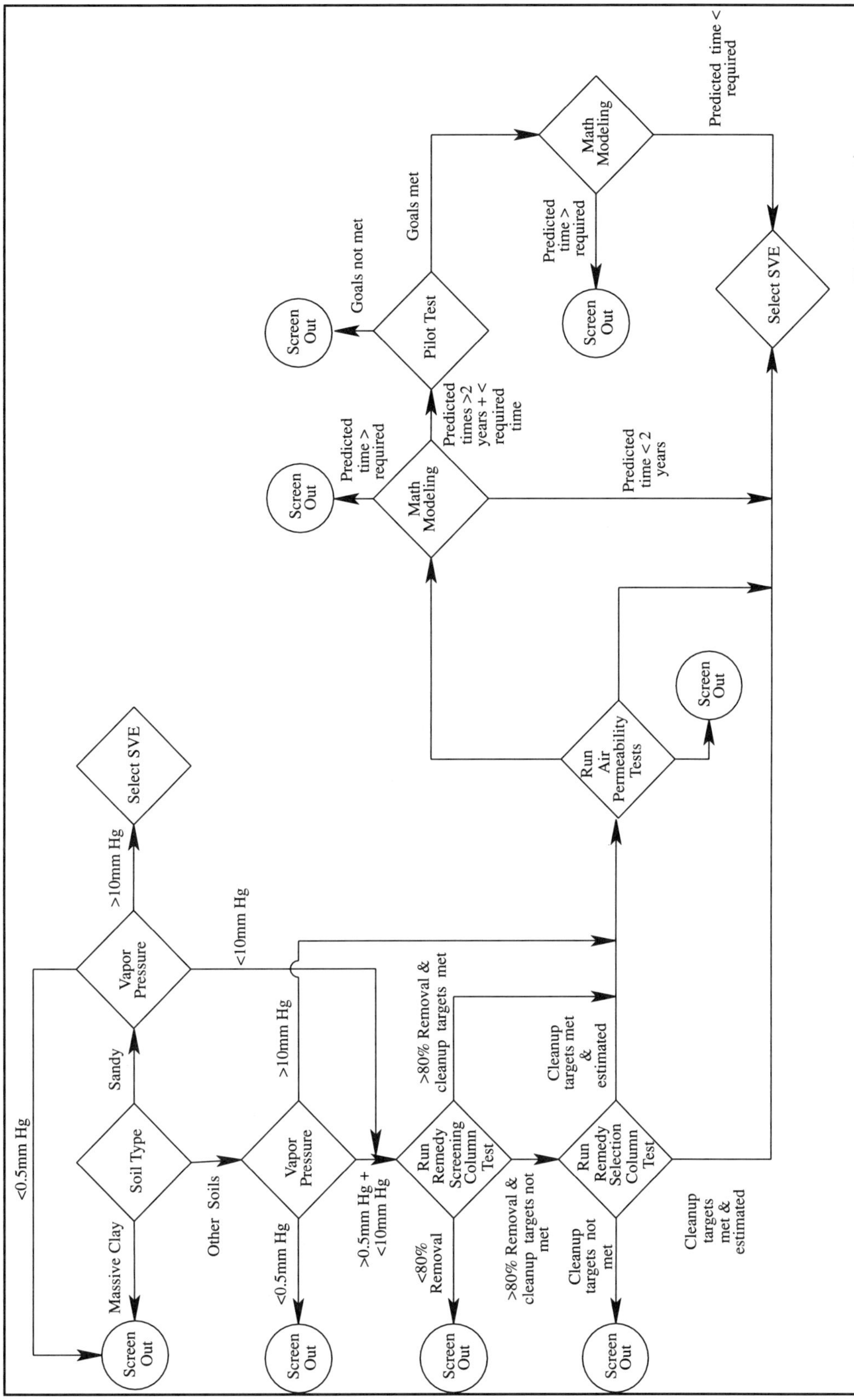

FIGURE 5.2
Treatability Flowchart For Evaluating SVE
Soil Vapor Extraction

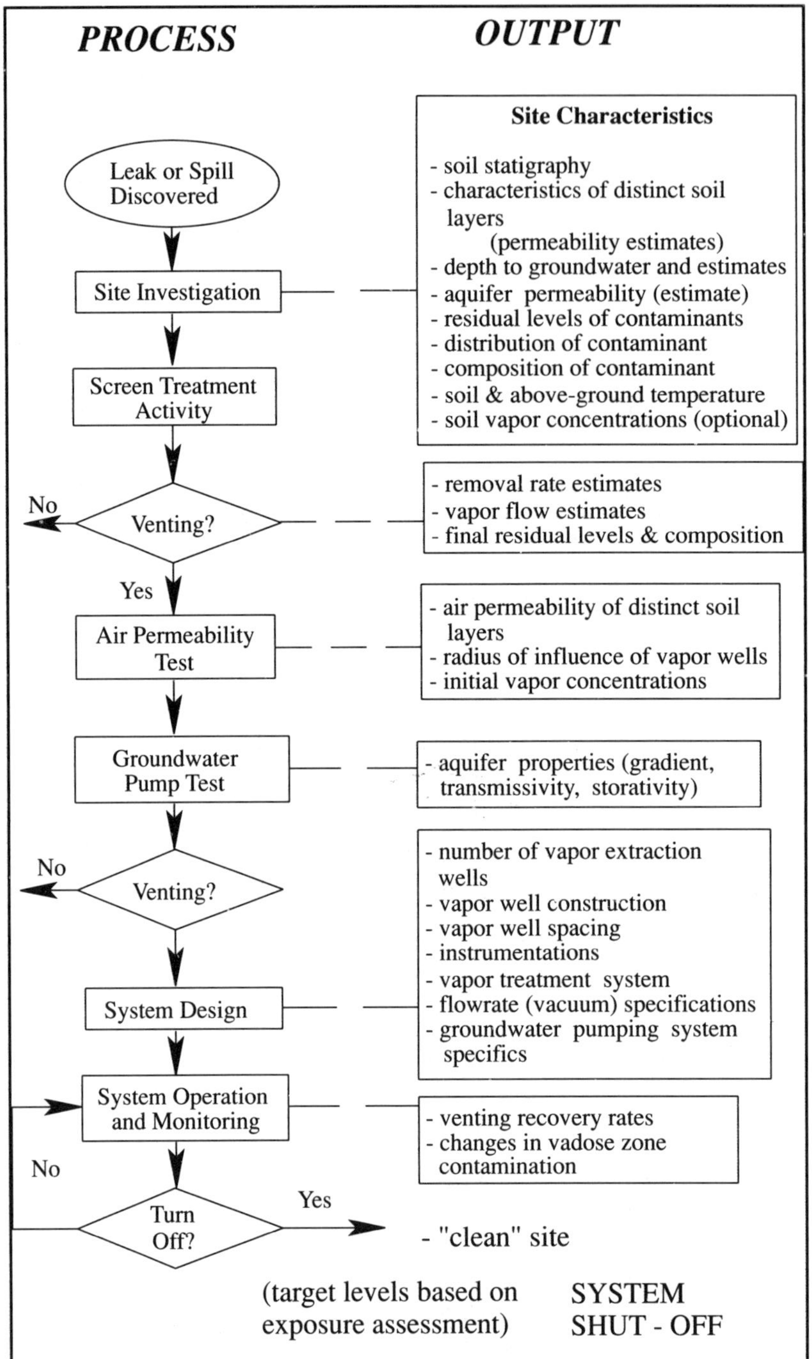

FIGURE 6.1
***In Situ* Soil Venting System Design Process**
Soil Vapor Extraction

6.2. Objectives

As with any remediation process, the objectives of the SVE process design are:

- To optimize the mass of contaminant removed
- To minimize cost of remediation
- To minimize time to remediate

This goal of optimum efficiency in SVE processes is dependent on:

- Vapor flow rate - (Intrinsic Air Permeability)
- Vapor flow path relative to contaminant distribution (Soil Heterogeneities)
- Composition of the contaminant

Design models relate the:

- Applied vacuum
- Intrinsic air/soil permeability
- Spill/contaminant composition
- Air flow boundaries

To the:

- Vapor flow rates
- Velocities
- Mass removal rates
- Residual composition changes with time

6.3. Alternatives

6.3.1. General

There are two different approaches to SVE process design. The first approach involves a 2-D radially symmetric analytical model [1, 5] and the second is the application of a numerical model [12], which requires detailed knowledge of the subsurface structure and spatial distribution of the contaminant. A detailed discussion of the analytical model is presented below.

6.3.2. 2-D Radially Symmetric Analytical Model (the following sections, utilizing this model, are adapted from references [5] and [1])

6.3.2.1. Assumptions

- Contaminant is uniformly distributed throughout soil at all times
- The vapor, a free-liquid, sorbed and dissolved phases are always in equilibrium
- Vapor behaves as an ideal gas

- Claussius-Clapeyron equation is valid
- The influence of surface tension on vapor pressures is neglected
- Activities are based on solubilities
- Homogeneous isotropic porous media
- Soil sorption coefficients are based on K_{OW} and f_{OC}
- Vapor flows are primarily orthogonal to gravity
- Changes in the apparent intrinsic air permeability are neglected
- The medium is porous and isothermal

6.3.2.2. Input Parameters

Required input data includes the:

- Pure component vapor pressure and concentration
- Sorption coefficient for each component
- Solubility of each component in water
- Molecular weight of each component
- f_{OC}
- K_{OW} of each component

6.3.2.3. Advantages

The model requires limited data input and is straightforward to utilize.

6.3.2.4. Limitations

The essential limitation of this model is its inability to explicitly describe heterogeneities in the porous media. This may affect results in a nonhomogeneous system by several orders of magnitude.

The specific problems are listed below:

- The horizontal vapor flow assumption may not be valid in cases where flow from the surface is significant
- Changes in air permeability, due to variable moisture and NAPL saturation in the unsaturated zone, may be significant, particularly when the liquid content of the soil is in excess of 20%
- Nonisothermal subsurface environments are common at SVE sites due to subsurface utility lines such as gas and water mains, electric conduits, and steam pipes
- Vertical heterogeneity is not accounted for in the radial flow two dimensional model. A three dimensional model is better suited to model heterogeneities
- Analytical solutions are obtained when the soil is assumed to be homogeneous. Heterogeneity may be accounted for by using a numerical scheme with appropriate boundary conditions [12]. A three dimensional model has been developed that allows for evaluating heterogeneities in the unsaturated zone [8]

A more detailed solution for determining venting removal rates, vapor concentrations, and residual composition changes can be developed by solving an advection-diffusion differential equation. However, a detailed knowledge of the subsurface structure and spatial distribution of contaminant is required.

6.4. Air Permeability

Because air permeability is a critical parameter in the design process, it is important to insure that the data to be used is representative of the site. It is common during feasibility studies for hydraulic conductivity values to be extrapolated to the intrinsic horizontal air permeability. More often than not, the assumptions upon which the extrapolation is based may be invalid. Errors in extrapolating hydraulic conductivity values to intrinsic horizontal air permeabilities may derive from any of the following [12]:

- Gas slippage is neglected
- Anisotropy is neglected
- Swelling soils are not addressed
- The variable water saturation in the unsaturated zone is neglected
- The presence of an oil phase is neglected
- The groundwater test may have occurred in different strata
- The scale of the groundwater test may invalidate the parameter evaluations

Air permeabilities are determined by removing gas at a constant rate from an extraction well while monitoring, with time, the transient subsurface pressure distribution at fixed points [5]. A spacial averaging of air permeabilities may be produced and used in the design process.

The relationship between air permeabilities, horsepower and vapor flow rates is presented in Figure 6.2 (adapted from reference [12]). As shown in this figure, the vapor flow rate is sensitive to the permeability of the soil, a parameter which is difficult to measure accurately.

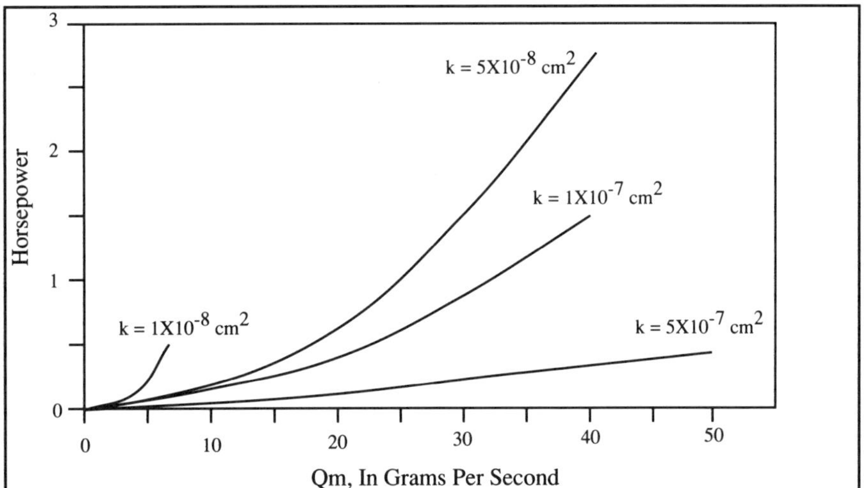

FIGURE 6.2
Relationship Between Air Permeability, Horsepower, and Vapor Flow
Soil Vapor Extraction

6.5. Design Equations

6.5.1. Basic Principles of Design (Vapor Flow and Vacuum-Induced Behavior)

Starting with Darcy's Law:

$$q = -K \frac{d\phi}{dx} \tag{17}$$

where:

$$\phi = z + \frac{P}{\rho g} \tag{18}$$

$K = \frac{k\rho g}{\mu}$ = hydraulic conductivity

k = intrinsic permeability

q = specific discharge = $\frac{Q}{A}$

Substituting:

$$q = -\frac{k\rho g}{\mu}\left(\frac{dz}{dx} + \frac{dP}{dx}\frac{1}{\rho g}\right) \tag{19}$$

Assuming flow orthogonal to gravity converting to three dimensions and replacing **q** with **u** (vapor velocity) to be consistent with the literature.

$$u = \frac{-k}{\mu} \nabla P \tag{20}$$

Assuming:

- Advective transport
- No changes in intrinsic permeability (k = constant)
- Isothermal porous media

Continuity equation can be written as:

$$\frac{\partial \theta_a \rho b}{\partial t} = -\nabla \rho b u \tag{21}$$

where:

g	=	gravity vector (cm/s^2)
k	=	soil/air permeability (cm^2) or (Darcy)
μ	=	vapor viscosity (f/cm-s) or (centipoise)
P	=	vapor phase pressure (g/cm^2-s^2) or (atm)
∇	=	gradient operator (cm^{-1})
ρ_{atm}	=	vapor density at the reference pressure, P_{atm}
b	=	stratum thickness (cm)
θ_a	=	air filled porosity

Employing the ideal gas law:

$$\rho = \rho_{atm} \left(\frac{P}{P_{atm}}\right) \tag{22}$$

Substituting into the continuity equation:

$$\left(\frac{2\theta_a \mu}{k}\right) \frac{\partial P}{\partial t} = \nabla^2 P^2 \tag{23}$$

Expressing the deviation from atmospheric pressure (i.e. vacuum applied) as ΔP. Then:

$$P = P_{atm} - \Delta P \tag{24}$$

Substituting:

$$\left(\frac{2\theta_a \mu}{k}\right) \frac{\partial (P_{atm} - P)}{\partial t} = \nabla^2 \left(P_{atm}^2 - 2P_{atm}\, \Delta P - \Delta P^2\right) \tag{25}$$

Assuming $\Delta P^2 << 2P_{atm}\,\Delta P$, then:

$$\frac{\theta_a \mu}{kP_{atm}}\left(\frac{\partial \Delta P}{\partial t}\right) = \nabla^2(\Delta P) \tag{26}$$

Or in radial coordinates [14]:

$$\frac{\theta_a \mu}{kP_{atm}}\left(\frac{\partial \Delta P}{\partial t}\right) = \frac{1}{r}\frac{1}{\partial r}\left(r\left(\frac{\partial \Delta P}{\partial r}\right)\right) \tag{27}$$

Based on considerations similar to aquifer drawdown with the following boundary conditions:

$$\Delta P = 0 \qquad r \rightarrow \infty$$

$$\lim_{r \rightarrow 0} \; r\left(\frac{\partial \Delta P}{\partial r}\right) = \frac{Q}{2\pi b\left(\frac{k}{\mu}\right)} \tag{28}$$

The above equation is a combination of Darcy's Law and continuity equation.

The solution is:

$$\Delta P = \frac{Q}{4\pi b\left(\frac{k}{\mu}\right)} \int_{\frac{r^2\theta_a\mu}{4kP_{atm}t}}^{\infty} \frac{e^{-x}}{x}dx \tag{29}$$

Details may be found in reference [15].

For small values of $\left(\frac{r^2\theta_a\mu}{4kP_{atm}t}\right)$, say < 0.01 the solution can be approximated by:

$$\Delta P = \frac{Q}{4\pi b\left(\frac{k}{\mu}\right)}\left(-0.5772 - \ln \frac{r^2\theta_a\mu}{4kP_{atm}t}\right) \tag{30}$$

or,

$$\Delta P = \frac{Q}{4\pi b\left(\frac{k}{\mu}\right)}\left(\ln \frac{2.25\ kP_{atm}t}{r^2\theta_a\mu}\right) \tag{31}$$

Although a variety of techniques exist to estimate k [16], the most appropriate is based on equation (31). To estimate k, plot transient pressure data at given locations, r.

$$\Delta P \text{ vs } \ln t$$

$$\Delta P = At + B \tag{32}$$

where:

$$A = \frac{Q}{4\pi b\left(\frac{k}{\mu}\right)} \tag{33}$$

$$B = \frac{Q}{4\pi b\left(\frac{k}{\mu}\right)}\left[-0.5772 - \ln \frac{r^2\theta_a\mu}{4kP_{atm}}\right] \tag{34}$$

If Q or b are known, then:

$$k = \frac{Q\mu}{4A\pi b} \tag{35}$$

If Q or b are not known, then:

$$k = \frac{r^2\theta_a\mu}{4P_{atm}} \exp\left(\frac{B}{A} + 0.5772\right) \tag{36}$$

The time to achieve steady state vs radial distance from the well for typical soil air permeabilities may be calculated from equation (29). When:

$$\left(\frac{r^2 \theta_a \mu}{4k\, P_{atm} t} \right) < 0.001$$

,

it is very close to the asymptotic limit and may be considered to be at steady state. Thus, the time required to achieve steady-state vapor flow is the time required for $\left(\frac{r^2 \theta_a \mu}{4k\, P_{atm} t} \right)$ to become less than 0.001.

6.5.2. Estimation of Vapor Flow Rates, Pressure Distributions and Vapor Velocities

The amount of time required and the cost of remediation is dependent on the vapor flow rate and off-gas treatment, which accounts for the largest percentage of the overall cost of remediation.

An increased flow rate will:

- Decrease vapor concentrations
- Increase volume of vapor.
  - Increase off-gas treatment costs
  - Decrease time of remediation.

An increase in the volume of air passed through the contaminated zone accelerates the remediation if the site is advection limited.

A decreased flow rate will:

- Increase vapor concentrations
- Decrease volume of vapor
  - Decrease off-gas treatment cost
  - Increase time of removal

Once k is known, vapor flow rates, pressure distributions, and vapor velocities may be obtained from the the steady-state solution to equation (23) when subject to the following boundary conditions [1]:

$$\begin{array}{ll} P = P_w & r = R_w \\ P = P_{atm} & r = R_I \end{array} \tag{37}$$

The analytical solution for a homogeneous or stratified soil is [1]:

$$P(r) = \left[P_w^2 + \left(P_{atm}^2 - P_w^2\right) \frac{\ln\left(\frac{r}{R_w}\right)}{\ln\left(\frac{R_I}{R_w}\right)} \right]^{\frac{1}{2}} \tag{38}$$

where:

P_w	=	pressure at the vacuum well
R_w	=	radius of the vacuum well
R_I	=	radius of influence
P_{atm}	=	ambient atmospheric pressure

The corresponding solutions for the radial darcian velocity distribution, u(r), and the volumetric vapor flow rate, Q, for a homogeneous soil system are [1]:

$$u(r) = -\left(\frac{k}{2\mu}\right) \frac{\left[\frac{P_w}{r \ln\left(\frac{R_w}{R_I}\right)}\right]\left[1 - \left(\frac{P_{atm}}{P_w}\right)^2\right]}{\left[1 + \left(1 - \left(\frac{P_{atm}}{P_w}\right)^2\right) \frac{\ln\left(\frac{r}{R_w}\right)}{\ln\left(\frac{R_w}{R_I}\right)}\right]^{\frac{1}{2}}} \tag{39}$$

$$Q = 2\pi R_w\, u(R_w)\, H = H\left(\frac{\pi k}{\mu}\right) \frac{P_w\left[1 - \left(\frac{P_{atm}}{P_w}\right)^2\right]}{\ln\left(\frac{R_w}{R_I}\right)} \tag{40}$$

To convert actual flow rates to standard flow rates:

$$Q^* = \left(\frac{P_w}{P_{atm}}\right) Q \tag{41}$$

R_I ranges for 9m to 30m; however the above equation is not sensitive to R_I. A value of R_I = 12m is typically used without loss of accuracy.

Figure 6.3 (adapted from reference [5]) illustrates steady-state flow for different permeabilities. Adjustments can be made with the following data:

R_w = 5.1cm (2 in)	R_I = 7.6m (25 ft)	- multiply Q^*/H	by 1.09
R_w = 5.1cm (2 in)	R_I = 23m (75 ft)	- multiply Q^*/H	by 0.90
R_w = 7.6cm (3 in)	R_I = 12m (40 ft)	- multiply Q^*/H	by 1.08
R_w = 10cm (4 in)	R_I = 12m (40 ft)	- multiply Q^*/H	by 1.15
R_w = 10cm (4 in)	R_I = 7.6m (25 ft)	- multiply Q^*/H	by 1.27

The standardized vapor flow rates, along with the change in vapor concentrations with time, can then be used to design appropriate off-gas treatment systems.

Vacuums that create an exceedingly high flow rate may:

- Cause diffusion to be the rate limiting step.
- Be less cost effective

High moisture content can decrease flow rate and has the following effects:

- Water occupies the pore spaces of the soil and causes a reduction in pore spaces for a given control volume of soil.
- A reduction in area requires greater flow (Q) values to achieve the same velocities. Greater flows can be achieved by inducing a higher vacuum.
- Dewatering may be utilized to decrease the moisture content. Increased horsepower or vacuum is an inefficient solution

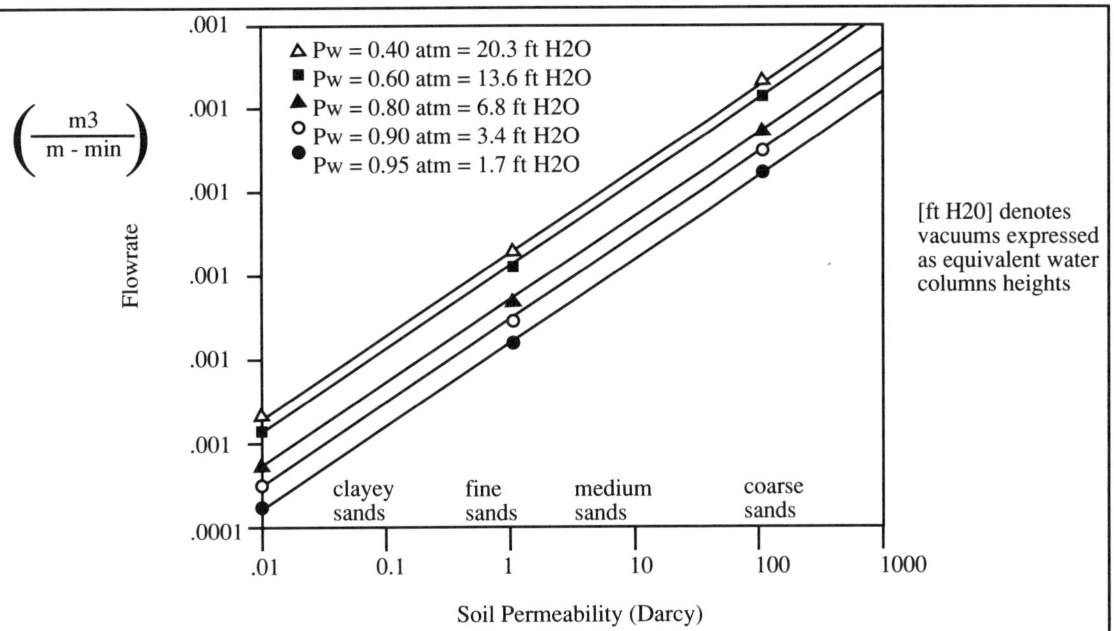

FIGURE 6.3
Predicted Steady-State Flowrates
(per unit well screen depth)
Soil Vapor Extraction

From equation (40), the required vacuum well pressure P_W, can be calculated. The amount of vacuum pressure required is dependent on the amount of vapor flow rate needed to meet remediation schedule requirements, while compensating for moisture content and avoiding an excessive flow rate. The vacuum pump can then be selected according to the vacuum well pressure required.

Once the number of wells, type of off-gas treatment, and type of pumps have been selected, costs of remediation may be determined.

6.5.3. Venting Removal Rates, Vapor Concentration, and Residual Composition Changes

Referring to Figure 2.2.

Location of contaminant can be in:

1.) Gas phase

2.) Sorbed

3.) Aqueous phase

4.) Free liquid

Mole balance for contaminant is:

Total mole of component i in soil	=	Number of moles of i in the vapor phase	+	Number of moles of i in the free liquid phase	+	Number of moles of i sorbed in the soil particles
	+	Number of moles of i dissolved in the soil moisture	+	Number of moles of i dissolved in the water liquid		

The last term in the equation above may be neglected when the contaminant is located in the unsaturated zone. Since most VOCs are slightly soluble in water, all water in the unsaturated zone may be accounted for in the third term as soil moisture. Rewritten:

$$M_i^T = M_i^z + M_i^x + M_i^y + M_i^s \tag{42}$$

where:

M_i^T = total moles of contaminant i

M_i^z = gas phase moles of contaminant i $= \dfrac{z_i P \theta_a V}{RT}$

M_i^x = free liquid moles of contaminant i $= x_i M^{HC}$

M_i^y = aqueous phase moles of contaminant i $= y_i M^{H_2O}$

M_i^s = sorbed moles of contaminant i $= k_i y_i \dfrac{M_{soil}}{M_{w,\,H_2O}}$

substituting,

$$M_i = \frac{z_i P \theta_a V}{RT} + x_i M^{HC} + y_i M^{H_2O} + k_i y_i \frac{M_{soil}}{M_{w,\,H_2O}} \tag{43}$$

where:

x_i	=	mole fraction of i in the free-liquid phase
z_i	=	mole fraction of i in the vapor phase
P	=	total pressure in the pore vapor
θ_a	=	void fraction occupied by vapor
V	=	volume of contaminated soil
R	=	ideal gas constant
T	=	absolute temperature in the soil's subsurface

M^{HC}	=	total number of moles in the free liquid phase
M^{H_2O}	=	total number of moles in the soil moisture
y_i	=	mole fraction of i dissolved in the soil moisture
k_i	=	the partition coefficient for species i [(mole i/mass soil)/(mole i/mass H_2O)]
M_{soil}	=	Total mass of contaminated soil
M_{w,H_2O}	=	molecular weight of water

One can sum individual mole balances to estimate the total contamination assuming:

- Vapor phase behaves as an ideal gas
- Free liquid hydrocarbon phase behaves as ideal mixture
- Soil moisture phase is non-ideal
- Local equilibrium

The following relationships can be derived

$$z_i P = x_i P_i^v \qquad \text{Raoult's Law}$$

$$= K_H y_i \qquad \text{Henry's Law}$$

$$= \alpha_i y_i P_i^v$$

$$= C_i RT \qquad \text{Ideal Gas Law}$$

where:

α_i = the activity coefficient for i in water

Writing a mass balance:

$$\frac{dM_i}{dt} = - QC_i - \text{Reaction} \tag{44}$$

where:

Reaction = biodegradation, etc.

Neglecting the reaction term, an iterative algorithm can be written:

$$M_i(t + \delta t) = M_i(t) - \delta t \frac{Qx_i P_i^v}{RT} \tag{45}$$

to determine concentrations in all phases.
Substitutions can be made relating M_i and either x_i, y_i, z_i, or C_i. For example:

$$x_i\{\frac{P_i^v\theta_a V}{RT} + M^{HC} + \frac{M^{H_2O}}{\alpha_i} + \frac{k_i M_{soil}}{\alpha_i M_{W,H_2O}}\delta(M^{H_2O})\} = M_i \tag{46}$$

where: $\delta(M^{H_2O}) = 0$ if $M^{H_2O} = 0$

(when soil moisture is dried out during the venting process)

$\delta(M^{H_2O}) = 1$ if $M^{H_2O} > 0$

The equilibrium distribution between all phases can be determined for any set of M_i values by subjecting equation (46) to the constraint that the sum of x_i values must equal one. If the sum is not one, they should be normalized to fit the constraint.

When no free liquid phase is present, equation (46) reduces to [1]:

$$\alpha_i y_i = \frac{M_i}{\left[\frac{P_i^v\theta_a V}{RT} + \frac{M^{H_2O}}{\alpha_i} + \frac{k_i M_{soil}\delta(M^{H_2O})}{\alpha_i M_{W,H_2O}}\right]} \tag{47}$$

If the sum of $\alpha_i y_i$ is less than one for a given set of M_i, the equilibrium distribution does not include a free-liquid phase.

A numerical scheme may be developed to solve equation (45) for a given initial condition (contaminant composition). A graphical representation of one numerical scheme is presented in Figure 6.4.

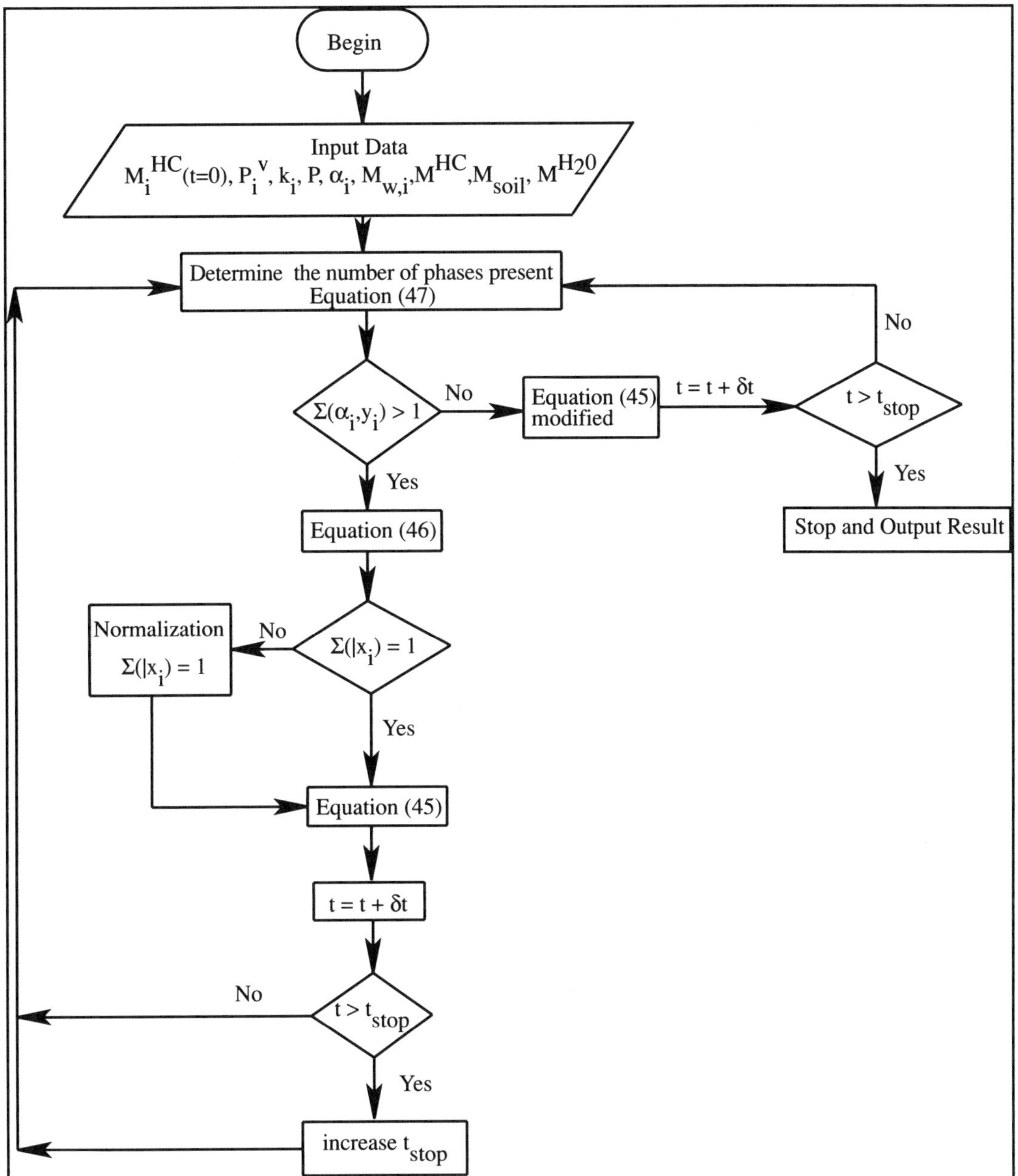

FIGURE 6.4
Flow Chart of Program Structure for the Design of Venting Removal Rates, Vapor Concentrations, and Residual Compositions Changes During Operation
Soil Vapor Extraction

6.5.4. Physical Properties and Parameter Estimation - Effects on Design

6.5.4.1. The Effect of Temperature on Vapor Flow

The viscosity of a gas at low pressures and temperatures is depicted in the following relationship [1]:

$$\mu(T) = \mu(T_0)\sqrt{\frac{(T + 273)}{(T_0 + 273)}} \tag{48}$$

Thus, based on equation (40)[1]:

$$\frac{Q(T_1)}{Q(T_2)} = \sqrt{\frac{(T_2 + 273)}{(T_1 + 273)}} \tag{49}$$

6.5.4.2. The Effect of Temperature on Vapor Pressure

A modification of the Claussius-Clapeyron equation [17] can be used to predict the vapor pressure dependence on temperature:

$$P_i^v(T) = P_i^v(T_0)\exp\left[\left(\frac{T_B T_0}{T_B - T_0}\right)\left(\frac{1}{T} - \frac{1}{T_0}\right)\ln\left(\frac{P_i^v(T_0)}{P_B}\right)\right] \tag{50}$$

where:

T_o = reference temperature at which the vapor pressure is known
T_B = boiling point at 1 atm
P_B = the pressure at the boiling point

6.5.4.3. The Effect of Capillary Forces on Vapor Pressure

The change in vapor pressure across a curved surface with a radius of curvature,r, may be described by the Kelvin equation:

$$P_i^v(\text{pore}) = P_i^v\exp\left(\frac{-2\sigma V_m}{rRT}\right) \tag{51}$$

where:

P_i^v(pore)= modified vapor pressure

σ = surface tension

V_m = molar volume of species, i

R = gas constant

However, changes in vapor pressure as a result of curvature are typically negligible.

6.6. Wells

6.6.1. General

The number and placement of wells is, perhaps, the most subjective part of SVE design. When one extraction well is sufficient it should be placed in the center of the contaminated zone, unless it is expected that vapor flow channeling along a preferred direction, more permeable, will occur [5]. When more than one well is employed, the effects of one well on the other wells must be considered [5]. The objective of well placement is to minimize:

- Occurrence of "Dead zones"
- Cost of remediation
- Time of remediation

6.6.2. Dead Zones

Where the radius of influences of two wells overlap, dead zones are created. This is caused by the vacuum forces of the two wells cancelling each other out. As there is no concentration gradient produced, there is no flow and the contaminant does not move. This is illustrated in Figure 6.5a. (adapted from reference [5]).

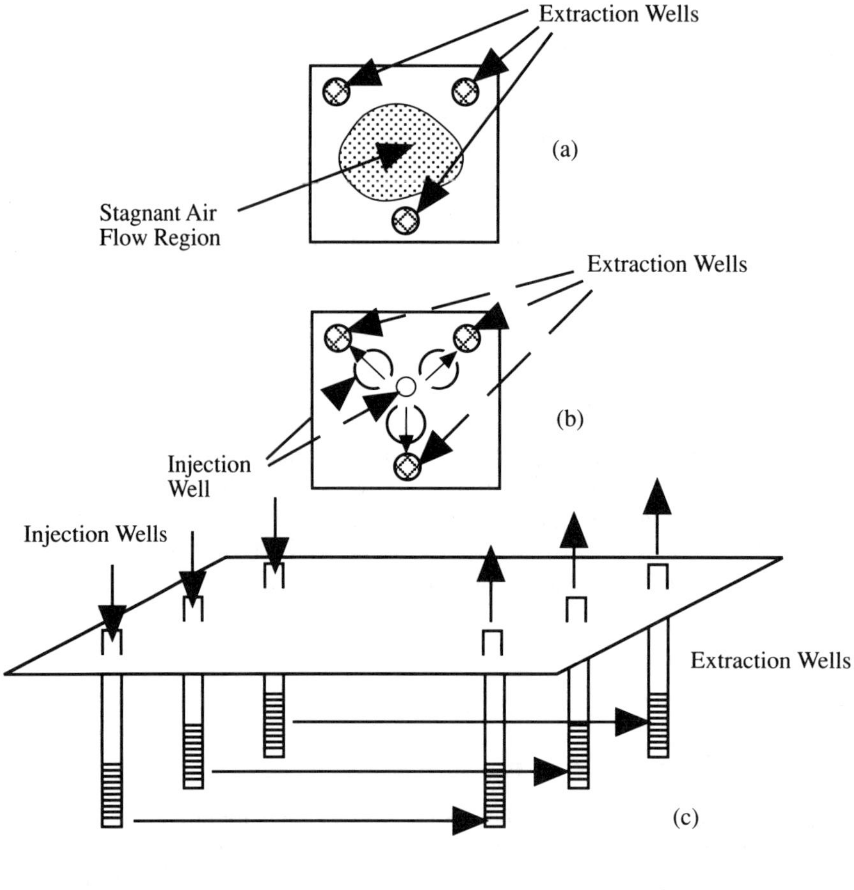

FIGURE 6.5 a,b,c
Injection & Extraction Well Configurations
Soil Vapor Extraction

Dead zones may be ameliorated through judicious placement of air injection wells or passive air wells. This allows air flow to both competing wells. This is illustrated in Figure 6.5b,c (adapted from reference [5]). In practice, vents are opened and closed to change air flow patterns, thus moving the location of stagnant regions.

6.6.3. Passive Air Wells

Passive air wells may be used for more than eliminating dead zones. They may be used to prevent contaminant migration from off-site sources. See Figure 6.6.

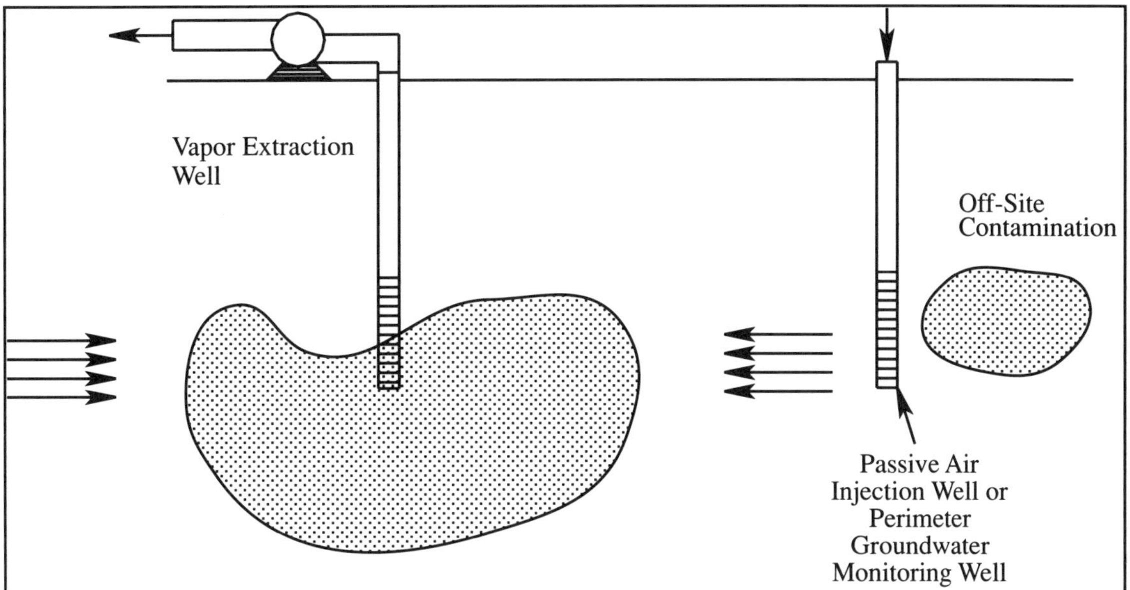

FIGURE 6.6
Use of Passive Vapor Wells to Prevent Migration of Off-Site Contaminant Vapors
Soil Vapor Extraction

6.6.4. Number of Wells

The number of wells can be estimated based on removal rates,

$$N_{WELLS} = \frac{R_{ACCEPTABLE}}{R_{ESTIMATED}} \qquad (52)$$

where:

$$R_{ESTIMATED} = C_{est}Q$$

$$R_{ACCEPTABLE} = \frac{M_{SPILL}}{\tau}$$

τ = maximum acceptable clean up time

$$C_{est} = \sum_i \frac{x_i P_i^v MW_i}{RT}$$

Or based on radius of influence,

$$N_{MIN} = \frac{A_{CONTAMINATION}}{\pi R_I^2} \tag{53}$$

R_I can be estimated from equation (38).

6.6.5. Well Screening and Construction

Wells should be screened only through the zone of contamination, unless the permeability of vapor flow is so low that removal rates would be greater if flow were induced in an adjacent more permeable soil layer [5].

Flow rates increase with increasing well diameter (15% when the well diameter is increased from 4" to 8")[5].

The slot size and number of slots should maximize the total open area of the pipe. The filter packing, placed in the annulus between the borehole and pipe, should be as coarse as possible [5]. Bentonite pellets and a cement grout are placed above the filter packing; care must be taken to ensure a good seal to avoid short circuiting [5]. See Figure 6.7 (adapted from reference [5]).

6.7. Nonideal Venting Conditions

Ideal case—when all vapor flows through contamination and there is no mass transfer limitation. Figure 6.8 (adapted from reference [5]) illustrates three non-ideal scenarios.

A. A fraction of the vapor, ϕ, flows through uncontaminated soil - estimated by assessing location of well relative to contaminant distribution.

$$R_{ESTIMATED} = (1 - \phi)\, Q\, C_{est} \tag{54}$$

B. Vapor flow parallel to, but not through, the zone of contamination. Therefore, vapor diffusive mass transfer limited. Based on a steady-state governing equation of

$$\frac{1}{r}\left(\frac{\partial ruC}{\partial r}\right) = D\left(\frac{\partial^2 x}{\partial z^2}\right) \tag{55}$$

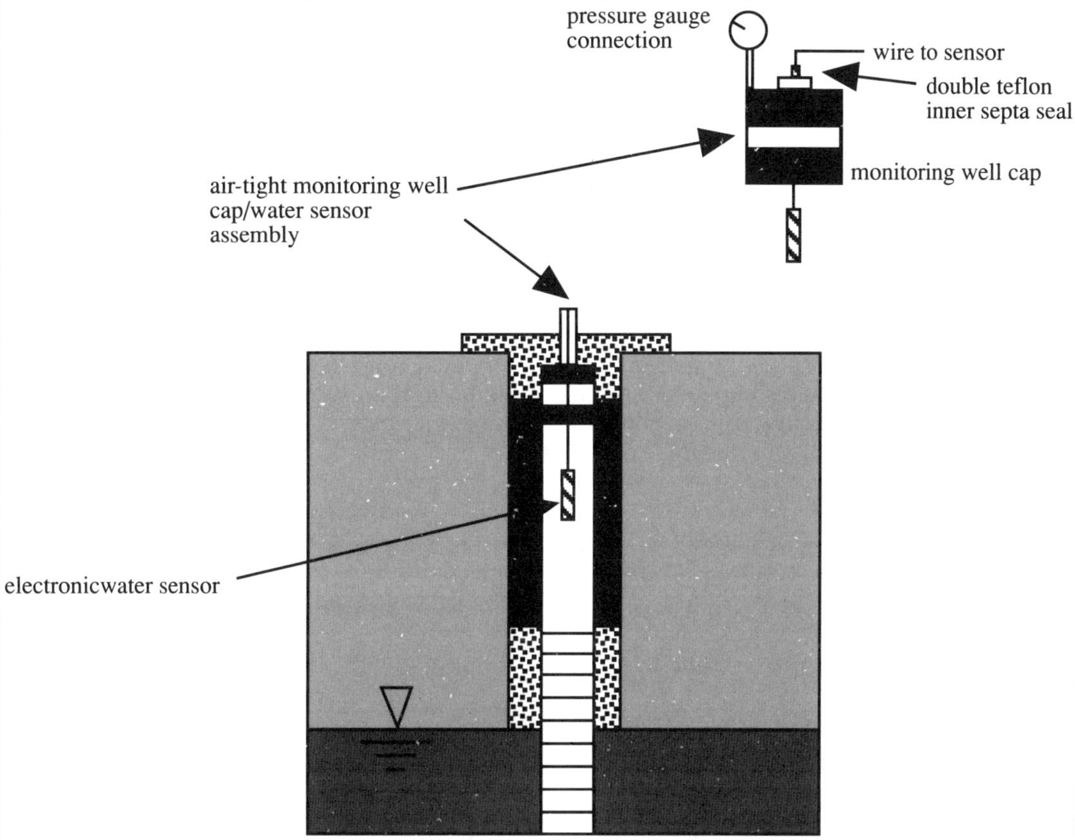

FIGURE 6.7
Monitoring Well Construction
Soil Vapor Extraction

$$R_{ESTIMATED} = \eta\, Q\, C_{est}$$

where:

$$\eta = \text{effectiveness factor} = \frac{\text{Removal rate for heterogeneous contaminant distribution}}{\text{Removal rate for homogeneous contaminant distribution}}$$

$$\eta = \frac{1}{3H}\left(\frac{6D\mu}{k}\right)^{\frac{1}{2}}\left[\frac{\ln\left(\frac{R_I}{R_W}\right)}{\left(P_{atm} - P_w\right)}\right]^{\frac{1}{2}}\sqrt{R_2^2 - R_i^2} \tag{56}$$

C. Vapor flow past rather than through contaminated soil—e.g. clay lens surrounded by sand. Diffusion through clay limiting. Maximum removal occurs when vapor concentration is low at permeable/impermeable interface. A contaminated free (dried-out) zone exists in low permeable soil.

$$R_{EST} = \pi\left(R_2^2 - R_1^2\right)\sqrt{\frac{C_{EST} D C_s \rho_b}{2t}} \tag{57}$$

where:

ρ_b = soil bulk density

C_S = residual level of saturation in a low permeability layer

6.8. Determining End of Remediation

Target soil clean up levels are often set on a site-by-site basis. They are based on the estimated potential impact that any residual may have on air quality, groundwater quality, or other health standards.

Generally soil borings, leaching tests (e.g. TCLP), soil-vapor surveys, or groundwater sampling may be required before closure is granted. Because these analyses may be expensive, system operation monitoring data may be used to determine when confirmation soil boring should be performed.

Valuable monitoring data include:

- Cumulative Mass of Contaminant Removed
- Extraction Well Vapor Concentration
- Extraction Well Vapor Composition

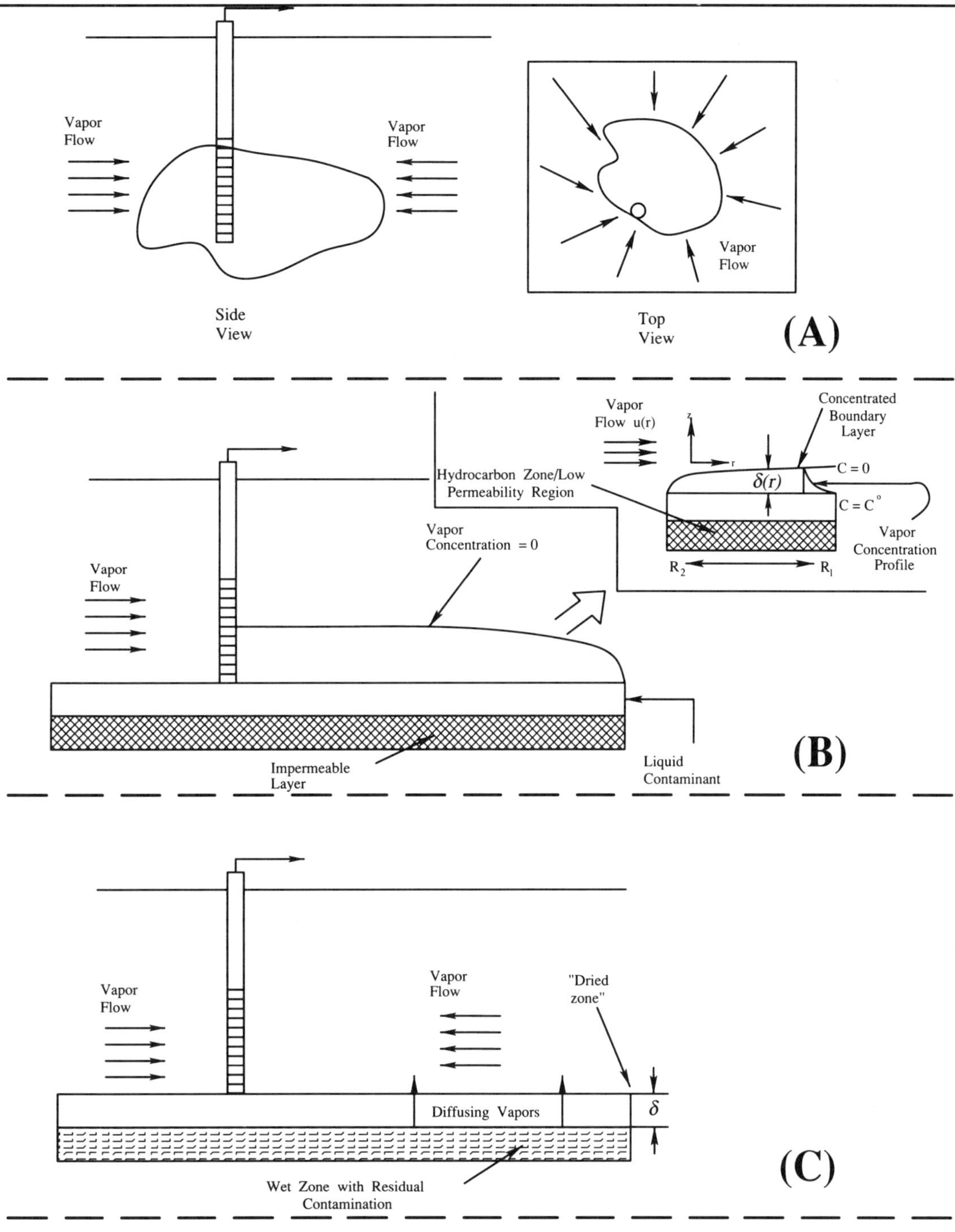

FIGURE 6.8
Scenarios For Removal Rate Estimates
Soil Vapor Extraction

6.8.1. Cumulative amount removed

The cumulative amount of contaminant removed is:

- Determined by integrating the measured removal rates (flow rate X concentration) with time.
- Not an appropriate method unless the original mass of the contaminant is known accurately. In most cases this information is not available. [5]

6.8.2. Extraction well vapor concentration

Well Vapor Concentration:

- Indicates of how effectively the venting system is working
- Decreases may not reflect actual soil concentration decreases
    - Decreases may be due to water table increases, increased mass transfer resistance due to drying, or leaks in the extraction system. [5]

6.8.3. Extraction well vapor composition

As the vapor extraction process progresses, composition is altered and relative amounts of:

- More volatile compounds will decrease
- Less volatile compounds will increase.[5]

The alterations in different compositions can be the result of different occurrences:

- If the total vapor concentration decreases without a change in composition, it is probably due to mass transfer resistance, water table upwelling, pore blockage or leaks
- If a decrease in vapor concentration is accompanied by a shift in composition toward less volatile compounds, it is most likely due to a change in the residual contaminant concentration.

6.8.4. Sampling Requirements

The most appropriate place to take boring samples and vapor surveys is mid-way between wells because this is the most likely place for "dead zones." At least two sampling events (2 to 3 weeks apart) are required to ensure representative concentrations of the contaminant in the soil and soil-vapor are obtained. After system closure is granted, a quarterly monitoring program is typically required to ensure that closure levels are maintained.

7. COSTS

7.1. Approximate cost

SVE costs are mostly dependent on the treatment processes required for air and water treatment. Wastewater treatment may add as much as 20% to the total. Site remediation may range from $10 to $50 per ton depending if off-gas or water treatment is required. Small sites are typically more expensive; as much as $150/ton. [6]

Economic advantages of the SVE system are most noticeable when the depth of the unsaturated zone is at least 10 ft deep. Beyond 10 ft the cost of excavation rises sharply, far exceeding the cost of an *in situ* SVE system.[6]

Most important factors affecting cost

- Site size
- Soil type
- Nature of contamination
- Amount of contamination
- Requirements for off-gas treatment
- Wastewater generation possibilities

See reference [13] for a good summary of costs.

REFERENCES

1. Johnson, P. C., M. W. Kemblowski and J. D. Colthart. "Quantitative Analysis of the Cleanup of Hydrocarbon-Contaminated Soils by In-Situ Soil Venting." *Groundwater*, **28**: 413-429, 1990.

2. Risk Reduction Engineering Laboratory and Office of Emergency and Remedial Response.*Guide for Conducting Treatability Studies Under CERCLA: Soil Vapor Extraction.* U.S. Environmental Protection Agency, Cincinnati, Ohio, March, 1991.

3. Hutzler, N. J., B. E. Murphy and J. S. Gierke. *State of Technology Review: Soil Vapor Extraction Systems.* Laboratory, R. R. E. EPA/600/2-89/024. U. S. Environmental Protection Agency, Cincinnati, August, 1990.

4. Office of Solid Waste and Emergency Response.*Technology Screening Guide for the Treatment of CERCLA Soils and Sludges.* EPA/540/2-88/004. U.S. Environmental Protection Agency, Washington DC, September, 1988.

5. Johnson, P. C., et al. "A Practical Approach to the Design, Operation and Monitoring of *In Situ* Soil-Venting Systems." *Ground Water Monitoring Review*, **10**: 159-178, 1990.

6. Micheals, P. A. *Terra Vac In Situ Vacuum Extraction System: Applications Analysis.* Laboratory, R. R. E. EPA/540/A5-89/003. U.S. Environmental Protection Agency, Cincinnati, July, 1989.

7. Ong and Lion. *Journal of Environmental Quality*, 1991.

8. Hoag, G. Director of Environmental Research Institute of University of Connecticut. Storrs, Connecticut, 1991.

9. Bruell, C. J. *The Diffusion of Gasoline-Range Hydrocarbon Vapors in Porous Media.* PhD Dissertation. University of Connecticut, Storrs, Connecticut, 1987.

10. Millington, R. J. and J. M. Quirk. "Permeability of Porous Solids." *Soc*, **57**: 1200, 1961.

11. Weber, W. J., Jr., P. M. McGinley and L. E. Katz. "Sorption Processes and Their Effects on Contaminant Fate and Transport in Subsuface Systems." *Water Research*, **25**: 499, 1991.

12. Baehr, A., G. Hoag and M. Marley. "Removing Volatile Contaminants from the Unsaturated Zones by Inducing Advective Air Phase Transport." *Journal of Contaminant Hydrology*, **4**: 1-26, 1988.

13. Pedersen, T. and J. Curtis. *Soil Vapor Extraction Technology.* EPA/540/2-91/003. U.S. Environmental Protection Agency, Cincinnati, February, 1991.

14. Bird, R., W. Stewart and E. Lightfoot. *Transport Phenomena.* Ed. Vol. John Wiley, New York, New York, 1960.

15. Bear, J. *Hydraulics of Groundwater.* Ed. Vol. McGraw-Hill, New York, New York, 1979.

16. Sellers, K. and T. Pedersen. "Soil Air Permeability Method Evaluation." In: *Hazardous Materials Control Research Institute Conference*, Ed. . Boston, Massachusetts, 1991.

17. Barrow, G. M. *Physical Chemistry*. Ed. Vol. McGraw-Hill, New York, New York, 1961.

G - Chemical Extraction / Soil Washing

CHEMICAL EXTRACTION/SOIL WASHING

Instructional objectives:

- To provide the theory necessary to understand chemical extraction and soil washing
- To provide a guideline for the design of chemical extraction or soiling washing unit operations.

1. GENERAL DESCRIPTION

1.1. Introduction

Chemical extraction (CHEX) and soil washing (SW) are volume reduction processes. Although often considered to be the same, the U.S. Environmental Protection Agency (EPA) has defined soil washing as a separation process that uses water as the solvent, and chemical extraction (for contaminated soils) as a process that uses additives (such as surfactants or chelating agents). However this module will present fundamental concepts common to both for the sake of unification.

The use of soil washing as a treatment technology is based on the following assumptions: (1) contaminants are typically concentrated in the fine fraction of the soil matrix, therefore separating the fine fraction of soil from the matrix reduces the volume of contaminated soil; (2) contaminants that are not chemically bound to the soil particles may be physically stripped from the soil matrix, thus further reducing the volume of contaminated soil.

The use of chemical extraction as a treatment technology is based on the assumption that contaminants are either reversibly sorbed to the soil or free floating. To remove sorbed contaminants, a non-destructive mass transfer process is employed. To aid in the desorption and solubilization of the contaminant, additives are typically used. The process takes advantage of the inverse miscibilitly of certain solvents, for example, triethyleneamine (TEA) is miscible with water below 65°F; above this temperature it is immiscible.

Soil washing and chemical extraction were first developed by the EPA in the late 1970's as a means of remediating beach sand contaminated from oil spills [1]. The first unit used a surfactant to aid in the remediation of the sand. Over the past ten to fifteen years, the distinction between soil washing and chemical extraction has been made [2].

Although design models are virtually nonexistent for soil washing and chemical extraction (for soils), this module presents a design protocol based on relevant theories and recent research.

Key factors to consider in the design of chemical extraction or soil washing units are:

- Contaminant solubility in extracting fluid
- Gradation of soil particles
- Contaminant sorption characteristics
- Hydrogeology (if *in situ*)

- Moisture content of the soil

1.2. Compounds Removed by Chemical Extraction and Soil Washing

Theoretically, it is feasible to treat a variety of contaminants as long as the correct solvent is used. Compounds that have been reported to be amenable to CHEX and SW include [2]:

Table 1
Compounds Amenable to CHEX and SW

Volatiles:	
2, 3, 7, 8, - tetrachlorodibenzo-p-dioxin	Xylene
Pentachlorophenol	Styrene
1,2,- dichloroethene	Phenol
Tetrachloroethene	Chloromethane
Semivolatiles:	
Anthracene	Total PAH
Trinitrobenzene	Total PCBs
Aromatic hydrocarbons	Pyrene
Bis- (2- ethylhexyl) phthalate	Napthalene
Pheneanthrene	
Metals:	
Arsenic	Lead
Cadmium	Nickel
Chromium	Zinc
Copper	

2. FUNDAMENTALS

2.1. Introduction

Mass transfer fundamentals presented in this section are applicable to soil washing and chemical extraction. However, it should be noted that soil washing is generally considered and designed as a mechanical separation process.

Equilibrium stage concepts and advective diffusive transport considerations are important fundamentals in the evaluation and application of chemical extraction processes. Other important fundamentals in the chemical extraction process are surfactant and coordination chemistry.

Important definitions:

Advection and diffusion - thoroughly discussed in the Soil Vapor Extraction module

Surfactant [3]- a material that exhibits the characteristics of modifying interfacial interactions by way of enhanced adsorption at interfaces

Chelating agent - is a polar molecule or anion that attaches to a central species (a Lewis acid) by coordinate covalent bonds, in which both of the electrons participating in the bond are derived from the chelating agent

2.2. General Theory

This section presents the theories applicable to the design of soil washing process and chemical extraction processes (*ex situ* and *in situ*).

Although soil washing is designed as a soil classification process, mass transfer fundamentals are also applicable.

There are two methods used in the design of a chemical extraction unit:

- Equilibrium stages (primarily for *ex situ*)
- Diffusion rate theory (primarily for *in situ*)

2.2.1. Volume Reduction

Soil washing is based on the following assumptions:

- Contaminants are typically concentrated in the fine fraction of the soil matrix, due to the high surface area to volume ratio
- Contaminants in the coarse and sandy fraction are typically physically bonded to the particles due to compaction and adhesion

Therefore, separating the soil fractions and stripping the contaminants from the larger size particles reduces the volume of contaminated soil that must be treated and/or landfilled. Soil washing does not rely heavily on the extraction of contaminants from the soil matrix unless the contaminant is water soluble. Figure 1 [4] is a schematic of a typical *ex situ* soil washing unit.

Soil washing may be applied *ex situ* or *in situ*:

Ex situ - the *ex situ* process is generally referred to as soil washing. It is the most common application

In situ - the *in situ* process is generally referred to as soil flushing. This application relies on elutriation or chemical enhanced pump-and-treat as the primary mechanism for contaminant removal

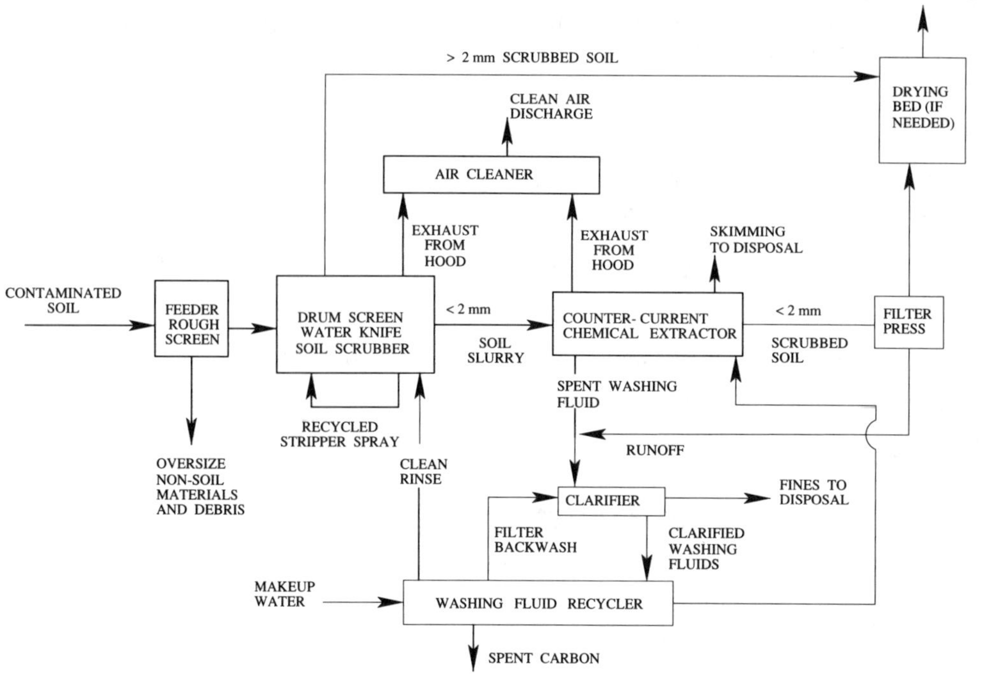

FIGURE 1
PROCESS FLOW SCHEME FOR SOIL WASHER
Chemical Extraction/Soil Washing

2.2.2. Phase Equilibria

When designing a chemical extraction unit for either *in situ* or *ex situ*, it is necessary to classify equilibria and determine the number of variables that must be fixed in order to design the unit.

If the two phases in a mass transfer operation come to equilibrium and the net transfer of material ceases, it is said that a limit of mass transfer is reached. Generally, equilibrium is avoided in mass transfer operations, however, in chemical extraction, it is a useful tool in system evaluation.

The phase rule classifies equilibria and establishes the number of independent variables or degrees of freedom available in a specific situation.

This phase rule can be stated as follows:

$$F = C - P + 2 \tag{1}$$

where:

- F = the number of degrees of freedom
- C = the number of components
- P = the number of phases

In chemical extraction processes, two simplified general situations could exist. One possibility is the solvent available is more than sufficient to dissolve all of the solute, and at equilibrium, all of the solute is in solution. With this situation there are two phases, the solid and the solution. The number of components is three (solid, solute and solvent). The variables are temperature, pressure, and concentration of the solution, each being independent.

The second possibility is that the solvent available is insufficient to dissolve all of the solute and the excess solute remains as a solid or liquid phase at equilibrium. Then the number of phases is three [solid (inert), solid solute or NAPL, and solvent], and the number of variables is two. The variables are temperature, and pressure. The concentration of saturated solution is dependent on these variables.

2.2.3. Hysteresis

The assumption that desorption is the reverse of adsorption will lead to erroneous predictions. This is due to the phenomena known as adsorption hysteresis.

Hysteresis is the failure of a material or property to return to its original conditions after a change, which is caused by an external agent that has been removed, has occurred. In the case of sorption, the desorption and adsorption isotherms do not follow the same path (see Figure 2 [5]). Although many theories have been presented for hysteresis, none have been widely adopted by the scientific community.

2.2.4. Equilibrium Stages—*Ex Situ*

A typical method for designing an *ex situ* chemical extraction unit is by the use of equilibrium stages.

A stage is an individual unit that is interconnected to other units or stages so that the soil and solvent being processed pass through each stage in turn. The term equilibrium stages is derived from the assumption that each stage is ideal, i.e., one where the streams exiting are in equilibrium (Figure 5 shows a typical equilibrium stage). It is the deviance from equilibrium of the entering streams that drives the mass transfer and the assumption of equilibrium of the existing streams that makes it possible to use graphical methods to determine the number of stages needed to achieve the needed degree of contaminant removal. Detailed discussion of equilibrium stage operations may be found in [6].

Before solid-liquid equilibria may be discussed, the value of a general theory must be considered. Because of the number variables affecting solid-liquid equilibria, a general theory of phase equilibria is of limited value. Reliance must be then be placed upon empirical correlations and experimental data. The methods outlined in the following sections must be validated experimentally.

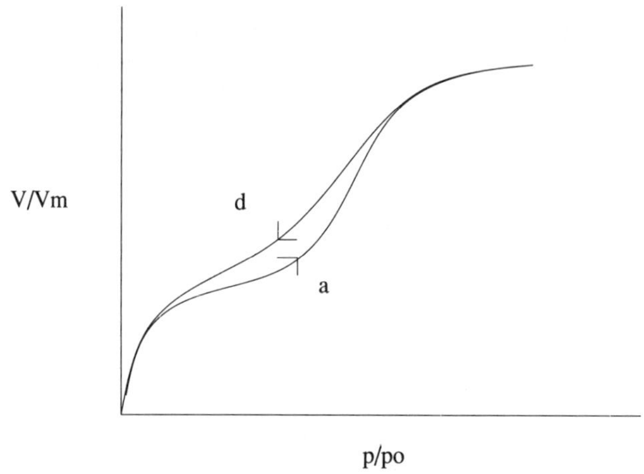

p = partial pressure
po = saturation pressure or normal vapor pressure of the material (atm)
V = total volume adsorbed
Vm = volume adsorbed at monolayer coverage

FIGURE 2
A Schematic Illustration of Hysteresis Between the Adsorption (a) and Desorption (d) Branches of an Experimental Isotherm
Chemical Extraction/Soil Washing

2.2.4.1. Terminology

Flow rates of the inert solid phase (soil) and the liquid phase (solvent) are referred to as L and V, respectively. The concentration of components in the L phase is x, and in the V phase is y. Therefore, the concentration of component A in the L phase is x_A and the concentration of component B in the V phase is y_B (see Figure 3).

In CHEX, there are two streams that enter and leave the reactor, making four terminal quantities. The four terminal quantities are identified by using subscript a to refer to the end of the process where the L phase enters and b to refer to the end where the L phase leaves. Figure 3 illustrates a schematic of a counter-current CHEX system and terminal quantities.

In chemical extraction systems, there are several concentrations that must be monitored. Each step in the system must allow for mixing of the solid with the liquid and subsequent separation. There are a variety of separation techniques, such as settling, centrifugation, or cake filtration. However, in each separation process, there will be one stream that contains primarily liquid and one stream that contains primarily solid. These streams are termed " overflow" and "underflow," respectively. As each stream will contain some concentration of solute and solvent, the amount of solute, solvent, and solids in each stream must be determined (see Figure 5 [6]).

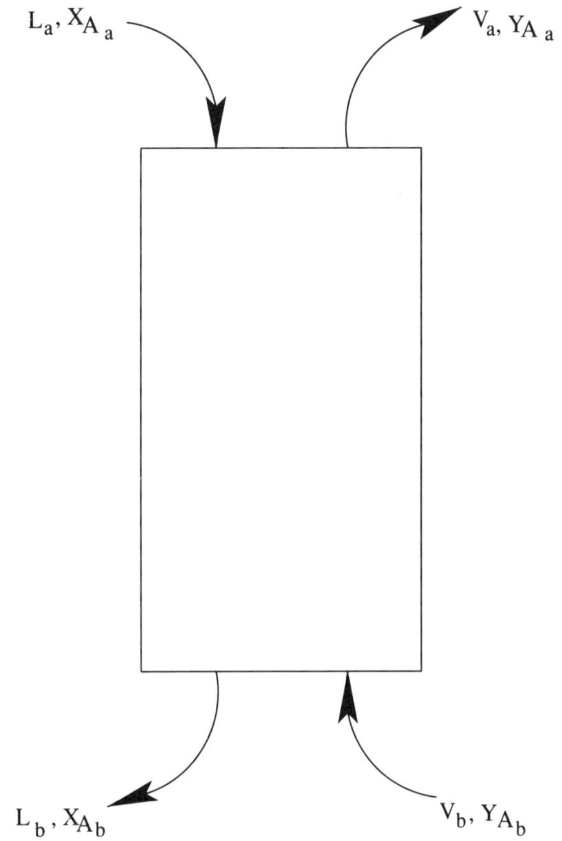

FIGURE 3
Terminal Quantities for Counter-Current Flow
Chemical Extraction/Soil Washing

2.2.4.2. Counter-Current Operations

Counter-current extraction is the most popular means of chemical extraction (see Figure 4). In counter-current extraction, the solids move in one direction and the liquid moves in the opposite direction. The thoroughness of extraction depends upon the solubility of the solute in the solvent, the amount of solvent, and the number of units. The solubility of the solute in the solvent will be discussed in the section titled " Surfactants and Coordination Chemistry."

2.2.4.3. Principals of Counter-Current Extraction

In counter-current extraction, the units are numbered in the direction of the flow of solids (L stream). Each unit receives two streams, the L stream and the V stream, and delivers them to the adjacent units. As the V stream moves in the counter direction of the L stream, it dissolves solute. Thus, fresh solvent is used to extract the nth stage L stream and the concentrated solution is used to extract the 1st stage of the L stream. This process ensures the largest driving force for transfer is realized (refer to Figure 4).

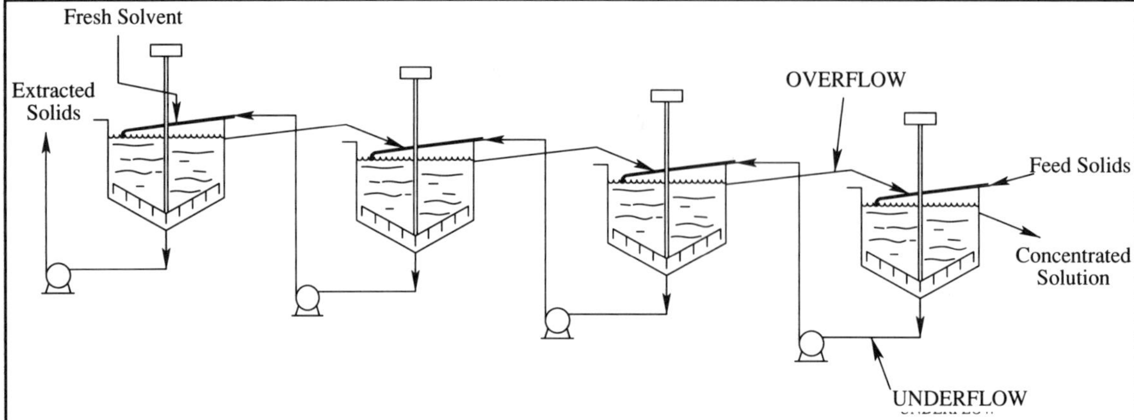

FIGURE 4
Counter Current Leaching Plant
Chemical Extraction/Soil Washing

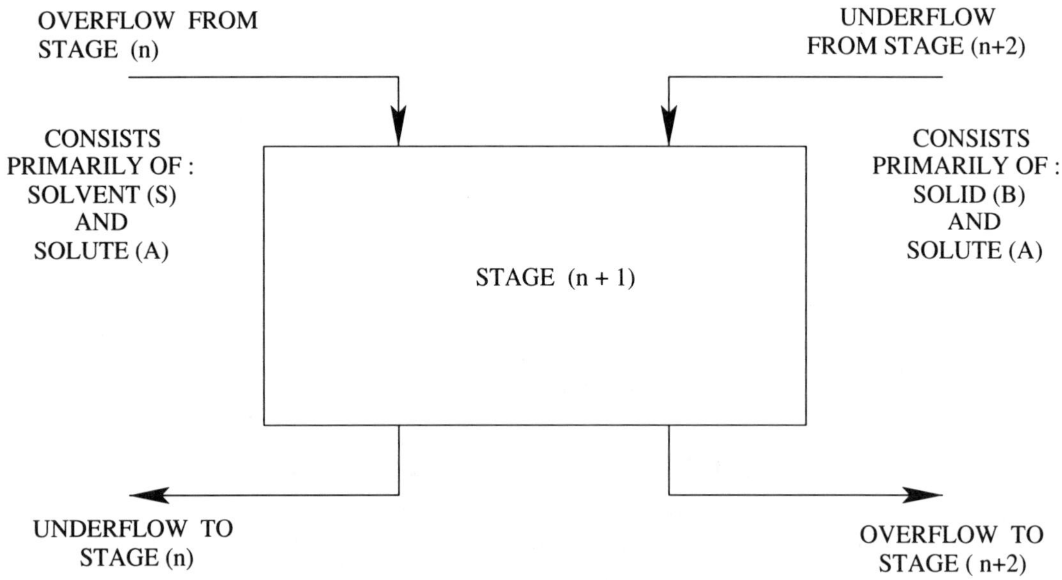

FIGURE 5
Schematic of Stage
Chemical Extraction/Soil Washing

It is common practice to assume that there will be sufficient solubility of the solute in the solvent, that there will be enough solvent to dissolve all the solute, and that there is no irreversible adsorption of solute by the solid. Thus, equilibrium is attained when the solute is completely dissolved and the concentration of the solution is uniform. However, these assumptions do not always hold. The exceptions will be discussed under "Diffusion Rate Theory." If diffusion is an issue in counter-current leaching, Fick's second law can be used to arrive at a stage efficiency (see ref.[6]).

If one uses the aforementioned assumptions, then the concentration of the liquid, which is a uniform mixture of dissolved solute and solvent, retained by the solid leaving any stage is the same as that of the liquid overflow from the same stage, thus the equilibrium relationship is simply $x_e=y_e$. To determine the ideal number of stages, the operating line and McCabe-Theile graphical method can be used.

2.2.4.4. Binary Systems

In a binary system, an operating line is used to determine the number of stages needed. An operating line is determined by an arithmetic graph with x (concentration of the solute in the underflow) as the abscissa and y (concentration of the solute in the overflow) as the ordinate. The operating line can be derived from the material balance on the total solution and the solute. The balances are:

For the total solution:

$$V_{n+1} + L_a = V_a + L_n \tag{7}$$

For the solute:

$$V_{n+1}y_{n+1} + L_a x_a = L_n x_n + V_a y_a \tag{8}$$

Solving for y_{n+1} gives the operating line as:

$$y_{n+1}=\frac{L_n}{V_{n+1}}x_n + \frac{V_a y_a - L_a x_a}{V_{n+1}} \tag{9}$$

where:

n = the number of the stage

The equilibrium line is the plot of the equilibrium values of x and y, where the equilibrium data is obtained either experimentally or via thermodynamic calculations (see Figure 6 for typical plot [6]).

There are two cases that must be considered when using the operating line to determine the number of stages needed in chemical extraction. The first case assumes that the amount of liquid retained by the solids in each stage is the same. This case is termed constant underflow. In this case, L_n is constant, thus the operating line is straight and the McCabe-Thiele diagram technique may be used [6]. However, if it is assumed that the equilibrium line is straight, equation (10) can also be used: (see Figure 6)

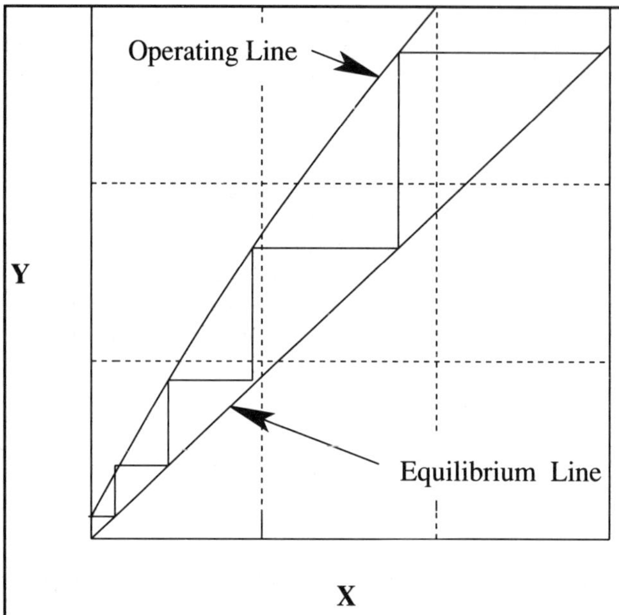

FIGURE 6
Typical McCabe-Thiele Diagram
Chemical Extraction/Soil Washing

$$N = \frac{\log[(y_b - y^*_b) / (y_a - y^*_a)]}{\log[(y_b - y_a) / (y^*_b - y^*_a)]} \tag{10}$$

where:

- N = the number of ideal stages
- y_a = the concentration of liquid leaving stage 1.
- y_b = the terminal concentration
- y^* = is used to indicate the concentration of the phase in equilibrium with a specified a phase.

The second case is termed variable underflow, because the under and overflow may vary from stage to stage. To determine the number of ideal stages for this case, the McCabe-Thiele graphical method is modified. These modifications include determining the terminal points by mass balance equations. Intermediate values of x_n are calculated by fixing L_n and V_n=1 for different stages of the operation. The terminal and intermediate points are plotted as in the constant underflow case to determine the operating line. The variance of L and V determine the number of intermediate points needed. If the variance is small enough so the operating line is close to the equilibrium line, then only one intermediate point is needed; if the variance is great, more intermediate points will be needed. A detailed description may be found in ref. [6].

2.2.5. Diffusion Rate Theory: *In Situ*

Although equilibrium conditions are typically assumed in the design of an *ex situ* CHEX unit, diffusion and desorption may be rate limiting steps in the design of an *in situ* CHEX unit and must be considered. The effect of mass transfer limitations can be shown by the following examples. If a cubic meter of soil with 35% porosity contains trichloroethylene (TCE) at 20% residual saturation, then approximately 270 pore volumes must pass through the soil and reach equilibrium with the TCE (1100mg/L) before the TCE is removed from the soil by dissolution [7]. Sandbox experiments with perchloroethylene (PCE) suggest that this equilibrium is achieved very quickly as the water passes through the fingers of residual contamination [7]. Longer periods of time are required to remove PCE if it is present in pools [7]. It has been shown that only the water that passes very close to the edge of the pool is likely to reach equilibrium concentrations with the PCE, while the concentrations of PCE in the water further above the pool will be significantly less due to the limitations of the rate of mass transfer from the pool to the bulk of the aquifer (see figure 7) [7]. The concentration of PCE, in a monitoring well placed just above the pool with a screen length of two meters, will increase across the length of the pool over which the groundwater has flowed (see figure 8) [7]. If the TCE (103 kg) in the example described above is distributed in a 20 cm thick pool below the cubic meter of soil, the average concentration of TCE existing from the soil is 28.6 mg/L and 10,200 pore volumes must pass through the aquifer before the TCE is removed [7]. This example site will take 39 times longer to remediate than the example site with residual contamination.

Another important concept to consider in the design of an *in situ* CHEX unit is the transport of the contaminant and solvent in the subsurface system.

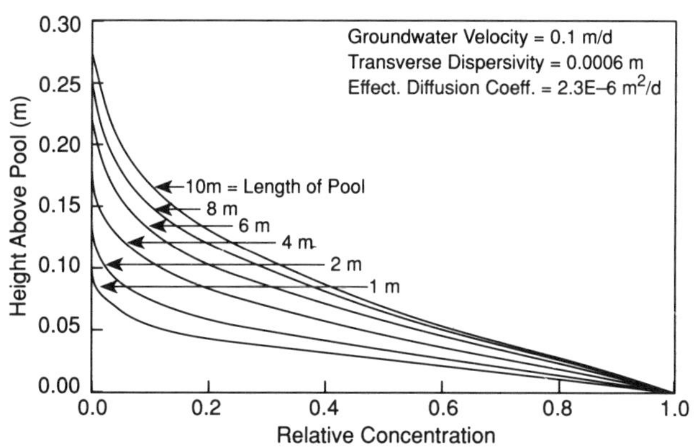

FIGURE 7
Concentration of a Contaminant at Different Elevations Above a DNAPL Pool For Different Pool Lengths
Chemical Extraction/Soil Washing

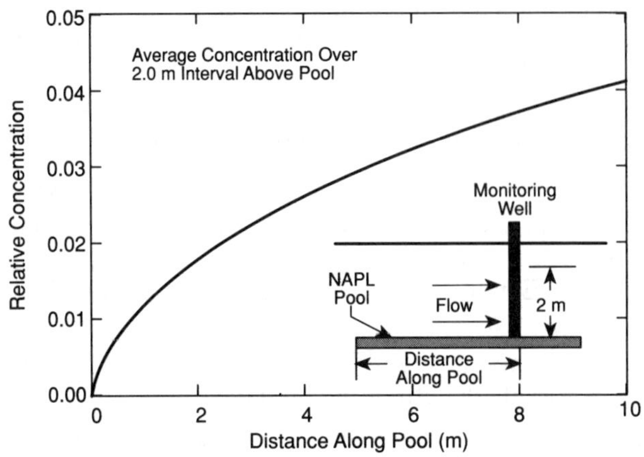

FIGURE 8
Average Concentration of a Contaminant Over a 2-m Interval Above a DNAPL Pool
Chemical Extraction/Soil Washing

2.2.5.1. Fick's First Law

Fick's first law describes the transport of a component into the bulk phase by diffusion due to a concentration gradient and may be used as a basis for diffusion limited systems.

$$J_A = -D_A \frac{\partial C_A}{\partial x} \tag{11}$$

where:

J_A	=	contaminant A flux (mass/area-time, $M/(L^2t)$)
D_A	=	diffusion coefficient for A (area/time, L^2/t)
C_A	=	contaminant A concentration (mass/volume, M/L^3)
x	=	distance in the direction of diffusion (length, L)

2.2.5.2. Fick's Second Law

Fick's second law is generally used to express diffusion in a soil matrix. Fick's second law may be used to describe the diffusion of a solute from the interior of the solid to the outside boundary layer. It may also be used to describe the diffusion from the interface to the bulk solution.

$$\frac{\partial C}{\partial t} = D' \frac{\partial^2 C}{\partial x^2} \tag{12}$$

where:

- D' = effective diffusivity (L^2/t)
- x = may be either the distance from the solid/liquid interface or distance measured in the direction of diffusion. (L)
- C = concentration of solute in solution (M/L^3)

Effective Diffusivity can be described by the Millington and Quirk relationship:

$$D' = \frac{\theta_a^{10/3}}{n^2} D_v \tag{13}$$

where:

- θ_a = solution content of soil
- n = soil porosity
- D_v = diffusion coefficient of chemical in free solution

A more convenient method of quantifying diffusion in the soil matrix is by means of a mass transfer coefficient. This type of analysis is generally termed a "lumped parameter model" to distinguish it from diffusion coefficient models termed " distributed-parameter models".

A generalized equation of a mass transfer coefficient is:

$$\begin{pmatrix}\text{amount of mass}\\ \text{transfered}\end{pmatrix} = k \begin{pmatrix}\text{interfacial}\\ \text{area}\end{pmatrix}\begin{pmatrix}\text{concentration}\\ \text{difference}\end{pmatrix}$$

k, the mass transfer coefficient, is the proportionality between the amount of mass transferred and the concentration difference and interfacial area.

$$J = k(C_{1i} - C_1) \tag{14}$$

where:

- J = flux at interface ($mol/m^2{*}hr$)
- k = mass transfer coefficient (m/hr)
- C_{1i} = the concentration at the interface (mol/m^3)
- C_1 = the concentration in the bulk solution (mol/m^3)

2.2.5.3. Desorption Characteristics

The soil/chemical interactions that affect sorption/desorption are complex. The desorption of a contaminant is related to sorptive capacity. Adsorption is better understood than desorption, and, although the sorption/desorption relationship is not well-established, the concepts of adsorption serve as a basis for understanding desorption.

Adsorption of a contaminant in soil is dependent on:

- Clay factor
  - — Clay content of the total soil volume
  - — Type of clay mineral in the soil
  - — Exchange capacity of the clay
  - — Specific surface area of the clay
- pH
  - — The impact of pH is dependent on the retention mechanism involved in its removal from the soil
- Organic content of the soil
  - — This is particularly important when the contaminant is organic
- Contaminant concentration factor
  - — If contaminant concentrations are high enough, sorption sites may be limiting
- Time
  - — Contaminant contact time defines the extent of reaction.

  Time is most important in cohesive or clay-like soils since the permeability is low in these soils, indicating diffusion limitations

2.2.5.4. Contaminant And Solvent Transport

In situ CHEX must take into consideration the fate of contaminants and solvents in subsurface systems. An approach is to write a mass balance equation for a cubic element of the soil matrix containing the contaminant. This mass balance assumes homogeneous isotropic porous medium and steady-flow.

Three items that must be considered in the derivation of the mass balance equation are:

- Flow, as described by average linear velocity, that carries the dissolved substance by advection
- Hydrodynamic dispersion caused by variations in microscopic velocity within each pore channel and from one channel to another
- Retardation of solute transport through adsorption, chemical reactions, biological transformations or radioactive decay

Consider the cubic element containing contaminant C (see Figure 9 [8]). There will be a flux in and out of the element. Letting the average linear velocity be $\bar{v}$ ($\bar{v} = v / n$ where v is the specific discharge and n is the porosity), C be the concentration of the contaminant (mass of solute per unit volume of solution); the solute per unit volume of porous media is:

$$nC \tag{15}$$

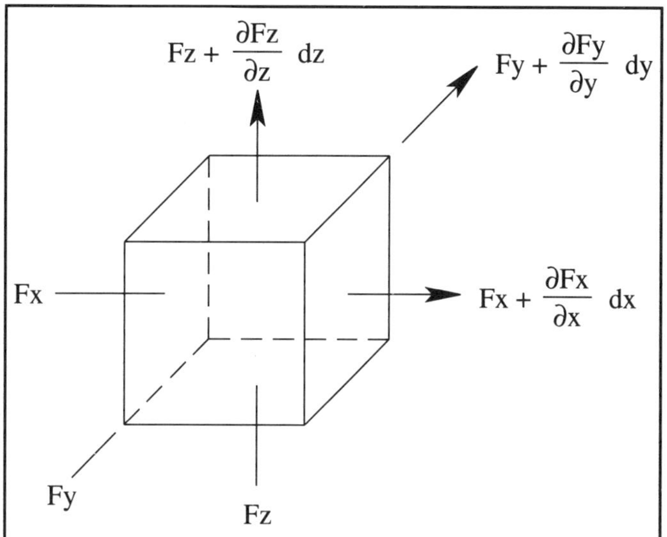

FIGURE 9
Cubic Diagram of a Subsurface Soil
Chemical Extraction/Soil Washing

The transport by advection is (in the x-direction):

$$\bar{v}_x n C \partial A \tag{16}$$

Transport by dispersion is (in the x-direction):

$$n D_x \frac{\partial C}{\partial x} \partial A \tag{17}$$

where:

D_x = the dispersion coefficient in x direction
∂A = the elemental cross-sectional area

From these equations it can be shown that the complete mass balance is:

$$+D_x \frac{\partial^2 C}{\partial x^2} - \bar{v}_x \frac{\partial C}{\partial x} = \frac{\partial C}{\partial t} \tag{18}$$

This equation can easily be expanded into three dimensions.

If adsorption is to be considered, a source-sink term must be added to equation (18). Typically, the general equation used in groundwater modeling for adsorption assumes local equilibrium and can be represented as:

$$\rho_b \frac{dS}{dt} \tag{19}$$

where ρ_b is bulk density of the porous medium.

S is equal to the mass of chemical constituent adsorbed per unit mass of the solid aquifer material. S is given by isotherm equations (For more information refer to [9]).

Desorption is significantly more complicated than assuming the reverse of adsorption. Hysteresis significantly impacts performance predictions.

Therefore, the final mass balance equation is:

$$+D_x \frac{\partial^2 C}{\partial x^2} - \bar{v}_x \frac{\partial C}{\partial x} + \frac{\rho_b}{n} \frac{\partial S}{\partial t} = \frac{\partial C}{\partial t} \tag{20}$$

S is typically represented in groundwater systems as a function of the partition coefficient (K_D) and the contaminant concentration C.

In conventional pump and treat operations the contaminant concentration in the extraction well may drop by several orders of magnitude but still remain at concentrations above the target clean-up level after an extended period of pumping (see figure 10) [7]. This is called "tailing". It is because of tailing that clean-up times are hard to predict. Tailing may be caused by physical or chemical processes. The primary physical process that causes tailing in pump and treat operations is variable residence times of contaminants in the stream tubes caused by pumping (see figure 11 [7]).

Chemical processes that may cause tailing are: (1) the contaminant mass may not be dissolved in the groundwater, but may be present as adsorbed species, precipitates, or NAPLS, (thus cannot be directly extracted by pumping, (2) the equilibrium concentration in the groundwater may be small compared to that on the soil, (3) the rate of mass transfer may be small relative to the groundwater velocity, and (4) there may retention of contaminants by microbial cell walls.

If chemical processes are the cause of tailing, then chemically enhanced pump-and-treat may be a viable alternative to expedite remediation.

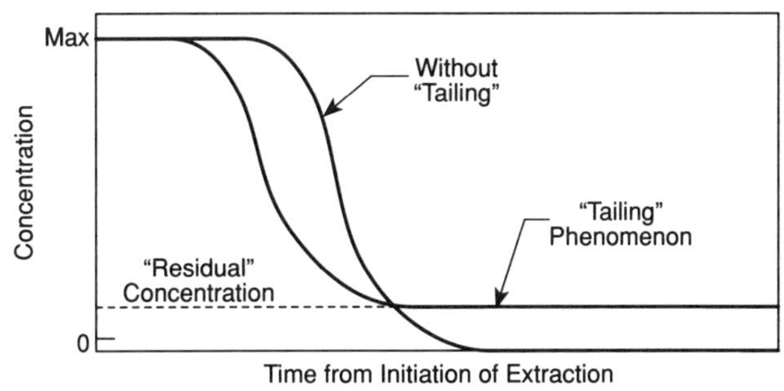

FIGURE 10
Concentration Versus Time Curve For
An Extraction Well With Continuous Pumping
Chemical Extraction/Soil Washing

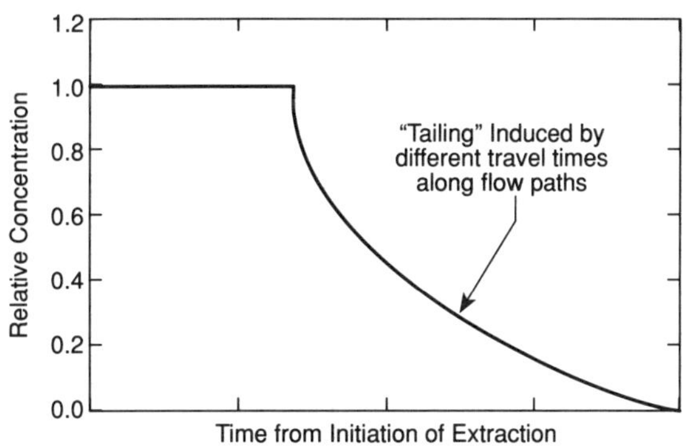

FIGURE 11
Concentration Time Versus Time Curve
The Tailing is Caused By The Differential Travel Times
Along the Individual Stream Lines
Chemical Extraction/Soil Washing

2.2.6. Surfactant and Coordination Chemistry

(This section adapted from references [10] and [3])

When designing either an *in situ* or *ex situ* CHEX unit, it is important to consider the use of either a surfactant or chelating agent to aid in the desorption and solubilization of the contaminants into the solvent. The use of a surfactant to aid in the solubilization of a contaminant is sometimes referred to as cosolvent chemical enhancement. This is because a surfactant is added to the solvent, increasing the contaminant's solubility in the solvent, hence cosolvent. Cosolvents may also be used as a substrate to increase microbial activity [7].

2.2.6.1. Surfactant Theory

A surfactant is a material that exhibits the characteristics of modifying interfacial interactions by way of enhanced adsorption at interfaces.

Surfactants have two distinct characteristics:

- Molecular components that have little attraction for the solvent or bulk material (called the lyophobic group).
- Molecular components that have a strong attraction for the solvent or bulk phase (called the lyophillic group).

(see Figure 12 [3])

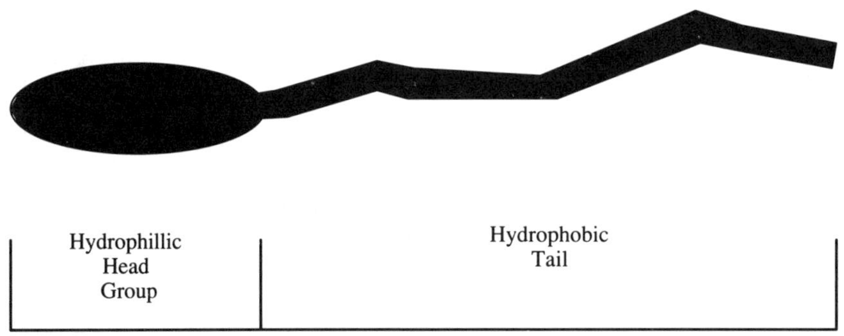

FIGURE 12
The Basic Chemical Nature of Surface Active Molecules
Chemical Extraction/Soil Washing

Because of the amphiphillic nature of surfactant molecules, they possess properties that make them appear "schizophrenic". Their chemical composition causes them to have a "love-hate" relationship with most solvents. This relationship causes a constant tug of war between the opposing forces. (adapted from [3])

The amphiphillic nature of the surfactant is responsible for the surfactant's tendency to concentrate at interfaces, reducing system free-energy. However, when all of the interfaces are saturated, other mechanisms may continue the overall energy reduction. The two main mechanisms are: (1)

the crystallization or precipitation of the solute from solution and (2) the formation of molecular aggregates called micelles. For the purpose of chemical extraction, the second mechanism is the most important and shall be discussed in detail.

Micelles are formed when the interfaces available in a system are saturated.
These molecular aggregates have geometrical configurations as spheres, oblate spheroids or prolate spheroids. The hydrophobic portion of each molecule in the micelle is directed inward, toward the center of the aggregate, forming a liquid core with a fairly smooth boundary, surrounded by a hydrophillic shell (see Figure 13 [3]). Micelles remain in solution, are thermodynamically stable, and have properties distinct from those of the monomeric solution.

The concentration of surfactant in solution at which micelle formation takes place is called the critical micelle concentration (cmc). Thus, at concentrations less than cmc, the surfactant molecules exist predominantly as monomers.

With micelle formation, there is an increased solubility of organic materials in aqueous surfactant solutions. This phenomenon is called micelle solubilization. A general definition of micelle solubilization is: the preparation of thermodynamically stable, isotropic solution of a substance (the additive) normally insoluble or only slightly soluble in a given solvent by the addition of one or more amphiphillic compounds at or above their critical micelle concentration [3] (see Figure 14 [3]).

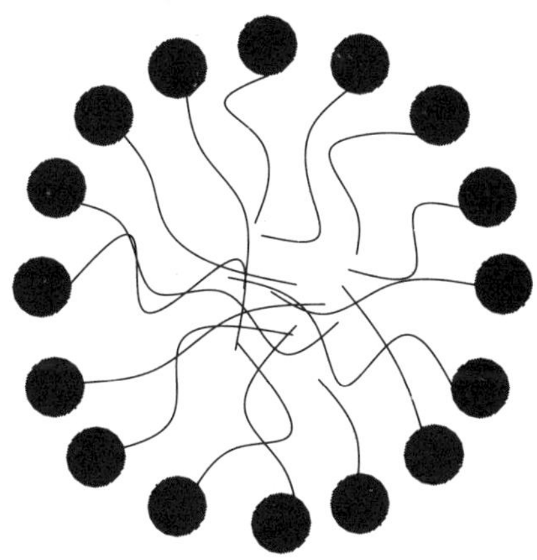

FIGURE 13
Cross Section View of Micelle
Chemical Extraction/Soil Washing

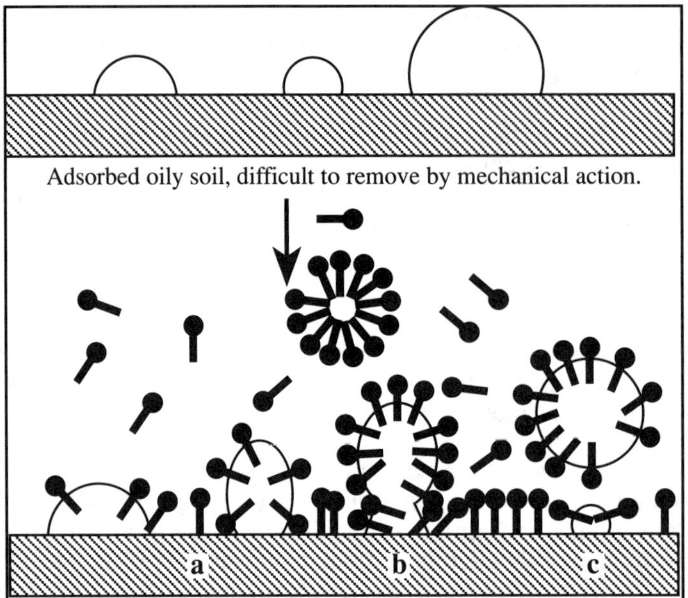

Surfactant adsorption at O/W and S/W interfaces acts to initiate drop deformation (a), followed by "necking" (b), and eventual partial oil removal (c). If the original contact angle of oil on the surface is less than 90°, a residual drop of oil may be left; if the angle exceeds 90°, complete soil removal may be attained.

The "roll-back" mechanism for oily soil removal.

FIGURE 14
Oil Removal Mechanism
Chemical Extraction/Soil Washing

The location of a solubilized molecule in a micelle relative to the structural components of the surfactant will be determined primarily by the chemical structure of the additive. In aqueous solutions, nonpolar additives, such as hydrocarbons, are intimately associated with the core of the micelle, while slightly polar materials, such as fatty acids, alcohols, and esters, will be usually located in what is termed the palisades layer and the core (see Figures 15a and b [3]). Additives may also be found entirely in the palisades region (the area between the hydrophobic core and the micellar surface) and the micellar surface (see Figures 15c and d [3]).

The orientation of such molecules is probably more or less radial with the hydrocarbon tail remaining closely associated with the micellar core. Other structural factors, such as charge on the surfactant head group, can significantly affect the locus of solubilization. Material containing aromatic rings may solubilize in or near the core of anionic systems, but may solubilize in the palisades layer of a cationic system due to the polarization interactions between the ring and the cationic head group.

As expected, the total mass of material solubilized increases as the size of the micelle increases. Therefore, it may be stated that an increase in the micelle size might also be expected to increase the solubilizing power of the system.

Another theory[3] states that the quantity of a substance that can be solubilized in surfactant micelles is centered around the relationship between molar volume of the additive and the maximum amount of material that can be incorporated in a given solution. It has been determined that there is a linear relationship between the logarithm of the volume of the material solubilized and its molar volume, although the slopes of the plots differed for the different classes of compounds.

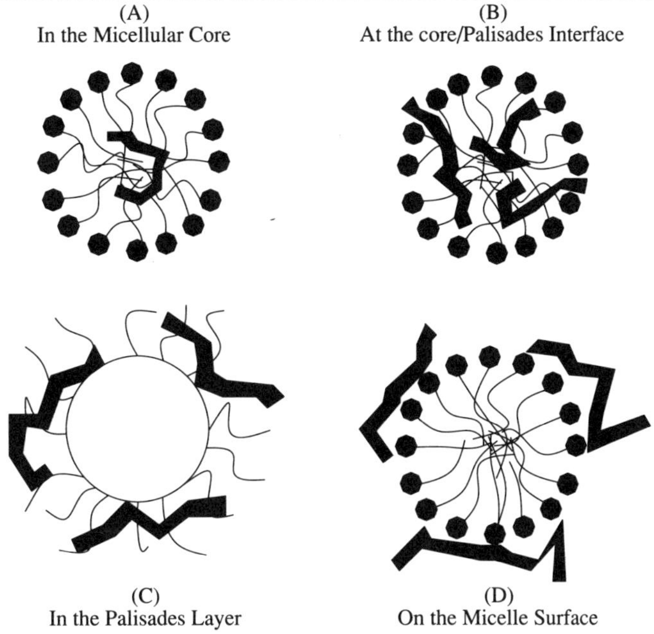

FIGURE 15
Possible Loci for the Solubilization of Additives in Micelles
Chemical Extraction/Soil Washing

2.2.6.2 Chelating Theory

When a metal ion or atom reacts with an electron-pair donor, it forms a coordination compound or complex ion. The electron pair donor is called a ligand. Ligands must have at least one pair of electrons that can bond with another atom. Most metals are Lewis acids and can form coordinate bonds with ligands (see Figure 16 and 17).

When a ligand has more than one electron donor site that is separated by several atoms, it may complex with a metal at each site, forming a five or six-membered heterocyclic ring called a chelate. Mechanistic explanations for coordination bonding have been developed and can be explained by valence bond theory, crystal field theory, or molecular orbital theory. Reactions are very fast and can be assumed to be at equilibrium.

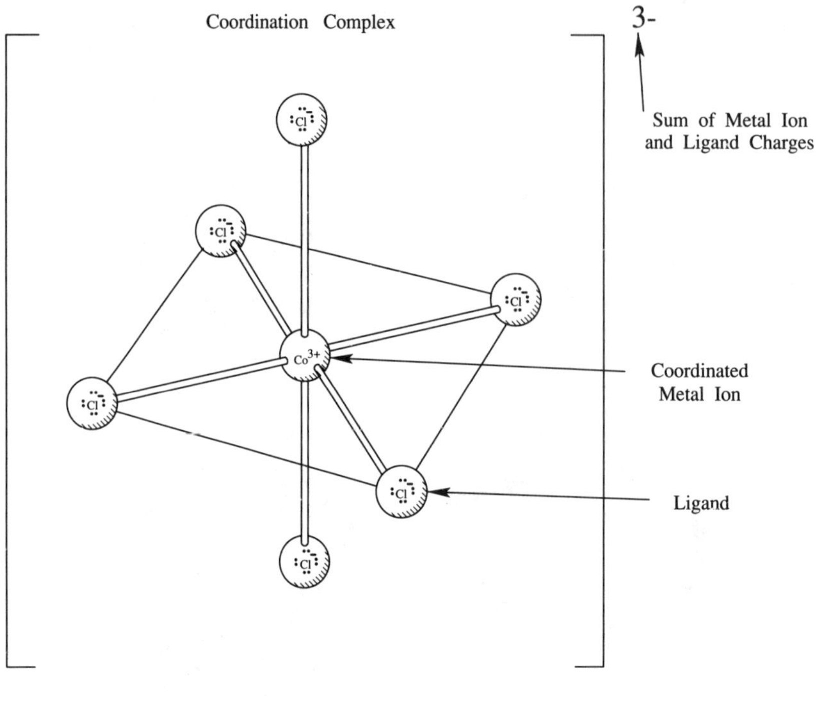

FIGURE 16
Coordination Complexes
Chemical Extraction/Soil Washing

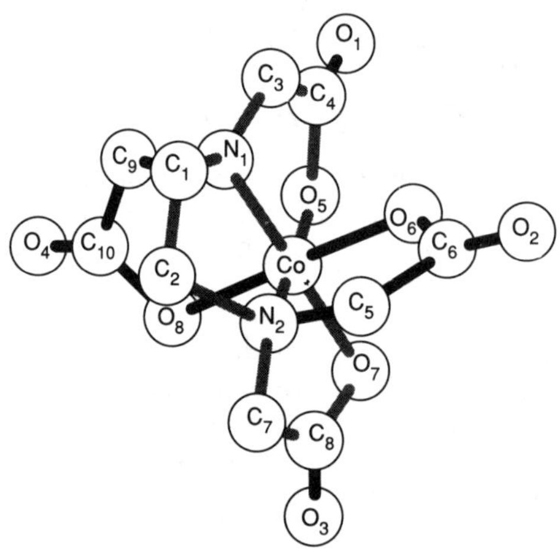

FIGURE 17
The Structure of [CoEDTA]- Showing Mode of Attachment of the Chelating Ion
Chemical Extraction/Soil Washing

3. APPLICATIONS

3.1. General Applications

Soil washing and chemical extraction have been used extensively in Europe and have had limited use in the United States. Currently, eight Superfund sites have incorporated soil washing and chemical extraction in their RODs [4].

In general, chemical extraction is most effective when a low solid/liquid partitioning coefficient can be induced through the use of additives and when diffusive or mass transport considerations are not limiting. Soil washing (soil classification) is most appropriate when a significant volume reduction can be achieved through soil gradation.

3.2. Advantages and Limitations

3.2.1. Advantages

Soil washing and chemical extraction have the following advantages:

- Clean-up level of <1 ppm achieved for some contaminants over a wide range of soil types and influent contaminant concentrations
- Favorably viewed by the public
- Relatively low-cost technology for waste separation and minimization of waste volume
- Can utilize a closed-treatment system, which permits control of volatile emissions
- Can effectively pretreat soils for bioremediation

3.2.2. Limitations

Limitations of soil washing and chemical extraction include the following:

- Contaminant not destroyed
- Fines and washing fluids require secondary treatment or must be disposed of as a hazardous waste
- Sludges from wastewater treatment must be treated
- Chemical specific
- Large organic fraction may require use of additional pretreatment steps.

4. PROCESS DESCRIPTION

Ex Situ

4.1. Introduction

The CHEX and SW system, typically consists of the following components:

- Size Reduction System
    - — Primary crushers
    - — Secondary crushers

- Screening Equipment
  - — Grizzly
  - — Trommel
  - — Screen
- Separation Equipment
  - — Classifiers
  - — Clarifiers
- Mixing and Extraction Devices
  - — Impellers
  - — Water Knife
- Washing Fluid Treatment
  - — Metal Precipitation
  - — Concentration/Separation of Contaminants
  - — Polishing
- Air Emissions
- Secondary Treatment of Fines

4.2. Size Reduction Equipment

Size reduction equipment is necessary when coarse particles greater than 2 -5 cm must be cleaned. This is done for ease of material handling in the soil washing process. A variety of size reduction equipment may be employed, including (see Figure18 [11]):

- Primary crushers (to +100 mm)
- Secondary crushers (100 mm to 5mm)

4.3. Screening Equipment

Screening equipment is used to separate coarse material from the soil and may be used when coarse material does not need to be treated. Screening equipment includes (see Figures 19,20 and 21 [11]):

- Grizzly, still or vibrating (debris)
- Screens (coarse particles)
- Sieve bend
- Trommel (revolving screen - coarse particles)
- Vibrators
- Shakers
- Rotary screens

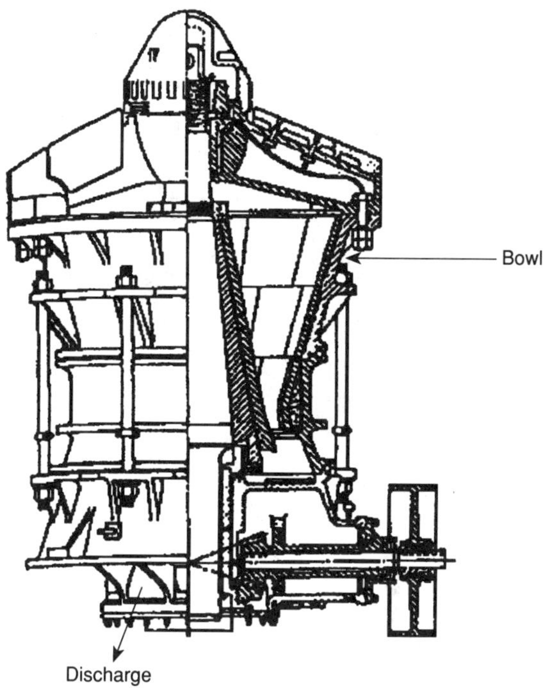

FIGURE 18
Primary Crusher
Chemical Extraction/Soil Washing

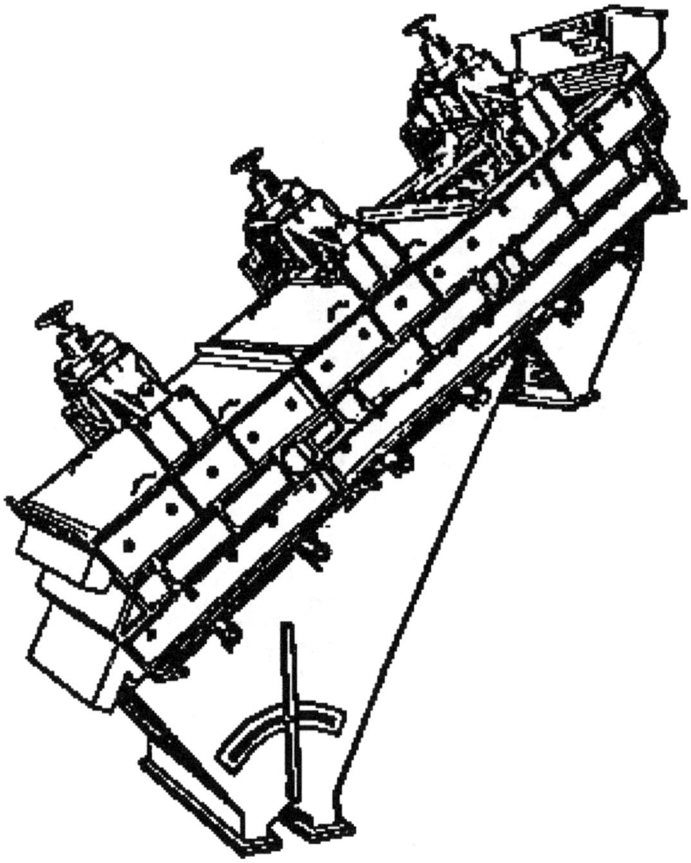

FIGURE 19
"Hum-mer" Screen"
Chemical Extraction/Soil Washing

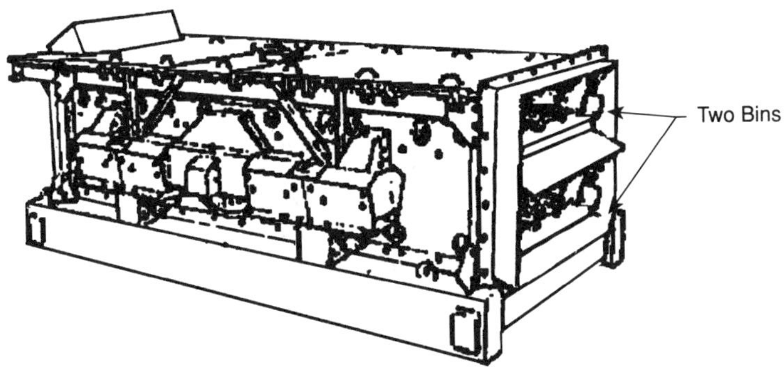

FIGURE 20
"Ty-rock" Screen With Air Seal Enclosure
Chemical Extraction/Soil Washing

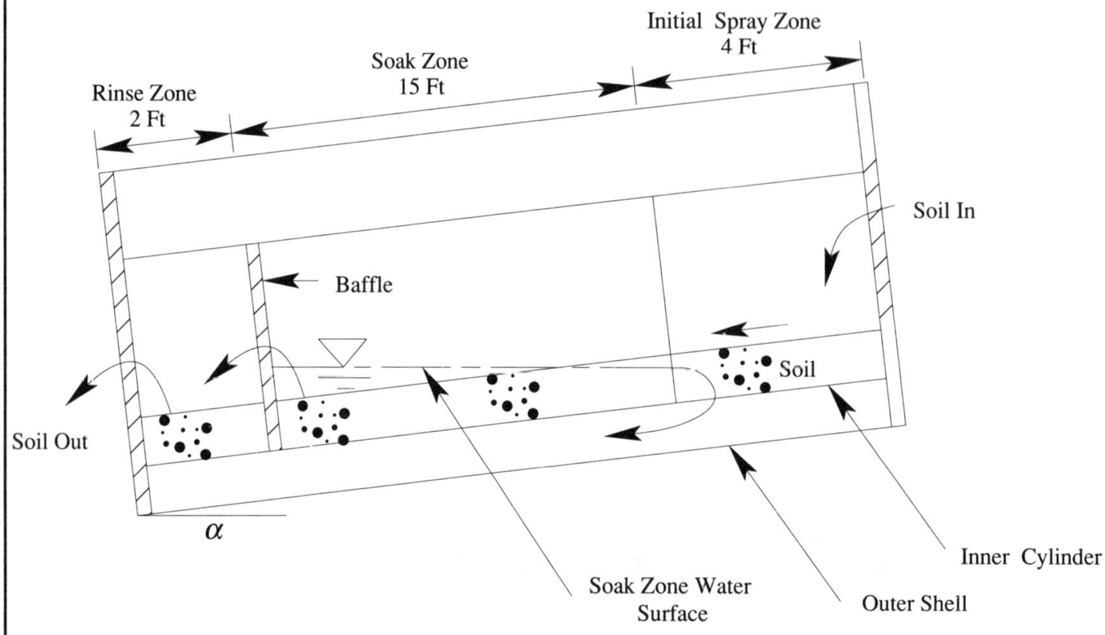

FIGURE 21
Rotary Drum Cross-Section
Chemical Extraction/Soil Washing

4.4. Separation Equipment

Separation equipment is used to separate fine to medium sized particles from water. Any settler which removes particles from a liquid is a clarifier. A device which separates solids into fractions, such as floating organics and settling fines, is a classifier. The same principles of sedimentation apply to both clarifiers and classifiers [6].

Classification equipment (size separation via settling rate)

- Sedimentation Tanks (medium particles)
- Lamella thickener (medium particles)
- Elutriator (upflow classifier)
- Inclined spiral classifier (medium particles)
- Hydrocyclone (fine particles)

  Hydrocyclones are less efficient than centrifuges but have a high capacity, lower energy requirements and low cost [12]. They utilize fluid pressure energy to create centrifugal fluid motion, permitting separation of suspended solids [12]. The solids concentration in the outlet streams vary with influent solids type and concentration, particle size, specific gravity, and fluid viscosity [12]

• Centrifuge (fine particles)

Centrifuges are often limited by solids loading and are extremely expensive, very large and often difficult to operate [12]. They are efficient, but not suitable, for multiple separations in a mobile unit [12].

Gravity Concentration Equipment - Centrifugal separations have to a considerable extent replaced gravity separators because of their greater effectiveness and smaller size [6].
Examples of gravity concentration equipment include:

• Shaking table

• Dense medium cyclone

Flotation (separation via density)

Dewatering (solid-liquid separation) - Dewatering is typically performed on the fines before disposal or after secondary treatment. A variety of methods may be used to dewater the fines.

• Thickening units produce sludges that contain 2 -15 % solids [13].

For fast settling particles a batch settling tank or continuous settling cone may be adequate. Other thickeners may be used to optimize settling of solids.

• Filtration may produce sludges that contain 20-45% solids [13].

A wide variety of filters are available for the dewatering of sludges, such as vacuum and pressure filters.

4.5. Mixing and Extraction Devices

These devices are used to contact the soil with the washing fluid. Again, a wide variety of equipment is available (see Figures 22 and 23 [4]):

• Impellers

• Water knife (60 psig)

• High pressure jet pipe (5000 psig water)

• Low frequency vibration unit

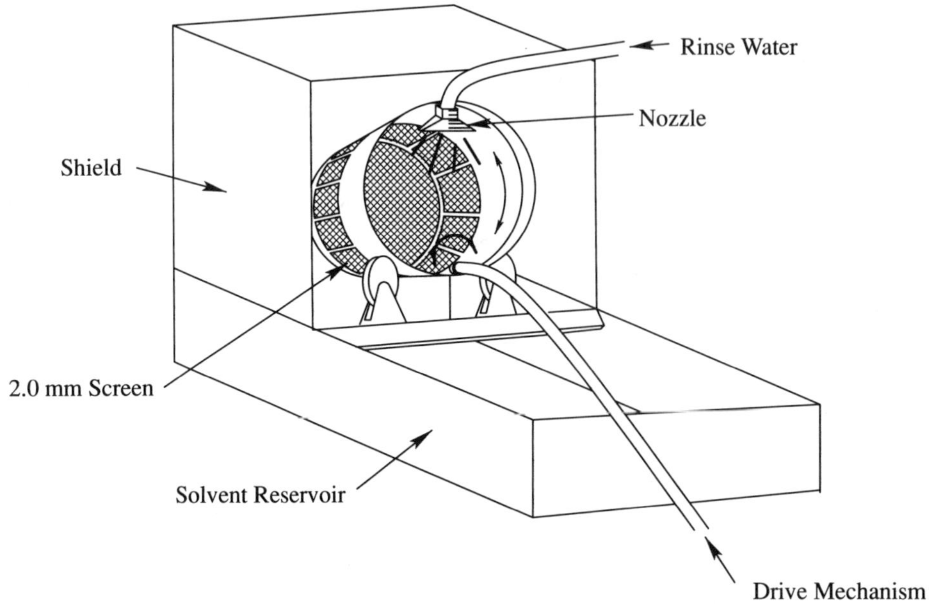

FIGURE 22
Schematic of Water Knife
Chemical Extraction/Soil Washing

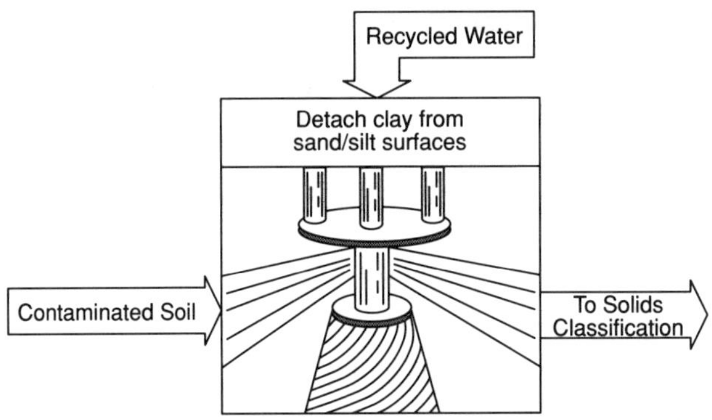

FIGURE 23
High Pressure Jet Pipe
Chemical Extraction/Soil Washing

4.6. Treatment of Washing Fluid

Treatment of washing fluid is necessary for soil washing and chemical extraction to be considered as a feasible alternative. Treated washing fluids may be recycled in the treatment process. Depending on the type of washing fluid to be treated, a number of treatment options are available:

- Metal precipitation
- Concentration/separation of contaminants
    - — Coagulation
    - — Flocculation
    - — Thickening
    - — Dewatering
    - — Filtration
- Biodegradation
- Polishing
    - — Ion exchange
    - — Activated carbon
    - — Microfiltration

4.7. Air Emissions

Air emissions may be of concern during:

- Excavation
    - — Contaminants may volatilize or become airborne as fine particulates.
- Size reduction process
    - — Contaminants may volatilize or become airborne as fine particulates.
- Particles size separation process (dry sieving)
    - — Contaminants may volatilize or become airborne as fine particulates.
- Treatment process
    - — Contaminants may volatilize from treatment units.

Air emissions are controlled by capture and treatment systems. This may be difficult during excavation. Treatment systems may include electrostatic precipitators, carbon systems, and wet air scrubbers.

4.8. Secondary Treatment of Fines

Some soil washing or chemical extraction systems, such as the EPA mobile soil washing unit, will perform secondary treatment of the fines [14]. This is typically comprised of:

- Secondary extraction
- Rinsing
- Dewatering

4.9. *In Situ* Equipment

4.9.1. Introduction

In situ systems generally consist of:

- Injection wells
- Extraction wells

- Wastewater treatment systems
- Pumps and instrumentation
- Containment systems

4.9.2. Injection Wells

Injection wells are used to inject the extraction fluid into the subsurface system. The depth and location are site specific.

4.9.3. Extraction Wells

Extraction wells are used to extract the used extraction fluid with the contaminant from the subsurface system. The depth and location are site specific.

4.9.4. Wastewater Treatment

The wastewater treatment system is dependent on the type of contaminant being extracted and the extraction fluid. Common wastewater operations are generally applied.

4.9.5. Containment Systems

A containment system is generally used to insure that contaminants that are mobilized during the extraction process do not migrate off site. Types of containment systems may be slurry walls, grout walls or prefabricated material walls.

5. FEASIBILITY

(This section has been adapted from reference [2])

The applicability of chemical extraction and soil washing is based on:

- Soil Characteristics
- Site hydrology (if *in situ*)
- Contaminants, i.e., solubility and adsorption affinity
- Extracting agent being used
- Mixing (time and efficiency)

<u>Soil characteristics</u>
It is important to know the soil characteristics of a site because of the following effects:

- Highly variable soil conditions can produce inconsistent flushing.
- High organic content soils can inhibit desorption of the contaminants and have a higher affinity for non-polar organics.
- Large organic fraction in the soil may require use of additional pretreatment steps
- Grain size, specific gravity, density, and water content determine available void volume, soil porosity, permeability, and the extent of volume reduction with soil washing.

- Low permeability (high salt or clay content) reduces percolation and leaching.
- Buffer capacity and soil pH affect neutralization and possibly precipitation reaction.
- A high cation exchange capacity (CEC) observed in some clays and fine silts may decrease contaminant mobility and can attenuate treatment of metals and metal salts.

Site Hydrology
It is important to know site hydrology because, in *in situ* applications, groundwater flow must be well defined to ensure recapture of soil flushing fluids.

Contaminants
Different contaminants have different affinities with soil or different soil types. The adsorption of the contaminant to the soil is related to the type of contaminant and the type of soil. The extent of adsorption/desorption can be defined using the aforementioned adsorption isotherms.

The solubility of the contaminant in the given solvent is also of great importance and must be determined either through a literature review or laboratory analysis. This solubility should be determined in multi-component systems, which typically exist in real world situations, since the solubility of a component may differ in multi-component systems as opposed to pure solutions.

Extracting agent
Extracting fluids should have the following characteristics:

- Favorable separation coefficient for extraction
- Low volatility under ambient conditions (to reduce air contamination effects)
- Low toxicity
- Safety and ease of handling
- Recoverability for reuse
- Treatability of washing fluid and
- High water solubility

If the extracting agent reacts with the contaminants, it should produce no toxic by-products. The chemical reaction rate varies with each extracting agent; a laboratory analysis should be conducted to determine the rate constant.

Mixing
Complete mixing of soil with the washing solution will enhance the contact between soil particles and contaminants, thus reducing the time required for extraction and increasing the efficiency of contaminant removal. It also will enhance separation of contaminants and accelerate the leachate rate. Consequently, diffusion limitation and flow pattern short-circuiting must be quantified.

6. DESIGN

6.1 Soil Washing Design Approach

For soil washing purposes, soils are classified as coarse, sandy, and fine [3]:

- Coarse fraction: typically greater than 2 cm in diameter

- — This fraction is tested to ensure that it is contaminant free before it is separated from the other fractions prior to the treatment process
- — If the course fraction contains an unacceptable level of contamination, it is subjected to a size reduction process prior to treatment

• Sandy fraction: typically between 2 cm and 63 µm in diameter
- — It is this fraction which is cleaned in the soil washing process.
- — The mechanisms of cleaning include elutriation and physical stripping of contaminants which are not chemically bound to the soil.

• Fine fraction: typically less than 63 µm in diameter
- — This fraction typically contains the highest concentrations of the contaminant and must be subsequently treated or land filled.
- — The mechanisms of cleaning include elutriation and physical stripping of contaminants which are not chemically bound to the soil.

An example of contaminant distribution among particle sizes is presented in Table 2 [4].

Table 2
Contaminant distribution

Particle size distribution	Oil & Grease mg/kg	Volume % of total	Contamination % of total
Total	36,000	100	100
> 2mm	40	5	0.1
2mm>x>0.06mm	300	80	0.8
< 0.06mm	35,700	15	99.1

When designing a soil washing process, mass balance equations (as discussed earlier) are used to size the unit operations.

The most important unit operation in the soil washing process is the stripper. This unit is of significant importance because it ensures that the fine fractions attached to the larger fractions and/or caught in the soil lumps are removed. Both contaminant stripping and lump breaking require a relatively high level of energy input for the greatest efficiency.

The most commonly used system to strip contaminants is a high energy system containing a rotary trommel screen and water knives (rotary drum/water knife) (see Figure 21 [4)]. The water knives are small flat blade nozzles that function by creating a high velocity stream of water in the shape of a long narrow line. The stripping is performed by directing the water stream at a moving layer of material. The trommel screen is used to rotate the material to ensure complete contact with the water knives and to remove the coarse fraction of the soil matrix. The volumes of water and pressure needed must be determined from laboratory testing [4].

There are many variations of the soil washing process. Two general schematics of the soil washing process are presented in Figures 24 and 25 [4].

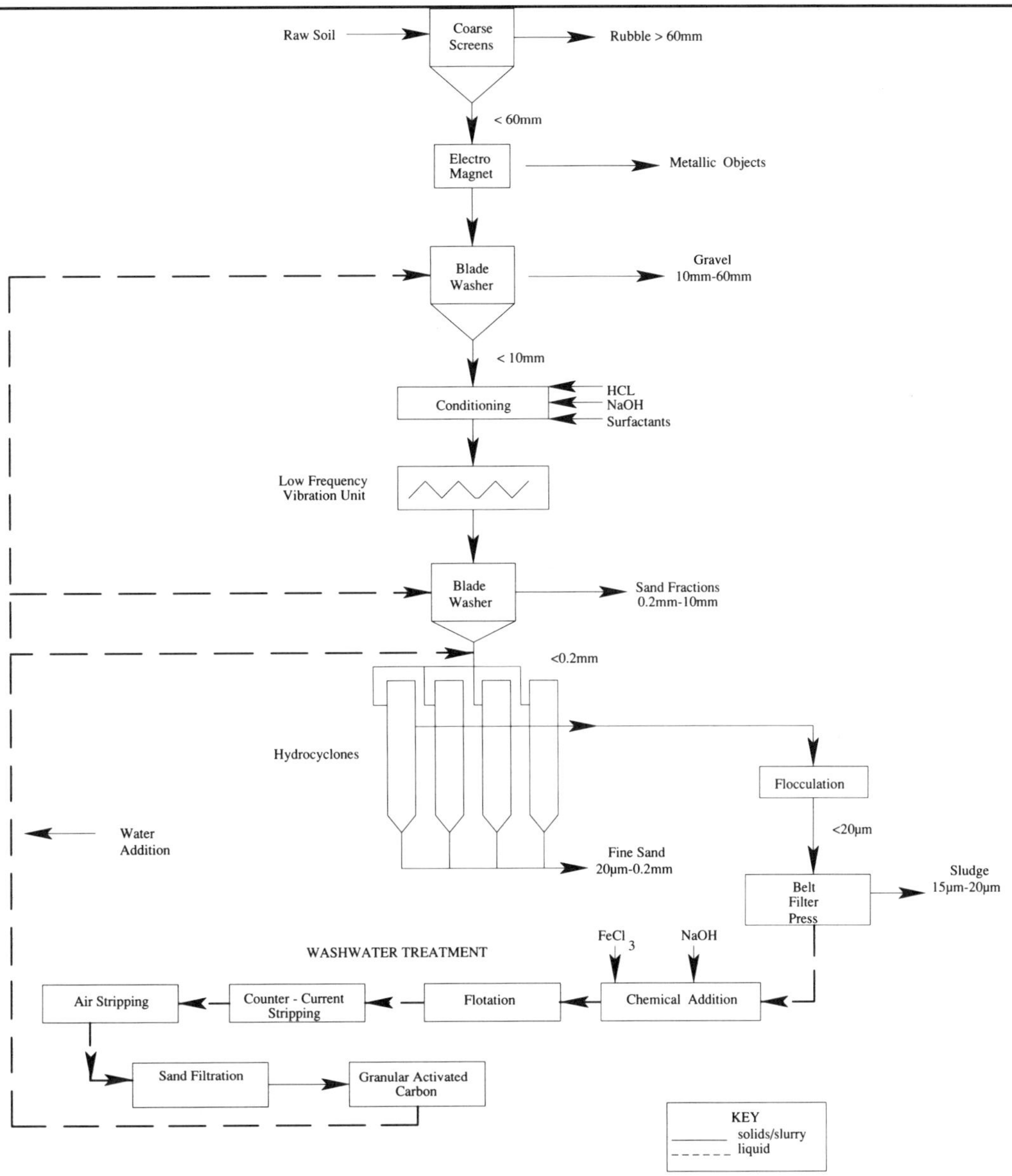

FIGURE 24
Flow Schematic of the Harbauer Soil Washing Installation
Chemical Extraction/Soil Washing

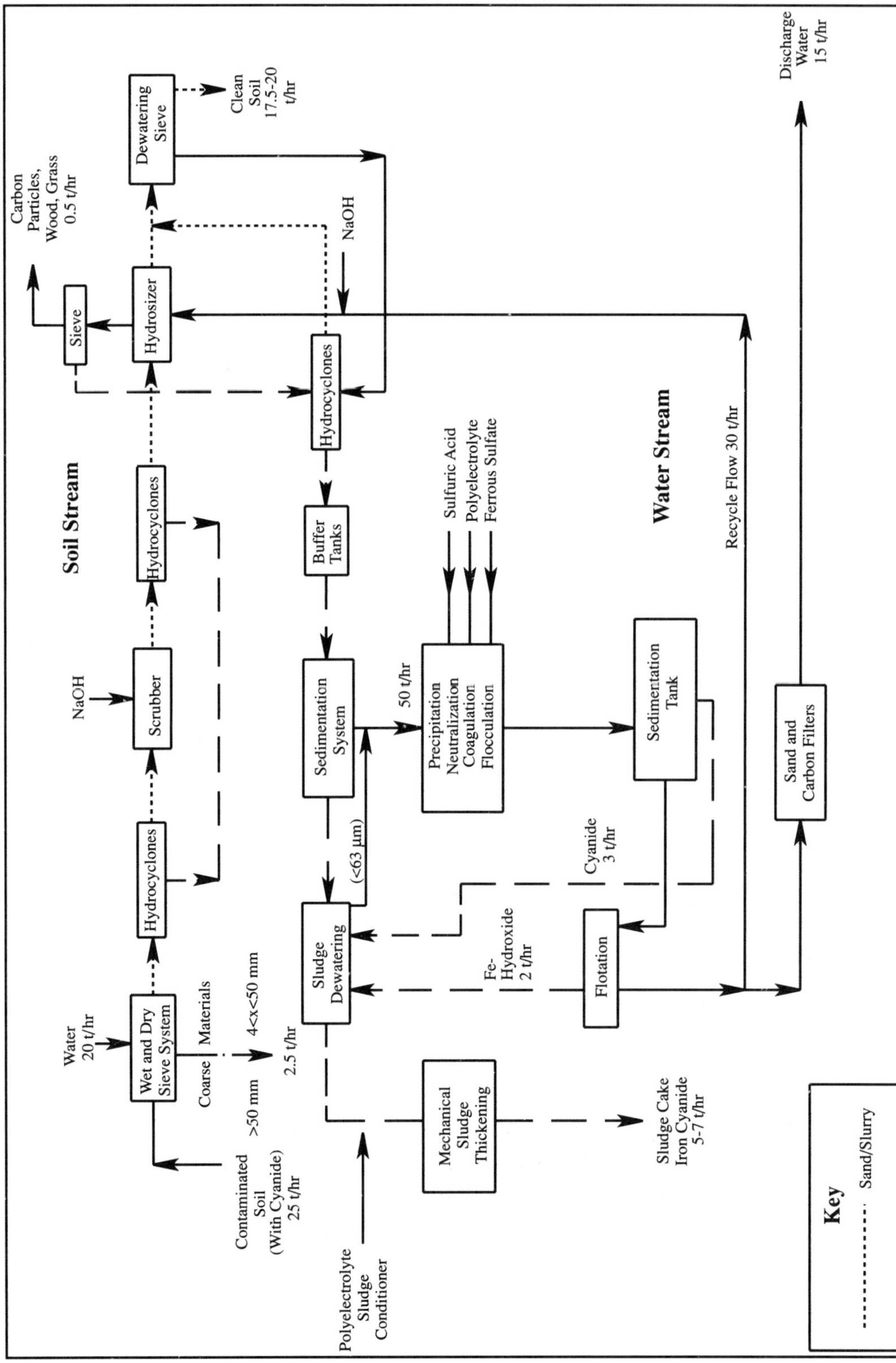

FIGURE 25
HWZ Soil Treatment Scheme
Chemical Extraction / Soil Washing

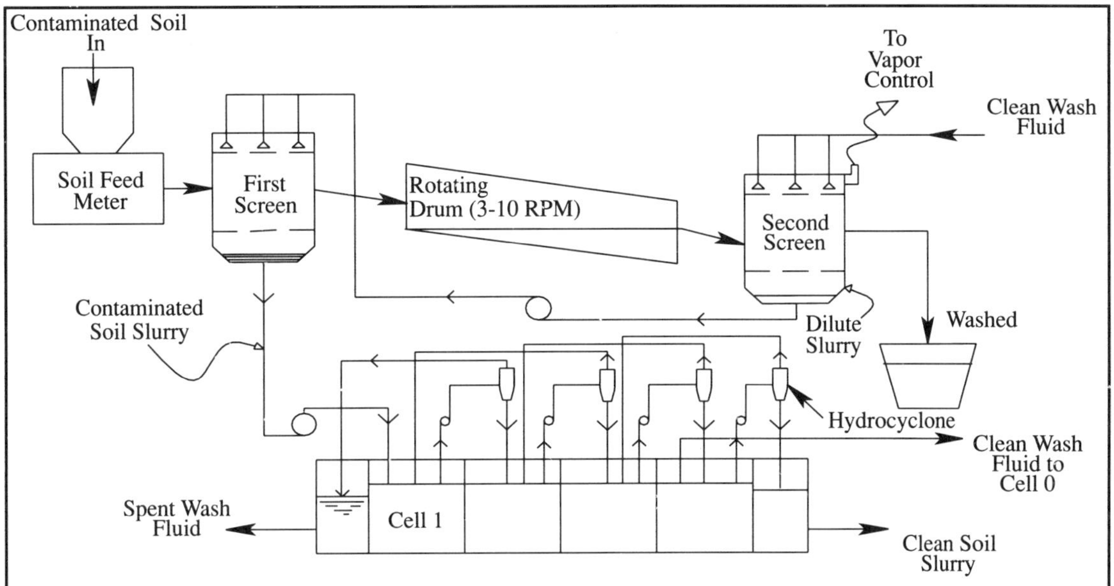

FIGURE 26
Schematic of the EPA Mobile Soil Washing System
Chemical Extraction/Soil Washing

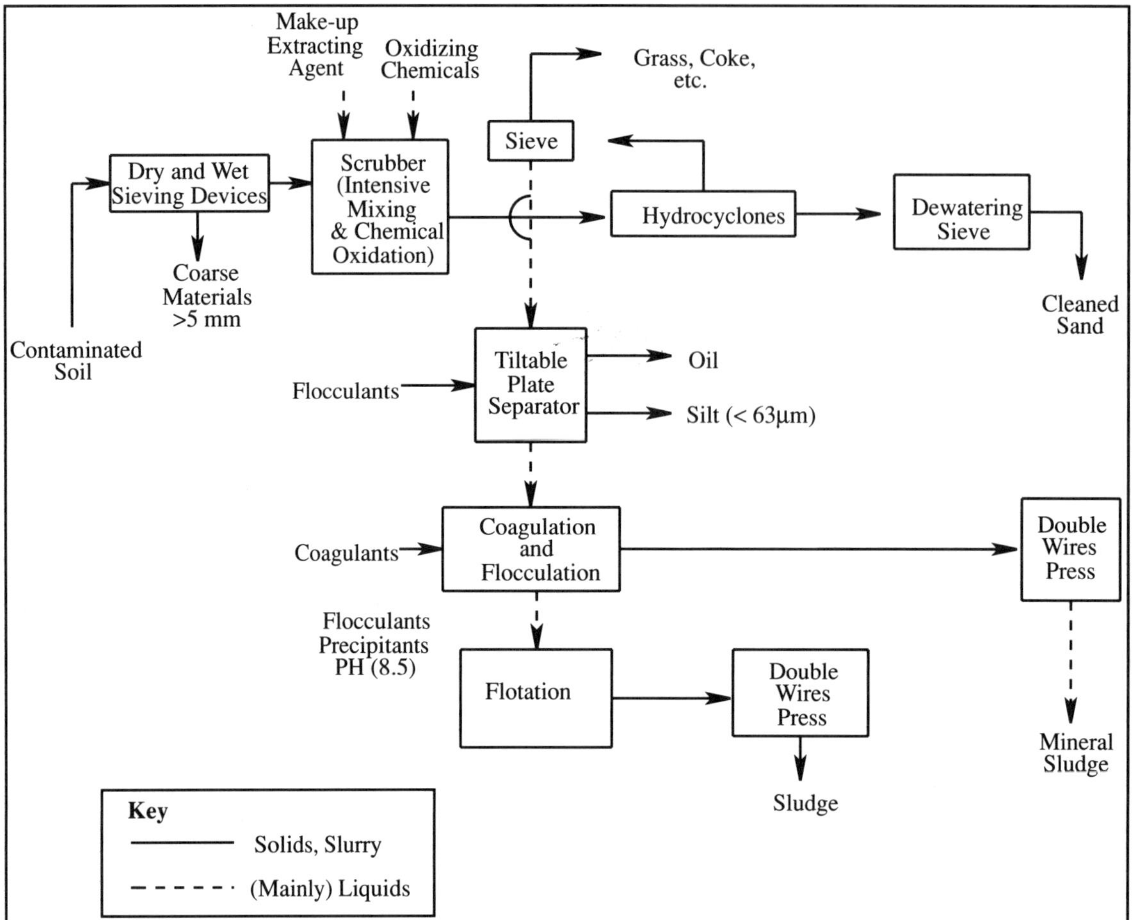

FIGURE 27
Process Scheme of Installation of Heijmans Milieutechniek
Chemical Extraction/Soil Washing

In the first schematic, the coarse material is separated from the fines and sand. The sand and fines are washed in a scrubber. After washing, a variety of methods may be used to separate the contaminated fines and the wash water from the washed sandy fraction. This schematic washes all particles less than 2 cm in diameter.

In the second schematic, the coarse material is separated from the sand and fines, then the fines are separated from the sand. The sand is then washed, and the fines are properly disposed.

To illustrate the variations in the soil washing systems, three schematics of three different soil washing systems are presented (source: reference [4]:

- Fig. 25 - HWZ Soil Treatment Schematic
- Fig. 26 - EPA Mobile Treatment Soil Washing
- Fig. 27 - Heijmans Soil Washing Schematic

6.2 Chemical Extraction Design

6.2.1 Introduction

In recent years, work has focused on exploring Chemical Extraction. The main areas of interest thus far have been contaminant transport, sorption and desorption of contaminants on the soil particles, developing design equations that are based upon mass transfer limitations, and micelle solubilization. This section will summarize recent results, their applications and limitations, and how they may be of use to the design engineer.

6.2.2. *In Situ* Chemical Extraction Example

6.2.2.1. Introduction

Recent work is expanding the basic equations for contaminant transport in subsurface medium to include chemical reactions taking place in these systems. This is of great importance in the design of an *in situ* chemical extraction unit, because one of the greatest difficulties in the design process is predicting how the extraction fluid will migrate and how it will react with the contaminant under subsurface conditions.

6.2.2.2. Contaminant Transport Model

Cederberg et al. (1985) devised a numerical model that includes equilibrium chemical reactions such as complexation, ion exchange, and competitive adsorption. In this model, the groundwater transport equation is:

$$\frac{dc_i}{dt} + \frac{\rho d\,\bar{c}_i}{\theta dt} = D_L \frac{d^2 c_i}{dx^2} - v\frac{dc_i}{dx} + S_{c_i} \qquad i=1,2,3...n_{aq} \tag{21}$$

where:

- c_i = aqueous phase concentration of species i (mole/L)
- $\bar{c}_i$ = sorbed phase concentration of species i (moles/g)
- n_{aq} = number of aqueous species to be considered
- S_{ci} = source-sink for species i due to aqueous phase complexation
- θ = medium porosity
- ρ = bulk density of medium (g/L)
- D_L = coefficient of hydrodynamic dispersion (m2/hour)
- v = average pore velocity (m/hour)
- x = distance (m)
- t = time (hour)

This equation uses the equilibrium chemistry separately from that of the mass transport equations. This leads to the coupling of a set of algebraic equations (chemistry) and a set of differential equations (mass transfer) called TRANQL. Cederberg used a simulation of cadmium transport in a one-dimensional-radial flow situation to test the performance of TRANQL. Although ion pair formation and transport were modeled the results may be indicative of inner-sphere complex behavior as well.

TRANQL defines a species as every chemical to be considered in the problem. Mass law equations and equilibrium stability constants are written for each entity in the group of species (see Table 3 and 4 [15])

Table 3
Reactions and Mass Law Equation: Cadmium Transport

Reactions	Species	$Log_{10}K$	Mass Law Equation						
1	$H^+ = H^+$	0						$[H^+]$	$K_1 = [H^+]$
2	$Cd^{2+} = Cd^{2+}$	0			$[Cd^{2+}]$				$K_2 = [Cd^{2+}]$
3	$Cl^- = Cl^-$	0	$[Cl^-]$						$K_3 = [Cl^-]$
4	$Br^- = Br^-$	0		$[Br^-]$					$K_4 = [Br^-]$
			Complexation						
5	$Cd^{2+} + Cl^- = CdCl^+$	1.8	$[Cl^-]$		$[Cd^{2+}]$				$K_5 = [CdCl^+]$
6	$Cd^{2+} + 2Cl^- = CdCl_2$	2.6	$[Cl^-]^2$		$[Cd^{2+}]$				$K_6 = [CdCl_2]$
7	$Cd^{2+} + Br^- = CdBr^+$	2.2		$[Br^-]$	$[Cd^{2+}]$				$K_7 = [CdBr^+]$
8	$Cd^{2+} + 2Br^- = CdBr_2$	3.0		$[Br^-]^2$	$[Cd^{2+}]$				$K_8 = [CdBr_2]$
9	$Cd^{2+} - H^+ + H_2O =$ $CdOH^+$	-12.69			$[Cd^{2+}]$			$[H^+]^{-1}$	$K_9 = [CdOH^+]$
			Dissociation of Water						
10	$- H^+ + H_2O = OH^-$	-13.91						$[H^+]^{-1}$	$K_{10} = [OH^-]$
			Sorption						
11	$SOH = SOH$	0				$[SOH]$			$K_{11} = [SOH]$
12	$SOH + H^+ = SOH_2^+$	7.4				$[SOH]$	$[e^{-F\psi/RT}]$	$[H+]$	$K_{12} = [SOH_2^+]$
13	$SOH - H^+ = SO^-$	-9.24				$[SOH]$	$[e^{-F\psi/RT}]$	$[H^+]^{-1}$	$K_{13} = [SO^-]$
14	$SOH - H^+ + Cd^2 =$ $SOCd^+$	-7 — -11			$[Cd^{2+}]$	$[SOH]$	$[e^{-F\psi/RT}]$	$[H^+]^{-1}$	$K_{14} = [SOCd^+]$

Table 4
Component Material Balance Equations

Components	Material Balance Equation
$T_1 = Cl_T$	$= [Cl^-] + [CdCl^-] + 2[CdCl_2]$
$T_2 = Br_T$	$= [Br^-] + [CdBr^+] + 2[CdBr_2]$
$T_3 = Cd_T$	$= [Cd^{2+}] + [CdCl^+] + [CdCl_2] + [CdBr^+] + [CdBr^2] + [CdOH^+] + [SoCd^+]$
$T_4 = SOH_T$	$= [SOH] + [SOH_2^+] + [SO^-] + [SoCd^+]$
$T_5 = T_{oT}$	$= [SOH_2^+] - [SO^-] + [SoCd^+]$
$T_6 = H_T$	$= [H+] + 2[SOH_2^+] - [CdOH^+] - [OH^-] - [SO^-] - [SoCd^+]$

T_o represents the charge on the surface, defined as the excess of positive groups over negative groups.

From these species, a set of components (T1, T2,...) are defined so that an equation for each species expressing each entity can be written as the product reaction involving only components. Cederberg develops separate interaction equilibrium equations from the mass transport equations.

To determine the equilibrium concentration of the species, Cederberg used the computer code MICROQL developed by Westall (1979). MICROQL uses the mass law equations and stability constant for each species and the material mass balance equations for each component to define the chemical equilibrium problem [15]. MICROQL is an iterative process, initialized by guesses at the free concentrations of components.

The transport equation for each species is reduced using partial differential equations, to fewer equations formulated around mass balance equations for the total concentration for each component. This keeps the interaction chemistry separate from the mass transport equations.

To investigate the behavior of TRANQL, Cederberg simulated the transport of a multicomponent solution of cadmium, bromide, and chloride in a one-dimensional laboratory column. Six test cases were run, to investigate the effect of the concentration of complexing ligands and the composition of sorbing substrate on the concentration of aqueous cadmium. Six test cases were run for both the upper and lower limit of the sorption stability constant.

Cederberg broke the test cases into two groups: group one considered the transport of cadmium in the presence of Cl_T and Br_T near background concentrations; group two considered the transport of cadmium in the presence of Cl_T and Br_T above background conditions.

The initial and boundary conditions are presented in Table 5 [15]. In each case, the column was considered to be in chemical equilibrium at time equal to zero. Cases 1 and 4 correspond to step input of Cd_T, with Cl_T and Br_T held constant. Cases 2 and 5 correspond to a step input of Cl_T and Br_T with Cd_T held constant. Cases 3 and 6 correspond to a step input of Cl_T, Br_T and Cd_T.

Table 5
Concentration of Components for Test Cases: Initial and Boundary Conditions

Case	Components			Conditions	
	Cd_T	Cl_T	Br_T		
	Group 1: Background Levels of Cl^- and Br^- (M)				
1	0.00001	0.0003	0.0001	$0 < x < \infty$	$t = 0$
	0.0001	0.0003	0.0001	$x = 0$	$t \geq 0$
2	0.0001	0.0003	0.0001	$0 < x < \infty$	$t = 0$
	0.0001	0.003	0.001	$x = 0$	$t \geq 0$
3	0.00001	0.0003	0.0001	$0 < x < \infty$	$t = 0$
	0.0001	0.003	0.001	$x = 0$	$t \geq 0$
	Group 2: Higher Than Background levels of Cl^- and Br^- (M)				
4	0.00001	0.003	0.001	$0 < x < \infty$	$t = 0$
	0.0001	0.003	0.001	$x = 0$	$t \geq 0$
5	0.0001	0.003	0.001	$0 < x < \infty$	$t = 0$
	0.0001	0.03	0.01	$x = 0$	$t \geq 0$
6	0.00001	0.003	0.001	$0 < x < \infty$	$t = 0$
	0.0001	0.03	0.01	$x = 0$	$t \geq 0$

Simulation results, in which the sorption equilibrium constant was at a maximum (i.e Log_{10} K_{SOCd+} = -7.0), are shown in Figures 28, 29, and 30 [15]. Significant sorption of cadmium was observed for these six examples. When the sorption equilibrium was set at the minimum (i.e. $log_{10}K_{SOCd+}$ = -11.0), the amount of cadmium sorbed was less than 0.05%.

When Cl_T and Br_T were held constant, sorbed cadmium [$SOCd^+$] decreases and Cd_{AQ} increases as Cl_T and Br_T increase, as can be seen in Figure 28. Figure 29 shows that [$SOCd^+$] decreases and Cd_{AQ} increases in the column as Cl_T and Br_T increase. The cadmium, which was initially sorbed onto the porous medium, desorbed, complexed Cl^- and Br^-, and entered the aqueous phase due to increases in Br_T and Cl_T concentrations.

In Figure 30, the step input of Cl_T, Br_T and Cd_T are used as boundary conditions, and Cd_{AQ} is increased as Cl_T and Br_T are increased. [$SOCd^+$] is smallest when Cl_T is at a maximum. Figure 30 shows that as Cl_T and Br_T are increased, cadmium becomes more mobile and is transported further down the column with respect to the predicted cadmium concentrations when Cl_T and Br_T are at a minimum.

As seen from these results, the partitioning of Cd is a strong function of the Cl_T concentration. As the concentrations of Cl_T and Br_T were increased for cases 1, 2 and 3, the percent of cadmium sorbed decreased (refer to Figures 28, 29, 30). When the concentrations of Cl_T and Br_T are increased, Cd_{AQ} is transported further down the column due to the equilibrium partitioning of cadmium. Cederberg observed that this behavior is consistent with results obtained in laboratory experiments.

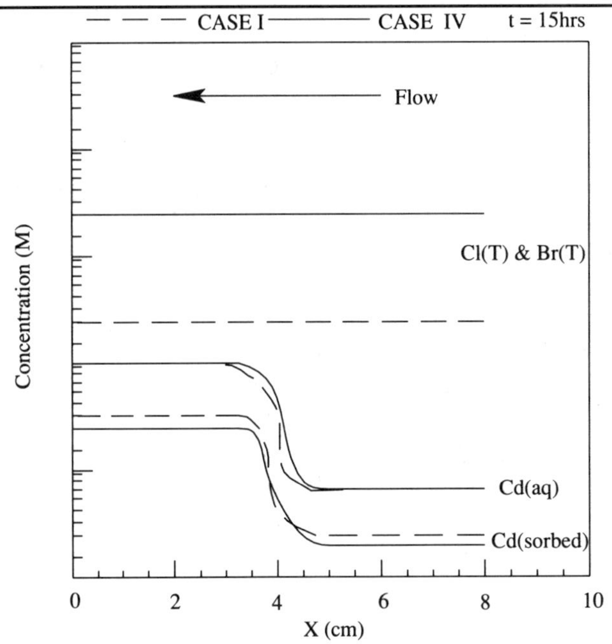

FIGURE 28
TRANQL Simulations of Cd(AQ) Transport in Solution of 3X10^-4 M Cl(T) and 1X10^-4 Br(T) (case 1) and 3X10^-3 M Cl(T) and 1X10^-3 M Br(T) (case 4)
Chemical Extraction/Soil Washing

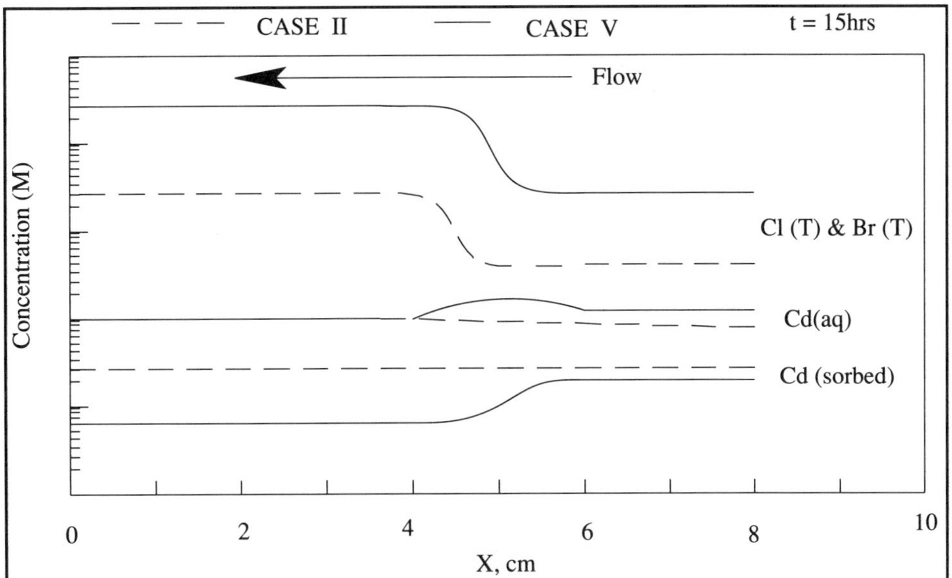

FIGURE 29
TRANQL Simulations of 3X10^-3 M (case2) and 3X10M^-2 (case 5)
$Cl_{(T)}$ Transport in Solution of 10-4 Cd(T)
Chemical Extraction/Soil Washing

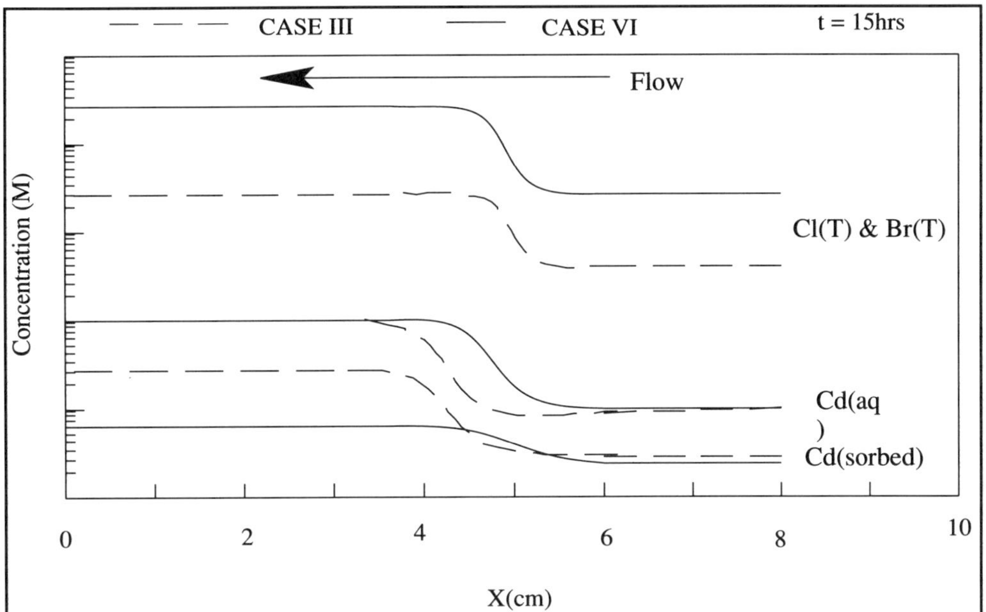

FIGURE 30
TRANQL Simulations of $Cd_{(AQ)}$ Transport and 3X10^-3M
(case 3) and 3X10M (case 6) $Cl_{(T)}$ Transport
Chemical Extraction/Soil Washing

Cederberg drew two conclusions from these simulations:

- Partitioning between the sorbed and fluid phase of particular components in a multicomponent solution cannot necessarily be represented in terms of a linear isotherm involving only the aqueous species. The K_D value is not spatially or temporally constant in a flow system and should not be used in transport models. The use of single K_D values will lead to erroneous results. (This is contrary to the earlier simplified presentation in section 2.2.5.4)
- Sorption equilibrium constants must be determined for the particular system under consideration.
- Control in the subsurface is difficult, therefore *in situ* treatment may accelerate migration off-site.

6.2.2.3. Applications of Model

This model can provide useful insight into the mechanics influencing *in situ* chemical extraction and soil washing. This model should not be used as the sole criteria of design, but if combined with the theory and application presented herein, it is a valuable resource.

6.2.2.4. Limitations of Model

There are two major limitations to this model:

- It assumes a homogeneous isotropic condition, which does not exist in real circumstances, therefore precise modeling is not possible
- It does not consider the chemical kinetics that influence the system. Equilibrium is an ideal situation

6.2.3. *Ex Situ* Chemical Extraction Example

6.2.3.1. Introduction

The need for quantitative methods of designing and modeling CHEX processes has become apparent in recent years. Because of this need, new theories and models are being developed.

Clarke, et al. [16] had developed a correlation to determine the micellar phase/water partition coefficient, and several models of surfactant extraction using a non-linear non-traditional isotherm developed by Wilson [17].

6.2.3.2. Micellar Phase/Water Partition Correlation

Clarke, et al. had developed a correlation between the octanol/water partition coefficient, K_{OW}, and the micellar phase/water partition coefficient, K_{SW}. The micellar phase/water partition coefficient is a useful tool in predicting the success of a surfactant extraction process. This correlation is:

$$\log_{10} K_{sw} = 1.12 \log_{10} K_{ow} - 0.686 \tag{22}$$

See Figure 31 [16].

Because octanol/water partition coefficients are known and/or can be determined with relative ease for most contaminants, this correlation is of great use.

6.2.3.3. Limitation of Correlation

In recent studies, it has been shown that the value of the octanol/water partition coefficient is not as accurate as previously thought, and that the use of it may lead to serious errors in model predictions. A new method has been developed that has a greater accuracy in determining partition coefficients. It is the determination of partition coefficients by reverse phase chromatography [18]. Therefore, while the micellar/water partition correlation is (based on octanol/water partition coefficient) useful, it may not be accurate.

6.2.3.4. Application of Correlation

This model may used a preliminary tool in determining if surfactant extraction is a viable remedial tool.

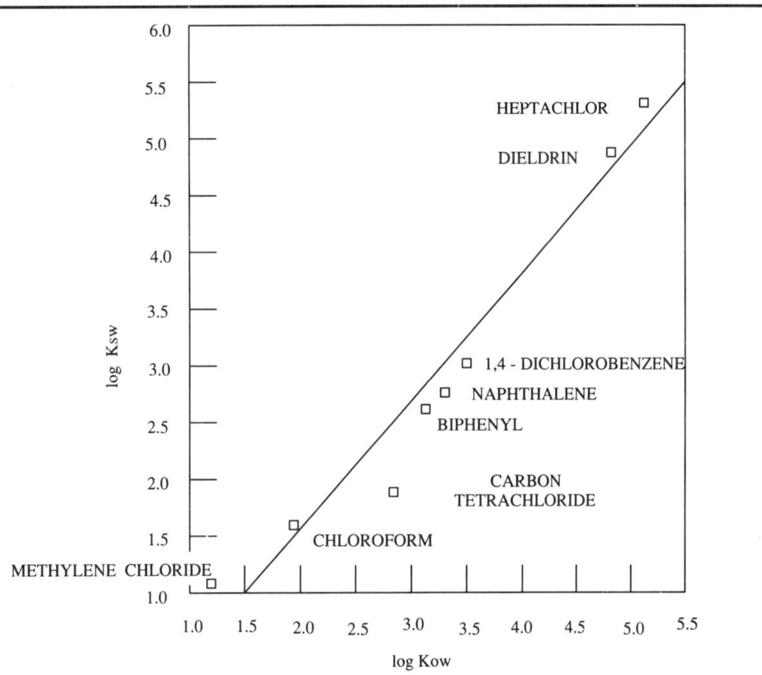

FIGURE 31
Correlation of Octanol/Water Partition Coefficient (Kow) with Sodium Dodocylsulfate (SDS) Surfactant/Water Partition Coefficient (Ksw)
Chemical Extraction/Soil Washing

6.2.3.5. Surfactant Extraction Model

Clarke, et al. developed three models that simulate surfactant chemical extraction based on an isotherm developed by Wilson [17]. These models are for batch processing, continuous flow surfactant processing, and counter-current flow processing.

Wilson's isotherm, which assumes local equilibrium, is:

$$c(m,C) = \left[c_o + K_D \ (C - cmc)\right] \frac{m}{m_{1/2} + m} \tag{23}$$

where:

c = concentration of contaminant in the surfactant solution after equilibrium, kg/m^3

C = Surfactant concentration. kg/m^3

m = Contaminant concentration in soil after surfactant solution has been allowed to equilibrate with the soil and is then drained away, kg/m^3

c_o = Contaminant concentration in water, kg/m^3

K_D = Distribution coefficient for contaminant in the surfactant being used, dimensionless

cmc = Critical micelle concentration of surfactant, kg/m^3

$m_{1/2}$ = Langmuir-type adsorption parameter, which is the soil concentration of the contaminant at which the aqueous concentration in equilibrium with the soil has dropped to half of its saturation value, kg/m3

Clarke modifies Wilson's isotherm for non-equilibrium conditions and for diffusion limitations, using igenvalues determined as a function of Millington and Quirk equation for effective diffusivity. Stating that if the soil is lumpy - i.e. contains pieces of porous rock, lumps of clay of low aqueous permeability etc., diffusion of the contaminant from the interiors of these blocks will be the rate-limiting step in the process.

The diffusion limited batch process model was run four times, varying the particle lump radius each time (see Tables 6-9, reference [16]). The results show that diffusion is the rate limiting step with particle lump sizes greater than 2mm. Therefore, if removal efficiency is sought, granular or lumpy soils must be broken-up before surfactant extraction.

Table 6
Diffusion-Controlled Model: Diameter Particle Lump = 1 x 10^{-4} m

Number of Washings	Effluent Soil Contaminant Concentration (kg/m^3)
0	17.00
1	12.01
2	7.268
3	3.164
4	0.7866
5	0.1368
6	2.176×10^{-2}
7	3.410×10^{-3}
8	5.331×10^{-4}
9	8.329×10^{-5}
10	1.301×10^{-5}

Table 7
Diffusion-Controlled Model: Diameter Particle Lump = 10^{-3} m

Number of Washings	Effluent Soil Contaminant Concentration (kg/m^3)
0	17.00
1	12.02

2	7.275
3	3.173
4	0.7918
5	0.1383
6	2.210 x 10^{-2}
7	3.478 x 10^{-3}
8	5.461 x 10^{-4}
9	8.569 x 10^{-5}
10	1.345 x 10^{-5}

Table 8
Diffusion-Controlled Model: Diameter Particle Lump = 2 x 10^{-3} m

Number of Washings	Effluent Soil Contaminant Concentration (kg/m^3)
0	17.00
1	12.85
2	8.855
3	5.187
4	2.328
5	0.8138
6	0.2549
7	7.710 x 10^{-2}
8	2.308 x 10-2
9	6.886 x 10^{-3}
10	2.053 x 10^{-3}
14	1.619 x 10^{-3}

Table 9
Diffusion-Controlled Model: Diameter Particle Lump = 5 x 10^{-3} m

Number of Washings	Effluent Soil Contaminant Concentration (kg/m^3)
0	17.00
1	15.76
2	14.53
3	13.32
4	12.12
5	10.94
6	9.783
7	8.656
8	7.570
9	6.534
10	5.563
67	1.107 x 10^{-5}

In the continuous flow surfactant process, the soil was batch processed while the surfactant was continuously passed upward through the soil (see Figure 32 [16]). This model was run varying surfactant flowrates, surfactant concentrations, and particle lump diameter (see Figures 33 - 35, reference [16]).

Figure 33 shows the effect of various surfactant flowrates on the contaminant removal rate. There is a significant increase in removal rate as the flow rate is increased from 0.5 L/s to 1.0 L/s, but little increase in removal rates at flowrates greater than 1.0 L/s. Clarke concluded that at flowrates greater than 1.0 L/s, diffusion from the interior of the particle lump was the rate controlling step.

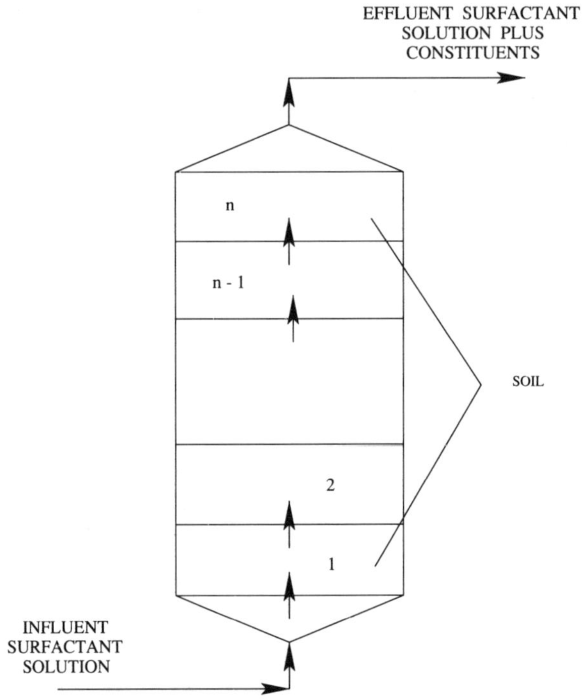

FIGURE 32
Schematic of the Semi-Continuous Flow Soil Washing Apparatus
Chemical Extraction/Soil Washing

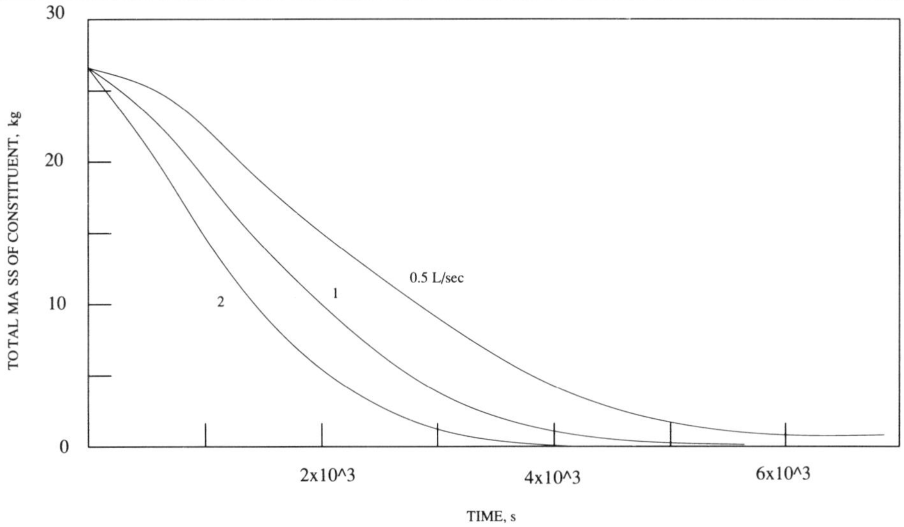

FIGURE 33
Plots of Total Contaminant Mass Versus Time Dependence on Surfactant Solution Flow Rate
Chemical Extraction/Soil Washing

The effects of increasing the surfactant concentration from 5 g/L to 10 g/L to 20g/L can be seen in Figure 34. The removal rate is nearly doubled when the surfactant concentration is increased to 10 g/L, however, the increase in removal rate is much less when the surfactant concentration is increased to 20 g/L. Therefore, Clarke concluded that diffusion is the rate limiting step at surfactant concentrations greater than 10 g/L.

Because it was evident from the first two runs that the model was diffusion limited, Clarke varied the diameter to the particles to be treated (see Figure 35). At diameters less than 1mm diffusion seemed unimportant, but at diameters greater than 1mm diffusion is an overwhelming factor. Thus, Clarke concluded that comminution is a must before surfactant extraction.

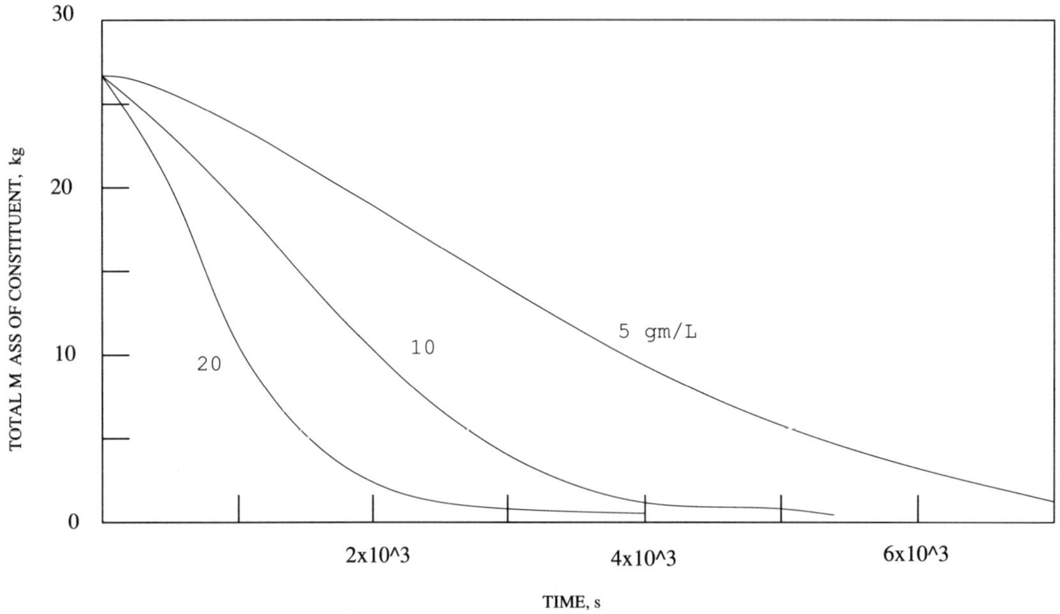

FIGURE 34
Plots of Total Contaminant Mass Versus Time; Dependence on Surfactant Concentration
Chemical Extraction/Soil Washing

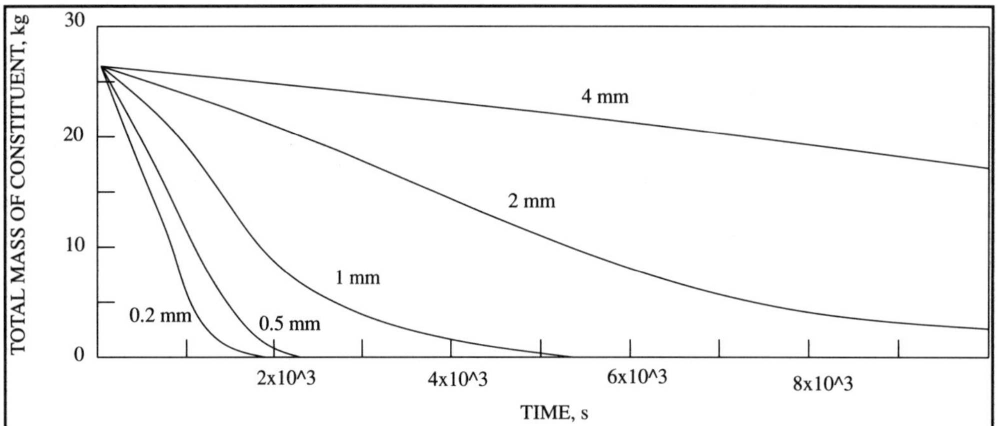

FIGURE 35
Plots of Total Contaminant Mass Versus Time; Dependence on Soil Lump Effective Diameter
Chemical Extraction/Soil Washing

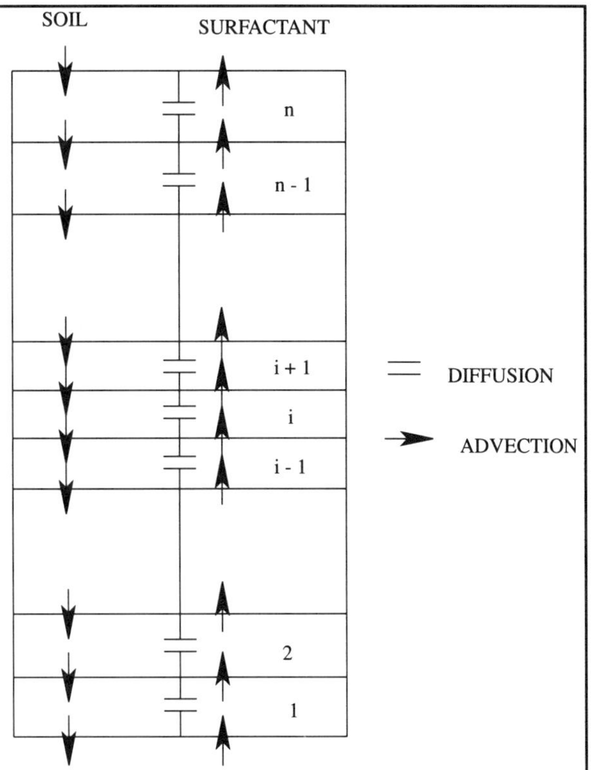

FIGURE 36
Schematic of the Counter-Current Flow Washing Column
Chemical Extraction/Soil Washing

In the counter-current flow process, the soil and extraction fluid are moved in opposite directions (see Figure 36,[16]). This model was run varying the particle lump diameter, the soil velocity, and the surfactant flowrate.

As was expected, the particle lump diameter had a great effect on the contaminant removal rate (see Table 10 [16]). Clarke again concluded that at particle lump sizes greater than 1mm diffusion was the rate limiting step.

Table 11 [16] shows the effect of soil velocity on the contaminant removal rate. At 4 cm/s, the soil effluent concentration is 0.796 mg/kg, while at 8 cm/s the effluent soil concentration is approximately 15 mg/kg. Clarke concluded that the linear velocity of the soil must be sufficiently small to achieve the desired removal efficiency.

Another critical variable in the counter-current process is the surfactant solution flowrate. Table 12 [16] shows that a 25% decrease in flowrate below 10 L/s will increase the effluent soil contaminant concentration by five orders of magnitude and a decrease to 5 L/s will result in only approximately 50% removal of the contaminant. Therefore, Clarke concluded that this parameter should be closely monitored to ensure removal efficiencies.

Table 10
Effect of Soil Particle Diameter on Effluent Soil Contaminant Concentration

a (cm)	Effluent soil contaminant concentration (mg/kg)
0.10	3.83×10^{-3}
0.15	6.61
0.20	654
0.25	2980

Table 11
Effect of Linear Velocity of Soil through the Column on Effluent Soil Contaminant Concentration

v_1 (cm/s)	Effluent soil contaminant concentration (mg/kg)
1	3.83×10^{-3}
2	4.19×10^{-2}
4	0.796
8	14.7
16	169
24	519

Table 12
Effect of Surfactant Solution Flow Rate on Effluent Soil Contaminant Concentration

Q_W (L/s)	Effluent soil contaminant concentration (mg/kg)
5	5370
7.5	97.4
10	3.83×10^{-1}

6.2.3.6. Application of Surfactant Extraction Model

This model may be used as a guideline in the design of the surfactant extraction process.

6.2.3.7. Limitations of Surfactant Extraction Model

Because this model is yet to be verified by field and bench-scale analysis, it should only be used as a preliminary guideline to the design of the process.

6.3 Design Protocol

6.3.1. Introduction

The most important factors that impact the design of a chemical extraction or soil washing process are:

- Type of contaminant

- Contaminant solubility
- Kinetics
- Solvent
- Heterogeneities in the site (if *in situ*) - site hydrogeology
- Soil gradation
- Contact time

Optimum efficiency in soil washing and chemical extraction is dependent on:

- Contaminant Solubility
- Mass Transfer and Chemical Kinetics
- Flow path of solvent vs. solute (if *in situ*)

6.3.2. Design Parameters

Initial parameters that must be determined prior to design are:

- Rate at which soil is to be treated
- Effluent concentrations
- Influent contaminant concentrations
- Sorption Isotherm (mass transfer)
- Kinetics (chemical)
- Washing fluid to solids ratio
- McCabe-Theile operating line
- Water knife pressure and water volume
- Volume of soil to be treated
- Soil gradation

6.3.3. Design Protocol

After the design parameters have been determined, the following steps should be taken to ensure a proper design.

Step 1
Using the site information (i.e site geology, hydrogeology, etc.), calculate contaminant concentrations, analyze the data, determine if *in situ* or *ex situ* treatment is appropriate, and identify and quantify soil fractions containing contaminant.

Step 2
Determine if the contaminant is water soluble, chelation susceptible or soluble in a surfactant (via handbooks, literature review or bench scale analysis). This will determine the type of extraction fluid necessary and, with consideration of the contaminant, what wastewater treatment is appropriate.

EX SITU
Step 3

With the basic operations schematic, perform a mass balance calculation on the system. Use the required wash fluid to soil ratio and volumetric flow rate of the soil, and the estimated removal rates. If a surfactant extraction is being performed, simulations using the Clarke model may be used to estimate removal rates and volumetric flow rates. For other types of extraction, a literature review and/or bench-scale analysis is required to determine system parameters.

The mass balance equation will determine the volume of wastewater to treated, the volume of extraction fluid that must be added to the system to ensure proper soil/extraction fluid ratios, and the volume of sludge to be treated and/or disposed.

Step 4
Using the rate of flow of the solids and solvent, the required effluent concentration (determined by permits), the McCabe-Theile operating line (determined experimentally (see [6]), construct a McCabe -Theile diagram (see [6] for complete instructions) and determine the number of stages needed in the chemical extraction unit. Check stage efficiency.

Step 5
Determine the volume of each unit operation. This is done by using the flow rate of the solids and solvent, the mass transfer coefficient, and the chemical kinetics constant.

The mass transfer coefficient either may be determined by bench-scale analysis or by correlation formulas (see [19] for detailed discussion). The chemical kinetics must be determined by bench-scale analysis. Compare each constant to determine the rate-limiting process. This will be the minimum residence time of the soil needed for each unit operation.

The total volume per unit operation is obtained by multiplying the flow rate by the residence time per unit operation. The general equation is:

$$\begin{matrix}\textbf{Volume per Unit}\\ \textbf{Operation}\end{matrix} = \left(\begin{matrix}\textbf{Soil and Extraction Fluid}\\ \textbf{Flow Rate}\end{matrix}\right)\left(\begin{matrix}\textbf{Residence}\\ \textbf{Time}\end{matrix}\right)$$

To determine the volume per stage in the chemical extraction unit operation, the general equation is:

$$\textbf{Volume per Stage} = \left(\frac{\begin{matrix}\textbf{Total Volume}\\ \textbf{Needed}\end{matrix}}{\begin{matrix}\textbf{Number of}\\ \textbf{Stages}\end{matrix}}\right)$$

Step 6
Design the wastewater treatment system per conventional standards.

Step 7
With the modified design, build a pilot-scale operation and test on site. With the new data, modify the design, if needed, and test again on site. Continue until reasonable scale-up similitude is achieved. Build full-scale operation.

IN SITU
Step 8
Employ coupled transport/chemistry groundwater models.

7. COST ANALYSIS AND COMPARISON

Chemical extraction and soil washing costs are most dependent on the post-treatment process. Estimates range from $150/ton to $350/ton depending on soil volume to be treated. Costs include pre-treatment costs, post-treatment costs, residual disposal, quality assurance and quality control (QA/QC), sampling and analysis, and health and safety costs.

A comparison of chemical extraction and soil washing is presented in Table 13. From this table it can be shown that CHEX and SW are comparable to immobilization for the treatment of metals. In some situations, CHEX and SW will be more advantageous because they allow the site to be reused.

TREATABILITY GROUP \ TECHNOLOGY	THERMAL DESTRUCTION	DECHLORINATION	BIOREMEDIATION [4]	LOW TEMPERATURE DESORPTION	CHEMICAL EXTRACTION AND SOIL WASHING	IMMOBILIZATION [4]
NON-POLAR HALOGENATED AROMATICS	●	◐	◐ [3]	● ◐	◐	◐
PCBs, HALOGENATED DIOXINS, FURANS AND THEIR PRECURSORS	●	◐	◐ [3]	◐ [1]	◐	◐ [1]
HALOGENATED PHENOLS, CRESOLS, AMINES, THCLs, AND OTHER POLAR AROMATICS	● [3]	◐	◐	◐	◐	◐ [3]
HALOGENATED ALIPHATIC COMPOUNDS	●	◐ [2]	◐ [2]	●	◐	◐ [2]
HALOGENATED CYCLIC ALIPHATICS, ETHERS, ESTERS, AND KETONES	●	◐ [1]	◐ [1]	○ [1]	◐ [1]	◐ [1]
NITRATED COMPOUNDS	●	○ [1]	◐	○ [1]	◐	◐ [1]
HETEROCYCLICS AND SAMPLE NON-HALOGENATED AROMATICS	●	○ [2]	◐ [2]	●	◐	◐ [2]
POLYNUCLEAR AROMATICS	●	○ [2]	◐	○	◐	◐
OTHER POLAR NON-HALOGENATED ORGANIC COMPOUNDS	●	○ [2]	◐ [2]	◐	◐	◐ [2]
NON-VOLATILE METALS	○ [1]	○ [1]	○ X [1]	○ [1]	◐	● [3]
VOLATILE METALS	X [1]	○ [1]	○ X [1]	○ [1]	◐	●

● DEMONSTRATED EFFECTIVENES

◐ POTENTIAL EFFECTIVENESS

○ NO EXPECTED EFFECTIVENESS (NO EXPECTED INTERFERENCE WITH PROCESS)

X NO EXPECTED EFFECTIVENESS (POTENTIAL ADVERSE EFFECTS TO ENVIRONMENT OR PROCESS)

1 DATA WERE NOT AVAILABLE FOR THIS TREATABILITY GROUP. CONCLUSIONS ARE DRAWN FROM DATA FOR COMPOUNDS WITH SIMILAR PHYSICAL AND CHEMICAL CHARACTERISTICS

2 HIGH REMOVAL EFFICENCIES IMPLIED BY THE DATA MAY BE DUE TO VOLATAZATION OR SOIL WASHING

3 THE PREDICTED EFFECTIVENESS MAY BE DIFFERENT THEN THE DATA IMPLY, DUE TO LIMITATIONS IN THE TESTS CONDITIONS

4 THESE TECHNOLOGIES MAY HAVE LIMITED APPLICABILITY TO HIGH LEVELS OF ORGANICS

TABLE 13
Predicted Treatment Effectiveness for Contaminated Soil
Chemical Extraction/Soil Washing

REFERENCES

1. Masters, Hugh. Personal Communication 1992.

2. Hsieh, H. and D. Raghu. "Site Cleanup with Soil Washing Technology." In: *U.S. EPA Review*, Ed. . U.S. Environmental Protection Agency, Cincinnati, Ohio, 1990.

3. Myers, D. *Surfactant Science and Technology*. Ed. Vol. VCH Publishers, New York, New York, 1988.

4. Everson, F. *Overview: Soils Washing Technologies, for Comprehensive and Liability Act Resource Conservation and Recovery Act Leaking Underground Storage Tanks, Site Remediation*. U.S. Environmental Protection Agency, Cincinnati, Ohio, June, 1989.

5. Hiezmenz, P. *Principals of Colloid and Surface Chemistry*. 2 ed. Ed. Vol. Marcel Dekker Inc., 1986.

6. McCabe, Smith and Harriot. *Unit Operations of Chemical Engineering*. 4 ed. Ed. Vol. McGraw-Hill, New York, New York, 1985.

7. Palmer, C. D. and W. Fish. Chemical *Enhancements to Pump-and-Treat Remediation*. USEPA/540/2-91/010. U.S. EPA, January, 1992.

8. Freeze, R. A. and J. A. Cherry. *Groundwater*. Ed. Vol. Prentice Hall Inc., Englewood Cliffs, New Jersey, 1979.

9. Weber, W. J., P. M. McGinley and L. E. Katz. "Sorption Processes and Their Effects on Contaminant Fate and Transport in Subsurface Systems." In: *Water Research or 1990 Distinguished Lecture Series*, Ed. . 1990.

10. Bell, C. F. *Principles and Applications of Metal Chelation*. Ed. Vol. Clarendon Press, Oxford, 1977.

11. Perry, R. H. and C. H. Chilton. *Chemical Engineering Handbook*. Ed. Vol. McGraw Hill Company, New York, 1973.

12. Scholz, R. and J. Milanowski. *Mobile System for Extracting Spilled Hazardous Materials from Excavated Soils*. Inc., R. EPA/600/2-83/100. U.S. Environmental Protection Agency, Cincinnati, Ohio, December, 1982.

13. U.S. EPA. *Survey of Materials-Handeling Technologies Used at Hazardous Waste Sites*. USEPA/540/2-91/010. US EPA, June, 1991.

14. Schultz, P. and J. Milanowski. *Mobile System for Extracting Spilled Hazardous Materials from Excavated Soils*. Contract no. 68-2696. U.S. Environmental Protection Agency, 1982.

15. Cederberg, G. and et al. "A Groundwater Mass Transport and Equilibrium Chemistry Model for Multicomponent Systems." *Water Resources Research*, **21**: 1095

16. Clarke, A. N., et al. "Soil Clean-up by Surfactant Washing. I. Laboratory Results and Mathematical Modeling." *Separation Science and Technology*, **26**: 310, 1991.

17 Wilson, D. J. "Soil Clean-up by *In Situ* Surfactant Flushing, 1-Mathematical Modeling", *Separation Science and Technology*, **24**: 863-892, 1989.

18. Weber, W., Y.-P. Chin and C. P. Rice. "Determination of Partition Coefficients and Aqueous Solubilities by Reverse phase Chromatography-I." *Water Resources*, **20**: 1433, 1986.

19. Cussler, E. L. "Diffusion." In: *Mass Transfer in Fluid Systems*, Ed. . Cambridge University Press, 1984.

H - Solidification / Stabilization

SOLIDIFICATION/ STABILIZATION

Instructional objectives:

- Describe fundamental principles of S/S processes
- Evaluate applicability of S/S processes
- Discuss leaching as applied to S/S
- Identify interference compounds
- Explain the design procedure

1. GENERAL DESCRIPTION

1.1. Introduction

Originally developed in the 1950's, solidification/stabilization (S/S) has more recently been used to treat industrial, hazardous, and selected radioactive wastes. The overall objective of S/S is to contain contaminants and minimize their release to the environment [1, 2, 3, 4, 5, 6, 7, 8, 9, 10, 11, 12, 13, 14].

More specifically, S/S can [1, 14]:

- Improve handling and physical characteristics of waste
- Reduce loss of contaminant by decreasing surface area
- Reduce contaminant solubility
- Further treat residues from physicochemical, biological, or thermal technologies
- Be conducted on-site or *in situ*

S/S has been selected as BDAT for several categories of waste (see Table 1.1 [5]), primarily consisting of metal finishing waste and ash residues [1].

EPA has defined stabilization and solidification as follows [9]:

<u>Stabilization</u> refers to those techniques that reduce the hazard potential of a waste by converting the contaminants into their least soluble form. The physical nature and handling characteristics of the waste are not necessarily changed by stabilization.

<u>Solidification</u> refers to techniques that encapsulate the waste in a monolithic solid of high structural integrity. The encapsulation may be of fine waste particles (microencapsulation) or of a large block or container of wastes (macroencapsulation). Solidification does not necessarily involve a chemical interaction between the wastes and the solidifying reagents, but may mechanically bind the waste into the monolith. Contaminant migration is restricted by vastly decreasing the surface area exposed to leaching and/or by isolating the wastes within an impervious capsule.

Table 1.1: Example of U.S. EPA RCRA Hazardous Wastes for which S/S is Being Evaluated as a Treatment Technology

Waste code	Description of waste	Pollutant of concern for S/S
K048-52	Dissolved air flotation (DAF) float from the petroleum refining industry	chromium(+3), lead
K061	Emission control dust/sludge from the primary production of steel in electric furnaces	chromium(+3), lead, cadmium
K046	Wastewater treatment sludges from manufacturing formulation and loading of lead-based initiating compounds	lead
F006	Metal finishing sludges	cadmium, chromium(+3), lead, nickel, silver
F012, F019	Metal finishing sludges	cadmium, chromium(+3), lead, nickel, silver
K022	Distillation tar (treated)	chromium(+3), nickel
K001	Wood preserving sludges (treated)	lead

In the solidification process, waste is combined with water and other reagents (binders) to form a solid that will be structurally stable under expected conditions at disposal sites. Load bearing strengths are typically greater than 50 psi. A variety of material strengths can be achieved through solidification processes. Resulting contaminant mobility and structural properties will ultimately dictate use. Possible products resulting from solidification processes include construction grade materials (this may be limited under "derived from" rule), landfill covers, backfill for landfill gas collection systems, and structural fill. S/S materials are often also disposed of in landfill environments. The term fixation or chemical fixation is often substituted for stabilization.

S/S processes, in addition to solidifying, produce nonhazardous or less hazardous wastes (as defined by leaching tests) by physicochemical transformations (e.g. encapsulation, sorption, precipitation) [9]

A typical S/S flow schematic is presented in Figure 1.1.

1.2. Compounds amenable to S/S

S/S processes are primarily applicable to the treatment of inorganics. However, certain proprietary processes claim the ability to treat organic compounds.

Inorganic compounds amenable to S/S processes include, but are not limited to:

<u>Metals</u>

Aluminum	Antimony	Arsenic
Barium	Beryllium	Cadmium
Chromium*	Cobalt	Copper
Lead	Mercury	Tin

<u>Other inorganics</u>

Fluorides	Cyanides	Sulfides

* May need chemical reduction

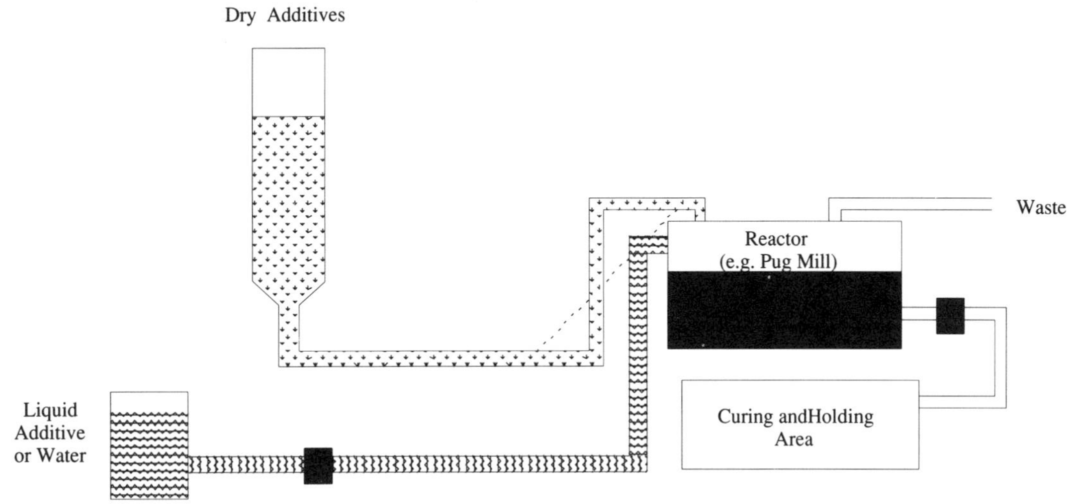

FIGURE 1.1
Typical S/S Flow Schematic
Solidification /Stabilization

2. FUNDAMENTALS

2.1. Introduction

S/S processes, unlike many technologies, do not destroy contaminants. The objective of this technology is to inhibit transport of contaminants through reactions designed to limit mobility. Theoretical prediction of mobility subsequent to S/S treatment is, at present, somewhat limited. Laboratory leaching and physical tests are employed to quantify compound mobility and hence S/S efficacy. Although presently employed tests (i.e. TCLP) do not emulate real world conditions, they are designed to simulate worst-case scenarios [9]. Should S/S waste be utilized in subsequent construction application where exposed to the ambient environment, attention should be given to the ultimate end of the life-cycle—disposal or reuse.

2.1. Leaching Behavior

The chemical testing of a waste is a function of both intrinsic properties of the waste form and hydraulic and geochemical properties of the site. Leaching behavior directly determines the performance of solidification/stabilization as a hazardous waste treatment for a particular process and waste. Controlled laboratory leaching tests may not be an accurate representation of actual field conditions.

Waste and leaching solution composition define the thermodynamics and kinetics of the chemical reactions that mobilize or immobilize contaminants in the solidified/ stabilized waste. When contaminants come in contact with a reagent solution, surface renewal rates, contact surface area, and volume affect contaminant concentration in the leaching media.

If solidified/ stabilized wastes have low permeabilities, contaminant transport is governed by molecular diffusion. Diffusion controlled transport is controlled by:

- Accumulation of waste species in the pore solution at the particle surface
- Concentration of reactive species (e.g. H^+ complexing agent) in the pore solution at the particle surface
- Contaminant free liquid diffusivity
- Polarity and ionic strength of the leaching solution and waste species
- Oxidation/ reduction conditions and competing reaction kinetics
- Porosity and tortuosity

See Figure 2.1 a-b.

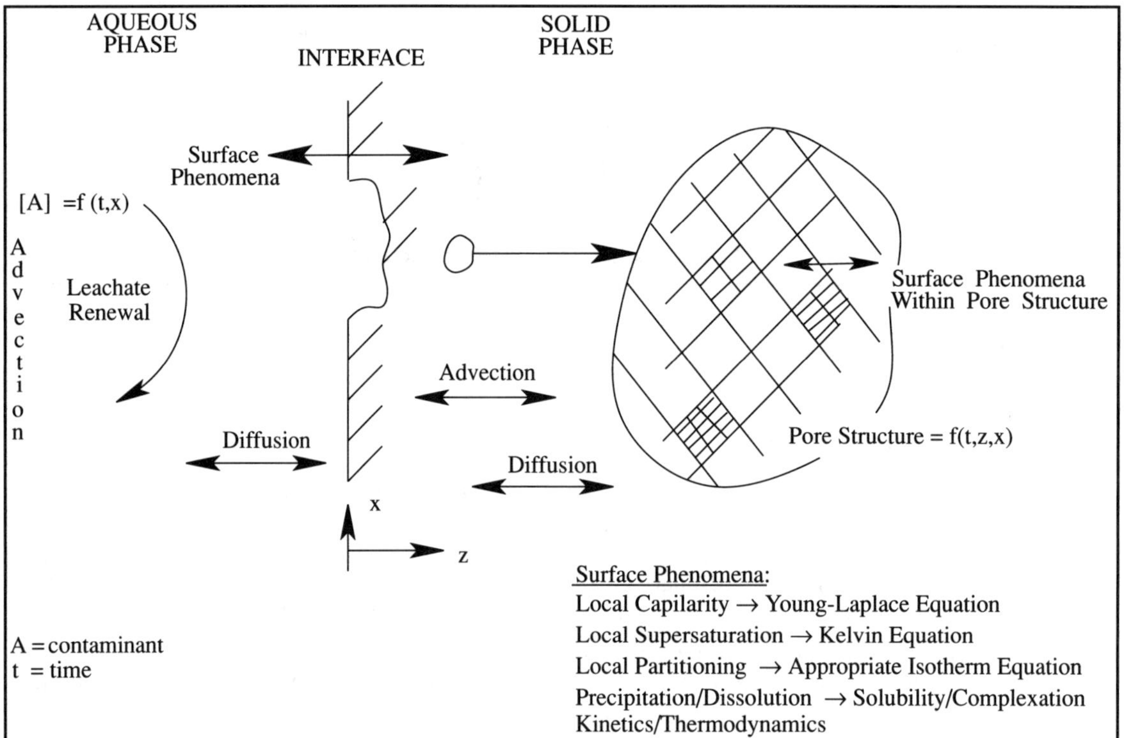

FIGURE 2.1a
Factors Influencing Leaching Behavior
Solidification/Stabilization

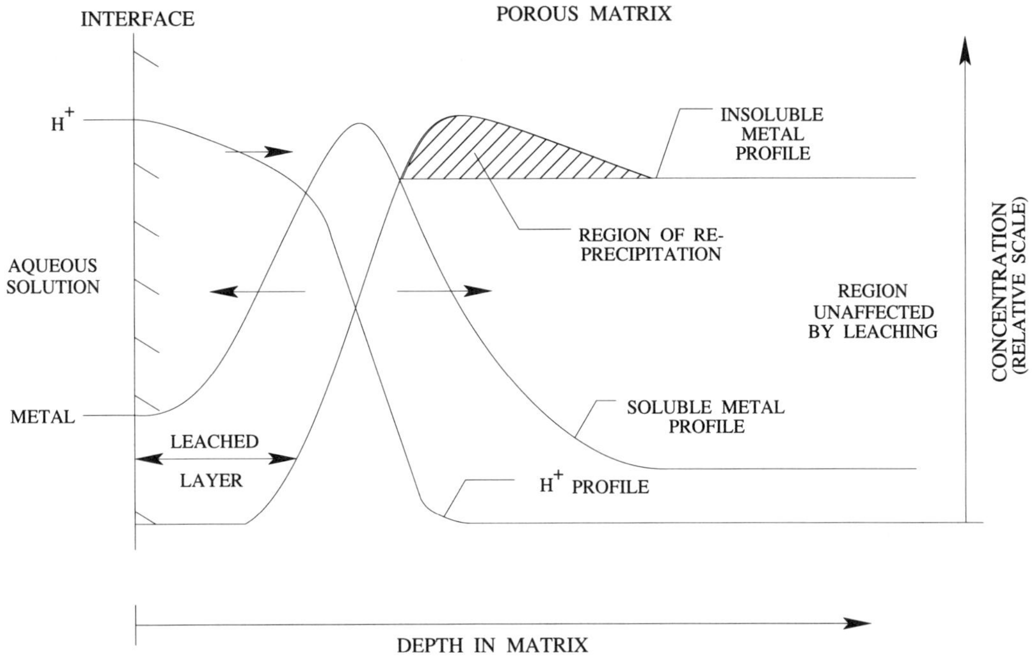

FIGURE 2.1 b
Concentration Gradients During Leaching
Solidification/Stabilization

2.2. Immobilization Mechanisms

There are a variety of operative mechanisms in S/S processes that can be exploited:

- Chemical Reaction
  - — Precipitation
  - — Redox
  - — Complexation
  - — Passivation
- Sorption
  - — Adsorption
  - — Ion exchange
  - — Diodochy
- Encapsulation

In the case of heavy metals, ligands present in the waste mixture may interfere with immobilization mechanisms through chelation reactions. Pre-treatment with strong oxidizing or reducing agents or appropriate reagents may be required.

2.2.1. Chemical Reaction

Chemical reactions can be subdivided into precipitation, complexation, and oxidation/ reduction. Redox reactions can be used in conjunction with precipitation reactions.

2.2.1.1. Precipitation

pH control is the most common and simplest type of stabilization (via precipitation of metals).

Most common alkali used in S/S processes are:

- CaO
- $Ca(OH)_2$
- Na_2CO_3
- $Mg(OH)_2$

Some solidification reagents are alkaline, such as:

- Portland cement
- Fly ash
- Lime kiln dust
- Sodium silicate

Classical precipitation may not be the only operative process. Bishop reported that after alkali had been leached from cement-based forms, lead and chrome leaching was lower than expected (metals probably bound to silica matrix).

Hydroxide precipitation - described in detail in Stumm and Morgan [15]

Note: pH dependent; Typically optimum pH exists

Carbonate precipitation - Again, see Stumm and Morgan [15].

Sulfide precipitation - typically less soluble species (than hydroxides). Three classes of sulfide precipitation reagents. See Tables 2.1 and 2.2.

Table 2.1: Sulfide Precipitation Reagents

Soluble inorganic sulfides	
Sodium sulfide	Na_2S
Hydrogen sulfide	H_2S
Sodium hydrosulfide (sodium sulfhydrate)	NaHS
Calcium sulfide (low solubility)	CaS
"Insoluble" inorganic sulfides	
Ferrous sulfide	FeS
Sulfur	S
Organosulfur compounds	
Dithiocarbamates	[R-NH-CS-S]
Thiorea	H_2N-CS-NH_2
Thioamides	R-CS-NH_2
Xanthates	[RO-CS-S]

Table 2.2: Comparison of Hydroxide and Sulfide Solubilities

Metal	Approximate Solubility (mg/l)		
	Hydroxide	Sulfide	Difference Factor
Iron	5×10^{-4}	1×10^{-4}	5×10^{3}
Cadmium	3×10^{0}	1×10^{-8}	3×10^{8}
Chromium	1×10^{-3}	None	None
Copper	2×10^{-2}	2×10^{-13}	1×10^{11}
Lead	2×10	6×10^{-9}	3×10^{8}
Mercury	6×10^{-4}	1×10^{-21}	6×10^{17}
Nickel	7×10^{-1}	6×10^{-8}	1×10^{7}
Silver	2×10	4×10^{-12}	5×10^{12}
Zinc	3×10^{2}	1×10^{-6}	3×10^{8}

$$\text{Difference factor} = \frac{\text{hydroxide solubility}}{\text{sufide solubility}}$$

All applications for organosulfur compounds have been with mercury. Figures 2.2 and 2.3 illustrate the pH dependence of both hydroxide and sulfide precipitation, respectively.

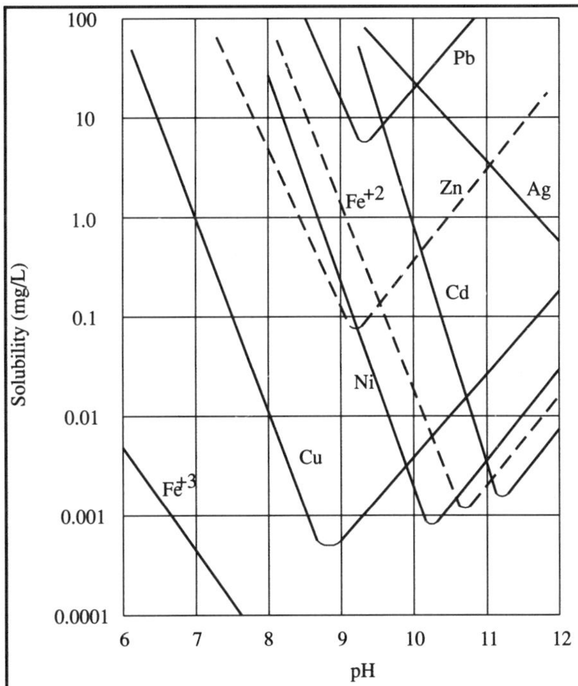

FIGURE 2.2
Solubilities of Metal Hydroxides as a Function of pH
Solidification/Stabilization

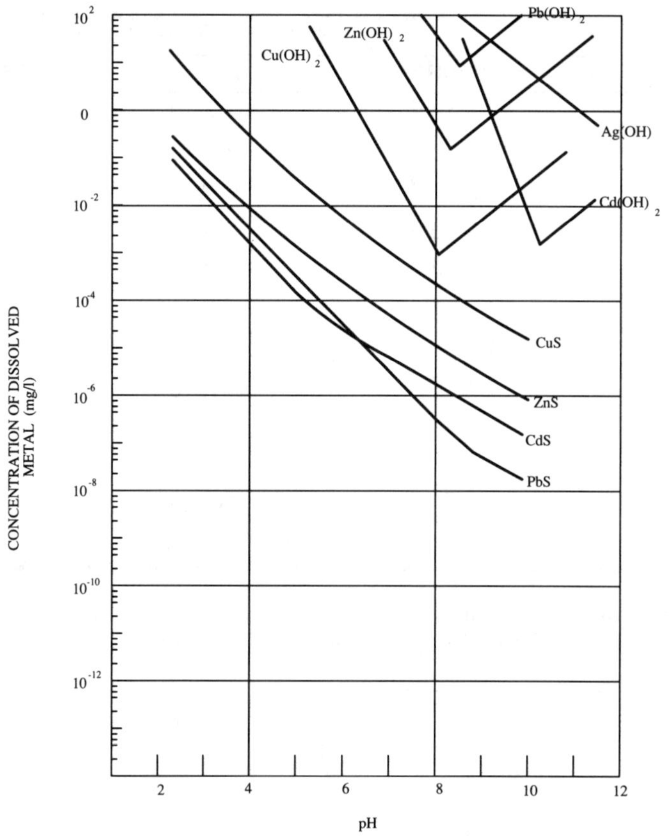

FIGURE 2.3
Solubilities of Metal Hydroxides and Sulfides
Solidification/Stabilization

Silicate Precipitation

Metal silicates are nonstoichiometric compounds in which a metal is coordinated with silanol SiOH groups in an amorphous silica matrix.

Insoluble silicates can dissolve under both alkaline or acidic conditions. Calcium silicates dissolve in alkaline environments created when cement is mixed with water, releasing $Ca(OH)_2$. Alkaline conditions do not result in the expected metal leaching behavior, possibly due to reprecipitation as silicates.

A variety of ligands may exist in complex waste mixtures that induce solubility behavior, which is inconsistent with that governed strictly by hydroxide solubility. See Figures 2.4 through 2.6 (adapted from reference [9]).

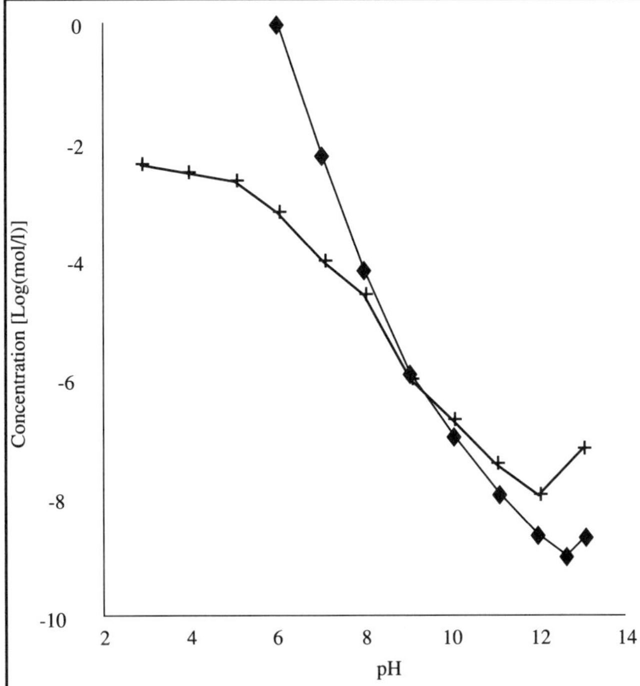

FIGURE 2.4
Solubility of Cadmium vs. pH in a Cement/Flyash S/S System
Solidification/Stabilization

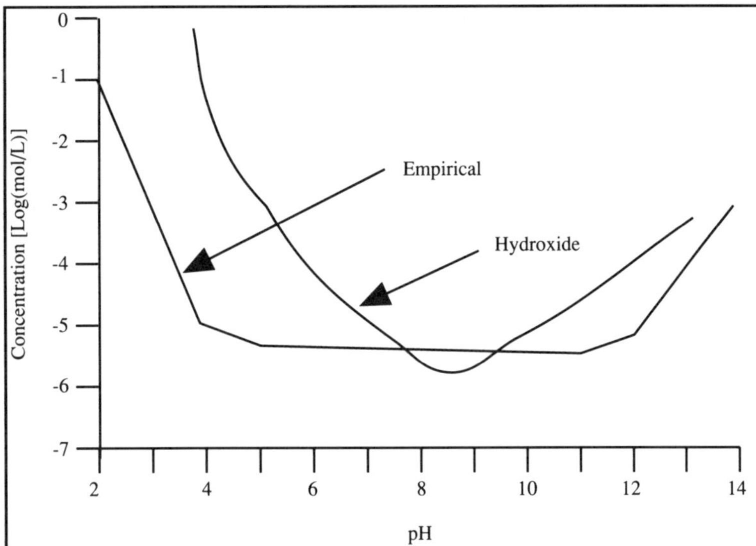

FIGURE 2.5
Solubility of Chromium vs. pH in a Cement/Flyash S/S System
Solidification/Stabilization

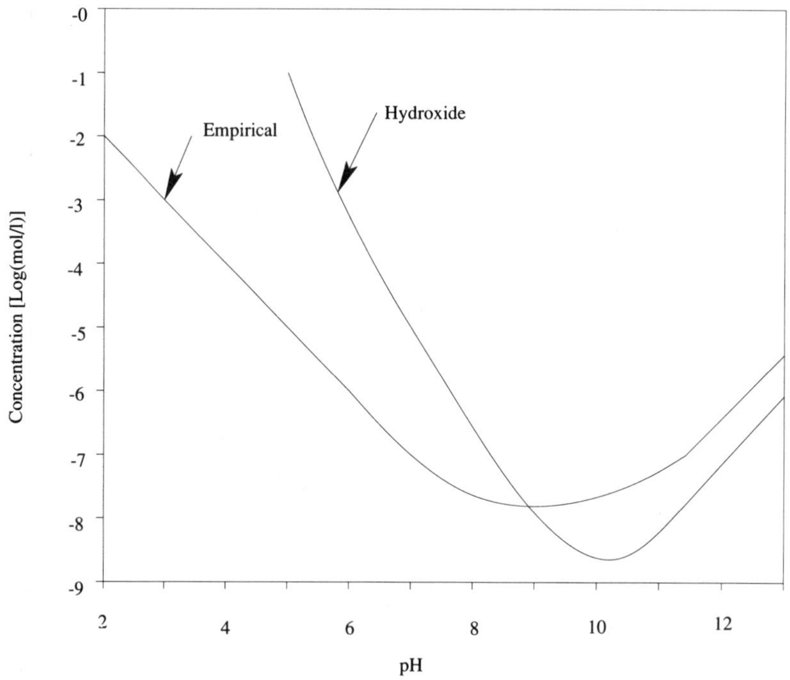

FIGURE 2.6
Solubility of Lead vs. pH in a Cement/Flyash S/S System
Solidification/Stabilization

Co-Precipitation

Metals can be co-precipitated or coagulated with iron salts. It has been proposed that ferrite crystals form, capturing metals in their lattice structure. Fe^{2+} and Fe^{3+} ratios of 1:1 to 1:2 have been reported to yield optimum results [9].

Passivation - metal ions dissolving from a solid surface may reprecipitate on the surface after contacting an anion in solution, forming a less soluble species.

2.2.1.2. Redox potential control

Redox reactions may be employed to change oxidation state of metals, rendering them less soluble. Common reducing agents are:

- $FeSO_4$ (primarily for Cr+6 bearing waste)
- $NaHSO_3$
- SO_2
- $NaBH_4$ (sodium borohydride)
- $Na_2S_2O_5$
- $Na_2S_2O_4$
- FeS
- N_2H_4 (hydrazine)

Oxidation is significantly less common in S/S processes.

2.2.1.3. Complexation

Although complexation can interfere and inhibit immobilization, certain complexes can prove to be insoluble. For instance, cadmium-tartrate complex is insoluble in water.

2.2.2. Sorption

Unlike the use of sorption processes in the water and wastewater applications, sorption processes in S/S operations are typically allowed to achieve equilibrium. Sorbent categories can be classified as follows:

- Metal oxides - iron, manganese, aluminum, etc.
- Clays - bentonite, illite, kaolinite, etc.
- Natural materials - zeolites, vermiculite, peat moss, etc.
- Synthetic materials - zeolites, fly ash, cullite, activated alumina, activated carbon, etc.

Sorption processes in S/S treatment can include [9]:

Ion exchange - replacement of adsorbed ions by similarly charged ions

Ion pairing - adsorption of surfactant ions from solution onto oppositely charges sites unoccupied by counter ions

Hydrogen bonding - hydrogen bond formation between substrate and adsorbate

Adsorption by pi-electron polarization - association of electron-rich aromatic substrate nuclei and adsorbent positive sites

Dispersion forces - induced dipole-dipole quadrapole-quadrapole interactions

Hydrophobic bonding - nonpolar and amphiphatic molecules seek nonpolar environments

Diodochy - substitution of one element for another of similar size and charge in a crystal lattice

It is likely that all solidification methods for metals involve ion exchange activity. Some of the more common ion exchange materials investigated for use with S/S processes include:

- Organic resins
- Natural and synthetic zeolites
- Soils/ clays
- Metal oxides

Metal oxides, clays, and activated carbon are the most common types of adsorbents used in S/S.

Clays - a significant amount of work has conducted done on the interactions of clays with metals and is presented in soil science literature

Activated carbon/ active oxides/ ion exchange - although a significant amount of work has been reported regarding these processes in the water and wastewater fields, this work may not be directly applicable to S/S processes.

Modified clays - quaternary ammonium compounds $[R_4N]^+$ are substituted for group IA and IIA metals (Li^+, Na^+, K^+, Mg^{2+}, Ca^{2+}, Ba^{2+}) present between silica and alumina layers in clay. This increases basal spacing and provides an organic (nonpolar) phase for stabilization of organic compounds [11]. See Figure 2.7.

2.2.3. Encapsulation-Isolation from Environment [9]

Microencapsulation - most S/S wastes are microencapsulated. Sizes range from small occlusions to agglomeration visible under low powered microscopes. Particles retain identity even after mixing. Mixture becomes homogeneous with time.

Macroencapsulation - solid form coated with or contained in an impervious layer to limit access of leachant

Embedment - incorporation of discrete waste masses of discernable size into a solid matrix.

Examples include:

- Hospital waste (e.g. syringe)
- Hazardous waste personal protective equipment

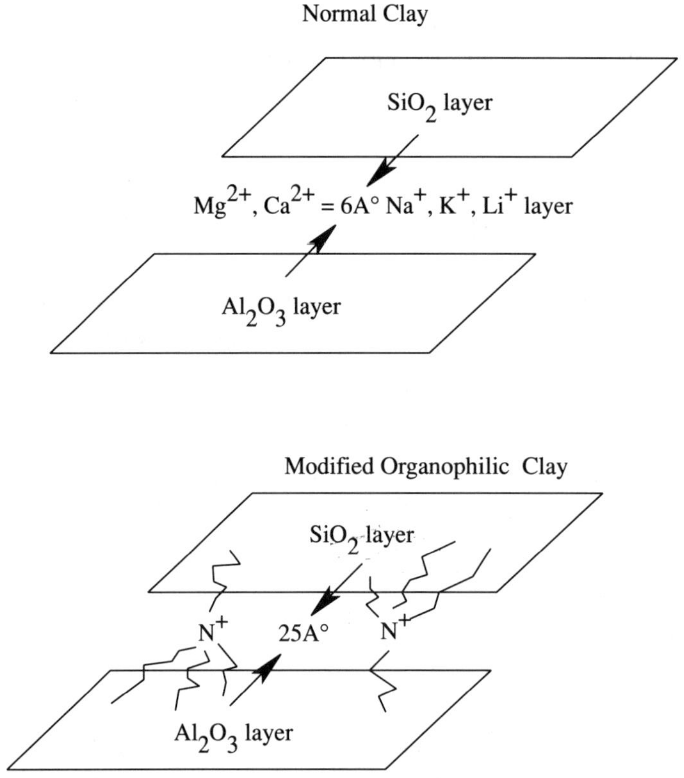

FIGURE 2.7
Modifying a Clay With a Quatermary Ammonium Compound
Solidification/Stabilization

2.3. Transport through Solid

(Abstracted from [9])
For the contaminant to be transported through the solid phase, it must first dissolve. Once dissolved (refer to section 2.1.), it can then be transported via molecular diffusion to the solid/liquid interface. Molecular diffusion is described by Fick's 1st and 2nd laws.

A proposed leaching model for S/S treated wastes is based on work conducted by the American Nuclear Society [12]. Effective diffusivity can be calculated as:

$$D = \pi \left[\frac{a_n / A_o}{(\Delta t)_n} \right]^2 \left[\frac{V}{S} \right]^2 T$$

where:

a_n	= contaminant loss during leaching period n, mg
A_o	= initial amount of a contaminant present in the specimen, mg
V	= volume of the specimen, cm^3
S	= surface area of the specimen, cm^2
t_n	= time to the end of the leaching period n, s
D	= effective diffusion coefficient, cm^2/s

$$T = \left[\frac{1}{2} \left(t_n^{\frac{1}{2}} - t_{n-1}^{\frac{1}{2}} \right) \right]^2$$ = Leaching time representing mean time of leaching interval, S.

$$\Delta t_n = t_n - t_{n-1}$$ = duration of nth leaching interval

Graphs to evaluate diffusivity for cylindrical structures are available [12].

Assumptions for this model are that the contaminant concentration in the initial solid is uniform, leaching is controlled by diffusion through the solid waste form (but not solid-state diffusion as the term is used here), and there is always zero surface concentration at the solid-liquid interface. The latter assumption may not be justified in real leaching situations.

Combining the above equation with the American Nuclear Society's (ANS) leaching protocol, which is a series of ten sequential batch leachings, are presented as the leachability index, L_i:

$$L_i = \frac{1}{10} \sum_{1}^{10} \left[\log \left(\frac{\beta}{D_i} \right) \right]_n$$

where β is a defined constant (1.0 cm^2/s).

The following can be derived,

$$\frac{\Sigma a_n}{A_o} = (1.128)\left(10^{-\frac{L_i}{2}}\right)\left(\Delta t_n^{\frac{1}{2}}\right)\left(\frac{S}{V}\right)$$

This equation is valid for $\frac{\Sigma a_n}{A_o} < 0.2$ otherwise see [12].

Most of the variables that effect metal leaching from S/S-treated waste can be accounted for by this equation. These include constituent speciation, particle size, and the initial concentration of the constituent. However, boundary layer effects and chemical reactions occurring during the leaching process can significantly affect the practical utility of such mathematical models.

A mathematical model such as the one proposed above can be used to determine leaching rates that, when coupled with data about the disposal site and groundwater, can be used to predict *in situ* leaching rates over extended time periods [9]. A number of other models have been developed and used for such predictions, including the EPA VHS (vertical and horizontal spread) model.

Diffusion through the leached layer. As more soluble constituents are leached from a relatively insoluble solid matrix, a layer deficient in the leached constituents develops [9]. Under low pH conditions, both H^+ and the leachable constituents must diffuse through this layer in opposite directions. The leaching rate in the leached layer should eventually be limited by diffusion of constituents since H^+ diffuses much faster than other species. However, this layer may not be rate-limiting in the overall leaching process. As constituents leach, the layer may become more porous compared to the unleached solid so that molecular diffusion in the pore water and boundary layer phenomena become the limiting factors.

Transport through the solid-liquid boundary layer. In many leaching scenarios, such as the electrochemical phenomena, the boundary layer at the solid-liquid interface may create the limiting condition for leaching. The speed with which a system comes to equilibrium is frequently limited by the thickness and nature of the boundary layer. In attempting to develop fast leaching assessment tests for routine quality control, for example, this limitation is thought to be controlling. The solid-liquid boundary can be described by double layer theory. Surface chemistry principles are important in the leaching of metals because of the electrostatic interactions that arise from their transport.

Figure 2.8 [1] illustrates an impact of solution velocity on the leaching rate of contaminants at the particle surface. Surface area renewal rate (v) is defined as:

$$v = \frac{V}{(SA)\,T}$$

where:

- V = volume of leaching solution contacted
- SA = surface area
- T = time

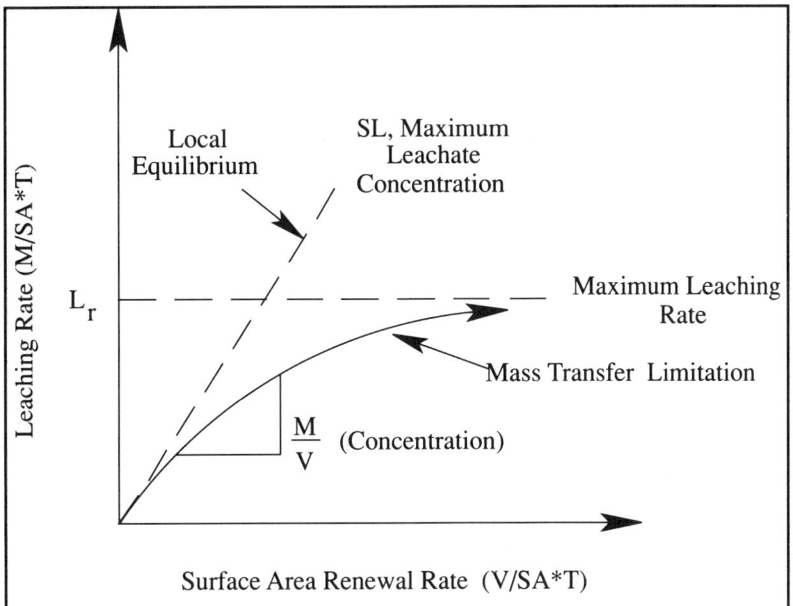

FIGURE 2.8
Relationship Between Velocity of Leaching Solution and the Leaching Rate
Solidification/Stabilization

The leaching rate (L) is defined as:

$$L = \frac{M}{(SA)\,T}$$

where:

M = mass of waste species

The slope of the curve, $\frac{dL}{dv}$, is the concentration of contaminant in the leaching solution $\left(\frac{M}{V}\right)$. As the velocity increases, the concentration decreases, approaching zero at a limiting leach rate, Lr. At low leach rates, local equilibrium is realized and the concentration approaches the saturation limit, SL. Clearly, the ANS model is a valid representation given local equilibrium. Should mass transfer limitations exist, local equilibrium may be somewhat over-predictive (conservative).

3. APPLICATIONS

3.1. General Applicability

S/S can be employed:

- For soils contaminated with heavy metals or other inorganics

- At sites that are easily accessible
- For contaminated soils with low organic content.

3.2. Advantages and Disadvantages

Advantages:

- Can immobilize certain compounds, thus allowing delisting
- Can utilize process by-products (e.g., flyash)
- Results in a stable structurally-sound end product

Disadvantages:

- Does not destroy compounds
- Can deteriorate over time
- Typically requires excavation and materials handling
- Requires curing period
- Contaminants will diffuse slowly (leaching)
- May increase volume

4. PROCESS DESCRIPTION

4.1. Introduction

S/S systems can be broadly grouped under organic and inorganic based on chemicals used. Inorganic systems are by far the most common. Organic systems maybe hydrophobic and may suffer long-term stability problems [9].
In order to capitalize on the immobilization mechanisms discussed in Section 2, S/S processes use binders. Binders cement together waste components into a solid mass. The most common S/S systems are [6]:

- Cement-based
- Pozzolanic processes
- Thermoplastic
- Organic polymer

4.1.1. Organic Systems

The two most common types of organic processes are polymerization and thermoplastic.

4.1.1.1. Polymerization [9]

Polymerization systems can work on organic solvent, OH, or water-based systems. Monomers, polymerized by use of catalysts, are mixed with waste. The most common polymer systems include:

- Urea-formaldehyde

- Polybutadiene
- Polyester-epoxy
- Acrylamid gel
- Polyolefin encapsulation
- Polyurethal

Water may interfere with these processes. Usually there is no direct reaction between waste and polymer. The primary operative mechanism is microencapsulation. Polymers typically have low permeabilities.

4.1.1.2. Thermoplastics [9]

Hot melt systems, primarily bitumen, work with either water-based or oil-based waste. However, water is typically evaporated prior to treatment. The presence of natural tars may induce microbial degradation of the S/S product.

A summary of organic systems is provided below:

- High cost
- Very low permeability
- Potential of using one system with a variety of wastes
- Wide-range volume increase
- Relative instability on the presence of ultraviolet light
- Questionable stability in the presence of microorganisms
- Potential for volatilization of contaminants
- Potential long-term degradation effects on the polymer matrix by certain waste constituents
- Unproven long-term stability
- Very limited commercial use experience except with radioactive wastes and no long-term commercial-scale field-test data
- Systems are more difficult to use and require more technical expertise than do the inorganic systems
- Some system components, especially catalysts, initiators, and accelerators, are hazardous

4.1.2. Inorganic Systems [5, 9, 10, 13, 14]

Inorganic systems can be divided into processes that require bulking agents and those that do not. Bulking agents are added to prevent settling out of suspended matter. Examples of the two groups include:

- Cement-based or cement soluble/ silicate systems (no bulking agents)
- Cement/ flyash, lime/flyash, cement/ clay (bulking agents)

There are two types of bulking agents:

- Inert
- Pozzolanic

  Pozzolan derives from the name "Pozzuolana" which refers to a class of volcanic materials first utilized as hydraulic cementing material in the town of Pozzuli, Italy.

A pozzolan is defined as a material that does not exhibit cementing ability when used by itself, but in combination with other material, such as Portland cement, exhibits cementitious reactions.

A summary of inorganic system characteristics is provided below:

- Relatively low cost
- Good long-term stability, both physically and chemically
- Documented use on a variety of industrial wastes over a period of at least ten years
- Widespread availability of the chemical ingredients
- Ease of use in processing (processed normally operated at ambient temperature and pressure and without unique or very special equipment)
- Wide range of volume increases
- Inertness to ultraviolet radiation
- High resistance to biodegradation
- Low water solubility
- Relatively low water permeability
- Good mechanical and structural characteristics

4.2. General Process

A typical S/S process involves the addition of reagents (dry or liquid) to the waste in a mixing vessel. Mixing typically takes place from 1 to 15 minutes in batch modes, or 1 to 3 minutes in continuous flow modes [9]. However, mixing times of up to 60 minutes also have been reported. Mixing time depends on reactor size and waste and reagent characteristics. Subsequent to mixing, the waste is transferred by either pumps or mechanical conveyors to curing areas. Pilot studies are essential in determining optimum process configuration.

4.3. Factors Affecting S/S Process

4.3.1. Physical

4.3.1.1. Particle Size and Shape

Particle content can impact handling and final product characteristics. Finer size particles have been reported [9] to both increase and decrease the leaching rate. Small particles can release alkalinity quicker, reduce diffusion limitation, and increase area available for sorption/ desorption. Particles are typically friable and can undergo size reduction during processing.

4.3.1.2. Free Water content

Only free water is available for setting reactions. Some bound water may become available during long-term curing [9].

Water may be immobilized in hydrating particle surfaces [1] (see Figures 4.1 a-b).

In a recent study, soil moisture content was reported as 10% prior to S/S and as 1% after S/S [1].

4.3.1.3. Solids Content

Depending on solids concentration:

- Solids may settle out resulting in two phase systems
- Final products may have undesirable strength characteristics

A low solids, low viscosity, high specific gravity waste requires either a fast setting S/S process or the addition of bulking agents [9].

It is virtually impossible to make strong, durable forms with an inorganic S/S system at low solids content.

4.3.1.4. Specific gravity

Large differences between waste and reagent specific gravities may result in phase separation. Most S/S reagents have specific gravity greater than 2.0 versus 1.0 to 1.5 for typical wastes [9]. Phase separation may take place until setting begins. Rheological agents can also be added to alter viscosity. These agents may subsequently interfere with setting reactions.

4.3.1.5. Viscosity

Viscosity may impact phase separation as discussed above as well as the kinetics of S/S processes. A useful rough classification of viscosity categories is given below [1]:

Consistency category	Characteristic Properties
Liquid waste	Less than 1% suspended solids*, pumpable liquid, generally too dilute for sludge dewatering operations
Pumpable waste	Less than 10% suspended solids*, pumpable liquid, generally suitable for sludge dewatering
Flowable waste	Greater than 10% suspended solids*, not pumpable, will flow or release free liquid, will not support heavy equipment, will undergo extensive primary consolidation
Nonflowable waste	Solid characteristics, will not flow or release free liquids, will support heavy equipment, may be 100% saturated, may undergo primary and secondary consolidation

*Suspended solid ranges are approximate

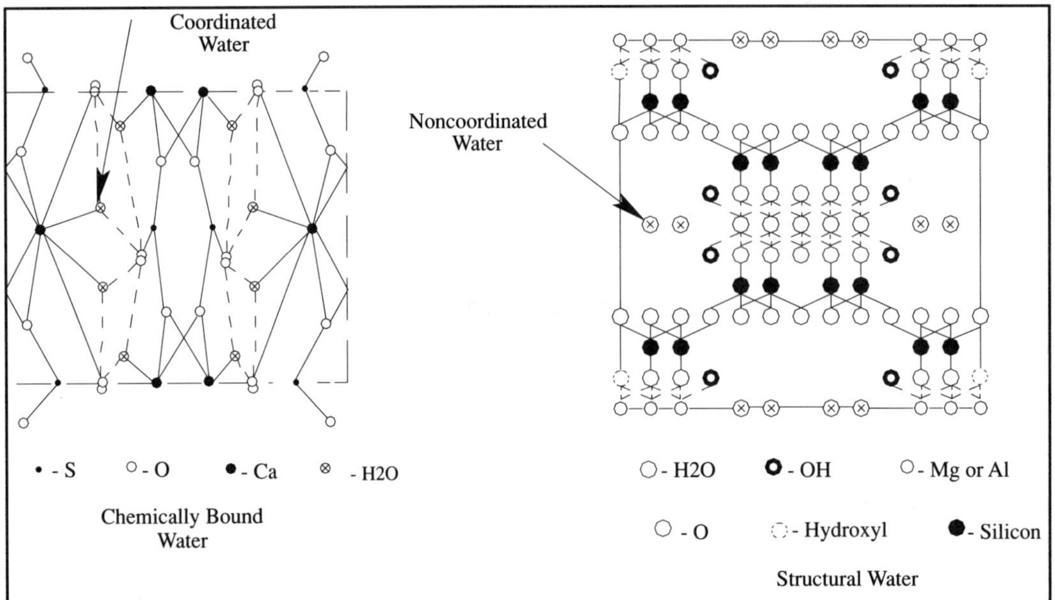

FIGURE 4.1a
Mechanisms Retaining Water and Ionic Materials on and in Solid Phases
Solidification/Stabilization

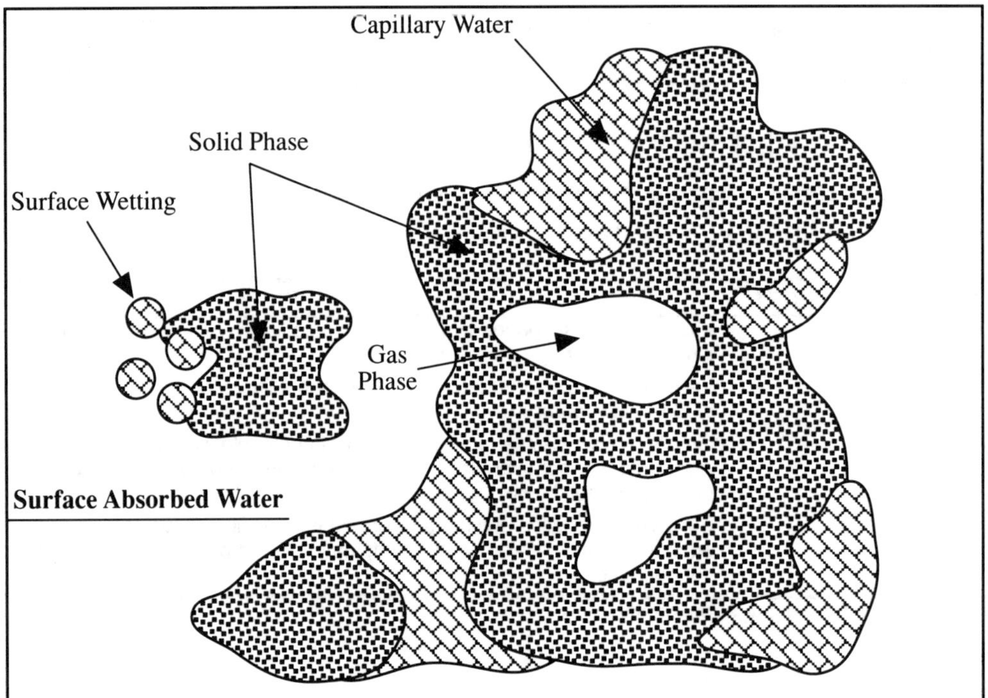

FIGURE 4.1b
Mechanisms Retaining Water and Ionic Materials on and in Solid Phases
Solidification/Stabilization

4.3.1.6. Wetting

Some additives and soils may be hydrophobic. Consequently, it may be necessary to add surfactants. However, interference with subsequent cementation reactions may occur.

4.3.1.7. Mixing

Some S/S processes are sensitive to energy input and shear rate. However, high-shear mixing may be required in certain instances, e.g. temporary viscosity changes of thixotropic materials. Over mixing may interfere with initial gel formation of cementitious S/S processes, causing delayed set, slow curing, and, possibly, loss of final physical properties. Intimate mixing on the microscopic scale may not be required for adequate process performance [9].

4.3.1.8. Temperature and Humidity

Typically, rates of reaction increase with temperature.

Below freezing, gel structures can be broken; at high temperatures, steam release can break up solid mass and reduce available water reactions. Some reagents, such as kiln dust and fly ash, contain CaO, which can hydrate violently, releasing heat.

To be assured that cementitious reactions will start, temperatures should be $\geq$ -3°C.

Ambient curing humidity should be kept high. Evaporation of water from the surface may inhibit solidification. Laboratory curing is typically conducted at 95% relative humidity.

4.3.2. Chemical

Chemical additives/constituents can retard, inhibit, and accelerate setting and curing in S/S systems. Final strength, permeability, and durability also may be impacted. A study [8] based on unconfined strength, cone index, and wet and dry tests reported the following conclusions:

- The metals, grease and oil, and phenol were generally deleterious to the strength and durability of all binder systems tested. The effects of these interferences generally increase with increasing concentration.
- Sulfate, hexachlorobenzene, and TCE had little measurable effect at any concentration
- Sodium hydroxide had mixed effects, increasing early strength (CI) at 2- and 5- percent interference addition, but decreasing early strength at the 8 percent NaOH level
- The strength and durability of the solidified/ stabilized products are strongly related, as might be expected.

During coring of S/S material, liberation of gases may increase effective porosity and decrease strength characteristics.

Table 4.1 [9] summarizes categories of additives or transformations with accompanying mechanisms.

A more detailed list of compounds or properties that may impact S/S processes is provided in Table 4.2 [9] and more quantitatively in Table 4.3 [9].

Table 4.1: Impact of S/S Additives

Method or Material	Mechanism
Flocculant	Aggregation of fine particles and filmformers
Wetting agent	Dispersion of oils and greases and fine particulates away from reacting surfaces
pH adjustment	Removal of interfering substances from solution; destruction of gels and filmformers
Fe+2/Fe+3 addition	Precipitation of interfering substances
Ion exchange	Removal of interfering substances from solution
Sorbent addition	Removal of interfering substances from reacting surfaces
Redox potential	Destruction/ conversion of interfering substances
Aeration	Alteration of biological status; removal of interfering volatiles
Temperature adjustment	Acceleration of reaction rate to counter retarding effect
Lime addition	Supplies additional calcium for reaction; reacts with certain interfering organics; pH adjustment
Sodium silicate	Reacts with interfering metals; causes acceleration of initial set
Calcium chloride	Accelerates set in Portland cement systems
Sodium Hydroxide	pH Adjustment; may solubilize silica for quicker reaction with calcium ion (low concentration)
Amines, other organics	Mechanism unclear
Metal ions	Mechanism unclear

Table 4.2: Factors Affecting Solidification

Compound or Factor	Effect	Mechanism Affected	S/S Processes
Fine particulates	I, P	P	PC, PZ
Ion exchange materials	I, A	I	AI
Metal lattice substitution	I, A	I	AI
Gelling agents		R, I, P–	P, I, M AI
Organics, general	I, P, R, I, D	AI	
Acids, acid chlorides	P–	I	C, Some O
Alcohols, glycols	R, P–	I, W	Some O
Aldehydes, ketones	P–	I	C, Some O
Amides	R	I, W	Some O
Amines	R, A	I, F	Some O
Carbonyls	R	I, D	AI
Chlorinated hydrocarbons	P–,R	I, M	PC, Some O
Ethers, epoxides	P–	I	Some O
Grease	I, P	P	PC, PC/PZ, L
Heterocyclics	P–	I	C
Hydrocarbons, general	P	I	C, Some O
Liginins	I	C	AI
Oil	I, P	P	PC, PZ
Starches	I	C	AI
Sulfonates	R	D	AI
Sugars	I, R	C	AO
Tannins	I	C	AI

Table 4.2: Factors Affecting Solidification (continued)

Compound or Factor	Effect	Mechanism Affected	S/S Processes
Organics, specific			
Adipic acid			
Benzene			
EDTA			
Ethylene glycol	P	I PC	
Formaldehyde			
p-Bromophenol			
Hexachlorobenzene		P–,P+ I PC	
		P– I	PC/PZ
		P+, P– I	L
Methanol			
NTA			
Phenol	P–	I	PC, PC/PZ, L
Trichloroethylene	P	I	PC, PC/PZ, L
Xylene			
Inorganics, general			
Acids	P–	I	PC, Some O
Bases	P–	I	PC/SS, C, Some O
Borates	R	M	PC, PZ
Calcium compounds			
Chlorides	R, P	I	AI
Chromium compounds	A	I	AI
Heavy metal salts		P–, A, R, I	AI, Some O
Iron compounds	A	F, M	PC, PZ
Lead compounds	R	M	PC, PZ
Magnesium compounds	R	M	PC, PZ
Salts, general	P–, A, R	I	AI, Some O
Silicas	R	F	PC, PZ
Sodium compounds	I	I	AI
Sulfates	R, P	I	AI
Tin compounds	R	M	PC, PZ

Table 4.2: Factors Affecting Solidification (continued)

Compound or Factor	Effect	Mechanism Affected	S/S Processes
Inorganics, specific			
Calcium chloride		A, R M	PC, PZ
Copper nitrate	P+	I	PC
		P+, P –I	PC/PZ
		P – I	L
Gypsum, hydrate	R	I	PC, PZ
Gypsum, semihydrate	A	I	PC, PZ
Lead nitrate		P –, P + I	PC
	P–	I	PC/PZ, L
Sodium hydroxide		P+, P – I	PC, PC/PZ, L
Sodium sulfate	P–	I	PC
	–	P+, P – I	PC/PZ, L
Tin			
Zinc nitrate		P+, P – I	PC
	P–	I	PC/PZ, L

Key: Effect: o = setting/curing inhibition (long term); A = setting/curing acceleration; R = setting/curing retardation (short term); P+ = alteration of properties of cured product, positive effect; P– = alteration of properties of cured product, negative effect. Mechanism: P = coats particles; I = interferes with reaction; C= complexing agent; M = disrupts matrix; F = flocculant; D = dispersant; W = wetting agent. Process: PC = Portland cement-based; PC/SS = Portland cement/ soluble silicate; PC/PZ – Portland cement/ pozzolan; PZ = pozzolanic (kiln dust, flyash); C = clay-based; L = lime-based; AI = all inorganic; O = organic.

Note: When the effect maybe positive or negative, depending on concentration, the first symbol listed represents lower concentration, the last higher concentration.'

Table 4.3: Twenty-Eight-Day Unconfined Compressive Strength as a Percent of Control Specimen[a]

	Portland Cement Binder I/B/S[b] Ratio			***Cement/ Flyash I/B/S Ratio***			***Lime/ Flyash I/B/S Ratio***		
Interference Chemical	0.02	0.05	0.08	0.02	0.05	0.08	0.02	0.05	0.08
Oil	–20	–38	–44	–8	–28	–42	–7	–27	–32
Grease	–12	–25	–45	–48	–40	–20	–7	–27	–54
Lead nitrate	–2	+18	+10	–51	–75	–97	–40	–77	–90
Copper nitrate	+91	+181	+4	+17	–84	–98	–48	–48	–98
Zinc nitrate	+3	–73	–86	–47	–93	–95	–53	–77	–88
Tricholroethylene	–28	–36	–27	–7	–33	–29	+51	–20	–34
Hexachlorobenzene	–5	–6	+15	–10	–9	–10	+6	+9	–1
Sodium hydroxide	+14	–33	–52	+23	+5	–20	+1	–7	–4
Sodium sulfate	–7	–13	–53	+23	+16	–64	+65	+38	–82
Phenol	–22	–26	–54	–49	–82	–92	–65	–88	–96

a All results reported as percent increase (+) or decrease from the control specimen rounded to nearest whole percent.

b Interference/ binder/ sludge ratio.

4.4. Inorganic Processes comparison [9]

Given that inorganic processes are the most common, this section compares strength and leachability characteristics among the most common inorganic processes.

Typical gel set times are given in Table 4.4.

Table 4.4: Gel Times and Physical States

	Set or Gel Time (hours) Solids		Physical State Solids	
Process	**Low**	**High**	**Low**	**High**
Cement-based	24-48	1-48	heavy slurry[a]	soft solid
Cement/ flyash	48+	1-48	heavy slurry[a]	soft solid
Cement/ soluble silicate	0.01-0.5	0.01-0.5	pumpable slurry	soft solid
Lime/ flyash	48+	1-48	soft	soft solid
Kiln dust	48+	24-48	soft	soft solid

[a] May be pumpable with positive displacement pumps if distance and head are not too great.

An EPA study evaluated S/S of four synthetic analytic reference matrices (SARM):

- SARM 1: High organic, low metal; 31.4 percent water
- SARM 2: Low organic, low metal; 8.6 percent water
- SARM 3: Low organic, high metal; 19.3 percent water
- SARM 4: High organic, high metal; 22.1 percent water

Strength and leaching characteristics for these waste types stabilized by a variety of processes are summarized in Tables 4.5 and 4.6.

Portland cement clearly yields the strongest products. In terms of leaching:

Cadmium and Nickel - all processes similar low leaching

Chromium - all processes increased leaching over raw waste

Arsenic, copper, lead, zinc - Portland cement was superior except for SARM 2 where lime/ flyash performed best on copper, lead and zinc.

Table 4.5: Comparison of Product Strengths from Various S/S Processes*

Sample Number	SARM Type	Binder Type	B/S[a] Ratio	W/TS[b] Ratio	Days After Mixing**			
					7	14	21	28
1	I	Portland Cement Type 1	0.7	0.40	NA	977	NA	1093
2	I		1.2	0.40	NA	>1000	NA	>1000
3	I		2.3	0.40	NA	>1000	NA	>1000
4	II	Portland Cement Type 1	0.7	0.40	NA	>1000	NA	>1000
5	II		1.2	0.40	NA	>1000	NA	>1000
6	II		2.3	0.40	NA	>1000	NA	>1000
7	III	Portland Cement Type 1	0.7	0.40	NA	28	NA	>1000
8	III		1.2	0.40	NA	99	NA	>1000
9	III		2.3	0.40	NA	71	NA	>1000
10	IV	Portland Cement Type 1	0.7	0.40	NA	15.8	NA	16.2
11	IV		1.2	0.40	NA	167	NA	160
12	IV		2.3	0.40	NA	177	NA	300
13	I	Kiln Dust	1	0.40	5	72.9	93	113
14	I		2	0.45	5	51.8	54	241
15	I		3	0.40	176	211	215	81.1
16	II	Kiln Dust	1	0.40	37.5	59.7	78.3	85.1
17	II		2	0.40	128	190	164	216
18	II		3	0.40	183	225	275	252
19	III	Kiln Dust	1	0.40	32.9	36.6	37.1	38.5
20	III		2	0.40	33	38.4	40.8	39
21	III		3	0.40	45.7	44.7	43.7	79.8
22	IV	Kiln Dust	1	0.42	27.9	28.1	26.8	32.2
23	IV		2	0.43	38.9	55.7	52.3	52.2
24	IV		3	0.40	35.7	38.2	33	40.1
25	I	Lime/Flyash	1	0.45	24.1	27.3	26	32.3
26	I		2	0.45	22.2	29	33.9	40.4
27	I		3	0.45	19.5	30.4	32.9	46.6
28	II	Lime/Flyash	1	0.45	9.9	17.1	17.3	28.8
29	II		2	0.48	17.2	24.2	26.9	62.4
30	II		3	0.49	19.4	31.4	41.2	73.4
31	III	Lime/Flyash	1	0.48	21.9	28.7	29.1	30.7
32	III		2	0.49	30.3	33	36.4	36.5
33	III		3	0.50	34.8	48.8	48.2	50.9
34	IV	Lime/Flyash	1	0.45	34.9	36	34.8	37.9
35	IV		2	0.40	29.8	38.7	36.3	40.5
36	IV		3	0.45	36.9	35.7	37.9	42.2

[a] Binder (reagent) to soil ratio. [b] Water to total solids ratio.

*All tests run at 90 - 100% relative humidity at 21°c ± 6°c **Results expressed in psi.

Table 4.6: Summary of TCLP Results for Metals

(SARM) Sample Number	Binder (Day)	Arsenic *a*	Arsenic *b*	Cadmium *a*	Cadmium *b*	Chromium *a*	Chromium *b*	Copper *a*	Copper *b*	Lead *a*	Lead *b*	Nickel *a*	Nickel *b*	Zinc *a*	Zinc *b*
I	RAW	ND		0.53		ND		0.61		0.49		0.27		9.2	
1	PC(14)	ND	—	ND	100	0.06	+	0.07	81	0.15	75	0.04	70	0.23	96
14	KD(14)	ND	—	ND	100	0.06	+	0.04	81	ND	100	ND	100	0.27	94
27	LF(14)	ND	—	ND	100	0.02	+	0.03	98	ND	100	ND	100	0.14	94
1	PC(28)	ND	—	ND	100	0.06	+	0.06	83	0.15	75	0.04	70	0.49	91
15	KD(28)	ND	—	ND	100	0.09	+	0.03	80	ND	100	ND	100	0.62	73
27	LF(28)	ND	—	ND	100	0.02	+	0.03	98	ND	100	ND	100	ND	100
II	RAW	ND		0.73		ND		0.89		0.7		0.4		14.6	
4	PC(14)	ND	—	ND	100	0.03	+	0.04	92	0.15	82	0.04	83	0.09	99
16	KD(14)	ND	—	ND	100	0.08	+	0.07	79	0.44	+	ND	100	0.25	97
30	LF(14)	ND	---	ND	100	ND	---	ND	100	ND	100	ND	100	0.22	99
4	PC(28)	ND	—	ND	100	0.03	+	0.06	89	0.15	83	0.04	83	0.54	94
16	KD(28)	ND	—	ND	100	0.05	+	0.09	89	0.37	+	ND	100	0.78	89
29	LF(28)	ND	—	ND	100	ND	—	0.03	90	ND	100	ND	100	0.02	100
III	RAW	6.39		33.1		ND		0.89		0.7		0.4		14.6	
7	PC(14)	ND	—	ND		0.07	+	0.15	100	0.63	95	ND	100	0.58	100
21	KD(14)	ND	—	ND	100	0.22	+	1.02	96	13.3	+	ND	100	4.38	95
33	KD(14)	0.81	52	0.02	100	0.03	+	2.96	87	51	+	ND	100	3.81	96
7	PC(28)	ND	—	ND	100	0.07	+	0.09	100	ND	100	ND	100	0.69	100
21	KD(28)	0.21	98	ND	100	0.12	+	0.85	96	18.3	+	ND	100	4.07	95
33	LF(28)	0.79	51	0.02	100	0.07	+	2.59	87	51	+	0.03	99	2.97	96
IV	RAW	9.58		35.3		0.06		10		70.4		26.8		396	
10	PC(14)	ND	100	ND	100	0.06	+	0.14	100	0.39	99	ND	100	0.39	100
23	KD(14)	0.16	95	ND	1--	0.11	+	1.88	97	12.4	43	ND	100	4.57	97
	LF(14)	1.61	50	ND	100	0.07	+	1.92	96	91.8	+	ND	100	3.22	96
10	PC(28)	ND	100	ND	100	0.06	+	0.17	100	0.37	99	ND	100	0.74	100
23	KD(28)	0.27	92	ND	100	0.12	+	1.647	97	21.4	9	ND	100	3.72	97
	LF(28)	0.98	59	0.02	100	0.07	+	2.18	95	65	+	ND	100	3.64	96
Detection Limit		0.15		0.01		0.01		0.02		0.15		0.04		0.01	

a. TCLP results in ppm *b*. Percent reduction corrected for dilution
ND = Below detection limit + = increase over raw SARM

Clearly, no one process is consistently superior.

In dynamic leaching studies (Figures 4.2-4.5), cement/fly ash proved most efficacious. The moderation of the initially high pH due to pozzolanic reactions that consume $Ca(OH)_2$ was credited for this behavior.

Dilution effects should be considered when evaluating process efficiency.

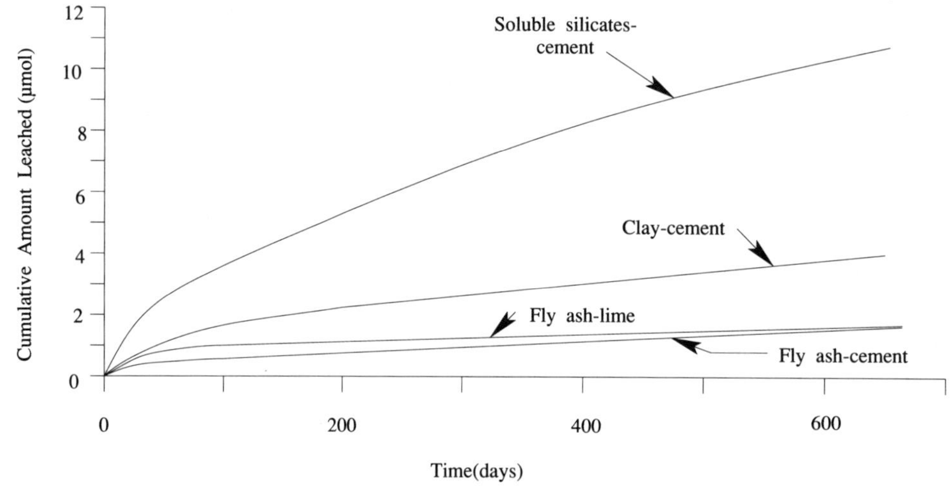

FIGURE 4.2
Cadmium Leaching vs. Time for Various S/S Processes
Solidification/Stabilization

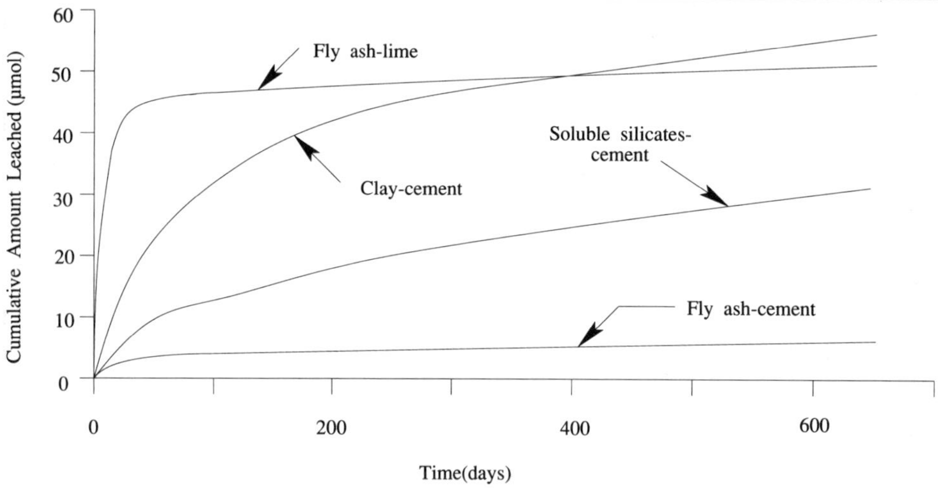

FIGURE 4.3
Lead Leaching vs. Time for Various S/S Processes
Solidification/Stabilization

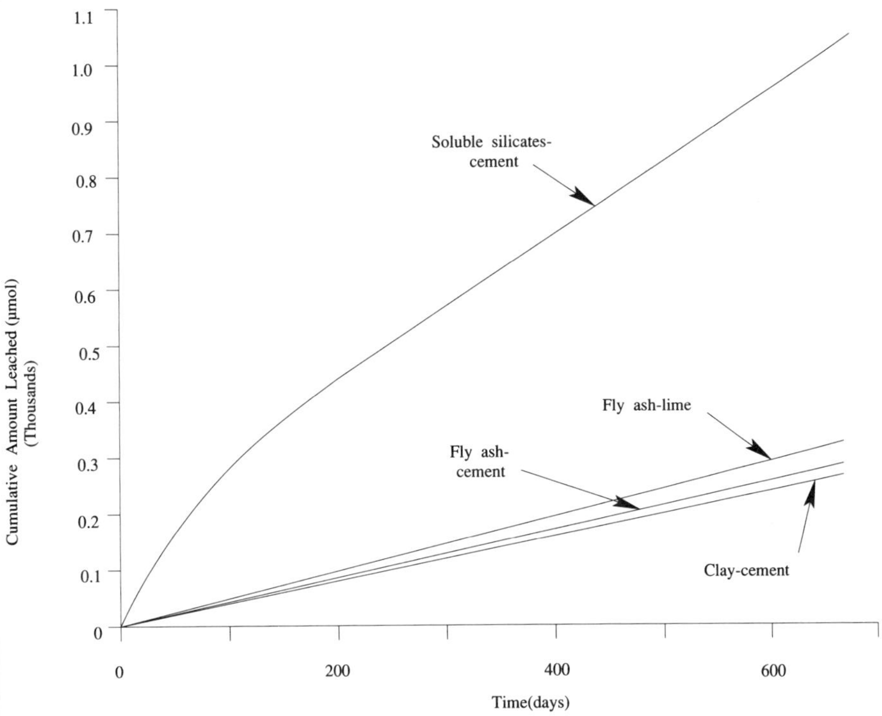

FIGURE 4.4
Arsenic Leaching vs. Time for Various S/S Processes
Solidification/Stabilization

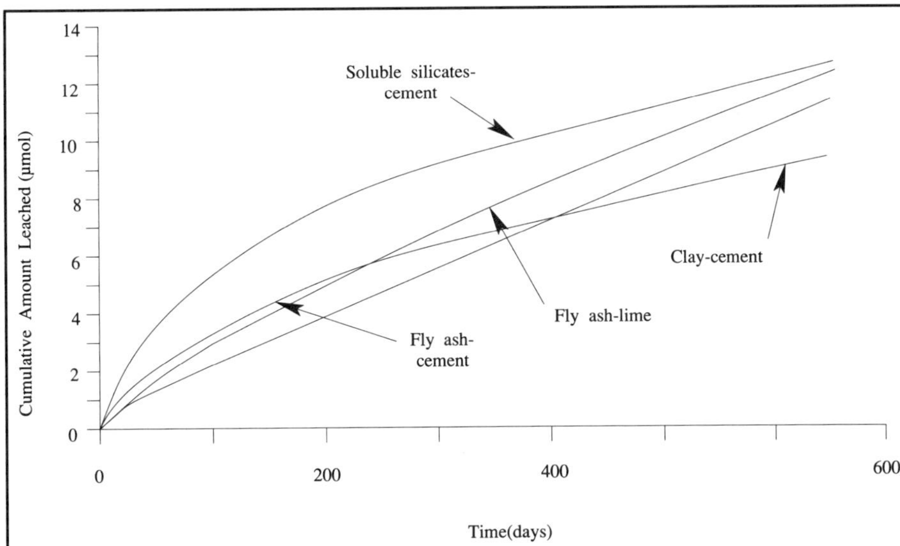

FIGURE 4.5
Chromium Leaching vs. Time for Various S/S Processes
Solidification/Stabilization

5. FEASIBILITY

5.1. Introduction

In general, the applicability of S/S processes is based on:

- Soil and site characteristics
- Contaminant location and properties

5.2. Prescreening [1]

Potential interferences between waste and binder should be addressed. Previous studies on similar waste should be reviewed. However, available information on interferences and application of S/S processes to hazardous wastes is often incomplete and sometimes conflicting. Potential pretreatment options should be identified.

Site-specific conditions may pose further restrictions on the applicability of S/S technologies.

Table 5.1 presents a listing of interferences and applications of S/S technologies for common hazardous waste material.

Table 5.1: Effect of Waste Components on Stabilization/ Solidification

Waste component	Cement-based	Pozzolan-based	Thermoplastic	Organic polymer
Organics				
Nonpolar as: oil and grease, aromatic hydrocarbons, halogenated hydrocarbons, PCBs	May impede setting. Decreases durability over a long time period.Volatiles may escape upon mixing. Demonstrated effectiveness under certain conditions.	May impede setting. Decreases durability over a long time period. Volatiles may escape upon mixing. Demonstrated effectiveness under certain conditions.	Organics may vaporize upon heating. Demonstrated effectiveness under certain conditions.	May impede setting. Demonstrated effectiveness under certain conditions.
Polar as: alcohols, phenols, organic acids, glycols	Phenol will significantly retard setting and will decrease durability in the short run. Decreases durability over a long time period.	Phenol will significantly retard setting and will decrease durability in the short run. Alcohols nay retard setting. Decreases durability over a long time period.	Organics may vaporize upon heating.	No significant effect on setting.

Table 5.1: Effect of Waste Components on Stabilization/ Solidification (continued)

Waste component	Cement-based	Pozzolan-based	Thermoplastic	Organic polymer
Inorganics				
Acids as: hydrochloric acid, hydrofluoric acid	No significant effect on setting. Cement will neutralize acids. Types II and IV portland cement demonstrate better durability characteristics than Type I. Demonstrated effectiveness.	No significant effect on setting. Compatible, will neutralize acids. Demonstrated effectiveness.	Can be neutralized before incorporation.	Can be neutralized before incorporation. Ureaformaldehyde demonstrated to be effective.
Oxidizers as: sodium hypochlorite, potassium. permanganate, nitric acid, potassium dichromate	Compatible	Compatible	May cause matrix breakdown, fire	May cause matrix breakdown, fire
Salts as: sulfates, halides, nitrated, cyanides	Increase setting times. Decrease durability. Sulfates may retard setting and cause spalling unless special cement is used. Sulfates accelerate other reactions.	Halides are easily leached and retard setting. Halides may retard setting, most are easily leached. Sulfates can retard or accelerate reactions.	Sulfates and halides nay dehydrate and rehydrate, causing splitting	Compatible
Heavy metals as: lead, chromium, cadmium, arsenic, mercury	Compatible. Can increase set time. Demonstrated effectiveness under certain conditions.	Compatible. Demonstrated effectiveness on certain species (lead, cadmium, chromium).	Compatible. Demonstrated effectiveness on certain species (copper, arsenic, chromium).	Compatible. Demonstrated effectiveness with arsenic.
Radioactive materials	Compatible	Compatible	Compatible	Compatible

References for Table 5.1

1. Federal Register. Vol. 51, No. 216, Friday, November 7, 1986, pp. 40643-40652.

2. JACA Corporation. 1985. Critical Characteristics and Properties of Hazardous Waste Solidification/ Stabilization. Unpublished Report. Prepared for U.S. Environmental Protection Agency, Hazardous Waste Engineering Research Laboratory, Cincinnati, Ohio.

3.Kolvites, B. and P. Bishop. Column Leach Testing of Phenol and Trichlorethylene Stabilized/ Solidified with Portland Cement". University of New Hampshire, Department of Civil Engineering, Presented at ASTM 4th International Hazardous Waster Symposium, Atlanta, Georgia. p. 27.

4. Kyles, J. H., K. C. Malinowski, and T. F. Stanczyk. 1987. Solidification/ Stabilization of Hazardous Waste - A Comparison of Conventional and Novel Techniques, Toxic and Hazardous Wastes. In: Proceedings of the 19th Mid-Atlantic Industrial Wastes Conference, June 21-23, 1897. Jeffery C. Evans, ed. pp. 554-568.

5. Musser, D., ENRECO, and R. Smith, Raba-Kistner Consultants. 1987. In Situ Solidification/ Fixation of Oil Field Production Fluids - A Novel Technique. In: Proceedings of the 39th Industrial Waste Conference, Perdue University, West Lafayette Indiana, 1984.

6.Tittlebaum, M. E., *et al.* 1985. " State of the Art on Stabilization of Hazardous Organic Liquid Wastes and Sludges." CRC Critical Reviews in Environmental Control, 15(2): 191-193.

7. U.S. Environmental Protection Agency. 1984 Case Studies - Remedial Response at Hazardous Waste Sites. EPA/540/2-84-002, February 1984. Tramel Crow Site and College Point Site.

8. U.S. Environmental Protection Agency. 1986a. Hazardous Waste Management System. 51 FR 46824, December 24, 1986.

9. U. S. Environmental Protection Agency. 1986b. Test Methods for Evaluating Solid Waste, Method 9095, SW-846. Third Edition. November 1986.

10. U.S. Environmental Protection Agency. 1986c. Prohibition on the Disposal of Bulk Liquid Hazardous Waste in Landfills - Statutory Interpretive Guidance. OSWER Policy Directive No. 9487.00 - 2A. EPA/530-SW-016, June 1986.

11. U.S. Environmental Protection Agency. 1986d. Hazardous Waste Management System; Land-Disposal Restrictions. 51 FR 40572, November 7, 1986.

12. U.S. Environmental Protection Agency. 1986e. Handbook for Stabilization/ Solidification of Hazardous Waste. EPA/540/2-86/001.

13. U.S. Environmental Protection Agency. 1986f. A Procedure for Estimating Monofilled Solid Waste Leachate Composition. Technical Resource Document SW-924, Second Edition. Hazardous Waste Engineering Research Laboratory, Office of Research and Development, Cincinnati, Ohio, and Office of Solid Waste and Emergency Response, Washington, D.C.

14. U.S. Environmental Protection Agency. 1986g. SW-846 Test Methods for Evaluating Solid Waste, Vol. 1C: Laboratory Manual Physical/ Chemical Methods, Third Edition. Office of Solid Waste and Emergency Response, Washington, D.C.

15. U.S. Environmental Protection Agency. 1986h. Mobile Treatment Technologies for Superfund Wastes. Office of Solid Waste and Emergency Response. EPA/540/2-86/003, September 1986.

16. Van Keuren, E., et al. 1987. Pilot Field Study of Hydrocarbon Waste Stabilization, Toxic and Hazardous Wastes. In: Proceedings of the 19th Mid-Atlantic Industrial Waste Conference, June 21-23, 1987. Jeffery C. Evans, ed. pp. 330-341.

5.3. Technology Screening [1]

Elements:

1. Waste Characterization
2. Feasibility Testing
3. Site Evaluation
4. Cost

5.3.1. Waste Characterization [1]

Complex waste mixtures should be analyzed to insure no "land ban" constituents exist. The presence of oils or grease may require the addition of proprietary binder additives. Up to 25% (by weight) oil and grease has been reported to be accommodated. Pretreatment to remove oil and grease may be required (oil-water separators, centrifuges, DAF).

Although organic content is typically reported as total organics, various classes behave differently. For instance, some organic chemicals can act as solvents for some binders. Other organics, such as phenol, can inhibit setting and curing reactions. Although EPA has reported up to 40% total organic content can be treated by inorganic binders, identification of individual chemical constituents is recommended.

Waste consistency may pose mixing problems. For example, creosote laden waste may be too viscous to mix. Mixing problems typically arise when materials with significantly different viscosities are combined.

Excess water may have to be removed through dewatering operations. Both anthropogenic and natural debris may have to be removed prior to treatment.

Spatial and temporal variability in both characteristics and quantities of waste at clean-up sites may result in somewhat variable process performance (e.g. strength, durability, leaching characteristics).

A variety of compounds can result in interference mechanisms. See Table 4.2 (section 4). Physical properties of the waste will also impact the selection process (see section 4).

5.3.2. Feasibility Testing

No set protocol has been established for determining the feasibility of S/S processes. Typically bench-scale tests are conducted.
Two primary parameters are evaluated:

A. Leachability

B. Durability

Physical and chemical testing methods used for S/S evaluations are summarized in Tables 5.2-5.4 [1].

Volume changes (increase or decrease) must also be quantified.

During testing, the following information is important for an evaluation of inorganic and organic binder technologies:

- Leaching tests appropriate for the waste site
- Durability tests as appropriate for the final disposition of the product (e.g., freeze-thaw testing has little relevance for a material that will be buried beneath the freeze line or that is in a part of the country where it does not freeze)
- Permeability tests of the final product, which should be two orders of magnitude below that of the surrounding materials
- The waste-to-binder ratio needed to achieve leaching and durability characteristics appropriate for the waste site
- Increase or decrease in the volume of the final product
- Evolution of gases during processing, curing, or preprocess drying
- The need for dewatering or adding waste to the inorganic waste/binder mixture
- Ability to mix the waste material and binder
- Sorption characteristics

Table 5.2: Physical Testing Methods

Test procedure	Reference	Purpose
Index property tests		
Particle size analysis	ASTM D422-63	To determine particle size distribution of material
Atterberg limits		
Liquid limit	ASTM D4318-84	To define the physical characteristics of a material as a function of its water content
Plastic limit	ASTM D4318-84	
Plasticity index	ASTM D4318-84	
Moisture content	ASTM D2216-80	Determines percentage of free water in a material
Suspended solids	USEPA Method 208 C	To determine the amount of solids that do not settle from a column of liquids
Paint filter test	USEPA Method 9005-SW846	Determines the presence of free liquids in a representative sample of bulk or noncontainerized waste
Density testing		
Bulk density - drive cylinder method	ASTM D2937-83	To determine the in-place density of soils or soil-like material
Bulk density - sand-cone method	ASTM D 1556-82	To determine the in-place density of soils or soil-like materials
Bulk density - nuclear methods	ASTM D2922-81	To determine the in-place density of soils or soil-like materials
Bulk density - stab. waste		Determines density of a monolithic stabilized waste
Compaction testing		
Moisture density relations of soil-cement mixtures	ASTM D558-82 ASTM 31557	To determine the relation between moisture content and density of a monolithic stabilized waste
Permeability testing		
Falling head permeability	USEPA Method 9100-SW846	To measure the rate at which water will pass through a soil-like material
Constant head	USEPA Method 9100-SW846	To measure the rate at which water will pass through a soil-like material
Strength testing		
Unconfined compressive strength of cohesive soils	ASTM D2166-85	To evaluate how cohesive soil-like materials behave under mechanical stress
Unconfined compressive strength of concrete specimens	ASTM D1633-84	To evaluate how cement-like materials behave under mechanical stress
Compressive strength of hydraulic cement mortars	ASTM C109-86	To measure the compressive strength of hydraulic cement mortars
Flexural strength	ASTM D1635-87	To evaluate a material's ability to withstand loads over a large area
Cone index	ASTM D3441-79	Evaluate a material's stability and bearing capacity
Durability testing		
Freeze-thaw durability	ASTM D4842	To determine how materials behave or degrade after repeated freeze-thaw cycles
Wet-dry durability	ASTM D4843	To determine how materials behave or degrade after repeated wet-dry cycles

[a]ASTM - American Society for Testing and Materials

Blank indicates no ASTM method or other reference method.

Table 5.3: General Chemical Test Methods with Application to Stabilized/Solidified Wastes.

Parameter	Test method	Applicability to untreated and stabilized /solidified wastes	Reference
pH	EPA Method SW-9045	Leachability of hazardous constituents (e.g., metals) may be governed by the pH of the solid	1
Oxidation/ reduction potential (E_H)	ASTM D 1498-76	Changes in E_H after treatment can change the leachability of many elements	2
Major oxides	ASTM C114	Mineralogy of the stabilized/solidified waste may aid in interpretation of leach test results	3
Total organic carbon (TOC)	Combustion Method	Used to approximate the nonpurgeable organic carbon in wastes and treated solids	4
Oil and grease	EPA Method 413.2	May be used to compare the leachable oil and grease from the treated and untreated wastes	5
Elemental analysis	EPA Method SW-846	Used to determine the fraction of metals leached to the total metals content of the untreated and stabilized/solidified wastes	1
Volatile organic compounds (VOCs)	EPA Method SW-846 (Methods 5030 and 8240)	Used to compare VOC concentrations in stabilized/solidified wastes and untreated wastes with the VOC concentrations in TCLP extracts to determine relative leachability of the treated and untreated wastes	1
Base, neutral, and acid compounds (BNA)	EPA Method SW-846 (Methods 3540, 3520,and 8270)	Used to compare BNA concentrations of leachates with respective concentrations in treated and untreated wastes to determine relative leachability of the treated and untreated wastes	1
Polychlorinated biphenyls (PCBs)	EPA Method SW-846 (Methods 3540, 3520, 680, and 8080)	Same as for VOC with respect to PCB leachability from treated and untreated wastes	1
Ion measurements	Std. Method No. 429	Used to determine leachate ionic species concentrations	7
Heat of hydration	ASTM C186-86	Measurement of temperature change during current mixing will allow prediction of VOC emissions in the field	8
Alkalinity	Titrometry	Alkalinity changes in leachates may be used to determine changes in stabilized/solidified waste form	9,10

a Reference:
1) U.S Environmental Protection Agency. 1986. Test Methods for Evaluating Solid Waste. Volumes 1A-1C: Laboratory Manual, Physical /Chemical Methods; and Volume II: Field Manuals, SW-846, Third Edition, Office of Solid Waste. Document Control No. 955-00100000-1.
2) ASTM D1498-76, Standard Practice for Oxidation-Reduction Potential of Water.
3) American Society for Testing and Materials. 1981. Standard Methods for Chemical Analysis of Hydraulic Cement, ASTM Committee c-1 on Cement, Philadelphia Pennsylvania. July 1981.
4) Perkin Elmer 240C Users Manual.
5) U.S. Environmental Protection Agency. !979. Methods for the Chemical Analysis of Water and Wastes. Office of Research and Development. EPA-600 4-79-020, March 1979.
6) Albord-Stevens, Bellar, Erchelberger and Bubble, Method 680, November 1985, Determination of Pesticides, PCBs in Water and Sediment by GC-MS, Office of Research and Development, U.S. Environmental Protection Agency.
7) Standard Methods for the Examination of Water and Wastewater, 16th Edition, APHA, AWWA, WPCF, 1985, Washington D.C.
8) ASTM C186-86. Standard Test Method for Heat of Hydration of Hydraulic Cement.
9) Bishop, P.L. 1986. Prediction of Heavy Metal leaching Rates from Stabilized/Solidified Hazardous Wastes. In Toxic and Hazardous Wastes Proceedings of the 18th Mid-Atlantic Industrial Waste Conference.
10) APHA-AWWA-WPCF. 1975. Standard Methods for the Examination of Water and Wastewater, 14th Edition. Method 403. pp. 278-281.

Table 5.4: Extraction or Leaching Tests

Test method	Leaching or extracting medium	Liquid/solid ratio	Maximum particle size	Number of extraction s	Time of extraction
TCLP	Acetic acid[1]	20:1	9.5mm	1	18 hours
EP Tox	0.04 M acetic acid (pH = 5.0)	16:1	9.5 mm	1	24 hours
Cal. WET[a]	0.2 M sodium citrate (pH= 5.0)	10:1	2.0 mm	1	48 hours
MEP[b]	Same as EP Tox, then with synthetic acid rain (sulfuric acid: nitric acid in 60:40 wt.% mixture)	20:1	9.5mm	9 (or more)	24 hours per extraction
MWEP[c]	Distilled/deionized water or other for specific si t	10:1 per extraction	9.5 mm or monolith	4	18 hours per extraction
Equilibrium leach test	Distilled water	4:1	150μm	1	7 days
Acid neutralization capacity	HNO_3 solutions of increasing strength	3:1	150μm	1	48 hours per extraction
Sequential extraction tests	0.04 M acetic acid	50:1	9.5mm	15	24 hours per extraction
Sequential chemical extraction	Five leaching solutions increasing in acidity	Varies form 16:1 to 40:1	150μm	5	Varies from 2 to 24 hours
Total waste analysis	Hexane or methylene chloride				

acetate buffered solutions: 1) pH = 3.0
a California Waste Extraction Test
b Multiple Extraction Procedure
c Monofill Waste Extraction Procedure

5.3.3. Site Evaluation

At some remediation sites, the original contaminated waste must be replaced with S/S material. Site specific parameters that should be evaluated include:

- Depth to groundwater
- Location of sensitive environmental receptors
- Equipment access
- Processing space
- Health and safety issues

Since the overall objective is waste containment, evaluation of sites should be conducted in light of these considerations. More specific site parameters that should be evaluated include:

- Area of the site
- Permeability of the area soils, both for a review of leaching capabilities and also for possible liner/cap material
- Existing groundwater contamination

- Velocity and direction of both groundwater and ambient air
- Site drainage
- Proximity to populated areas
- Access routes to the site
- Available work area/stockpiling area on the site
- Final disposal options and their site-specific implications
- Post remediation use of the site
- Sensitive environmental areas within the work site, such as flood plains or marshes
- Waste product volume increase and its implications for the capacity of the site to contain final product if on-site disposal is required
- Dusting and material stockpiling
- Ability to mix the materials adequately on the site
- Availability of the binder/additives in the amounts required for the entire site

5.3.4. Cost Evaluation

Cost estimates should include costs of the following (where applicable):

- Binder
- Pretreatment
- Liners
- Monitoring wells
- Excavation and moving equipment
- Additives
- Labor
- Caps

Figure 5.1 provides a treatability flowchart.

6. DESIGN

There is no well established design protocol for S/S processes.

Once an appropriate formulation has been selected during feasibility testing, the remainder of the design process focuses on the delivery system. The primary elements of an effective S/S design include [1]:

1. Waste removal
2. Untreated waste transportation/storage
3. Reagent storage
4. Mixing
5. S/S waste transportation/replacement

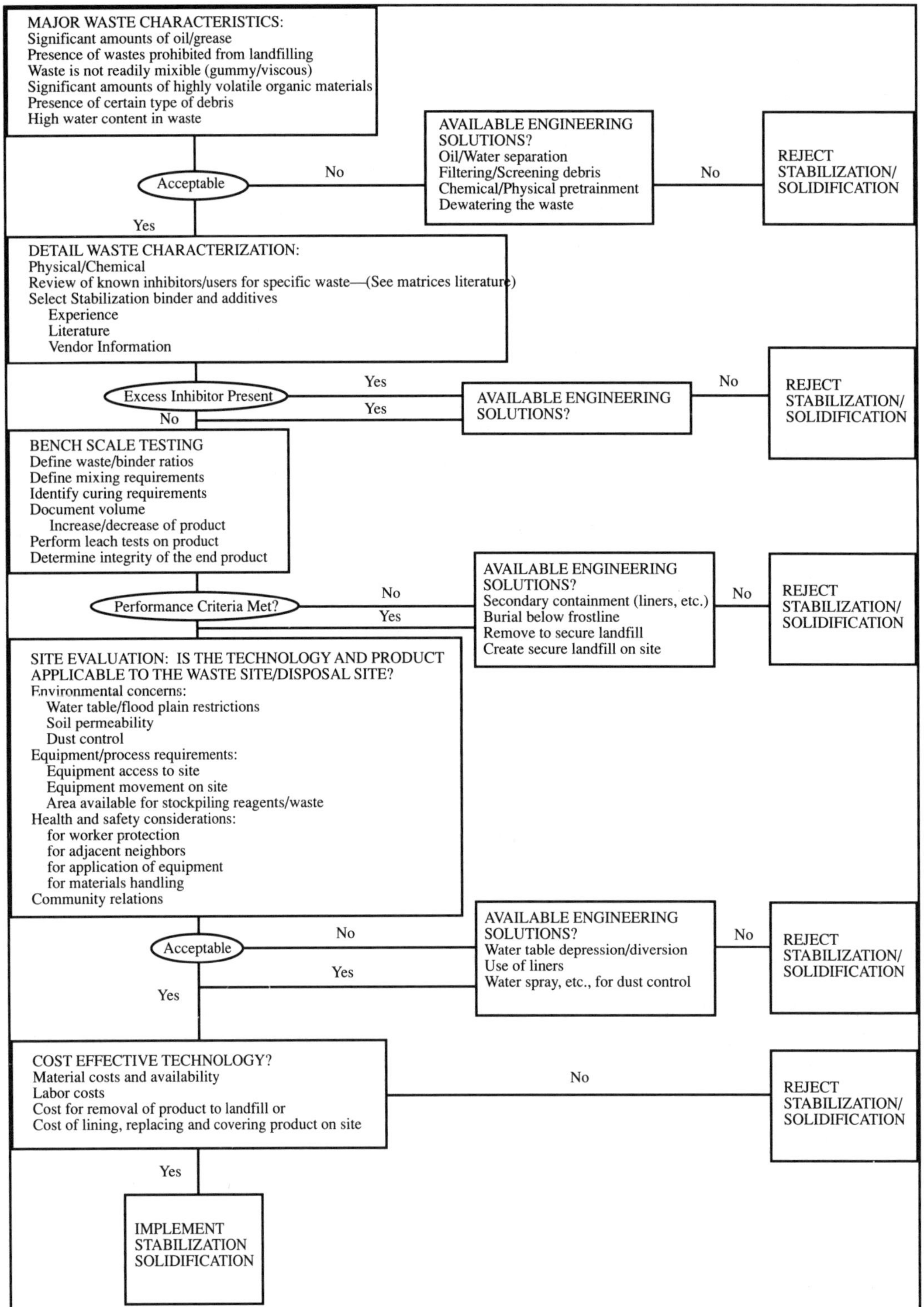

FIGURE 5.1
Treatability Flowchart
Solidification/Stabilization

6.1. Waste Removal

Although earth moving equipment and processes are well documented in the construction literature, hazardous waste characteristics must be considered during this process. For example, corrosive waste will damage equipment constructed of steel. The soil moisture content will also play a significant role in the removal operation. Operator safety is an additional concern. Typically operator spaces are completely enclosed.

6.2. Untreated Waste Transportation/ Storage

Waste can be transported by truck, pumped, or via conveyor belts or screw augers. Sludge with high moisture content can be easily pumped. For soils with low or high moisture content, conveyors can be used when large amounts of waste are to be moved over a fixed distance for long periods of time. Should the waste require storage prior to stabilization, the following considerations should be addressed:

- A 2 to 3 percent sloping of the storage area to provide for collection of liquids
- Lining of the storage area with a high-density polyethylene flexible membrane liner and sand to prevent contact between liquids and the soil and to allow for easy drainage of the liquids
- Preparation of the underlying soils to remove rocks, cobbles, and vegetative matter that could puncture the liner
- Installation of drainage pipes and a sump for final collection of liquids
- Operation of a water-treatment system to remove hazardous constituents (from the collected liquids) and to reduce the volumes of liquids that must be stored
- Covering of the wastes to prevent their dispersion
- Berming of the sides of the area to resist slumping of the waste
- Installation of a cement loading ramp for the unloading of waste from trucks
- Provision of a rubber-tired front-end loader to move wastes within the storage area

6.3. Reagent Storage

Most S/S reagents are dry. Consequently, storage facilities should be provided to keep the reagent dry. Bins and hoppers are the most common reagent storage facilities.

6.4. Mixing

Mixing is an important aspect of the S/S process. Mixing can be conducted via "area mixing" where the reagent is delivered to an area to be stabilized and mixed directly with the waste or the waste can be removed and mixed in a specially designed vessel.

Area mixing is typically conducted with backhoes. However, *in situ* stabilization also has been reported (see Figure 6.1) through the use of cassion/ auger system. A major shortcoming of these systems is the potential for incomplete mixing. If waste is excavated, land disposal restrictions may apply for ultimate disposal.

If the waste is excavated, mixing vessels, such as pug mills, ribbon blenders, extruders, or screw conveyors, can be used for mixing. The waste and reagents can be metered into the system for quality control purposes.

6.5. S/S Waste Transportation/ Replacement

Transportation of stabilized waste can be conducted in a similar fashion to untreated waste. Trucks or conveyors can be used. Typically, stabilized waste is easier to transport due to the lower water content but is usually greater in volume. Cement-like material should be placed in a disposal area immediately because it sets-up and hardens.

Placement of soil-like materials should be performed in accordance with procedures used in road-building industry (i.e. stabilized waste is placed in lift of 8 to 10 inches with graders or bulldozers). After placed, the waste is compacted.

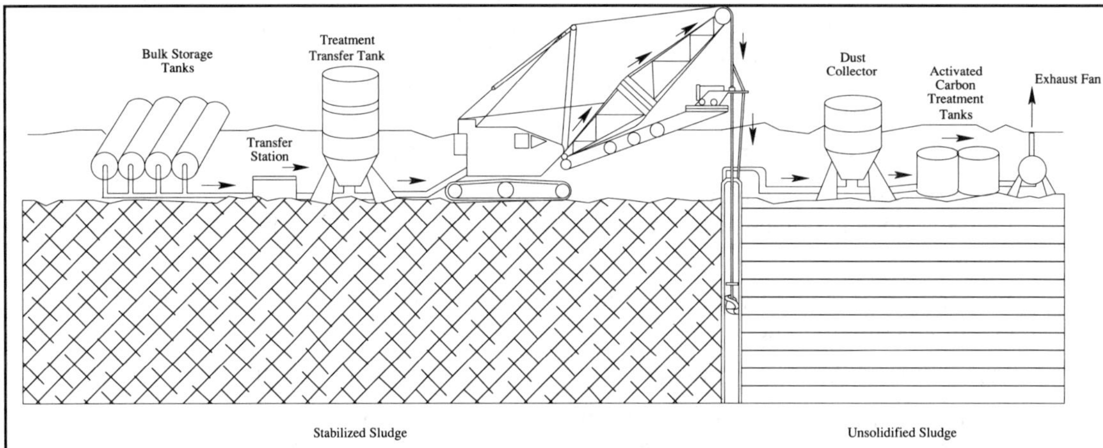

FIGURE 6.1
Crane-Mounted Mixing System Advancing Through Unstabilized/Unsolidified Sludge Layer
Solidification/Stabilization

7. COSTS

Estimated costs for a Portland cement sodium silicate process are given below [13]:

Fixed cost category	Estimated cost (1990$)
Site preparation	$ 50,000
Permitting and regulation	$ 25,000
Equipment installation and startup	$8,500
Trailer transport to site	$12,500
Mobilization/ demobilization	$ 12,000
Equipment repair and replacement	NA[c]
Effluent treatment and decontamination	$ 2,500
Total fixed costs	*$ 110,500*
Total fixed cost per ton	*$ 3*
Variable cost category	**Cost per ton of waste treated**
Capital equipment rent/lease[d]	$ 5
Labor	$ 27[e]
Chemfix reagents	$ 30
Utility and miscellaneous	$ 3
Analytical	$ 5[f]
Total variable cost per ton	*$ 70*
Total treatment cost per ton of raw waste	*$ 73*[g]

[a] Based upon 30,000 cubic yards (37,000 tons) processed at 160 tons per day.

[b] Excludes excavation costs.

[c] Assumed to be included in the cost of renting/leasing equipment from Chemfix.

[d] Based on the rent/lease cost of equipment from Chemfix. Does not include costs for purchasing or renting screening devices, size reduction equipment or waste/product storage containers.

[e] Includes cost for an operations crew.

[f] Excludes the cost of raw waster or environmental sampling.

[g] Excludes the cost of possible post-treatment, curing, storage, excavation, and final disposal.

Potential additional costs include:

Options	Cost (per ton)
Screening	$ 3 to 7
Grinding	$ 0.5 to 2
Soil washing	$ 100 to 400
Organics removal	$ 250 to 600
Land disposal	$ 50 to 250

REFERENCES

1. USEPA. *Stabilization/Solidification of CERCLA and RCRA Wastes: Physical Tests, Chemical Testing Procedures, Technology Screening, and Field Activities.*(U. S. Environmental Protection Agency, Cincinnati, May, 1989).

2. F. K. Cartledge, H. C. Eaton and M. Tittlebaum. *Morphology and Microchemistry of Solidified/ Stabilized Hazardous Waste Systems.* (U.S. Environmental Protection Agency, Cincinnati, May, 1987).

3. USEPA.*International Waste Technologies/Geo-Con In Situ Stabilization/Solidification: Applications Analysis Report.* (U. S. Environmental Protection Agency, Cincinnati, August, 1990)

4. L. Jones. *Interferences Mechanisms in Waste Stabilization/ Solidification Processes.* (U.S. Environmental Protection Agency, Cincinnati).

5. USEPA.*Immobilization Technology Seminar: Speaker Slide Copies and Supporting Information.*(U. S. Environmental Protection Agency, Cincinnati, October, 1989).

6. U.S. Army Engineer Waterways Experiment Station: Environmental Laboratory.*Guide to the Disposal of Chemically Stabilized and Solidified Wate.*(U.S. Environmental Protection Agency, Cincinnati, September, 1980).

7. USEPA.*Hazcon Solidification Process Douglassville, PA: Application Analysis Report.*(U. S. Environmental Protection Agency, Cincinnati, May, 1989).

8. R. M. Bricka and L. W. Jones. *An Evaluation of Factors Affecting the Solidification/ Stablilzation of Heavy Metal Sludge.* (U.S. Army Engineer Waterways Experiment Station, Vicksburg.)

9. J. R. Conner. *Chemical Fixation and Solidification of Hazardous Wastes.* (Van Nostrand Reinhold, New York, New York, 1990).

10. J. D. Hamernik and G. C. Frantz. *Properties of Municipal Solid Waste Fly Ash and Their Use in Concrete.* Technical Report No. ERI 90-1. (Environmental Research Institute, Storrs, January, 1990).

11. Soundarajan, R., E.F. Barth and J.J. Gibbons. *Using an Orgainc Clay to Chemically Stabilize Waste Containing Organic Compounds.* Custom, 1990.

12. American National Standard ANSI/ANS. **16**: 1. 1986.

13. USEPA.*Chemfix Techonolgies Solidification/ Stabilization Process.*(U.S. Enviromental Protection Agency, Cincinnati, 1991).

14. USEPA.*Technology Evaluation Report: Chemfix Technologies, Inc. Solidification/ Stabilization Process Clackmas, Oregon. Volume I.*(U. S. Environmental Protection Agency, Cincinnati, September, 1990).

15. W. Strumm and J. J. Morgan. *Aquatic Chemistry: An Introduction Emphasizing Chemical Equalibria in Natural Waters.* (John Wiley & Sons, New York, New York, 1981).

I - Chemical Destruction

CHEMICAL DESTRUCTION

Instructional Objectives

- Understand the chemical and physical properties associated with chemical destruction processes.
- Develop an understanding of how to apply basic concepts of chemical kinetics, equilibrium, and mass transfer to the design of chemical destruction processes.
- Develop a general design approach for hydrolysis, dechlorination, and chemical oxidation.

1. GENERAL DESCRIPTION

1.1. Introduction

Chemical destruction processes chemically alter a contaminant's structure to form new, less toxic, compounds. Three remedial technologies that apply this method are hydrolysis, dechlorination, and chemical oxidation.

2. HYDROLYSIS

2.1. General Description

2.1.1. Introduction

Hydrolysis destroys organic contaminants by chemically altering their structure, to form new, less toxic, compounds. Hydrolysis is the general term used for reactions in which hydrogen-oxygen bonds are broken and new compounds formed. Hydrolysis should not be confused with other organic-water reactions, such as acid:base reactions, hydration of carbonyls, addition to carbon-carbon bonds, or elimination.

The hydrolysis process is typically an *ex situ* operation. Key factors involved in hydrolysis design include, but are not limited to,:

- type of contaminant
- gradation of soil particles
- solvent solution

2.1.2. Compounds Removed by Hydrolysis

Hydrolysis is a family of reactions involving organic compounds, such as:

- Alkyl halides
- Organophosphates
- Epoxides
- Carboxylic acid esters
- Carbamates
- Nitriles

There are many organic functional groups that are completely or relatively inert with respect to hydrolysis. Tables 2.1 and 2.2 give a brief list of those compounds generally inert and potentially susceptible to hydrolysis, respectively.

Table 2.1
Types of Organic functional Groups That Are Generally Resistant To Hydrolysis

Alkanes	Aromatic nitro compounds
Alkenes	Aromatic Amines
Alkynes	Alcohols
Benzenes/Biphenyls	Phenols
Polycyclic Aromatic Hydrocarbons	Glycols
Heterocyclic Polycyclic Aromatic Hydrocarbons	Ethers
Halogenated Aromatics/ PCBs	Aldehydes
Dieldrin/Aldrin and Related Halogenated Hydrocarbon Pesticides	Ketones
Sulfonic Acids	Carboxylic Acids

Table 2.2
Types of Organic Functional Groups That Are Potentially Susceptible to Hydrolysis

Alkyl Halides	Nitriles
Amides	Phosphonic Acid Esters
Amines	Phosphoric Acid Esters
Carbamates	Sulfonic Acid Esters
Carboxylic Acid Esters	Sulfuric Acid Esters
Epoxides	

2.2. General Theory

2.2.1. Introduction

Fundamentals that are relevant to hydrolysis are nucleophilic substitution, and mass transfer (diffusion). Contaminants that are treated by hydrolysis generally are either sorbed to the soil or are free floating in the soil matrix.

Glossary:

Elimination - the loss of an atom

Optically active substance - a substance that rotates a plane of polarized light

Levorotatory - rotation is to the left

Dextrorotatory - rotation is to the right

Carbocation ion - (Carbonium ion) a group of atoms that contains a carbon atom bearing only six electrons

Good leaving group - generally a weak base

Isomer - different compounds that have the same molecular formula, (i.e., they contain the same kinds and numbers of atoms but are arranged differently.)

Nucleophile - "nucleus loving;" a reactant that has a partial or complete negative charge that enables it to bond with an electron-deficient atom

Electrophile - a reactant that has a partial or complete positive charge and will bond with an atom that has an available electron pair

S_N1 - substitution nucleophile unimolecular

S_N2 - substitution nucleophile bimolecular

Racemic - a chemical compound containing equal quantities of dextro-rotatery and levorotatery isomers so that it does not rotate the plane of incident polarized light

Rearrangement - the migration of a group from one atom to the very next atom, in order to create a more stable molecule, generally occurs with carbocations

Steric - spatial arrangement of atoms in molecule

Dielectric constant - the ratio of electric flux density produced in a vacuum by the same field

Inverted configuration - a product of reaction whose configuration is opposite of that of the reactant

2.2.2. Fundamentals

This section presents the general principles used in the design of a hydrolysis destruction unit operation. There are two principles that must be considered in the design of a hydrolysis unit operation:

- nucleophilic substitution
- diffusion rate theory

2.2.2.1. Nucleophilic Substitution

Hydrolysis occurs when a nucleophile attacks an electrophile and displaces a leaving group. The general reaction is:

$$\underset{\text{substrate}}{R\text{-}X} + \underset{\text{nucleophile}}{H_2O} \rightarrow R\text{-}OH + \underset{\text{leaving group}}{X^-} + H^+ \qquad [1\text{-}1]$$

The carbon compound that undergoes substitution is the substrate and the group that is displaced from the carbon is the leaving group. This reaction may follow either the S_N1 or S_N2 mechanism. There is a general rule of thumb to determine if a hydrolysis reaction follows a S_N1 or S_N2 mechanism: if an Alkane is substituted with a tertiary alkyl, a benzyl, or an allylic group, it is likely to follow a S_N1 mechanism. In most other cases, if a good leaving group is present, it will follow a S_N2 mechanism.

2.2.2.1.1. S_N1 Mechanism

The S_N1 mechanism is unimolecular because the rate determining step in the kinetics only involves one molecule (see 1-2). In general, S_N1 reactions follow 1st order kinetics.

$$RX \rightarrow R^+ + X^- \qquad \text{Slow}$$
$$R^+ + H_2O \rightarrow ROH + H^+ \qquad \text{Fast}$$

[1-2]

S_N1 reactions are kinetically characterized by:

- a rate dependence on the concentration and nature of the nucleophile
- the formation of racemic products from optically active material
- the enhancement of rate by electron-donating substituents on the central atom

S_N1 mechanisms are favored by:

- R-systems that form stable carbonium ions
- X-systems that are good leaving groups (e.g. halide ions)
- high dielectric constant solvents such as water

2.2.2.1.2. S_N2 Mechanism

The S_N2 mechanism is bimolecular because the rate-determining step involves the collision of two particles. (see [1-3]). In general, S_N2 reactions follow second order kinetics.

$$H_2O + R\text{-}X \rightarrow [H_2O \cdots R \cdots X] \rightarrow H^+ + HO\text{–}R + X^- \qquad [1\text{-}3]$$

S_N2 reactions are kinetically characterized by:

- a rate that depends on the concentration and identity of the nucleophile and substrate

- an optically active starting material gives a product of inverted configuration

The S_N2 mechanism is the general mechanism in the hydrolysis of carboxylic acid esters, amides, or organophosphorus compounds. In general, the S_N2 case is favored by:

- R-systems with low steric hindrance and low carbonium ion stability
- X-systems that are poor leaving groups (e.g. NH_2^- or $CH_3CH_2O^-$)
- organic solvents

2.2.2.2. Chemical Kinetics

Hydrolysis processes are generally modeled as having first-order or pseudo-first-order kinetics. In hydrolysis, the rate of disappearance of the substrate RX is proportional to the concentration of [RX].

$$\frac{d[RX]}{dt} = -\mathbf{k_T}[RX] \quad [1\text{-}4]$$

where:

$\mathbf{k_T}$= hydrolysis rate constant

This assumption of first-order rate kinetics is important because it implies that the hydrolysis half-life of RX is independent of the concentration of [RX].

$$t_{\frac{1}{2}} = \frac{0.6931}{\mathbf{k_T}} \quad [1\text{-}5]$$

Thus, it also implies that results of high concentrations of [RX] can be extrapolated to low concentrations of [RX] assuming constant conditions (e.g. pH, temperature).

The general rate expression:

$$\frac{d[RX]}{dt} = -\mathbf{k_T}[RX] \quad [1\text{-}6]$$

is over simplified. $\mathbf{k_T}$ is a rate constant that may include contributions from nucleophilic substitution and acid:base catalyzed hydrolysis. In order to account for these possibilities, the following equation may be used:

$$\mathbf{k_T} = k_H[H^+] + k_0 + k_{OH}[OH^-] + \sum k_{HA_i}[HA_i] + \sum k_{B_j}[B_j] \quad [1\text{-}7]$$

where:

$\mathbf{k_T}$	=	Total hydrolysis rate constant
k_H	=	Rate constant for specific acid-catalyzed hydrolysis
k_0	=	Rate constant for neutral hydrolysis. Equals $K_{H2O}[H_2O]$
k_{OH}	=	Rate constant for specific base-catalyzed hydrolysis
k_{HA}	=	Rate constant for general acid-catalyzed hydrolysis

k_B	=	Rate constant for general base-catalyzed hydrolysis
$[H^+]$	=	Hydrogen ion concentration
$[OH^-]$	=	Hydroxyl ion concentration
$[HA]$	=	General acid concentration
$[B]$	=	General base concentration
i,j	=	Indices to identify different acids and bases that may be present

$k_H[H^+]$ represents a specific acid-catalyzed hydrolysis. The catalyst is the hydronium ion. This catalysis is common in both S_N1 and S_N2 reactions of RX compounds.

The term k_0 represents the neutral hydrolysis. This term can be written as a second order rate constant, however, for S_N2 reactions, it is generally treated as a psuedo-first-order rate.

The third term, $K_{OH}[OH^-]$, represents specific base-catalysis by the hydroxide ion, which is observed generally for all substrates that undergo S_N2 reactions.

The last two terms for general acid and base catalysis, other than by H^+ and OH^-, are generally neglected because it is impossible to predict the types and concentrations of acidic and basic species that may be present in aquatic environments.

It should be noted that the assumption of psuedo-first order reaction rates may cause significant error in the prediction of hydrolysis efficiency.

2.2.2.3. Factors that influence the rate of hydrolysis

The rate of hydrolysis is influenced by:

- pH
- temperature
- solvent composition
- catalyst

2.2.2.3.1. Effect of pH

pH will have a great effect on the overall rate of hydrolysis constant, especially for acid-or-base catalyzed reactions. Figure 2.1 [1] shows the pH dependence of the hydrolysis rate. There are complex pH rate profiles for organic species that undergo acid:base dissociation in the environmental pH range of 5-9 [1].

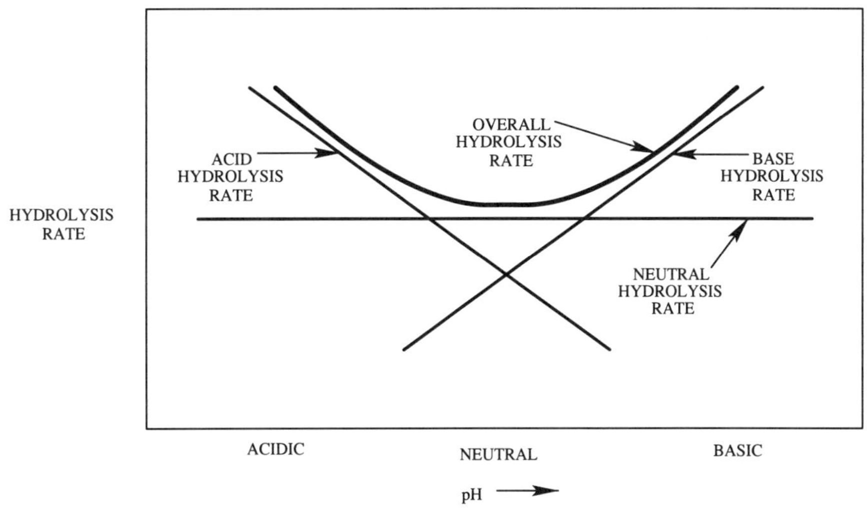

FIGURE 2.1
pH Dependence on Hydrolysis Rate by Acid, Neutral, and Base Promoted Processes
Chemical Destruction

2.2.2.3.2. Effect of Temperature

The rate of hydrolysis is also dependent upon temperature. As temperature increases, the rate of hydrolysis for organic chemicals also increases. The Arrhenius equation is generally used to describe the relationship between temperature and the rate of hydrolysis.

$$k = Ae^{-E_A/RT}$$ [1-8]

where:

E_A	=	The Arrhenius activation energy (Kcal/mol)
R	=	Gas constant (1.987 cal/deg•mol)
T	=	Temperature (K)
k	=	Rate of hydrolysis (general) ($time^{-1}$)
A	=	Pre-exponential factor (same units as k)

The log of A is the intercept of the plot k versus 1/T where the slope is equal to:

$$\frac{E_A}{2.303\ R}$$ [1-9]

Another equation used to show temperature dependence is derived from the Eyring rate theory. It is:

$$k = \frac{kT}{h} e^{-\Delta H^{\neq}/RT} e^{\Delta S^{\neq}/R}$$ [I-10]

where:

k	=	Boltzmann's constant
h	=	Planck's constant
$\Delta S^{\neq}$	=	Entropy of activation of transition complex
$\Delta H^{\neq}$	=	Enthalpy of activation of transition complex

$\Delta H^{\neq}$ can be calculated from the slope of the plot of k/T versus 1/T and $\Delta S^{\neq}$ can be calculated from the intercept of the following equation:

$$\log\frac{k}{T} = \log\frac{\mathit{k}}{h} - \frac{\Delta H^{\pi}}{2.303RT} + \frac{\Delta S^{\pi}}{R} \qquad [1\text{-}11]$$

These aforementioned equations are applied independently to k_H, k_0, and k_{OH} instead of $\mathbf{k_T}$, the overall rate of hydrolysis constant. This is because the catalyzed and uncatalyzed reaction pathways are usually different in their temperature dependencies.

2.2.2.3.3. Effect of Solvent on k

Until recently, little consideration was given to the effect of solvent composition on the rate of hydrolysis. It is now known that hydrolysis rates are much greater in water than in a water and organic solvent mixture. Therefore, it is important to note what solvent was used in the determination of the rate of hydrolysis when reviewing literature. Hydrolysis rates for organic solvent mixtures may be taken as conservative for a compound dissolved in water.

When using water as the solvent, ion strength impacts the rate of hydrolysis. For ionic strengths greater than 0.1 M, increases in ionic strength may either accelerate or retard hydrolysis.

2.2.2.3.4. Effect of Catalyst

Mabey and Mill (1978) postulate that alkaline earth and heavy metal ions can catalyze hydrolysis reactions, thereby increasing the rate of hydrolysis. However they note that more research must be done to determine the actual effect of heavy metal ions on the rate of hydrolysis. Yet, this potential effect should be noted when extrapolating data.

2.2.2.4. Estimation of Rate of Hydrolysis

Because experimental determination of the rate of hydrolysis is not always feasible or practical, different methods have been developed to estimate the rate of hydrolysis. However, it is important to note that estimation methods are only valid when the hydrolysis pathway is reasonably simple and straightforward and the product(s) can be predicted.

A general procedure for estimating the rate of hydrolysis of an organic chemical has been compiled by Lyman, *et al*[2].

1) Categorize the organic chemical in terms of functional groups present; identify hydrolyzable groups
2) Determine whether k_H and/or k_{OH} in the hydrolyzable groups are potentially significant in the pH range of 5-8

3) If k_H is required, estimate it from correlations

4) Estimate k_0 from correlations

5) If k_{OH} is required, estimate it from correlations

6) If k_H, k_0, and/or k_{OH} refer to temperature T_2 other than 298K (25°C) convert to 25°C by:

$$\log k_{25°C} = \log k_{T_2} - 3830 \left(\frac{T_2 - 298}{298\ T_2} \right)$$

[1-12]

where T_2 is in Kelvin, and an average $\Delta H^{\neq}$ or E_A of 17.5 Kcal/mol has been assumed

7) Calculate $\mathbf{k_T}$ for pH(s) of interest

$$\mathbf{k_T} = \left(k_H \times 10^{-pH}\right) + k_0 + \left(k_{OH} \times 10^{pH - 14}\right)$$

[1-13]

8) Calculate the hydrolysis half-life ($t_{1/2}$) by:

$$t_{1/2} = \frac{0.693}{\mathbf{k_T}}$$

[1-14]

There are many different correlations, each to be used with specific types of compounds. Lyman, *et al*, have created Table 2.3, showing estimation methods that may be used for different compounds. These methods are outlined in[2].

2.3. Applications

2.3.1. General Applications

Hydrolysis may be applied to:

- soils contaminated with nitriles
- soils contaminated with epoxides and esters

An example of the application of alkaline hydrolysis is presented by Robuck *et al* (1989), where iron-complexed cyanide compounds found in industrial waste are hydrolyzed under alkaline conditions at elevated temperatures and pressures. These complexes are generally resistant to destruction by conventional technologies used to treat cyanide-bearing waste. Robuck et al found in this study that the rate of hydrolysis may be estimated with the Arrhenius equation, this is shown in Figure 2.2 [3].

Table 2.3
Characteristics of Estimation Methods

Estimate[a]	Basis	Chemical Classes Covered[b]
k_H	Hammet Correlation	Ring-substituted benzamides; ethyl benzoates
k_H	Taft Correlation	Ortho-substituted Benzamides
k_0	Hammet Correlation	Benzyl Halides; Dimethyl benzyl halides; Benzyl tosylates. (All in mixed organic/ aqueous solvents)
k_{OH}	Hammet Correlation	Benzene ring-substituted compounds based on $ArCOOCH_3$, $ArCOOCH_2CH_3$, $ArCH_2COOCH_2CH_3$, $ArCH{=}CHCOOCH_2CH_3$, $ArCONH_2$, $ArOCOCH_3$, $ArCH_2OCOCH_3$, $ArCON(CH_3)_2$, $ArCONHCH_3$, $ArCH_2Cl$, and $ArOSi(CH_2CH_3)_3$. (All in mixed organic/aqueous solvents.)
k_{OH}	Taft Correlation	Dialkyl phthalate esters
k_{OH}	Correlation with pK_a of Leaving Group	Aryl esters of methylphosphonic acid ($(CH_3)_2CHOP(O)(CH_3)OAr$); carbamates of the form: (1) $(C_6H_4)NHCOOAr$ (2) $CH_3N(C_3H_5)COOAr$; (3) $CH_3NHCOOAr$; or (4) $(CH_3)_2NCOOAr$.

2.2.3. Mass Transfer Limitations

In the remediation of contaminated soils, the desorption of the contaminant from the soil particle and/or the diffusion of the contaminant into the liquid phase, may be the rate limiting step in hydrolysis processes. Therefore, it is pertinent to the design of a hydrolysis unit that mass transfer considerations be applied. The Chemical Extraction/Soil Washing module has an extensive overview on mass transfer, which should be reviewed.

2.3.2. Advantages and Limitations

2.3.2.1. Advantages of Hydrolysis

Hydrolysis offers the following advantages:

- primary reagent is water
- alkaline conditions can be produced inexpensively

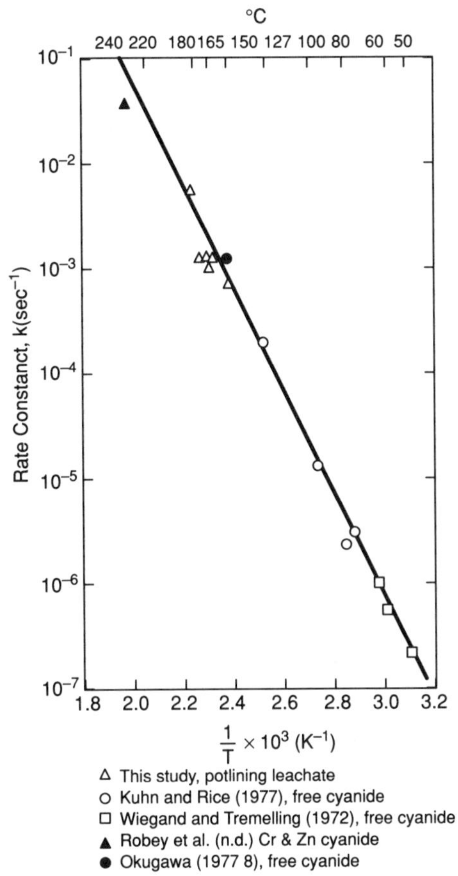

FIGURE 2.2
Rate of Alkaline Hydrolysis
Chemical Destruction

2.3.2.2. Limitations of Hydrolysis

Hydrolysis may be limited due to:

- mass transfer limitations
- pH conditions that may leach minerals out of soil

3. DECHLORINATION

3.1. General Description

3.1.1. Introduction

Dechlorination is a destruction process that uses a chemical reaction to remove chlorine atoms from chlorinated molecules. The purpose of dechlorination in remediation is to make a toxic substance less toxic and more water soluble in order to expedite the remediation process. Several U.S. patents may be applicable to the technology. Key factors involved in dechlorination design include:

- type of contaminant
- gradation of soil
- reagent used

The contaminated soil is mixed with an alkaline chemical reagent generically referred to as "APEG" (alkali metal polyethylene glycolate). The reagent is generally a mixture of alkali metals or their hydroxides with polyethylene glycols or their derivatives. The actual reagent is proprietary. The chlorinated contaminant is then stripped of its chlorine molecules.

3.1.2. Compounds Destroyed by Dechlorination

There are many compounds that may be destroyed by dechlorination. Table 3.1 gives a brief list of compounds potentially susceptible to dechlorination.

Table 3.1
Compounds Susceptible to Dechlorination

PCB's	Carbon tetrachloride
Polychlorinated dibenzo-p-dioxins (PCDD's)	Chloroform
Polychlorinated dibenzofurans (PCDF's)	Dichloromethane (methylene chloride)
Chlorobenzenes	Ethyl dibromide
Organochlorine pesticides	

3.2. General Theory

3.2.1. Introduction

Fundamentals that are relevant to dechlorination are elimination, nucleophilic substitution mechanisms, and diffusion. Contaminants that are treated by dechlorination are generally either sorbed to the soil or are free floating in the soil matrix.

The soil properties that affect dechlorination are:

- Soil moisture

  Alkali polyethylene glycoate reagents are deactivated by water bound in the soil. Therefore low moisture content of the soil is a necessity.

- Soil gradation

  Contaminants may be bound lightly to fine clay particles, and/or are trapped in soil particle clumps.

3.2.2. Elimination

Elimination is the removal of atoms or groups from two adjacent carbons. There are two idealized mechanisms that this reaction can follow. These mechanisms are:

- E_1 (unimolecular elimination)
- E_2 (bimolecular elimination)

E_1 is the elimination mechanism that involves one molecule in the rate determining step. E_2 is the elimination mechanism that involves two molecules in the rate determining step. These mechanisms are relatively similar to their nucleophilic substitution counterparts S_N1 and S_N2.

E1
Unimolecular
elimination

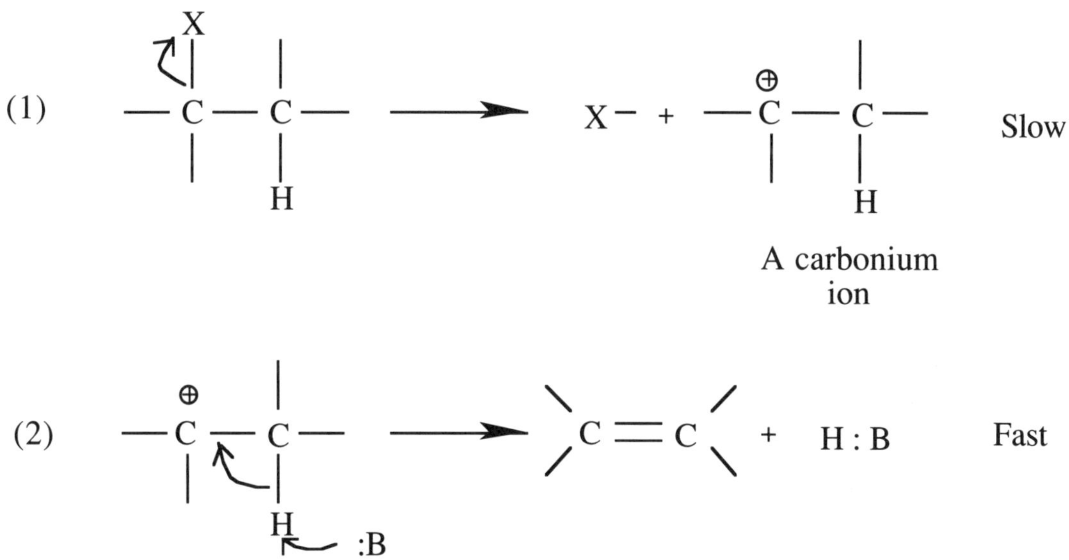

[3-1]

E2
Bimolecular elimination

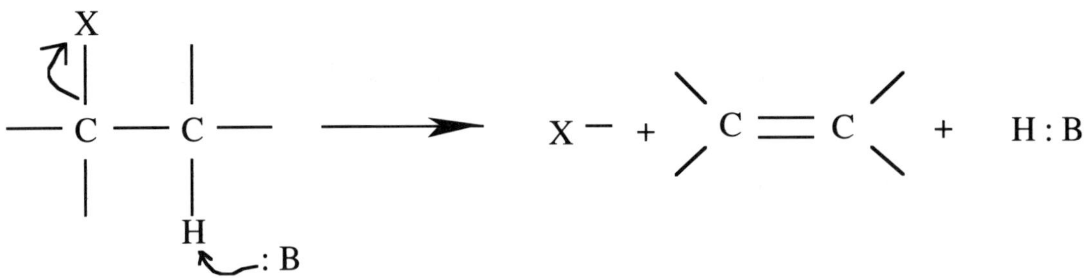

[3-2]

3.2.2.1. E_1 - Unimolecular Elimination

E_1 reactions are characterized by:

- first-order kinetics
- showing the same effect of structure on reactivity as in S_N1 reactions
- a rate that is independent of base concentrations
- rearrangement, where structure permits

The E_1 mechanism is generally favored:

- in solutions of low base concentrations
- by tertiary halides

3.2.2.2. E_2 - Bimolecular Elimination

The E_2 mechanism is the <u>principal</u> path for elimination.

E_2 reactions are characterized by:

- a rate that is dependent on base concentrations
- second-order kinetics
- no rearrangement
- no hydrogen deuterium exchange
- showing a large element effect

  <u>Large element effect</u> - the larger the halogen element (by mass) to be eliminated, the faster the reaction will proceed (R-I > R-Br > R-Cl > R-F)

3.2.2.3. Dehydrohalogenation

Dehydrohalogenation is the likely form of elimination in the dechlorination process. Dehydrohalogenation is the removal of a hydrogen from the carbon adjacent to the halide, as well as the removal of the halide to form a double bond.

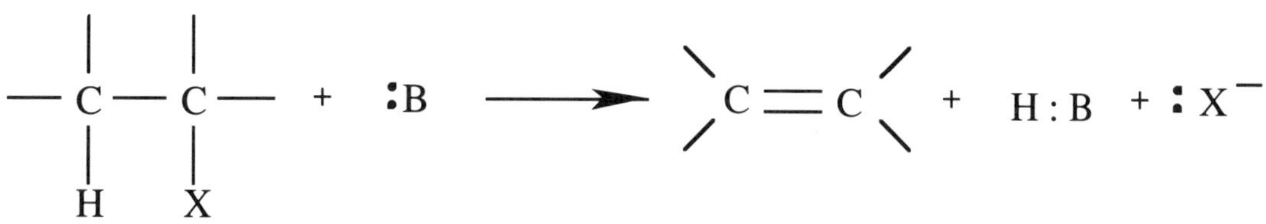

[3-3]

In dehydrohalogenation, the function of the hydroxide ion is to pull away the hydrogen ion from the carbon; simultaneously the halide separates and a double bond forms (Refer to [3-3] [4]). Dehydrohalogenation is the principle mechanism for alkyl halides.

Whether or not a reaction is going to proceed via elimination or substitution depends on the functional group and the designation of the halide. Refer to source [5] for further discussion.

3.2.3. Nucleophilic Aromatic Substitution

In dechlorination, aromatic compounds follow the nucleophilic aromatic substitution mechanism. This mechanism is similar to the nucleophilic substitution mechanism discussed in the section pertaining to Hydrolysis.

Nucleophilic aromatic substitution has two different pathways it may follow:

- bimolecular displacement mechanism (equation 3-4)
- elimination-addition mechanism (equation 3-5)

$$Ar{:}X + {:}Z \longrightarrow Ar{:}Z + X^-$$

[3-4]

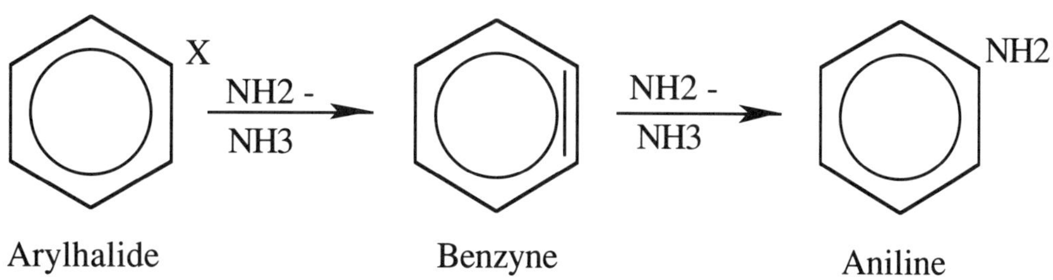

[3-5]

Bimolecular displacement is the mechanism that aryl halides will follow if they are activated by electron-withdrawing groups (:X⁻). If the ring does not contain an electron-withdrawing group, it is not activated. Substitution can take place, however, by the use of a strong base. This reaction follows the elimination-addition mechanism.

For the destruction of PCBs and other activated aryl halides, bimolecular displacement is the general path.

3.2.3.1. Bimolecular Displacement

There are two steps in bimolecular displacement nucleophilic aromatic substitution.

1) The nucleophilic reagents attack the ring to form a carbanion

2) The expulsion of halide ion from this carbanion

(See [3-6])

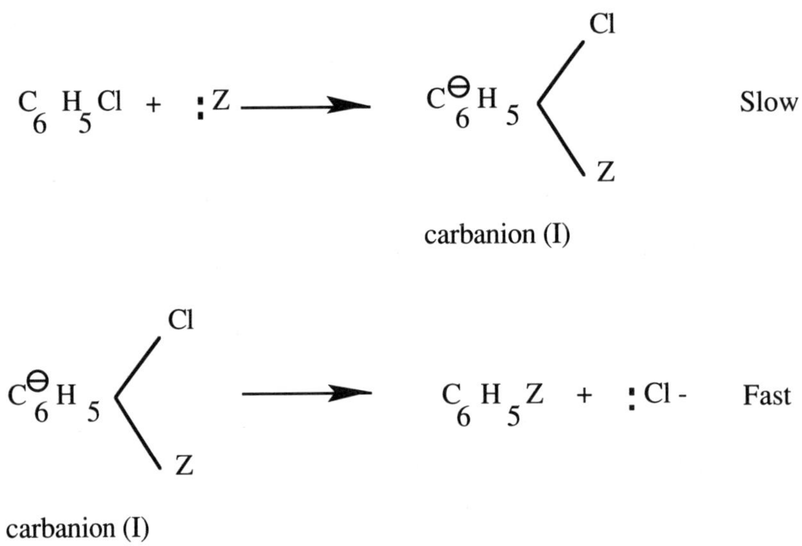

[3-6]

This carbanion is stable because it is a hybrid (see [3-7]), although it is sometimes represented by the single structure [3-8]. Thus, the first step is the rate-determining step of the overall reaction because once the carbanion is formed, it reacts rapidly to form the final product.

For a more detailed analysis of dechlorination chemistry of aryl halides, refer to references [6] and [5].

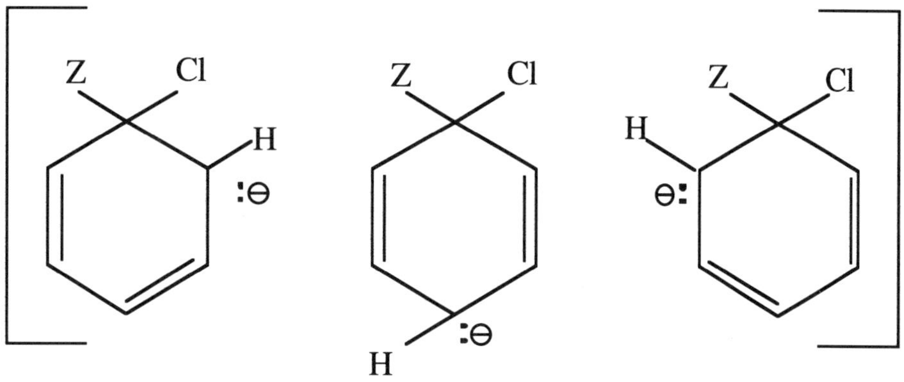

II III IV

[3-7]

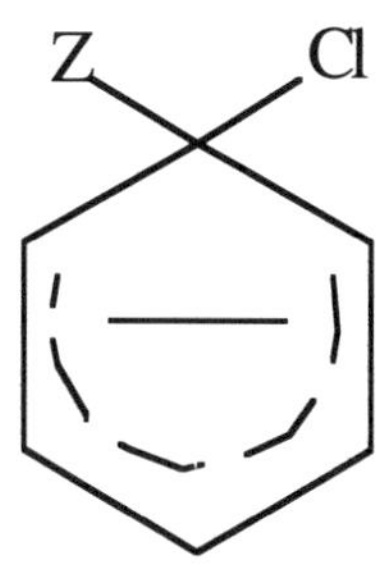

V

[3-8]

3.3. Applications

3.3.1. General Applications

Dechlorination can be applied to [7]:

- soil contaminated with activated halogenated aromatic compounds
- soil contaminated with halogenated aliphatics
- volumes greater than $1000m^3$

Dechlorination is generally an *ex situ* process.

3.3.2. Advantages and Limitations of Dechlorination

3.3.2.1. Advantages

Dechlorination offers the following advantages:

- treats soils previously considered untreatable
- lower cost for removable of PCBs

3.3.2.2. Limitations

Dechlorination is limited due to:

- mass transfer limitations
- low permeability of treated soil
- limited field data
- moisture content of the soil

4. CHEMICAL OXIDATION

4.1. General Description

4.1.1. Introduction

Chemical oxidation (CHOX) is a chemical process that chemically alters the toxic compound by oxidation to form a new less toxic compound. Chemical oxidation is part of a process known as oxidation-reduction (redox). Chemical oxidation is a traditional process that is well developed and documented. Consequently, only a brief discussion will be presented here.

Many different oxidants may be used, but the three major oxidants are hydrogen peroxide, ozone, and hypochlorites. Key factors to consider in the design of a chemical oxidation unit are:

- contaminant chemistry
- gradation of soil

4.1.2. Compounds Destroyed by CHOX

Different oxidants destroy different compounds. A general list of compounds destroyed by chemical oxidation is presented in Table 4.1.

Table 4.1
Compounds Susceptible to Chemical Oxidation By Hydrogen Peroxide and Ozone

Hydrogen Peroxide	Ozone
Alkyl Halides	Ethers
Alcohols	Esters
Glycols, Epoxides	Alcohols
Aldehydes, Ketones	Aldehydes, Ketones
Amides	Amines
Phenols	Halogenated Aromatic Compounds
Nitriles	Sulfides

4.2. Fundamentals

4.2.1. Introduction

Chemical oxidation is generally controlled be reaction kinetics, however, thermodynamic principles and mass transfer limitations may also apply.

Chemical oxidation has been applied to the treatment of water and wastewater for many years and is well documented. The principles applied to treat water and wastewater also apply to the treatment of hazardous waste. Refer to Aquatic Chemistry by Stumm and Morgan for theory and principles of chemical oxidation.

Mass transfer limitations may also apply to chemical oxidation due to the sorption of contaminants to the soil. Chemical extraction/soil washing presents, in detail, applicable theory.

4.3. Applications

4.3.1. General Applications

The applicability of chemical oxidation is controlled by the oxidant used.

In general, it can be stated that chemical oxidation can be applied to soils contaminated with:

- unsaturated aliphatic hydrocarbons
- halogenated phenolic compounds
- halogenated aromatic compounds
- sulfides and other organic contaminants

4.3.2. Advantages and Limitations

Advantages and limitations of this process are proprietary of the oxidant used.

5. BY-PRODUCT FORMATION

When considering any chemical destruction process, the by-products of the reaction must be identified and evaluated for environmental impact. This is very important, especially in the case of multi-constituent contamination, because reactions between the contaminants, by-products and reagent may cause additional by-products to be formed that are hazardous. Therefore, it is imperative that bench-scale and pilot-scale analysis be performed.

6. FEASIBILITY OF CHEMICAL DESTRUCTION

6.1. Introduction

In general, the applicability of a chemical destruction process is based on:

- soil and site characteristics
- contaminant type and concentration

6.1.2. Prescreening

A detailed information and guidance on the development and screening of remedial alternatives is provided in the RI/FS guidance (EPA 1988a). However, a brief outline of the necessary data needed to prescreen chemical destruction processes as a treatment is provided in the next section.

6.1.2.1. Necessary Data

- Matrix type (i.e. soil, sludge, sediment)
- Volume of media
- Contaminant type
- Treated product characteristics
- Past performance on similar waste contaminants and matrices

7. GENERAL PROCESS DESCRIPTION

7.1. Introduction

In general, chemical destruction processes typically consist of the following components:

- Soil pretreatment system
- Reactor tank
- Wash tank
- Reagent recovery system
- Control building

7.2. Soil Pretreatment System

Depending on the gradation of the soil to be treated and the amount of miscellaneous material present (tree limbs, barrels), pretreatment of the soil may be necessary.

This treatment usually consists of screening devices to remove unwanted particles and a crusher to break down soil particles into sizes that can be handled by the rest of the system.

7.3. Reactor Tank

The reactor tank is the most important unit in the chemical destruction operation. The tank or reactor may be of any design configuration (i.e. batch, counter-current, etc...), although it must insure proper mixing between the soil and reagent. Short circuiting of the soil must be avoided.

7.4. Wash Tank

The wash tank process separates the treated soil from the reagent and washes (rinses) the soil to remove any residual reagent.

7.5. Reagent Recovery System

The reagent recovery system depends on the reagent and the destruction by-products. Many conventional treatment processes used in water and wastewater treatment may be applicable to this process.

7.6. Control Building

The control building will contain all electrical systems necessary to operate a chemical destruction operation and an analytical laboratory with necessary applications to monitor the chemical destruction operation.

8. GENERAL DESIGN

8.1. Design Protocol

8.1.1. Introduction

The most important factors that impact the design of a chemical destruction process are:

- Type of contaminant
- Solvent
- Kinetics
- Heterogeneities
- Contact time

Optimum efficiency in chemical destruction is dependent on:

- Mass Transfer and Chemical Kinetics

8.1.2. Design Parameters

Initial parameters that must be determined prior to design are:

- Rate at which soil is to be treated
- Effluent concentrations
- Influent contaminant concentrations
- Sorption isotherm (mass transfer)
- Kinetics (chemical)
- Volume of soil to be treated
- Soil gradation

8.1.3. Design Protocol

Step 1
Using the list of contaminants present, determine the most appropriate destruction method

Step 2
Using the site information (i.e. site geology, hydrogeology, etc...) and contaminant concentrations, analyze the data and determine if proposed treatment is appropriate. Identify and quantify the soil fraction containing contaminant.

Step 3
Determine if the process will be mass transfer limited or chemically kinetic limited. This is done by determining the mass transfer coefficient, which may be determined by bench-scale analysis or by the correlation formulas (see [8] for detailed discussion), and the chemical kinetic constant. The chemical kinetic constant may be determined by correlations presented in this module and in other handbooks or by bench-scale analysis. If possible, bench-scale verification of the chemical kinetic constant is recommended. Compare transport and transformation rates to determine the rate-limiting process. This rate-limiting process will determine the minimum residence time of soil in the operation.

Step 4
Because the reagents used in chemical destruction processes are usually toxic or hazardous, care must be taken to ensure safety on site. Equipment used for reagent storage, handling, and processing must be selected to be compatible with specific reagents. The equipment required for reagent storage, handling, and processing generally consists of:

- main storage tanks, bins, or silos
- temporary holding tanks and batching vessels
- feed systems
- required process control equipment

Step 5
With the basic operations schematic, perform a mass balance calculation on the system. Use the required solvent to soil and volumetric flow rate of the soil and the estimated removal rates. A literature review and/or bench-scale analysis is required to determine system parameters.

Mass balance equations determine the volume of solvent that must be added and expected reaction times ensure proper soil/solvent fluid ratios.

Step 6

the total volume per unit operation is obtained by multiplying the flow rate by the residence time per unit operation. The general equation is:

$$\left(\begin{array}{c}\textbf{Volume per Unit}\\ \textbf{Operation}\end{array}\right) = \left(\begin{array}{c}\textbf{Soil and Extraction Fluid}\\ \textbf{Flow Rate}\end{array}\right)\left(\begin{array}{c}\textbf{Residence}\\ \textbf{Time}\end{array}\right)$$

Step 7

Design the wastewater treatment system per conventional standards.

Step 8

With the modified design, build a pilot scale operation and test on site. With the new data, modify the design and, if needed, test again on site. Continue until reasonable scale-up similitude is achieved. Build full-scale operation.

REFERENCES

1. Amdurer, M., *et al.* Systems to Accelerate In Situ Stabilization of Waste Deposits. EPA/540/2-86/002. 1986.

2. Lyman, W. J., W. F. Reehl and D. H. Rosenblatt. Handbook of Chemical Property Estimation Methods. Ed. Society, A. C. Vol. Washington, D.C., 1991.

3. Robuck, S. J. and R. G. Luthy. "Destruction of Iron-Complexed Cyanide by Alkaline Hydrolysis." Walee Science Technology, **21**: 547, 1989.

4. Stumm, W. and J. J. Morgan. Aquatic Chemistry. Ed. Vol. Wiely - Interscience, New York, 1981.

5. Morrision, R. T. and R. N. Boyd. Organic Chemistry. Ed. Vol. Allyn and Bacon, Inc., Boston, 1987.

6. Brunelle, D. J. and D. A. Singleton. "Destruction/Removal of Polychlorinated Biphenyls From Non-Polar Media. Reactions to PCB with Poly(Ethylene Glycol)/KOH." Chemosphere, **12**: 183, 1983.

7. PEI Associates, I. Guide for Conduction Treatability Studies Under CERCLA: Chemical Dehalogenation. EPA Risk Reduction Engineering Laboratory, November, 1990.

8. McCabe, W. L., J. C. Smith and P. Harriott. Unit Operations of Chemical Engineering. Ed. Vol. McGraw-Hill, New York, 1985.

J - Bioremediation

BIOREMEDIATION

Instructional objectives:

- Discuss basic microbial processes occurring during bioremediation
- Quantify mass transport effects in bioremediation
- Discuss kinetics of bioremediation
- Identify optimum conditions for bioremediation
- Determine feasibility of alternative bioremediation methods
- Develop a systematic approach for using bioremediation

1. MICROBIAL PROCESSES OCCURRING DURING BIOREMEDIATION

1.1. Overview

Bioremediation is defined as the use of microorganisms to mediate the transformation of hazardous chemicals to less toxic and environmentally acceptable compounds.

Bioremediation is a process that utilizes microbial organisms, or their metabolic products to convert complex substances (substrates) to simpler products, typically reducing their toxicity to the environment. The process breaks down the substrate and utilizes the chemical energy liberated to drive the thermodynamically non-spontaneous process, such as synthesis of the cell components. The process converts complex substances, using them as an energy source or carbon source, to simpler compounds. Bioremediation typically effects in an increase in the redox potential of the substrate. When an organic compound is used as a carbon source, and oxygen is used as the terminal electron acceptor (aerobic processes), carbon dioxide is typically liberated as a byproduct of the process.

Biological treatment can be an efficient and cost effective way to remove hazardous substances from:

- Liquid
- Vapor
- Solid Phases

A myriad of bioremediation technologies have been explored, ranging from attempts to utilize bacterial cultures typically found in the environment to sophisticated genetic engineering applications that produce tailor made organisms capable of degrading specific compounds. Among the innovative or emerging methods are:

- Plasmid-assisted molecular breeding
- Gene-engineering
- Enhancement of indigenous microorganisms.

Xenobiotic compounds are typically somewhat recalcitrant to biodegradation. Xenobiotic compounds are chemical structures that involve novel chemical groups or substitutes found rarely in nature. Hence, these compounds undergo slow biological transformations, tending to persist and accumulate in the environment.

The fate and transport of xenobiotic compounds introduced into the environment are impacted by:

- Sorption
- Coagulation
- Precipitation
- Biotransformation
- Photolysis
- Hydrolysis
- Oxidation / Reduction

Naturally existing microorganisms adapt to degrade xenobiotic compounds either by:

- Utilizing existing catabolic enzymes, or
- Acquiring new metabolic pathways

Bioremediation can be a feasible remedial technology if the microorganisms are capable of acquiring new metabolic pathways and synthesizing appropriate enzymes to achieve the desired degradation within a reasonable period of time.
The generalized chemical reactions for a microbial growth can be depicted as follows:

Substrate + TEA + ATP ----> Oxidized substrate + Reduced TEA + more ATP

ATP + Substrate/Intermediate products ----> Microbial cells

TEA = Terminal Electron Acceptor
ATP = Adenosine Triphosphate (Energy currency of the cells)

1.2. Microorganisms

Microorganisms fall under the kingdom of Protists. Because Protists consist of the fundamental and earlier forms of life, they have simple metabolic and physiological requirements. The Protists are further divided into two categories, Eucaryotes and Procaryotes.

Procaryotes - consist of organisms like bacteria, blue green algae and a few others having rather "simple" cell structures. The typical length of procaryotic cells is 5 μm and weighs around 10 pico grams.

Bacteria, classified as prokaryotic organisms, generally reproduce by the process of binary fission, in which a cell enlarges due to new cell-wall growth and then divides, producing two daughter cells of equal size.

Eucaryotes - consist of protozoa, algae, fungi etc. having larger and more advanced cell structure.

Chemical formulas have often been used to represent the proportions of various elements in a bacterial cell. The most commonly used formula is $C_6H_7NO_2$. Typical bacteria cell composition consists of 50% carbon, 20% oxygen, 10-15% nitrogen 5% hydrogen and a small fraction of other elements like phosphorus and sulfur. Since carbon constitutes the major elemental fraction of a cell, carbon sources are important factors for cell growth. Carbon is required for the synthesis of amino acids, enzymes, nucleic acids, vitamins etc. Cells vary in their ability to synthesize materials they require to thrive and reproduce.

1.3. Classification

All microorganisms require a carbon source for cell growth and an energy source to sustain the metabolic functions required for growth. The nature of microbial metabolism is diverse and therefore the source of carbon and energy used by the microbes define their classification. The following definitions illustrate some of the various categories.
Microbe classifications are based on:

C source:

- Heterotrophic - use organic compounds (e.g. glucose) as a carbon source
- Autotrophic - use inorganic carbon compounds (e.g. carbon dioxide) as a carbon source

Energy source:

- Chemotrophic - chemical reactions provide the energy for growth. If inorganic compounds provide the energy they are called Lithotrophic; if the source of energy is from an organic compound, they are known as Organotrophic
- Phototrophic - use sunlight or other form of light as the energy source

TEA:

- Aerobes - oxygen is the terminal electron acceptor
- Anaerobes - a compound other than oxygen is the terminal electron acceptor
- Facultative - can degrade both in absence or presence of oxygen, though the growth yield is normally higher in presence of oxygen

These terms are often used in combination with each other.

> Example: A human is a Hetero-organotroph, meaning he/she uses organic carbon for cell growth and organic chemicals for the energy required for growth. Similarly plants are auto-phototrophs and require an inorganic carbon source, carbon dioxide, and sunlight as energy sources. Plants are auto-phototrophs in the presence of sunlight as explained above; in the absence of sunlight, however, plants may be classified as Hetero-organotrophs.
>
> Sometimes terms like "obligate anaerobe" are used for classification, which indicates that the microorganism can not grow in the presence of air. Therefore, the species of this microbe would be more abundant in the oxygen depleted zones.

Aerobic microorganisms are often used to treat water streams and other media exposed to air. Aerobic treatment can also be used to bioremediate subsurface environments as long as an oxygen source is provided. Anaerobic organisms, while useful for treatment of organic sludges and slurry organic waste, may not be advantageous for subsurface bioremediation in view of their low growth rate and slow response in subsurface environments. Applicability of anaerobic treatment in subsurface environments is currently under research.

Having knowledge of growth characteristics of these organisms assists in predicting their habitats and feasibility for use in environmental remediations; likewise, knowledge of environmental conditions assists in predicting the type of species and degradation pathways likely to be present.

For example, when glucose is degraded aerobically, it can be represented by the equation given below:

$$C_6H_{12}O_6 + 6O_2 \text{ -----> } 6CO_2 + 6H_2O + \text{ENERGY.}$$

Similarly, anaerobic degradation of glucose in the presence of nitrate may be written as:

$$C_6H_{12}O_6 + 4NO_3 \text{ ---------> } 6CO_2 + 6H_2O + 2N_2 + \text{ENERGY}$$

In view of the lower position of oxygen in the electron tower, as compared to nitrate, more energy is liberated under aerobic oxidation of glucose. i.e. when oxygen is used as the terminal electron acceptor as compared to anaerobic oxidation with nitrate as the TEA. Due to the lower energy liberated in anaerobic oxidation, the rate of cell synthesis is lower in anaerobic environments as compared to aerobic ones.

In addition to carbon and energy sources, microorganisms also require nutrients for growth and for metabolic activities.

1.4. Definitions

Catabolic Process - the biochemical process involved in the breakdown of complex compounds to simpler molecules, usually leading to the production of energy; e.g. Glycolysis breaks down glucose to pyruvate, yielding 2 moles of ATP for every mole of glucose.

Anabolic Process - the biochemical process involved in the building up of cell components and other macro molecules, usually requiring energy; e.g. protein synthesis (carried out in ribosomes) synthesizes proteins from amino acids.

Oligotrophic - describing a body of water in which nutrients are in low supply

Xenobiotic - a completely synthetic chemical compound not naturally occurring on earth

Macronutrients - elements needed in major quantities - C, K, Mg, Ca, H, N, O, S, P, Na, Cl

Micronutrients - elements required in small quantities - Co, Ni, Fe, Zn, Mo, Wb, Se

Neutralism - when there is no competition for substrate or any interaction between two species in a given microbial community

Mutualism - when two species are mutually benefitted by each other in a microbial community. For example, in a pond, the bacteria produce carbon dioxide from a carbon source only in the presence of oxygen. The carbon dioxide helps algae synthesize cell mass in the presence of sunlight, releasing the needed oxygen as a by-product

Antagonistic - when the growth of one species hampers or inhibits the growth of another species. This can occur in several ways: competition for resources, amensalism (release of chemicals having deleterious effects on other organisms), and predation

Comensalism Metabolism- when one species is benefited by the growth of another while the other is unaffected, e.g.. growth factor secretion, cellulose degradation resulting in free monomers that may act as substrates for others

Gratuitous - when enzymes for biodegradation are induced due to similarity of biodegradation between compounds, e.g.. Cresol is degraded by the microbes degrading Toluene due to the similarity in the structure of the two compounds

Cometabolism - transformation of a non-growth substrate in the presence of another transformable compound, e.g.. TCE is degraded in the presence of methane and oxygen but remains unaffected otherwise

Further information on the basic mechanism of biodegradation is provided in the appendix.

2. MASS TRANSPORT ISSUES DURING BIOREMEDIATION

Bioremediation processes can be used to treat contaminants present in the liquid phase, vapor phase and/or solid phase. The microorganisms needed for bioremediation can either be present as a suspension in the liquid phase or attached to a solid surface. Mass transport in bioremediation refers to the transport of the reactant (contaminant, nutrients) to the microorganisms, which is necessary for biodegradation of the contaminant to occur.

Biodegradation occurs in the liquid phase when the microorganisms are suspended in water and assimilate chemical compounds dissolved in the liquid phase. The chemical compounds diffuse to the cell wall of the microorganisms and are then eventually transported into the cell for biological growth and energy. The potential for microorganisms to consume or biodegrade a chemical compound is directly dependent on the chemicals ability to dissolve in water (solubility) and diffuse through the cell wall (diffusion coefficient in water).

If a solid phase is present, such as soil, debris, etc., the microorganisms may attach to the solid phase, forming a biological film on the solid surface. In this case, the chemical compound diffuses through the biological film covering a solid surface, and then adsorbs on the cell walls. It then undergoes two types of transport:

- Diffusion through the cell wall into the cell, wherein it is consumed (biodegraded) for growth and energy
- Diffusion to the walls of other adjacent cells, i.e. diffusion through the biofilm, and then subsequent biodegradation by the other cells constituting the biofilm

The net result is the gradual diffusion of the compound through the biofilm and simultaneous biodegradation by the microorganisms.

As a result of this biodegradation and consequent growth of the cell population, the biofilm thickness increases, which increases the resistance to diffusion. Eventually, a biofilm thickness is achieved, where the microbial cells attached to the solid surface on the inner surface of the biofilm are able to completely consume the amount of the chemical compound reaching the cells through the diffusion process; hence, no additional compound enters the solid phase. This results in a balance between the amount of the compound diffusing into the biofilm and the amount biodegrading in the biofilm (steady-state).

The actual thickness of a biofilm depends on several factors:

- The rate of diffusion of the chemical compound to the biofilm (external diffusion) and the rate of diffusion through the biofilm (internal diffusion)

- The rate at which the cells are removed from the biofilm by death, decay and/or predation by other species and also by hydrodynamic shear.

2.1. Mass Transport in The Liquid Phase

There are two mechanisms by which a chemical can be transported through a liquid phase:

- Diffusion due to molecular motion, known as "Fickian Diffusion"
- Transport by eddies formed in the liquid phase due to local turbulence caused by mixing, known as "Eddy Diffusion"

Diffusion can be characterized by a diffusion equation defined as follows:

$$R = (D_m + D_e)\left(\frac{dC}{dx}\right) \tag{1}$$

where:

R	= the rate of transport of contaminant or chemical per unit area per unit time
D_m	= the diffusion coefficient due to molecular motion (Fickian diffusion)
D_e	= the diffusion coefficient due to movement of eddies (Eddy diffusion)
dC/dx	= the concentration gradient in the liquid phase

Fickian diffusion coefficients, D_m, are measured experimentally for a given chemical and liquid, and tabulated in handbooks. They can also be estimated, if not measured experimentally.

Eddy diffusion coefficients depend on the type of system, amount of mixing, liquid phase properties, etc.

Since eddy diffusion coefficients are not known generally, the term $(D_m+D_e)(dC/dx)$ in equation (1) is simplified as follows:

$$(D_m+D_e)\left(\frac{dC}{dx}\right) = k(C1 - C2) \tag{2}$$

where:

k= the "mass transfer coefficient"

C_1,C_2 = concentrations in the liquid phase

The mass transfer coefficient, k, is calculated from empirical equations based on experimentally determined correlations.

2.2. Mass Transport in Biofilms

When a solid phase is present in the process, cell attachment to the solid surface and subsequent cell growth results in the formation of a biofilm on the solid surface. In this case, biodegradation occurs in this biofilm through the process of simultaneous diffusion and biodegradation.

Fixed bed processes behave similar to plug-flow reactors, where the liquid phase flows through the system without mixing. Fluidized bed processes behave like well-mixed reactors where the liquid phase is continuously mixed by the fluidized beads.

The mass transport in biofilm processes can be quantified similarly to the liquid phase processes. The rate of transport through the biofilm can be written as follows:

$$R = D_f\left(\frac{dC_f}{dz}\right)$$

where:

R = rate of transport per unit area per unit time through the biofilm
D_f = the Fickian diffusion coefficient of the chemical in the biofilm
dCf/dz = the concentration gradient of the chemical in the biofilm

Note that only the Fickian diffusion coefficient is used here since the biofilm is stagnant, i.e., it is attached to the solid surface.

Furthermore, at steady-state, when the rate of transport of chemicals through the biofilm is balanced by its rate of degradation, we can write:

$$R = \text{rate of chemical biodegradation in the biofilm}$$

The rate of biodegradation, presented later, will be quantified using equations of chemical kinetics.

2.3. Mass Transport in The Solid Phase

Bioremediation in solid phase such as in soils and sediments, involves the transport of the contaminant from the solid phase to the microorganisms. The contaminant desorbs from the solid surface. The rate of desorption depends on the type of chemical and the nature of its interaction with the solid phase. Usually, the solid phase contains pores, and the contaminant diffuses out of the pores into the liquid or gas phase. The microorganisms can be present as a suspension in the liquid phase within saturated pores or attached to the solid phase either inside or outside the pores.

In principle, diffusion rates can be quantified by equation where there is no eddy diffusion, i.e., D_e=0, and the Fickian diffusion coefficient, D_m, is replaced by a diffusion coefficient representing diffusion in the pores of the solid phase. However, diffusion in the pores of a solid phase, such as soil or sediment, has several complexities. Some of these are:

- The diffusion coefficient is confined to the pores of the solid phase, since the contaminant does not diffuse through the solid phase itself-therefore the path is tortuous and unpredictable due to heterogeneous distributions of pore sites
- The effective diffusion coefficient is not independent of concentration
- Adsorption and desorption on the solid phase surface occurs
- The diffusion coefficient depends on the properties of the solid phase (soil) such as mineral composition, organic matter content, bulk density, water content, and temperature

3. BIOREMEDIATION PROCESS APPLICATIONS

Bioremediation processes can be classified into three main types:

- Above-ground treatment of contaminated soils
  - Biopiles/landfarming/composting

  - Slurry reactors
- Above-ground treatment of contaminated water
  - Bioreactors
- *In situ* treatment
  - Saturated zone
  - Unsaturated zone

3.1. Aboveground Treatment of Soils

Aboveground or ex-situ treatment of contaminated soils can be achieved using biopiles, landfarming, or composting.

3.1.1. Biopiles

A biopile is created by spreading contaminated soil on a lined treatment bed. Air and/or nutrients dissolved in water are passed through the soil pile to enhance biodegradation of the contaminants present in the soil.

There are two types of biopiles:

- Oxygen and nutrients are delivered to the bacteria by water movement through the soil pile. The water is added by an irrigation system and the bed is equipped with a drainage collection system. The drainage from the soil pile, or leachate, is treated in a liquid phase bioreactor and recirculated through the soil pile. The soil pile may be totally enclosed if volatile- emissions control is necessary. Figure 3.1 shows aboveground solid-phase bioremediation using an irrigated biopile
- Nutrients are mixed with the soil when the soil pile is created and oxygen is delivered by pumping air through the soil. The air is delivered by pipes placed at the bottom and middle of the soil pile. A plastic liner is placed on top of the pile to prevent rainwater from permeating into the pile

3.1.2. Landfarming

Landfarming is the practice of spreading contaminated soils or wastes on the surface of the ground to facilitate natural microbial degradation of contaminants. Landfarming was the first method used for bioremediation of soils and sludges and has been successfully applied by the petroleum industry for disposal of petroleum-refinery wastes. Landfarming is typically used in areas where air emission would not pose unacceptable health risks and would not require collection.

3.1.3. Composting

Composting is an established technology for management of municipal wastewater treatment sludges and is an emerging technology for treatment of hazardous wastes. Biodegradation of organic compounds occurs within a compost matrix consisting of contaminated soils and/or sludges mixed with organic carbon sources and bulking agents such as straw, bark, wood chips, sawdust, etc. The compost piles, which are three to six feet in height, are placed on lined treatment beds and the microbial degradation is enhanced by maintaining optimum temperature, moisture content, oxygen, and nutrient levels within the pile.

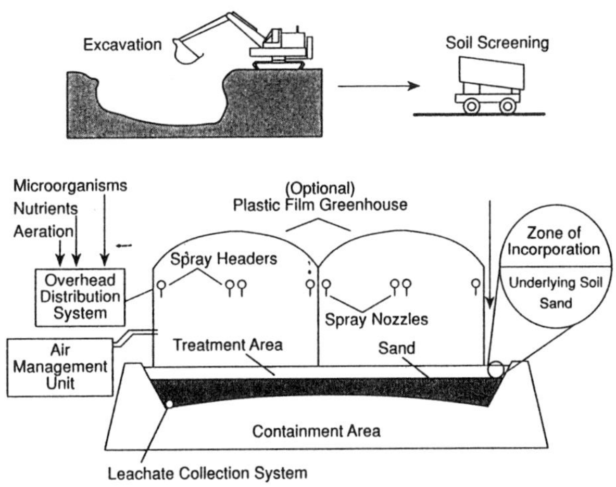

FIGURE 3.1
Schematic of an Aboveground Solid Phase Bioremediation Process Using an Irrigated Biopile
Biological Remediation

3.1.4. Slurry reactors

In a slurry phase reactor, soil or sludge and nutrient-containing water are combined to form as aqueous slurry wherein biodegradation of the contaminants occurs. Mixing must be sufficient to keep the solids in suspension and oxygen must be supplied throughout the slurry-phase to promote aerobic microbial activity.

Slurry-phase biotreatment may be performed either in bioreactor vessels or in lined lagoons.

3.2. Aboveground Treatment of Water

Aboveground treatment of contaminated water can be achieved in either suspended growth reactors or fixed film reactors. In suspended growth reactors the microorganisms are kept suspended in the aqueous phase. In fixed film reactors the microorganisms are grown in biofilms covering solid surfaces, and the contaminated water is circulated to maximize contact with the biofilm.

3.2.1. Suspended Growth Reactors

Suspended growth bioreactors are of two types:

- Completely-mixed systems, in which the microbial cells are suspended in the liquid phase, and the suspension is stirred to obtain a mixed state between the cells and the liquid phase
- Plug flow systems, in which the microbial cell suspension flows without axial dispersion in a tubular reactor

If oxygen is required for the cells, the water is either saturated by oxygen before it enters the reactor, or air containing oxygen is introduced to the liquid phase (air sparging). The chemical compounds and nutrients, dissolved in the water, diffuse into the cells and are biodegraded into carbon dioxide or transformed into metabolite products.

The well-mixed liquid phase process can be operated in either a "batch" or "continuous" mode. In the batch operating mode, the liquid phase and microbial cells are charged into the reactor, mixed, and when the compounds present in the liquid phase have been biodegraded, to target concentrations, the reactor contents are discharged. As oxygen is slightly soluble in the culture medium , a continuous oxygen supply is typically needed for aerobic processes. This is generally achieved by aerating the medium.

The main advantages of batch or discontinuous operation are:

- Lower capital cost for equipment and automated systems
- Greater flexibility by varying the reaction time and operating conditions, such as temperature, pH, concentrations, etc.
- Greater adaptability to variations in influent contaminant concentrations and waste stream volume.

The main disadvantages of batch operation are:

- Non-productive idle time during filling, heating/cooling, and decanting operations
- Higher operating expenses, since pumps have to be operated to fill and empty the reactor, and controls are needed for maintaining reactor liquid level, requires analytical testing of each batch.

In the continuous mode, there is a continuous flow of liquid and cells into and out of the reactor.

The main advantages of continuous operation are:

- Lower reactor volume for a given production rate
- Consistent product quality, i.e., extent of biodegradation can be controlled by maintaining the operating conditions

The main disadvantages of continuous operation are:

- Lower flexibility in handling changing influent characteristics, since it is difficult to change operating conditions
- More process control to maintain operating conditions
- Higher capital costs

In the plug flow or tubular flow bioreactors the reactor contents are not well-mixed but rather flow in plug-flow fashion through the reactor. A concentration profile develops between the inlet of the plug-flow reactor, where the concentration of the contaminant is high, and the outlet of the reactor where the contaminant concentration is low. Due to the presence of this concentration profile in the reactor, the rate of biodegradation varies from the inlet to the outlet, which is different from the completely mixed or continuous stirred-tank system where the reaction rate is constant.

Table 3.1 gives a comparison of total reactor volumes required for achieving a specified removal efficiency for a first order biodegradation reaction:

Table 3.1
Reactor Volumes Required to Achieve Required Removal Efficiency

Number of reactors in series	90% removal efficiency	98% removal efficiency
1 CSTR	9.00	49.00
2 CSTRs	4.32	12.14
8 CSTRs	2.64	5.04
1 PFR	2.30	3.91

Note: Volume of each reactor expressed in units of Q/k
CSTR - Continuous Stirred-Tank Reactor
PFR - Plug-Flow Reactor

It should be noted that as the number of continuous stirred-tank reactors connected in series increases, the total reactor volume approaches the volume of a single plug-flow reactor.

3.2.2. Fixed Film Reactors

Fixed film reactors can be classified into two main types:

- Stationary fixed film reactors
- Moving fixed film reactors

In fixed film reactors the microorganisms are immobilized on the surface of a solid surface. The immobilized microorganisms, called a biofilm, are contacted with the liquid phase containing the contaminants. The contaminants diffuse into the biofilm, wherein the contaminants undergo degradation.

In the stationary fixed film reactor the solid phase is fixed and the liquid flows through the reactor in contact with the solid surface. In some cases, the biofilm is immobilized on the inside surface of hollow fibers or outside surfaces of solid tubes, and the liquid flows through the reactor contacting the biofilm.

In a stationary fixed film reactor the liquid phase can be completely-mixed or flow through the reactor in plug-flow fashion. The level of mixing in the liquid phase depends on reactor design.

A moving fixed film reactor consists of solid particles with the biofilm attached to it and suspended in the liquid phase. The liquid phase can be mixed mechanically using a mixer or the particles can be fluidized in the liquid phase, wherein flow turbulence causes the particles to mix in the reactor. The solid particles are settled from the liquid phase after exiting from the reactor and recycled back to the reactor inlet.

In a stationary fixed film reactor the biofilm stays in the reactor until it decays and sloughs off the solid phase and exits with the liquid phase. In a moving fixed film reactor, the biofilm stays in the bioreactor for a fixed time, depending on reactor design.

Examples of fixed film reactors includes:

- Trickling filters
  - Biofilm immobilized on solid phase, and the liquid phase containing the contaminants trickles down through the packed bed
- Rotating biological contactors
  - Biofilm immobilized on rotating plates that alternately contact that water containing the contaminants and air as they rotate

- Submerged fixed film reactors
  - — Biofilm immobilized on plastic packing and packed in a bed with the liquid phase containing the contaminants completely filling the void spaces bed may be considered fluidized.
  - — Air may be pumped from the bottom of the bed
- Powdered activated carbon treatment systems
  - — Biofilm immobilized on powdered activated carbon, which is mixed with liquid containing the contaminant

3.3. In-Situ Treatment

In-situ treatment can be applied to both inorganic and organic contaminants. In in-situ bioremediation, four main requirements must be satisfied as illustrated in figure 3.2:

- Presence of bacteria
- Availability of Terminal Elector Acceptor (TEA)
- Availability of nutrients
- Favorable environmental factors (e.g. pH, temp, moisture)

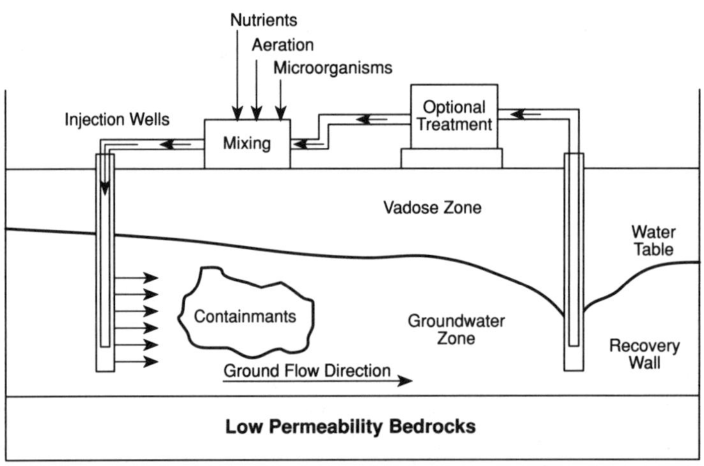

FIGURE 3.2
Schematic of an In Situ Bioremediation Process for Contaminated Groundwater
Biological Remediation

When indigenous bacteria are not present, suitable bacterial populations must be developed to effect biodegradation. TEA and nutrients have to be supplied for effective aerobic bioremediation. Distribution of the TEA, nutrients and microbial seeds is dependent on the soil property. A granular or lumpy soil results in the transport of these species to be diffusion limited resulting in a small growth rate. The problem is made worse in cases of soils with low permeability. Thus, in order to simulate microbial growth in subsurface environments, proper distribution of the TEA, nutrients and the microbial seeds must be ensured. Other environmental factors, such as temperature, pH, and soil conditions, must be maintained in proper range to sustain biodegradation.

In-situ treatment can be applied to both saturated and unsaturated (vadose) zones. If the desired biodegradation is an aerobic process, oxygen must be supplied to the microorganisms. In the saturated zone oxygen can be supplied by saturating the water with oxygen and then injecting it into the ground. Although this technique has been demonstrated to some success for previous soils the method does not work effectively for soils with low permeability. Other means of supplying oxygen include addition of hydrogen peroxide, which decomposes to oxygen and water. If the process is anaerobic, other terminal electron acceptors such as nitrate or sulfate can be supplied to enhance biodegradation. An effective way to remediate soils in the vadose zone or in the top layers of the saturated zone, is to depress the water table to create a thicker unsaturated zone. It is easier and more efficient to remediate the residues on soil in the vadose zone as compared to the saturated zone, in view of the higher oxygen trapped in the soil pockets of the vadose zone as compared to the dissolved concentration in the saturated zone. In addition, contaminant mobility in the vadose zone is less stringent compared to the saturated zone. In the unsaturated zone, water pumping is not required as part of clean-up. Enhancement of biological activity in the unsaturated zone is achieved by passing water with dissolved oxygen or air through the zone.

Figure 3.2 shows the process for in-situ bioremediation of groundwater.

4. KINETICS OF BIOREMEDIATION

4.1. Bioremediation Phases

The various microbial growth and decay stages occurring during bioremediation can be visualized using a batch model, wherein the cell mass concentration as a function of time, may be represented by Monod kinetics. Figure 4.1 represents the biodegradation growth stages associated with biological remediation. Seven different growth phases can be distinguished:[1, 2, 3]

Starting (lag) phase - When microorganisms are initially exposed to a chemical, it takes them some time (time lag) to get acclimated to the environment in which the chemical is present

Acceleration phase - In this phase, when the cell has adjusted to the environment and starts to propagate, the microbes synthesize the growth factors and enzymes required for the degradation

Exponential phase - Both the cell concentration-time curve and cell-mass concentration-time curve depict an exponential function. This exponential relationship results from the basic fact that cell growth occurs by cell division, called binary fission, as mentioned earlier. Hence, if the initial cell density is n_0 at time t_0, then after m time divisions, the cell density becomes

$$n = n_0 * 2^m \tag{4}$$

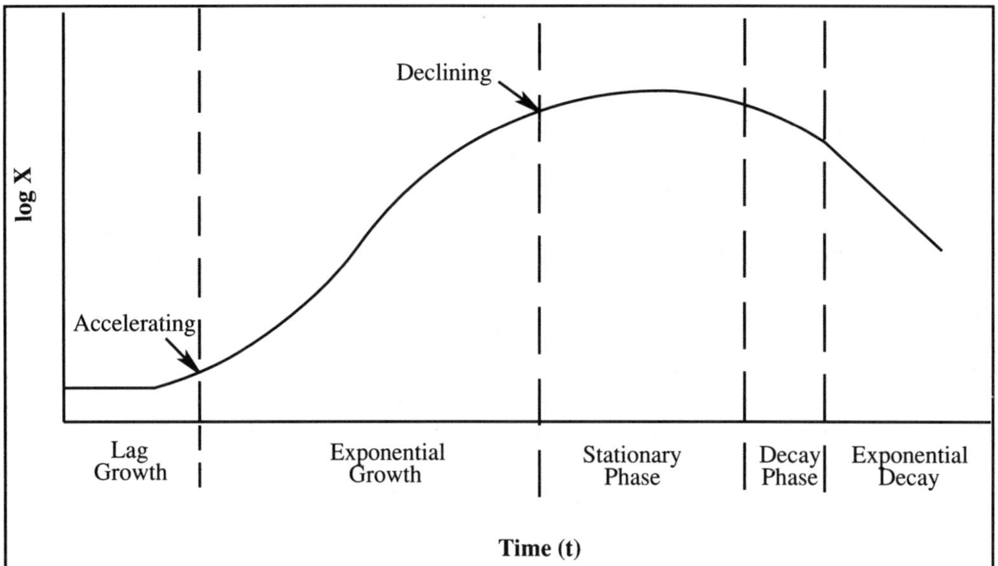

FIGURE 4.1
Biodegradation Growth Stages
Biological Remediation

Equation (4) can be rewritten as:

$$m = \frac{[\log(n) - \log(n_o)]}{(\log 2)} \tag{5}$$

The rate of increase in cell-mass concentration (x) can be written as follows:

$$R_x = \frac{dX}{dt} = \mu X \tag{6}$$

where:

R_X = the growth rate and μ is the specific growth rate

Equation (6) can be rewritten as:

$$\mu = \left(\frac{dX}{dt}\right)\left(\frac{1}{X}\right) \tag{7}$$

The specific growth rate, μ, is maximum when (dX/dt) is maximum, and can be expressed as:

$$\mu_m = \left(\frac{dX}{dt}\right)_{max}\left(\frac{1}{X}\right) \tag{8}$$

Note that μ_m is constant during the exponential phase.

$$n = n_0 * 2^m$$

$$\log n = \log n_0 + m \log 2$$

Where; $m = t / k$
t = total time elapsed since the beginning of division
k = time required for one cell division

$$\log n = \log n_0 + \frac{t}{k} \log 2$$

Differentiating the above expression with respect to t gives:

$$\frac{1}{n}\frac{dn}{dt} = \frac{\log 2}{k}$$

n = the number of cells in a unit volume
Replacing n by x (the microbial density)

$$\frac{1}{x}\frac{dx}{dt} = \frac{\log 2}{k}$$

Defining the Specific Growth Rate μ as:

$$\mu = \frac{\log 2}{k}$$

$$\frac{dx}{dt} = \mu x$$

Equation (8) can be integrated using the boundary condition that $X = X_o$ when $t = 0$

$$X = X_o \exp(\mu_m t) \tag{9}$$

The cell-mass doubling time, t_d, can be determined from equations (9):

$$t_d = \frac{\ln(2)}{\mu_m} \tag{10}$$

Declining phase - In this phase, the growth starts retarding, which is mainly due to the fact that the chemical used for growth has been depleted or the metabolite products are inhibiting further growth.

Stationary phase - Cell propagation is balanced by cell decay. The cell concentration reaches a plateau value.

Decay phase - The cell decay is higher than the propagation rate.

Exponential death phase - The number of living cells decreases exponentially, when cell growth or propagation has ceased completely.

4.2. Rate Constants and their Applicability

Standard methods for determining the biodegradability of a compound in the aquatic environment produce results in terms of percentage removal after a fixed time period.[4] Although this can be useful for determining the "biodegradability" of a compound in the environment, it cannot be used for relative quantifying the rate of biodegradation. A quantitative expression for the rate of biodegradation is needed to calculate and predict the fate of the chemical in the environment and determine its distribution and concentration in the environment in the presence of other competing abiotic processes, such as adsorption and volatilization.

Quantitative measurements of intrinsic biodegradation kinetics, independent of the test system, are useful for developing reliable quantitative structure-biodegradability relationships, which enable biodegradation kinetics to be predicted for new chemicals and allow synthetic chemicals to be modified for enhanced biodegradability.[5, 6]
The kinetics of biodegradation have been described by a variety of mathematical expressions, ranging in complexity and applicability. A review of some of the commonly used expressions is given below.[3]

First-order reactions:
If biodegradation is a first-order reaction, then the rate of substrate depletion (R) is proportional to the substrate concentration:

$$R = -\frac{d[S]}{dt} = k_1 [S] \tag{11}$$

where:

$-d[S]/dt$	= the rate of disappearance of the biodegrading chemical
t	= time
k_1	= the first-order rate constant
$[S]$	= the concentration of the test chemical in the test system at any given time t

The rate of product (carbon dioxide) formation, can be mathematically described as follows:

$$R(CO_2) = \begin{cases} 0 & \text{for } t < t_d \\ a\left(1 - e^{\left\{-k_l\left(t - t_d\right)\right\}}\right) & \text{for } t > t_d \end{cases} \qquad (12)$$

where:

$R(CO_2)$ = the rate of carbon dioxide generation due to biodegradation
a = the extent of biodegradation
t_d = the lag time prior to onset of biodegradation due to acclimation.

Care should be exercised when using the half-life equation for describing biodegradation, since at low concentrations of the chemical, the rate of chemical transport can become limiting, and the reaction expression may not be sufficient to predict the temporal profiles of concentration. In such cases, substrate depletion, for example, may be governed by diffusive characteristics of the environment rather than the rate of reaction.

The rate of biodegradation, in the environment, also depends on the concentration of the microorganisms, responsible for biodegradation. However, if the concentration of microorganisms can be assumed to be constant in the test system, i.e., if the test chemical concentration is so small that growth of microorganisms can be ignored, then the rate constant, k_l, in equation (12) incorporates the constant biomass concentration value. Variations in the quantity of biomass can change the rate constant, k_l, and this is usually the reason for large variations in the reported values of the rate constant, k_l, between different test sites. The use of first-order kinetics can be verified by checking whether k_l varies (> 10 fold)over a wide range of test chemical concentration.

Representative values of the first-order rate constant, k_l, are given in table 4.1.

Table 4.1
Representative First - Rate Order Constants

Chemical	k_l (day^{-1})	Sample	Reference
Benzene	0.12	River	Lee and Ryan (1979)
Chlorobenzene	0.07-0.3	River	Bartholomew and
	0.01	Marine	Pfaender (1983)
4-Chlorophenol	0.035	River	Lee and Ryan (1979)
Methyl parathion	0.30	River Sediment	Cripe et al. (1987)

Second-order reactions:

When the concentration of the microorganisms does not remain constant in the test system, due to growth, then the rate of biodegradation may be written as follows:

$$R = -\frac{d[S]}{dt} = k_2[S][X] \tag{14}$$

where:

k_2	= the second-order rate constant
[S]	= the concentration of the test chemical in the system
[X]	= the concentration of the microorganisms in the test system

The above second-order equation implies that the rate of biodegradation increases with the concentration of the test chemical and the biomass present. However, the rate of biodegradation reaches a plateau with increases in the test chemical concentration. A possible explanation of this phenomenon is, that as the substrate concentration is increased, it reaches a point where the environment becomes to toxic for micro - organisms to reproduce. Hence, the above second-order equation is valid only in a narrow range of test chemical concentration.

It should be noted that the second-order equations involve the concentration of the microorganisms, [X]. This microbial concentration can be obtained from either plate count tests or other enumeration methods. It is generally assumed that:

- Plate counts (or other enumeration methods) detect similar percentages of the total microbial population in different samples;
- All, or a constant proportion, of the enumerated bacteria are transforming the chemical; and
- Transformation rates per cell are approximately the same for all sites examined.

The second assumption is usually valid for initial hydrolysis of the chemical, but generally does not hold for the "ultimate" biotransformation to carbon dioxide. For example, in the hydrolysis of pesticides, such as malathion, the enzymes involved in the hydrolytic biotransformation are present in most bacteria, and, hence, the concentration of bacteria as measured by enumeration is proportional to the hydrolytic enzyme concentration. The variation in the second-order kinetic constant, k_2, as measured experimentally at different sites for the hydrolysis of malathion is small. Similar small site variations in k_2 values have been reported for primary biodegradation of phenol, para-substituted phenols, and various esters of 2,4-D in North American fresh waters.
Generally, all the bacteria present in natural populations are not equally active in the degradation of a specific test chemical. Therefore, the use of a second-order equation, wherein [X] is used as the concentration of the entire microorganism population is not always valid, and may lead to large site variations for the kinetic constant k_2.

Monod Growth and Second-Order Kinetics

If the concentration of the microorganisms increases continuously due to growth resulting from the biodegradation of the test chemical, then the rate of biodegradation can be described by the following equation, first proposed by Monod (1949):

$$R = -\frac{d[S]}{dt} = -\left(\frac{1}{Y}\right)\left[\frac{[S][X]}{[K_s + [S]]}\right] \quad (15)$$

$$\frac{d[X]}{dt} = (\mu_{max})\frac{[S][X]}{|K_s + [S]|} \quad (16)$$

where:

[X]	=	the concentration of the microorganisms
μ_{max}	=	the maximum specific growth rate
Y	=	the yield coefficient (usually about 0.5 for aerobic cultures)
[S]	=	the concentration of the test chemical
K_S	=	a constant numerically equal to [S] at 0.5 u_{max} (i.e., the test chemical concentration at which growth occurs at half the maximum rate)

The Monod equations, as applied to enzyme-catalyzed reactions, were first proposed by Michaelis and Menten in 1913.

The specific growth rate of the microorganisms, μ, can be written as follows:

$$\mu = \left(\frac{1}{[X]}\right)\frac{d[X]}{dt} \quad (17)$$

The Monod equations have been successfully used to describe the rate of removal of synthetic chemicals in laboratory studies where the test chemical was the sole source of carbon and energy. However, it is important to note the limitations of the Monod equations.
The limitations of the Monod equations are as follows:

- It is assumed that the microbial growth occurs on the test chemical only and that the test chemical is the single growth-limiting chemical in the system. Thus, in in-situ applications where other chemicals present in the environment may compete for the growth substrate, Monod equation may not represent the microbial growth that accurately.
- The value of the yield coefficient, Y, is a constant. However, there are data which indicate that at low concentrations of the test chemical, a large amount of the energy derived from the biodegradation of the test chemical is used for maintenance energy requirements and not for growth
- Laboratory studies have also shown that many synthetic chemicals can be transformed by microorganisms which cannot use them as the sole source of carbon and energy. This phenomenon, called co-metabolism, is defined as the transformation of a non-growth chemical in the obligate presence of a growth chemical. The non-growth chemical is a compound that is unable to support cell replication

The Monod equation cannot be applied for the biological transformation of a non-growth chemical, since μ_{max} and Y are zero.

4.3. Applicability of Various Kinetic Models

It has been proposed that the variety of substrate disappearance curves encountered in biodegradability testing is the result of interactions between substrate concentration, [S], and population density, [X], (Simkins and Alexander, 1984). To accommodate changes in population density, a mass balance equation is used to relate the population density and the substrate (test chemical) concentration.

$$[So] + \frac{[Xo]}{Y} = [S] + \frac{[X]}{Y} \tag{18}$$

where:

$[S_o]$ = the initial concentration of the substrate (test chemical)
$[X_o]$ = the initial concentration of bacteria
$[S]$ = the test chemical concentration at a given time
$[X]$ = the bacteria concentration

Combining the above equation with the Monod equations, yields a general differential equation, as follows:

$$R = -\frac{d[S]}{dt} = \mu_{max}[S]\left(\frac{(S_0 - S)Y + X_0}{\{Ks + [S]\}}\right) \tag{19}$$

This equation can be simplified into simpler forms, as shown in Table 4.2. The conditions under which these models apply is also included in this table.

Table 4.2
Kinetic - Models for Biodegradation

Model	R = -d[S]/dt =	Conditions
First-order	k_l [S]	[So] < Ks [X] = [Xo] = constant
Second-order	k_2[S][X]	[So] < Ks [X] = [So] + [Xo] - [S]
Monod (no growth)	k_o[S]/{Ks + [S]}	[X] is constant [S_o] > Ks
Monod (growth)	equation (19)	[X] = [So]+[Xo]-[S] [So] > Ks
Zero-order	k_o	[S] >> Ks bacteria saturated with test chemical [X] is constant
Logarithmic	u_{max} [X]	[S] >> Ks [X] is not constant

At substrate concentrations below K_S (as would be the case for most synthetic chemicals in the aquatic environment), degradation can be described by either first-order equation, if the bacterial concentration is constant, or by a second-order equation, when the bacterial concentration is varying linearly, with time due to growth. The Monod (no growth) equations can be used when the initial concentration of the test chemical is higher than K_S, but if the bacterial yield coefficient, Y, is very low, the increase in bacterial concentration is small, i.e., the bacterial concentration can be assumed to be constant. The Monod (no growth) equation is the same as the Michaelis-Menten kinetic equation. The Monod (growth) equations are used when the initial test chemical concentration is higher than Ks and the bacterial concentration is varying due to growth. Zero-order kinetics can be used when the initial test chemical concentration is at least 50 times higher than Ks, and bacterial growth is small. Logarithmic kinetics is used when the initial test chemical concentration is significantly higher than Ks and bacterial growth occurs during biodegradation. Table 4.3 provides overall guidelines for the applicability of the various kinetic models, for various situations.

Table 4.3
Guidelines for Kinetic Model Applications

Chemical Concentration (M)	Situation	Model Type
< IOE-8	Environmental concentration	First-order
IOE-6 - IOE-5	Optimum analytical concentration	First-order Second-order Monod
> IOE-2	Spill concentration	Zero-order

4.4. Determination of Biodegradation Rate Constants

There are three types of experimental reactors that can be used to measure the kinetics of biodegradation. The three types of reactors are as follows:

- Batch reactor
- Continuous stirred tank reactor
- Fed-batch reactor

Batch Reactor System

In a batch reactor system nutrients needed for bacterial growth and water are well mixed in a closed reactor system. The concentration of the test chemical is monitored by analyzing the liquid contents periodically.[5] If it is difficult to measure the concentration of the test chemical in the liquid phase, the test chemical is obtained as a uniformly ^{14}C labeled compound. The concentration of the chemical is monitored by measuring the radioactivity due to the ^{14}C using a scintillation counter. The rate of biodegradation can be calculated using the equation:

$$R = \left(\frac{[S_o]}{t}\right)\left(\frac{\text{dpm in cells} + \text{dpm } CO_2^{14}}{\text{dpm originally added}}\right)$$

where:

- R = the rate of biodegradation
- [So] = the initial concentration of the test chemical
- t = the biodegradation time
- dpm = the activity measured in disintegrations per minute (dpm) as measured by scintillation counting

The major problems in measuring biodegradation kinetics using a batch reactor are:

- Requirement of measuring liquid concentration of the test chemical periodically
- Difficulty in providing oxygen for aerobic cultures and removing the carbon dioxide that is produced

To circumvent this problem, attempts have been made to measure the oxygen uptake and to derive the biodegradation kinetics from the oxygen uptake measurements without measuring the liquid concentration. This has been achieved by using:

- Manometric respirometry
- Electrolytic respirometry

In manometric respirometry, the carbon dioxide produced by the biodegradation process is absorbed in soda lime pellets that are kept in a container located in the headspace of the batch reactor. As oxygen is used from the headspace, and the carbon dioxide produced is absorbed in the soda lime pellets, the total pressure in the batch reactor decreases, which is monitored manometrically. A large flask is usually used so that the oxygen available in the head space is sufficient for the entire biodegradation of the test chemical. The bio-kinetics using the kinetic equations is then derived from the oxygen consumption measurement.

In electrolytic respirometry, the batch reactor flask is connected to another flask, where a copper sulfate solution in water is electrolytically decomposed to produce oxygen. A pressure differential cell monitors the pressure in the batch reactor. When a finite pressure decrease occurs in the batch reactor, the electrolytic flask is triggered to produce oxygen, which then flows into the reactor flask. The pressure decrease in the reactor flask occurs due to the absorption of the carbon dioxide produced due to biodegradation, in the soda lime pellets that are kept in the head space of the reactor flask. The voltage needed to produce the oxygen in the electrolytic flask is monitored continuously, and this information is used to calculate the oxygen uptake in the reactor flask.

The main advantages of using a batch reactor for measuring biodegradation kinetics are:

- Ability to handle volatile organic compounds, since the reactor system is closed
- Can be automated, as in the case of manometric or electrolytic respirometry

- Easy to conduct the experiment, since the reactor system is simple to operate
- No need for sampling and chemical analysis, as in the case of respirometry or when using a radio labeled compound

Measurement of oxygen consumption is one of the oldest means of assessing biodegradability. Time consuming manual measurement of oxygen uptake was gradually replaced by manometric respirometry and then later by electrolytic respirometry. A rather comprehensive review of the use of respirometers for the study of sewage and industrial waste biodegradation was published by.[6, 7]

Continuous Stirred Tank Reactor Systems
In a continuous stirred tank reactor system, a constant volume reactor is used with a continuous feed entering the reactor and a continuous waste stream leaving the reactor system. The reactor is usually well-mixed and samples are taken from the inlet and outlet liquid streams to monitor the rate of biodegradation. Oxygen needed for aerobic cultures is provided by saturating the influent stream with oxygen and the carbon dioxide produced in the reactor leaves in the dissolved state in the effluent stream.

Measurement of the test chemical concentration in the influent and effluent provides a direct measure of biodegradation. In some cases, the dissolved oxygen concentration can be measured at the inlet and outlet streams to measure the oxygen consumption in the reactor.

Continuous stirred tank reactor systems are more difficult to operate since a constant feed and effluent rate has to be maintained. This requires flow controllers, feed tank and effluent tank, level control, etc., making the entire system fairly difficult to operate. Furthermore, the test chemical can volatilize into the head space or adsorb on the surface of the microorganism cells, and these abiotic mechanisms have to be quantified before the rate of biodegradation can be measured.

Fed-Batch Reactor Systems
In a fed-batch reactor system, there is no constant influent and effluent flow, but rather the system is very similar to a batch system, except that a constant volume of feed stream is added to the reactor system at constant intervals of time, and an equivalent volume of the reactor contents is removed. This makes the fed-batch reactor system operate at a constant volume. Fed-batch reactor systems are easier to operate than continuous stirred tank reactors.

The rate of degradation determines, in part, the length of time needed to bioremediate a site. The extent to which a contaminant is degraded will determine if remediation goals are achievable. Bioremediation may not be a feasible option if the extent and/or the rate of degradation is low relative to the remediation objectives, and if it is not feasible to enhance the rate or extent by site modification. As discussed in the kinetics of bioremediation section, the rate of degradation is a function of contaminant concentration, microbial concentration, temperature, and other environmental factors.

5. SYSTEMATIC APPROACH FOR BIOREMEDIATION IMPLEMENTATION

5.1. Determining Contaminant Biodegradability

5.1.1. Screening for Biodegradability

In order to employ biodegradation as a remediation technology, the first step is to investigate the biodegradability of the compound of interest. A compound could be easily biodegradable resulting in complete mineralization, it could be persistent requiring specific conditions and presence of specific species and chemicals, or it could be recalcitrant indicating no degradation at all.

Simple experiments can be performed to test the biodegradability of the compound.

- A solution containing the test compound, as the sole carbon, energy source, and nutrients is inoculated with an acclimated microbiological culture.
- Temperature, pH, oxygen concentration and other important parameters are adjusted for the optimal yield and the culture is observed for microbial growth
- Results:
  - If the growth increases (usually turbidity is used as an indication of growth) substantially over a period of time then it may be inferred that the compound is biodegradable
  - If the turbidity does not increase for a considerable period of time a second assimilable carbon source is added
  - If growth is observed accompanied by reduction in the amount of test compound it implies that co-metabolism may be required for the successful biodegradation of the compound. *(A reduction in the quantity of a substance does not necessarily imply biodegradation. Other possible reasons for the decrease in quantity could be evaporation, sorption to cells, or apparatus etc. Thus all these possibilities should be examined before confirming biodegradation.)*
- A microbial culture may take months for acclimation and therefore sufficient time must be provided for observation before any conclusions are made.

The rates of mineralization (biodegradation) of some chemicals are directly proportional to their concentrations. Below a certain threshold concentration, usually mineralizable molecules are not degraded [2] Microorganisms may not assimilate carbon from chemicals present in trace amounts in natural environments and will not grow or produce large populations needed for enhanced biodegradation. The minimum concentration of a chemical necessary to support microbial growth must be ascertained.

5.1.2. Estimating Biodegradation Potential for Contaminant

The chemical structure of the contaminant has a profound effect on its biodegradability and biodegradation kinetics. The environmental degradation of organic compounds involves a complex interplay of organism type, degradation environment, chemical structure, and biochemical reactions. Hence, isolating the affect of chemical structure on degradation is usually difficult, except when strict experimental protocols are used to minimize the impact of the other parameters.

Qualitatively, compounds containing highly branched alkyl chains degrade slowly when compared with straight chains. Functional groups, such as hydroxyl and carboxylate groups on benzene rings, increase biodegradability, while halogen, nitro, and sulfonate groups decrease biodegradability. Ortho-substituted aromatic compounds are slightly more biodegradable than para-substituted compounds, and the meta-substituted compounds have the least biodegradable substitution pattern. The information also suggests that compounds containing either pattern are resistant to biodegradation. Other qualitative relationships imply that, as a general rule, water soluble chemicals are usually more biodegradable than insoluble chemicals. These qualitative relationships have facilitated the synthesis and development of compounds which may be discharged to the environment with some degree of safety. For example, the replacement of alkyl benzene sulfonate surfactants with linear alkyl sulfonate surfactants in detergents has been based on biodegradability considerations.

Quantitative relationships between the structure of the chemical and its biodegradation characteristics appear rarely in the literature. In recent years, molecular descriptors, such as molecular surface area and volume, have been used in the prediction of biological properties. This is a result of the bioconcentration phenomena being partially related to the area of interface between the hydrophobic molecules and water. Early studies on quantitative relationships involved the correlation of biological activity with physiochemical properties, such as octanol-water partition coefficient, aqueous solubility, density, boiling point, and molecular weight. The processes by which the chemical becomes distributed and available in water are dependent on measures of its solubility. Hence, when solubility becomes rate limiting, correlations based on solubility become good indicators of biodegradability.

Recently attempts have been made to correlate quantum chemical descriptors with biodegradation kinetics. The ready availability of computing techniques has enabled the easy calculation of some quantum chemical parameters, such as atomic charge, electronic energy, ionization potential, and dipole moment [6].

Chemical structure of the contaminant is important in assessing its biodegradability. In a study conducted to develop an expert system for biodegradation assessment, the order in which the various groups contribute to biodegradability was determined, as is shown in the following list [8]. The groups appear in a decreasing order of biodegradability:

Esters	Amide
Anhydride	Hydroxyl
Carbonyl	Epoxide
Site of unsaturation	Benzene ring
Methyl	Methylene

Hydrolyzable groups, azo bonds, halogens, and nitro groups are sites of anaerobic attack. Among the negative influences on aerobic biodegradability were molecular mass, branching, halogenation, and nitrogen heterocycles.

Catabolite repression, or substrate inhibition, occurs when a cell is faced with more than one utilizable substrate. Enzymes could be formed to catabolize all substrates, but this would be uneconomical for bacteria. Instead, enzymes are made to utilize the best substrate and only after exhaustion of the primary substrate other enzymes are formed which catabolize the "poorer" carbon sources. The most common type of catabolite repression involves the inhibition of formation of certain enzymes (i.e., repression) by the catabolite products of a readily utilizable carbon source. Catabolite repression has been historically known as the "glucose effect", because glucose is an easily assimilable substrate, and catabolite repression of other substrates was found to occur in the presence of glucose. However, other easily utilizable substrates, such as acetate and citrate can also repress the utilization of other substrates.

The potential for both induction of enzymes and catabolite repression in a cell insures that inducible enzymes are produced only in the presence of substrate. If several substrates are present, only the enzymes acting on the most favorable substrate will be produced.

5.1.3. Possibility of Co-Metabolism

Co-Metabolism is the biotransformation of a compound by a microorganism that gains little or no energy for growth from the compound. It is a fortuitous transformation brought about by non-specific enzymes produced by the organism for other purposes. Examples are oxidation by co-metabolism of chlorinated aliphatic compounds, such as trichloroethylene, 1,2-dichloroethane, and vinyl chloride by methane oxidizers that possess methane mono oxygenases. A typical mechanism of co-metabolism is given in Figure 5.1 Other co-metabolizing microorganisms that possess broad substrate specific enzymes are toluene oxidizers, nitrifying bacteria, and ethane, propane, and ethylene oxidizers. The oxygenases produced by these microorganisms can initiate oxidations of a broad range of aromatic and aliphatic compounds under aerobic conditions. Degradation of TCE can also be brought about anaerobically. However, a degradation by product is vinyl chloride, which is more toxic than TCE itself. A typical mechanism for the anaerobic degradation of TCE is presented in Figure 5.2 [9]. The addition of a carbon source is imperative when the degradative process is co-metabolic. However, more than one type of microorganism and/or carbon source can be used to co-metabolize a certain chemical reaction (e.g. trichloroethylene is co-metabolized when either methane or toluene is supplied as a primary carbon source.) In a field demonstration at Moffett Naval Air Station, methane addition was required for biodegradation of dissolved chlorinated ethenes introduced into an injection well. The concentration of trichloroethylene was diminished by 20 - 30% and vinyl chloride was reduced by 90 - 95% at a two meter distance downstream. When methane addition was ceased, 95% of the injected chlorinated ethene was detected in observation wells.

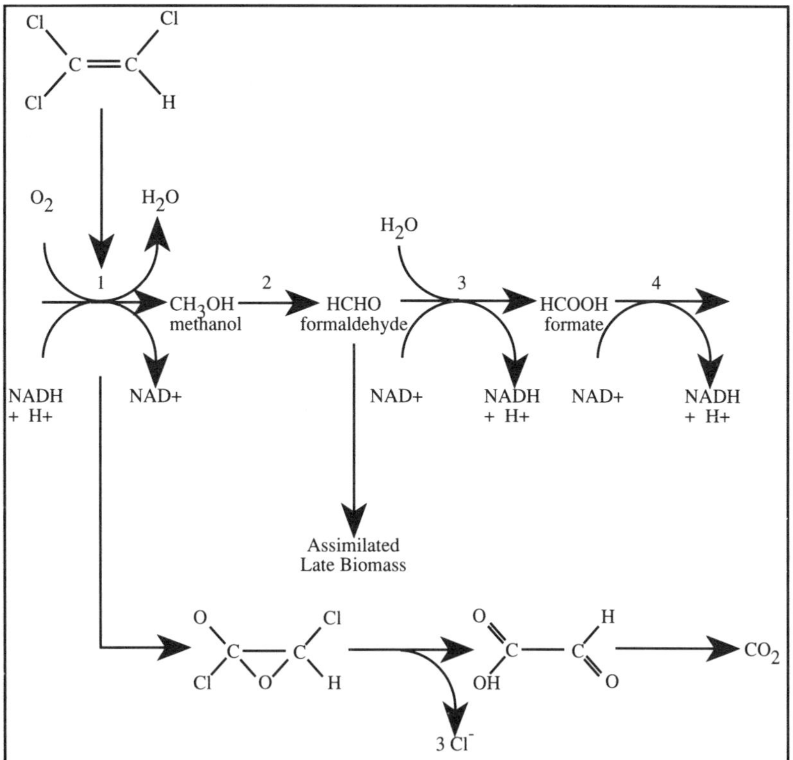

Co-Metabolism of TCE in the presence of methane. Enzymes involved: (1) methane monooxygenase; (2) methanol dehydrogenase; (3) formaldehydedehydrogenase; (4) formate dehydrogenase. Adapted from Dalton and Stirling (1984) and Little et al. (1988)

FIGURE 5.1
Mechanism of Co-Metabolism
Biological Remediation

5.1.4. Enhanced Biodegradation

5.1.4.1. Nutrient Addition

In general, it is essential to maintain a minimum level of any required nutrient in order to support a desired microbe growth rate and sustain biodegradation. However, there is also a maximum nutrient level, above which substrate inhibition occurs. Hence, it is important to define these restrictive nutrient levels. Quantification of nutrients mineral needs is especially difficult because of the occurrence of metal complexes, some cellular and other non-cellular in origin. Excessive amounts of essential nutrients such as PO_4^{3-} and SO_4^{2-} also can exert catabolite repression that affects specific metabolic pathways while not affecting growth.

A first approximation of the minimum requirements in nutrient design is based on the stoichiometry of the overall biodegradation process:

$$\text{C-source} + \text{N-source} + O_2 + \text{Minerals} + \text{Specific Nutrients} \longrightarrow$$
$$\text{Cell Mass} + \text{Products} + CO_2 + 2H_2O$$

Various empirical formulas have been proposed to represent the organic composition of microbial cells. One of the oldest and most widely accepted in the field of wastewater treatment is $C_5H_7O_2N$. Another formula which includes phosphorus is $C_{60}H_{87}O_{23}N_{12}P$. When microbial cells are incinerated, the ash left after incineration, includes potassium, calcium, magnesium, sodium, iron, and trace amounts of manganese, cobalt, copper, molybdenum, and zinc. Although these elements are known to play a role in metabolism, the specific contribution of each element is unknown, and elemental constants varies widely.

Generally nutrients can be classified into Macronutrient minerals, which include potassium, magnesium, calcium, iron and sodium, and Micro-nutrients minerals, which includes cobalt, zinc, molybdenum, copper and manganese.

Using the above empirical formulas for biomass, the amounts of Nitrogen source and Phosphorus source can be calculated using the overall stoichiometry equation. Generally the amount of phosphorus required is about one-fifth of the nitrogen requirement.

Nitrogen can be introduced in the nutrient media in either inorganic or organic forms. The most common inorganic forms are as ammonium or nitrate ions. Ammonium ions can be assimilated directly by the cell through amination of a keto acid to form glutamic acid and other amino acids, which can be incorporated into proteins and nucleic acids. When nitrate is supplied as the only nitrogen source, it must be first reduced to ammonia before the bacterium can use it.

Phosphorus is an essential component of nutrient media, since insufficient phosphorus will limit growth and prevent the complete removal of organic substrate in a manner similar to insufficient nitrogen. Usually the phosphorus is added in the form of inorganic orthophosphate.

Biodegradation can be limited by the availability of nutrients and hence the application of nutrients (sufficient nitrogen, phosphorus and mineral media) can enhance the rate and extent of biodegradation.

Mineral nutrients, in particular, nitrogen and phosphorus, are often limiting in the subsurface environment and must be supplied to ensure biodegradation of a large contaminant mass. For example, Mulkins-Phillips and Stewart found that phosphorus limited the rate and extent of growth of a Norcadia sp. on 1% (volume/volume) Bunker C fuel oil. Prior to beginning an *in situ* bioremediation project in Ambler, PA, laboratory experiments indicated that the native microorganisms could be stimulated by the addition of inorganic nitrogen, phosphate salts, and air. The carbon:nitrogen:phosphorus ratio necessary to enhance bioremediation can vary between 100:10:1 to 100:1:0.5, depending on the methodology used and the phase of the contaminant (liquid or solid phase).

Addition of nutrients is not always necessary. For example, in bioventing applications, nutrient addition is not necessary when the soil matrix contains adequate amounts of nitrogen and phosphorus. Nutrients may be needed to sustain increased rate of degradation as bacteria populations increase.

5.1.4.2. Application and Feasibility of Alternate Terminal Electron Acceptors

An organic substrate is not directly oxidized to carbon dioxide and water in a single chemical step because there is no energy conserving mechanism that could trap so large an amount of energy. Consequently, biological oxidation occurs in small steps. Basically, oxidation requires the transfer of an electron from the substance being oxidized to some acceptor molecule which will subsequently be reduced. In most biological systems, each step in the oxidation of the substrate involves the removal of two electrons and the simultaneous loss of two protons. The electrons and protons are transferred to an acceptor molecule. The acceptor molecule will not accept the protons until it has accepted the electrons, and thus it is referred to as an electron acceptor.

In the biodegradation of organic substrates, the terminal electron acceptor is an inorganic molecule, such as oxygen, nitrate, sulfate or carbonate. Depending on the type of electron acceptor, the degradation can be classified as either aerobic or anaerobic.

Aerobic Respiration:

$$\text{Organic substrate} \xrightarrow{\text{carbon flow}} CO_2$$

$$\downarrow \text{ electron flow}$$

$$O_2$$

Anaerobic Respiration:

$$\text{Organic substrate} \xrightarrow{\text{carbon flow}} CO_2$$

$$\downarrow \text{ electron flow}$$

$$NO_{3^-} \text{ or } SO_4^{=} \text{ or } CO_3^{=}$$

Depending on the conditions under which the biodegradation has to be conducted, alternate electron acceptors, such as nitrate, sulfate or carbonate, can be added to replace oxygen. In subsurface environments, where oxygen availability is limited, biodegradation using alternate electron acceptors can be conducted (for example, microbial degradation of acenapthene and napthalene has been demonstrated using NO_3^- as the TEA in soil-water systems [10].

Aerobic biodegradation is always preferred because it is faster than anaerobic processes. (Bioremediation is cost-effective only if it occurs rapidly, due to increased operation and maintenance cost over extended time. Regulatory agencies will also require rapid remediation to remove the source of continuing contamination.) For example, the biodegradation rate constant for carbofuran in soil is 0.047/day aerobically and 0.026/day anaerobically. Methods for increasing oxygen availability include air sparging, addition of pure oxygen, and addition of hydrogen peroxide during *in situ* bioremediation. The low solubility of oxygen in water limits the amount of oxygen that can be provided during *in situ* bioremediation.

The use of alternate electron acceptors has also been used in bioremediation. Nitrate addition is a method of supplying both a mineral nutrient and an electron acceptor.

5.1.4.3. Effect of pH

The pH, a measure of the hydrogen ion (H^+) activity, is particularly important as a parameter of microbial growth. The rate of biodegradation is influenced by the pH, the optimum pH and range of pH over which growth can occur being determined by the nature of the organisms present in the system. It is possible to make some generalizations that are useful in defining a chemical environment for the cell. Most bacteria have a pH optimum (i.e., that pH at which growth rate is maximum) around 6.5 - 7.5. Most environments have been found to have a pH in the physiological range (i.e. between 5 - 8.5). Most microorganisms do not grow effectively in acute acidic (pH<5) or basic (pH>8.5) conditions.

Changes in pH occur due to microbial activities: a hydrogen ion is generated during uptake of ammonium ion ($NH^{+}{}_4$), and a hydrogen ion is consumed during NO_3^- metabolism, etc.

While control of pH is important to the maintenance of an optimal environment for microbial degradation, monitoring of pH can yield useful information about the rate and extent of degradation.

5.1.4.4. Effect of Temperature

Microbial growth rate, as with all chemical reactions, is a function of temperature. In general, microorganisms will grow over a temperature range of 25 to 30° C. However, it is important to note that there are microorganisms in the environment, that will grow at temperatures below 0° C and above 93° C — the primary requirement is availability of liquid water.

Figure 5.3 shows the effect of temperature on three groups of microorganisms: the psycrophilic, mesophilic, and therophilic.

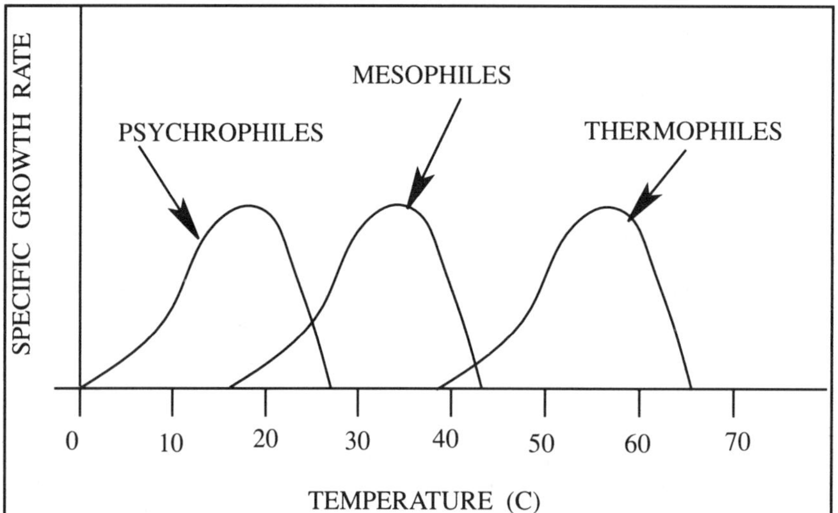

FIGURE 5.3
The Effects of Temperature on the Specific Growth Rate of Microorganisms
Biological Remediation

There are three generalizations that can be made concerning the effect of temperature on growth rate, and hence the rate of biodegradation. First, growth occurs only over a 30° C range for a microorganism: second, growth rate increases slowly with increasing temperature until the maximum growth rate is reached; and third, above the temperature for the maximum, growth rate falls rapidly with further increase in temperature.

The effect of temperature on the specific growth rate (refer to equation (7) in section 5.1.8) can be written as follows:

$$\mu = A \exp(-Ea/RT)$$

where:

A	= the Arrhenius constant
Ea	= the activation energy (kcal/mole)
R	= the universal gas constant (1.987 cal-mol)
T	= the absolute temperature

Typical values for activation energy for microbial growth are 13-17 kcal/mole. Above the maximum temperature for growth, the overall growth rate falls very rapidly, and is characterized by activation energies of 70-90 kcal/mole. The value of Arrhenius constant A, varies for different microorganisms. For the case of E. coli for example, for growth rates between 18 and 46 °C the value of the Arrhenius constant has been reported to be 2.92 E12[11]. For higher values of activation energy in temperature ranges above the maximum possible for growth research indicates, that for the same increase in temperature, the rate of inactivation is faster than the growth rate.

In *in situ* bioremediation, it is critical to know the environmental temperature and its variability since the efficacy of using microbial degradation processes varies with temperature. Usually the temperature at a given site may vary from freezing temperatures in the winter to about 30° C in the summer season. Accurate prediction of microbial growth rate requires as analysis of the monthly variation in the temperature, and the temperature variation with depth below the surface. Usually in the Northeast, bioremediation is dormant during the winter and only a nine month period is effectively available. In view of these observations an average growth rate may not be representative of the true growth rate. A first cut estimate, however, can be obtained by evaluating the growth rate for the average year round temperature.

5.1.5. Determining the Limiting Criteria

For determining the rate of biodegradation, the rate limiting step has to be investigated. In a typical subsurface, factors like dissolved oxygen and the availability of substrate to the microorganisms may be limited. Depending on the conditions of the system, different processes like diffusion, adsorption-desorption, or dissolution-precipitation may control the reaction. Biodegradation modeling is required to predict the rate of attenuation under the specific conditions. The model should be developed for the conditions which can be generated in the subsurface environment. A verification of the model predictions should be made by generating the model conditions in the subsurface environments and then measuring the rate of biodegradation.

One such model is developed by Rifai and Bedient, which compares a oxygen transport limiting rate condition and dual substrate Monod kinetic relationship to calculate biodegradation. The first condition assumes an instantaneous reaction between organics and oxygen, thus, assuming that the reaction is oxygen transport limited. The second condition, which is more accurate but at the same time more complex, utilizes a dual substrate Monod kinetic relationship to predict the biodegradation rate. Results of this analysis indicate that the instantaneous reaction is an adequate assumption when non-dimensionalised reaction rates are large.

Substrate availability could be the rate limiting step in some cases. In order to convert the substrate to biomass, it must be transported to the microbial cell. If the substrate is soluble, then it can be transported through the aqueous phase. If, however, the substrate is not soluble, then physical contact with the microbial cell is essential. In this case, diffusion becomes the operative mechanism. In some cases, an extra cellular enzyme is secreted by the microbes which emulsifies the substrate assisting diffusion through the cytoplasmic membrane. Thus, the form in which the substrate is available to the microbes is important. Additional forms could be as emulsion, vapor, or availability through an inert carrier.

5.1.6. Biodegradation Products

The products of biodegradation are not always carbon dioxide and water. TCE, for example, degrades to vinyl chloride which is more toxic than TCE itself. Biodegradation is a multi-step process, and the fate of the products formed in each step depends on their biodegradability and toxicity.

It is important to know the identity and amount of by-products that may be produced due to biodegradation. In certain cases the degradation products can pose a greater environmental hazard than the parent contaminant itself. If the by-products can not be treated either in sequence or concurrently with the bioremediation process, it may not be feasible to use bioremediation.

5.2. Developing the Microbes

5.2.1. Microcosm Stimulation

Adaptation of microorganisms to the environment is essential for ultimate bioremediation of the contaminants. This is usually achieved by applying essential nutrients and oxygen (for aerobic cultures) to the system in order to stimulate the natural microbiota. Both nutrients and oxygen are needed for acclimation and growth of the microorganism culture. When the culture is stimulated, the cells undergo growth through cell division.

5.2.2. Enrichment, Isolation, and Estimation of the Desired Species

Microorganisms having desirable characteristics, such as acclimation to certain contaminants, may be isolated from natural sources. Trained microbiologists are required to isolate, purify, screen, and test a culture from a natural source.

Culture enrichment involves the transfer of soil, sewage, or some other materials with a large and diverse population of microorganisms into a selective medium followed by incubation of the culture under such conditions of temperature, aeration, and pH, among others, that the growth of desired forms is favored. A small amount of the initial culture is transferred to a second that is set up in the same manner as the first, and this procedure is continued until the desired flora predominates and can be isolated in pure culture. This culture enrichment technique is based on free competition between the different organisms in liquid media. If a liquid selective medium is inoculated with a variety of organisms and incubated under any chosen set of conditions, these organisms for which the environment is most suitable will outgrow all others and eventually become predominant.

As listed in Table 5.1, for the enrichment of many organisms a basal medium of the following composition can be applied:

Table 5.1
Composition of Basal Medium

Basal Medium	
K_2HPO_4	1 g
$MgSO_4.7H_2O$	0.2 g
$FeSO_4.7H_2O$	0.05 g
$CaCl_2$	0.02 g
Trace element solution	0.1 mL
Water	1 liter

The requirements for the carbon, nitrogen and energy sources depend on the type of organism to be enriched. The trace element solution is usually applied to prevent any mineral deficiencies, and usually consists of very dilute quantities of trace salts, such as $AlCl_3$, KI, KBr, etc. Carbon dioxide is not only needed by those organisms that use it as the sole carbon source, but also by various chemoheterotrophs. It is not usually added to aerobic cultures, but should be included in growth media for anaerobes. It is generally added as $NaHCO_3$.

In addition, growth factors are added as supplements to the enrichment media for a large variety of microbes. These growth factors include vitamins, such as B_{12}, complex mixture of vitamins, and amino acids. Growth-factor requirements can often be met by including a small amount of yeast extract (0.001 to 0.05%) to the enrichment medium.

The enrichment culture is one of the most powerful techniques available to the microbiologist. Almost infinite number of permutations and combinations of the various environmental variables (nutrient composition of the medium, aeration, pH, temperature, etc.) can be used for the isolation of specific microorganisms from natural sources.

Isolation is the separation of a particular microorganism from the mixed populations that exist in nature. Isolation involves not only the separation of the pure culture from a mixed microbial population but also its maintenance in an artificial environment where it is protected from contamination by other microorganisms. The isolated culture vessel (usually a test tube, Petri dish, or flask) is initially sterilized (free of any living microorganism) and kept in a sterile room.

Isolation of the pure culture is generally achieved by:

- Plating methods
- Dilution methods
- Single cell separation

Plating methods: This method involves the separation and immobilization of individual organisms onto a solid nutrient medium solidified with agar or some other appropriate jelling agent. Each organism type grows to a visible colony from which organism can be readily isolated. The agar concentration used in plate culture is generally 1-2%.

The streaked plate generally is the most useful plating method. A sterilized bent wire is dipped into a diluted mixed suspension of organisms and is then used to make a series of parallel, non-overlapping streaks on the surface of the solidified agar plate. The inoculum (the microbial material used to seed or inoculate a culture vessel) is progressively diluted with each successive streak, so that the latter streaks can result in well-isolated colonies. The solid medium prevents migration of organisms and helps in forming isolated pure colonies.

Alternatively, isolations can be made with poured plates: successive dilutions of the inoculum are placed in sterile Petri dishes and mixed with the molten but cooled agar medium, which is then allowed to solidify. Colonies subsequently develop embedded in the agar.

Dilution Method in a liquid medium: Some of the larger-celled bacteria have not been successfully cultivated on solid media, and many protozoa and algae can only be grown in a liquid medium. The simplest procedure for isolation in a liquid medium is the dilution method. The inoculum is subjected to serial dilution in a sterile medium, and a large number of tubes of medium are inoculated with portions of each successive dilution. The purpose of this operation is to inoculate a series of tubes with a microbial suspension so diluted that the probability of introducing even one individual microorganism into a given tube is very small: a probability of the order of 0.05. The fraction of tubes that receives no microorganisms is therefore 0.95, the fraction receiving one microorganism is 0.048, the fraction receiving two organisms is 0.0012, the fraction receiving three organisms is 0.00002, etc. As a result, if a tube shows any subsequent growth, there is a very high probability that this growth has resulted from the introduction of a single organism. The dilution method has a major disadvantage that is that it can only be used to isolate the numerically predominant member of a mixed microbial population.

Single-cell isolation: When neither plating nor dilution methods can be applied, microscopically controlled isolation of a single cell or organism from a mixed population is used. The technical difficulty of single-cell isolation is inversely related to the size of the organism which one wishes to isolate. In the case of large microorganisms, purification involves the capture of a single individual in a fine capillary pipette and the subsequent transfer of this individual through several washings, in relatively large volumes of sterile medium, to eliminate microbial contaminants of smaller size.

For smaller microorganisms, a micromanipulator, which is a mechanical device, is used in conjunction with very fine glass operating instruments. Microbe cultures developed for injection for remediation applications are commercially available.

5.3. Introduction of the Microbes to Site

Depending on the type of site, the microbes can be introduced into the site in many ways. In the case of soil bioremediation, the microbes can be added to the nutrient media which is then sprayed evenly on the site or injected into the subsurface vadose zone through a grid of injection ports. In these cases, the transport of the microbes is dependent on the nutrient flow in the site. The transmissivity of the soil plays a significant role in the nutrient distribution and transport of the microbes to the site contamination. Low soil transmissivity limits the feasibility of soil bioremediation. If the site is a liquid mass, such as a lake or lagoon, then the microbes can be added directly and distributed very easily by natural convection or pump-around systems.

5.4. Optimum Conditions for Bioremediation

The essential requirements for bioremediation are:

- A viable inoculum
- An energy and carbon source
- Essential nutrients
- Absence of growth inhibitors
- Appropriate physico-chemical conditions

Normally, a viable inoculum exists in any natural environment. The indigenous microorganisms present in the natural environment may not be acclimated to the contaminants, but are usually viable. Appropriate physico-chemical conditions, such as temperature, pH, etc. may have to be adjusted to allow biological growth to occur. Usually, it becomes essential to add the required nutrients, supply oxygen (for aerobic degradation) and some additional carbon source for initial stimulation of the microorganisms and allow acclimation to occur.

Defining the optimum conditions for effective biodegradation usually involves selecting the proper nutrients, maintaining necessary supply of oxygen for the aerobes, pH adjustments, and applying the needed carbon and energy sources. The basic objective is to maintain the growth of the microorganisms based on the degradation of the contaminants, at its maximum value. In Table 5.2 the necessary conditions for microbe growth are collected.

Table 5.2
Specific Requirements of Microbes for Soil Bioremediation [13]

Environmental conditions necessary for bioremediation:	
pH	5.5 - 8.5
Moisture	25 - 85% of water holding capacity
Oxygen	Greater than or equal to 0.5 ppm
Temperature	10 - 45°C for most applications
Nutrients	Ratio of Carbon:Nitrogen:Phosphorus 120:10:1

5.5. Selection of Bioremediation Process

Bioremediation processes can be classified into four types. In the following sections, each bioremediation process, it feasibility, and its cost will be discussed in detail. Factors to be considered in determining whether or not bioremediation is feasible at a site include:

- Contaminant characteristics
- Microbial characteristics
- Site conditions and contaminated medium physical characteristics
- Regulations governing the implementation of bioremediation
- Public health and safety issues
- Time required to achieve remediation

5.5.1. Above Ground Treatment of Contaminated Soils (Landfarming/Composting)

5.5.1.1. Land Farming

5.5.1.1.1. Feasibility and Conceptual Approach

Organic contaminants in soils are susceptible to aerobic digestion, and this process occurs naturally to a limited degree. The rates may be greatly enhanced by fertilizing, irrigating, and tilling the soil to increase the availability of nutrients, moisture, and oxygen to the soil microorganisms. This process of fertilizing, irrigation and tilling the soil to enhance bioremediation is called land farming [12].

In land farming, the micro-organisms used are most often the indigenous population, unless the environment is too severe or biodegradation of specific contaminants requires specially-developed microbes. Through scheduled tilling and maintenance of the moisture content, pH, and nutrients, decomposition and immobilization of the contaminants occur within both the upper soil layers (zone of incorporation) and the underlying layers. Usually, landfarming is effective for 18-24" thick layer of contaminated soil. Treatment of layers >24" by landfarming is rare and has limited success.

Land farming has been used successfully throughout the United States, especially at petroleum refinery sites and also with creosote contaminated sludges and soils.

5.5.1.1.2. Factors Affecting Land Farming

Factors affecting land farming can be classified into three categories:

- Substrate limiting factors
- Environment limiting factors
- Biological consortia limiting factors

Substrate limiting factors are as follows:

- Substrate biodegradability
  - Organic substrate must be biodegradable, under aerobic conditions. Some compounds, such as PCBs are extremely bioresistant, and hence are not suitable for land farming techniques. For contaminated soils containing bioresistant compounds, pretreatment is required. Pretreatment schemes include supercritical extraction, solvent washing, and chemical dechlorination, as in the case of PCBs.
- Substrate toxicity
  - Some organic compounds are toxic to the microorganisms above a limiting concentration level. In some the cases, substrate toxicity can be controlled by dilution or pretreatment.

Environment limiting factors are as follows:

- Temperature
  - Temperature is an important environmental limiting factor. For successful land farming, temperatures favorable for microbial growth (15 - 45°C) are needed
- pH
  - pH needs to be maintained between 5 - 8 for effective bioremediation. pH control is needed where acidic (i.e. soils are present)
- Nutrients
  - For land farming operations, a C:N:P ratio of 100:2:0.2 has been found to be successful

Biological consortia limiting factors include:

- Diversity
  - The diversity of microorganisms present allow a wide variety of organics to be degraded successfully. The diversity of microorganisms present at the land farming site will control the rate of bioremediation and the kind of chemical contamination that can be treated

- Acclimation
  - Acclimation processes will delay the onset of biodegradation. However, natural evolutionary processes will result in acclimation of the indigenous microbes

5.5.1.1.3. Advantages/ Disadvantages

The major advantages of land farming are as follows:

- Simple to implement
  - The land is merely covered evenly with the contaminated soil, periodically dosed with nutrients, and mixed.
- Less expensive bioremediation technology.
- Applicable for slowly degrading contaminants, which would render bioreactor approaches infeasible due to large bioreactor sizes and costs.
- Usually employed for moderate to low contaminate concentrations.

The major disadvantages of land farming are as follows:

- Volatile contaminants, such as solvents, etc., must be pretreated, since they would evaporate into the atmosphere, causing air pollution.
- Particulate matter is also a concern because it may causing a dust generation problem.
- Presence of toxic metal ions requires pretreatment to avoid toxicity to the microbes and possible leaching from the contaminated soil into the ground.
- Land farming requires proper control of pH, water content, and nutrients.
- Can only be used where environment temperatures are favorable during most of the year.
- Can only be used for aerobic treatment of the organic waste.
- Requires a large area for treatment if significant volume of contaminated soils are present.
- May require bottom lining of large area if concern that leaching of contaminates may further degrade ground water.
- Less rapid method of remediation compared to other bioremediation techniques.

5.5.1.1.4. Site Characteristics For Selection of Land Farming

Land farming can be used when environmental conditions are suitable such as favorable temperatures for most of the year, reasonable rainfall, to provide moisture to the site, and availability of land.

5.5.1.1.5. Approximate Costs

Conventional land treatment costs range from \$50 to \$80/cubic yard. For 6,000 cubic yards of a soil contaminated with diesel fuel, land treatment in an unlined system, requiring no excavation, costs about \$33/cubic yard. These are the cost from a specific case study. In general for a variety of petroleum contaminated sites the costs are usually between \$30 - 50/cubic yard for sites containing greater than 1000 cubic yards of contaminated soils.

5.5.1.1.6. Additional Issues and Research Needs

Land farming is a well-established bioremediation technique that has been successfully used for a variety of applications. Experience and empirical relationships usually are relied upon for design. More research is needed toward the scientific development of landfarming techniques and broader applications. The major issues and research needs that need to be addressed are:

- Possible contamination of the environment from the degradation by- products
- Quantification of degradation kinetics
- Enhancement of degradation capabilities by addition of plasmids, to enable fast acclimation
- Monitoring of the land farming site, to obtain quantitative progress measurements

5.5.1.2. Composting

Composting is defined as the biological decomposition and stabilization of organic substrates under conditions which allow development of thermophilic temperatures as a result of biologically produced heat, with a final product sufficiently stable for storage and application to land without adverse environmental effects. Thus composting is a form of biological waste stabilization, but one that requires special conditions of moisture and aeration to produce thermophilic temperatures [12].

Composting is usually applied to solid or semisolid materials. (Approximately 800 million tons of solid and semisolid wastes are generated in the United States per year, and over 130 million are actually collected.) The process can be conducted under aerobic conditions, i.e., in the presence of air or oxygen, with carbon dioxide, water and heat being the main products of biological metabolism. Alternately, composting can be conducted under anaerobic conditions, with the metabolic products being methane, carbon dioxide, and numerous intermediates, such as low-molecular weight organic acids.

The objectives of composting have traditionally been to biologically convert putrescible organics to stabilized form and to destroy organisms pathogenic to humans. Composting can also effect considerable drying, which is of particular value with wet substrates, such as municipal and industrial sludges. Decomposition of substrate organics together with drying during composting can reduce the cost of subsequent handling and increase the attractiveness of compost for reuse or disposal.

Compost can be disposed of more easily then untreated solid waste. If the product is reusable, it can accomplish a number of additional purposes including. (However, when applying stabilized compost to soil one must be careful about plant uptake of undergraded residuals.):

- To serve as a source of organic matter for maintaining of building supplies of soil humus, necessary for proper soil structure and moisture holding capacity
- To improve growth or vigor of crops in commercial agriculture or home-related uses
- To reclaim and use valuable nutrients including nitrogen, phosphorous, and a wide variety of essential trace elements

Composting systems can be classified into two types: reactor systems and nonreactor systems.

Reactor composting systems:
Reactor systems can be further classified into vertical, horizontal and inclined flow reactors. Vertical flow reactors can be further defined based on the bed conditions in the reactor. Some systems allow for agitation of solids during their transit through the reactor, and are called agitated bed reactors. In other systems, the composting mixture occupies the entire bed volume and is not agitated during its residence time in the reactor. These are called moving packed bed reactors and can be fed on either a continuous, intermittent, or batch basis.

Horizontal and inclined flow reactors have been divided into those which employ a rotating drum (tumbling solids bed reactor) and those which use a bin structure and varying method of agitation, using either forced aeration or mechanical agitation.

Nonreactor composting systems:
Nonreactor composting systems are those that do not use a reactor, and can be further classified into agitated solids bed and static solids bed. An agitated solids bed disturbs or breaks up by some manner the composting mixture during the compost cycle. This may be achieved by periodic turning, tumbling, or other methods of agitation. The windrow system is an excellent example of a nonreactor agitated solids bed system. Mixed compost material is placed in rows and turned periodically, usually by mechanical equipment. The height , width, and shape of the windrows vary depending on the nature of the feed material and type of equipment used for turning. Oxygen is supplied by gas exchange during turning and by natural ventilation resulting from buoyancy of the hot gases in the windrow. In some systems, forced aeration is used for enhanced oxygen transfer.

The major limitation of composting for treating hazardous wastes is the lack of experimental data for a wide variety of toxic organics. Composting has been applied to wastes, such as TNT, petroleum sludges, and certain pesticides.

Major nuisance problems associated with composting are odors and dusts. Dust is also a problem for other landfarming techniques. Concern over odors has in recent years become of paramount importance in siting composting facilities and process selection.

5.5.1.2.1. Approximate Costs

Windrow composting, using recycled compost product, can usually be conducted for less than \$20-40/dry ton, including land purchase, site preparation, mobile equipment, concrete pad, fixed roof cover, controlled aeration system, and operation and maintenance expenses. Static pile composting with wood chips is considerably more expensive, primarily because of the cost of wood chips. Assuming chip costs of \$13/m^3, the cost of the new chip replacement alone is about \$42/dry ton of sludge. Total system costs are usually in the range of \$60-120/dry ton, the major cost items being the bulking agents and labor.

5.5.2. Above Ground Slurry/Soil Treatment

5.5.2.1. Feasibility and Conceptual Approach

A typical soil-slurry treatment reactor system is shown in Figure 5.4 The process consists of the following steps:

1. Feed addition system: The contaminated soil mixed with water, nutrients, and other pH adjusting solutions, is pumped into the bioreactor
2. Aeration system: An aeration blower is used to pump air through the slurry phase.

3. Emission control system: The air leaving the bioreactor has to be treated to remove volatile contaminants. Typical air emission treatment systems include carbon adsorption and/or biofilters
4. Bioreactor system: Consists of a closed tank, where the slurry is mixed using a mechanical mixer. The liquid slurry leaving the reactor is partly recycled, and the remaining flow is sent to other downstream slurry-bioreactor. The mixer design is critical to prevent settling of the soil in the bioreactor
5. Clarifier system: A suitable coagulant is added to the effluent of the final bioreactor, and a clarifier is used to separate the liquid from the treated soil and biomass. Part of the solid phase is recycled back to each bioreactor, to supply acclimated microorganisms and provide some dilution of the contaminated solid (soil) feed to each bioreactor. The liquid effluent is sent to a wastewater treatment plant

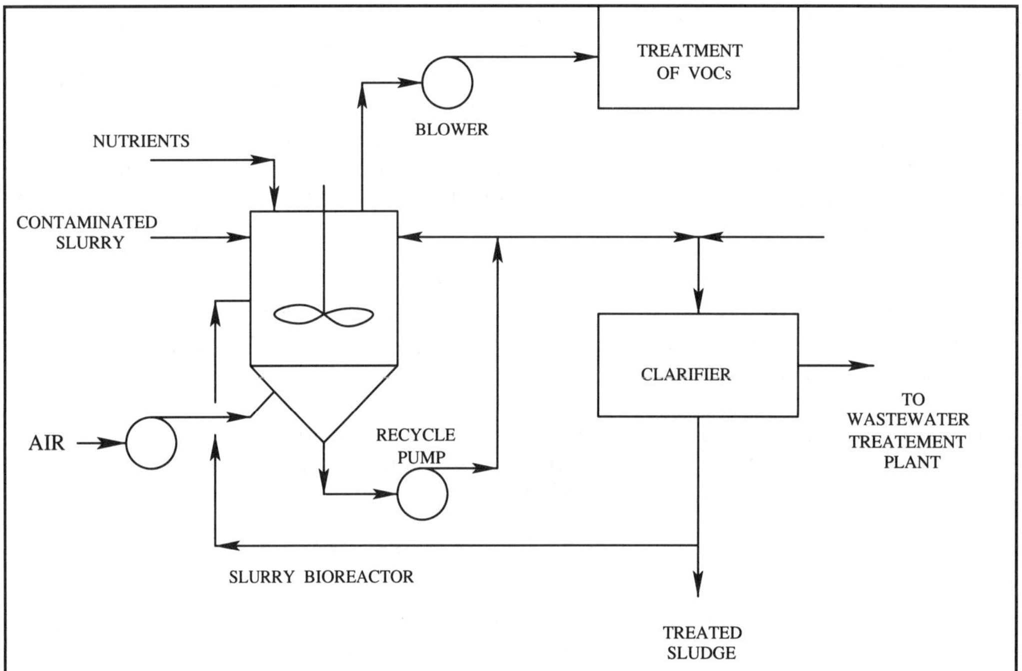

FIGURE 5.4
Schematic of a Soil-Slurry Bioreactor System
Biological Remediation

5.5.2.2. Factors Affecting Off-Site Treatment

The major factors affecting soil-slurry treatment are as follows:

- Solubility of the contaminant in water
  - — One of the mechanisms for the biodegradation of the contaminant is its solubilization in the water and subsequent degradation by the microbial population in the water. The extent of the contaminant's solubilization in the water phase depends on its aqueous solubility

- Adsorption/Desorption characteristics of the contaminant
  - All contaminants exhibit to varying degrees adsorption on the surface of the soil matrix. These contaminants remain on the soil particles in the slurry reactor and hence can only be biodegraded by the microorganisms which are attached to the soil particle surface. Some contaminants have a lower affinity for the soil surface, and hence desorb in aqueous solution. The extent of biotreatment in a slurry reactor depends on the adsorption/ desorption characteristics of the contaminant
- Aging of the soil since the onset of contamination
  - When soil is contaminated by an organic contaminant, usually, the organic chemical does not diffuse into the interior of the soil micropores, but rather remains on the outside surface of the macropores. However, as time elapses, the organic contaminant diffuses slowly into the interior of the soil micropores, which is referred to as "soil aging." Furthermore, some chemicals also undergo slow chemical transformations, usually catalyzed by the inorganic constituents of the soil matrix, which leads to chemical binding of the chemical to the soil matrix
  - The soil aging process makes the chemical difficult to treat, since for biological degradation, the chemical has to diffuse out from the interior of the micropores, which is inherently a slow process, and depends on the diffusivity of the contaminant
- Concentration of the microbial population in the reactor
  - The microbial population in the reactor is maintained by proper addition of nutrients, oxygen, and maintaining pH and reactor temperature
- Screen out gravel and coarser sized particles
- Plasticity of soil to accommodate thorough mixing in reactor (minimal clumping)

5.5.2.3. Advantages/Disadvantages

The major advantages of soil-slurry treatment are as follows:

- Proper operating conditions for effective biotreatment can be achieved in the slurry reactors. This allows the achievement of specified (limits set by designer) contaminant exit concentrations, and effective monitoring of the biotreatment process
- Treatment costs using slurry bioreactors are over 50% less than alternative technologies, such as incineration
- Slurry treatment provides effective control of volatile emissions and eliminates some statutory concerns related to contaminant migration

The major disadvantages of soil-slurry treatment include:

- Excavation of the site and transportation of the soil, which is an expensive process
- Slurry treatment requires material handling equipment with possible contaminated dust generation, emission of volatile compounds etc. When volatile compounds are present, bioventing and vacuum extraction methods may have to be used as pretreatment before off-site soil-slurry treatment can be used
- The water decanted from the clarifier system has to be subsequently treated in a treatment plant
- Effective slurry reactor treatment requires that the chemical can be easily desorbed from the soil matrix and solubilized in the water phase. Furthermore, the chemical should biodegrade at reasonable rates, so that the size of the bioreactor is not large

- Requires energy consumption, for mixing and pumping of the slurry, pumping of air, etc.

5.5.2.4. Site Characterization Factors for Using Soil-Slurry Treatment

Usually slurry-reactor treatment is used for somewhat water soluble contaminants. Soil-slurry biotreatment can be used if the contaminated soil can be excavated , when the contamination is only present in surface soil, and the extent of contamination is small.

5.5.2.5. Approximate Costs

Treatment costs using slurry reactors range form $100 to $150/cubic yard. Costs ranging from $125 to $160/cubic yard are incurred when the slurry-bioreactor off gas has to be further treated, due to the presence of volatile compounds.

5.5.2.6. Additional Issues and Research Needs

Additional issues and research needs are as follows:

- Effect of soil aging on the effectiveness of slurry-reactor treatment
- Development of more efficient soil/slurry agitation methods.
- Adsorption/desorption kinetics of multiple contaminants from the soil matrix
- Biodegradation kinetics for strongly adsorbed contaminants
- Effect of soil properties, such as particle size, pore size distribution, etc. on biodegradation kinetics
- Use of other bioreactor designs. such as plug flow, as compared to completely-mixed systems or combination of reactor systems

5.5.3. *In Situ* Bioremediation

5.5.3.1. Overview

Biodegradation finds wide applicability for in-situ bioremediation. Low capital and treatment cost can make it preferable over other *ex situ* treatment technologies. As an *in situ* treatment, bioremediation is particularly advantageous when a contaminant cannot be pumped out from an aquifer. The technology, however, may not be applicable in all the situations. *In situ* applications are limited to environments where conditions are amenable to easy dispersion of oxygen and nutrients, as well as injected microbes. In - situ bioremediation is also better suited for shallow contamination.

As mentioned above, it is a case specific technology and its applicability depends on: [12, 13, 14, 15]

- Contaminant characteristics
- Microbial requirements
- Hydrogeology of the site

The following parameters affect the applicability of *in situ* bioremediation and require further discussion:

- Biodegradability
- Inducibility and stimulation
- Feasibility in combination with other technologies
- Contaminant characteristics

5.5.3.2. Biodegradability

In order to employ *in situ* biodegradation as a remediation technology, the first step is to investigate the biodegradability of the compound of interest. Based on bench and pilot-scale tests with indigenous soils and contaminants, it may be ascertained whether the contaminant is biodegradable by indigenous microbes or special microbes are required. A compound could be easily biodegradable resulting in complete mineralization; it could be persistent requiring specific conditions; or it could be recalcitrant indicating no degradation at all. In some cases, as explained below, biostimulation and biotransformation of the contaminant can result in efficient disposal of the compound.

5.5.3.3. Inducibility and Stimulation

Once the biodegradability of the compound is established the next step is to find the optimal conditions for the process.
The growth requirements of the microorganisms must be considered, such as:

- Is TEA required?
- What nutrients are required?
- What is the favorable temperature range?
- What is the appropriate pH range?

In situ bioremediation may require:

- Pumping the microorganisms into the sub-surface
  - Especially if the sub-surface environment does not have an enriched population of the microbes
- Adding nutrients
  - Oxygen or nitrate, depending on the terminal electron acceptor
- Adding co-substrates
  - If co-metabolism is necessary to induce biodegradation
- Selecting a species appropriate to a site's temperature and pH

Stimulation of appropriate microorganisms can cause biodegradation of ordinarily non-degradable, or recalcitrant, compounds under given site conditions. The microbial population created can be used directly, or indirectly, to biodegrade the compound as a result of co-metabolism or biotransformation.

> Example: Chlorinated aliphatic compounds, for example have been have been transformed to stable nontoxic products through a co-metabolic process by

stimulating growth of indigenous methane oxidizing bacteria (methanotrophs) at the site.

Microbial activity in an environment can be greatly influenced by the site characteristics. Groundwater, for example, shows a higher microbial population and a significantly greater utilization of the dissolved organics by the bacteria than in the case of stream water.

If a site does not have a good microbial population, it can be induced artificially by "seeding" the site with the desired strains of the microorganisms.

This process is advantageous in providing:

- The option to choose an appropriate microorganism
- The possibility of engineering the microorganism being induced

These advantages, however, are greatly off-set by:

- The difficulty of the microorganisms to adapt and survive in the new environment
- The failure of technology to transfer from laboratory conditions to field conditions
- The unpredictability of survival of induced populations due to extraneous factors in the environment, such as:
    — Competition with existing species
    — Effects of metabolically and co-metabolically degraded substrates on the induced and existing populations
    — Possible interactions and products of biodegradation
- The inability to predict the rate of degradation from lab results due to inconsistency with conditions
    — No constant can be set, so lab results should be considered only as rough estimates

5.5.3.4. Feasibility in Combination with Other Technologies

In situ biodegradation can be enhanced through its combination with other technologies appropriate to the characteristics of the site and the contaminant. Approaches being reviewed for combination with biodegradation are [16].

- Ozonation
- Hydrogen peroxide treatment
- Genetic engineering
- Vapor gas extraction (bioventing)

5.5.3.5. Factors Affecting *In Situ* Bioremediation

The major factors that affect *in situ* bioremediation are as follows:

- Availability of oxygen. Oxygen diffusion through the soil can become rate limiting, especially in the subsurface environments
- Availability of nutrients. Nitrogen and phosphorous are required for microbial growth. Nutrients may be present in the soil matrix, or may have to be added externally

- pH of the soil matrix. pH control is necessary to support microbial growth, and hence *in situ* bioremediation in low pH soils is difficult
- Temperature. Microbial growth rate is affected by temperature. As temperature decreases, microbial growth slows down considerably
- Nature of the contaminant. The biodegradability of the contaminant depends on its chemical structure, i.e., size of carbon backbone, presence of functional groups, etc.
- Heterogeneous subsurface conditions limiting thorough dispersion of microbes, nutrients, and oxygen.

5.5.3.6. Advantages/ Disadvantages

The major advantages of *in situ* bioremediation are as follows:

- Lower capital cost for implementation
- Does not require excavation and transportation of the soil
- Does not cause cross-media contamination, since the soil is not disturbed
- Can be used for bioremediation of large-scale contamination of subsurface soils

The major disadvantages of *in situ* bioremediation are as follows:

- Difficult to monitor and quantify the extent of bioremediation
- Can produce biodegradation products, which may remain at the site
- Can take a very long time, depending on the type of contamination, and environmental conditions (thus involving considerable operation and maintenance costs)
- Requires favorable environmental conditions, such as temperature, moisture, pH of the soil matrix, etc.
- May be required to be combined with containment technology to avoid leaching and further contamination spread while remediation is in progress.

5.5.3.7. Site Characterization Factors for Selection of Techniques

Accessibility of the site can be important in determining feasibility of bioremediation. Typical site conditions that are important are:

- Location of contaminant
- Availability of land
- Building or utility locations
- Local population

Location of contaminant is important in determining whether nutrients, TEA, microbes etc. can be applied to the contaminated portion of the site. Biodegradation of a chemical depends on the spatial characteristics of the site. If the site characteristics vary with position, the biodegradation characteristics of the contaminant would also vary. A study was done to evaluate the effect of spatial variation of the contaminant on biodegradation at three sites representing a range of environments from fresh water to estuarine to marine systems. The heterotrophic rates of uptake at these sites differed, with the highest rate at the fresh water site and the lowest rate at the marine site. These rates were also a function of water temperature and this seemed to affect the fresh water and estuarine environments more than the marine environment. Thus, it may be inferred from this study that biodegradation rates are a function of both time and spatial distribution of the substrate.

Land area or building location may prohibit the building of a bioremediation treatment system, such as bioreactors, aboveground piles, or distribution systems for air/water for in-situ bioremediation. Finally, political and other issues concerning the local population may render bioremediation impractical. Local perceptions may be very important element in determining whether bioremediation will be an acceptable treatment process [12]. Bioremediation has been accepted by EPA on a number of Superfund sites. At Superfund sites selection criteria partly depends on public acceptance of the alternative.

The surface area and depth of contamination at the site are important factors for in-situ bioremediation. Other factors are soil structure and site geology. Permeability of the soil phase is critical for application of in-situ bioremediation. Low permeabilities generally indicates that the mass transfer of both TEA and nutrients may be severely impeded, making aboveground bioremediation more applicable. Other geological consideration include groundwater level and flow. Soil pH affects sorption of ionizable compounds in addition to limiting the type of suitable microorganisms. Temperature has a major impact on bioremediation rates. Concentrations of iron and manganese are important because high concentrations of these metals can cause precipitation under aerobic conditions due to infiltration of oxygen. Other heavy metals are important because at toxic levels they can inhibit the activity of microorganisms.

5.5.3.7.1. Hydrogeology (Saturated Soil)

Aspects of hydrogeology which affect biodegradation are:

- Specific yield and transmissivity
- Depth and thickness of the contaminated aquifer
- Contaminated aquifer heterogeneity
- Direction of ground water flow
- Recharge of contaminated areas
- Seasonal fluctuations of water table

The site should have the following characteristics:

- The conductivity of the soil should be high enough to allow addition of the nutrients, microorganisms etc.
    - Should be higher than 10^{-4} cm/s
- The heterogeneity of the subsurface should be limited, so that , irregularities can be accommodated in plans for distribution of microbes and nutrients.
    - For example, model should account for variations in hydraulic conductivity

5.5.3.7.1.1. Soil Properties (Vadose and Saturated Zone)

Soil properties such as water content, organic carbon content, and porosity affect the rate of biodegradation. If, for example, a soil has high porosity and low water content it will favor diffusion of oxygen, thus promoting aerobic reactions. A high organic carbon content of the soil may provide a carbon and energy source to the microorganisms and in some instances can result in co-metabolic degradation of the contaminant. Soil particles also provide surfaces for bacterial attachment and play an important role in microbial degradation. Studies show that the rate of biodegradation is directly proportional to the exposed surface area for the microbial attachment.

5.5.3.7.1.2. Groundwater Quality

5.5.3.7.1.2.1. Inorganic Nutrient Levels

Nutrient levels must be sufficient and constant for an efficient and uninterrupted process. If the site is not sufficiently rich in nutrients, it should be pumped periodically depending on the specific problem. Dispersion is the main mechanism for spreading the nutrients in groundwater. It is essential to restrict the area of the plume to avoid excess spreading of the plume resulting in wastage of nutrients. Interceptor walls, trenches, and extraction wells are commonly used to contain a plume.

5.5.3.7.1.2.2. Precipitation of Inorganic Nutrients

Compatibility between the site geo-chemical characteristics and the nutrient solution being pumped inside is very important. If the nutrients precipitate inside the soil, it could block the pores thus cutting off both water and oxygen supply. Other soil chemistry factors that are important are pH, and organic content.

5.5.3.7.1.2.3. Dissolved Oxygen Content

Dissolved oxygen in the sub-surface is the most important rate limiting factor for aerobic biodegradation. For a rapid and steady biological reaction, a constant and uniform supply of dissolved oxygen is essential. Oxygen can be supplied in the form of:

- Air
- Pure oxygen
- Hydrogen- peroxide

  Use of Hydrogen-peroxide can create some problems:
  - Is toxic to some of the microbial species
  - Should be introduced at the level of 50 ppm and gradually increased to the required level (should be around 750-1000 ppm)
  - Is very unstable and degases easily to water vapor and oxygen; therefore care must be taken while using it.

5.5.3.7.1.2.4. Presence of Other Terminal Electron Acceptors.

Some microbes do not require oxygen as the terminal electron acceptor (TEA). In such cases it should be ensured that the other required TEA is in sufficient quantity at the site. Nitrate is a common terminal acceptor used by some microbes. It is pumped into the ground when not available in sufficient quantities.
Adding Nitrate:

- Offers a cost effective technology
- Relieves the problem of solubility
- Raises problems as a drinking water contaminant

5.5.3.8. Approximate Costs

The cost of treatment depends on the nature of the contaminability (the solubility, volatility, and sorptive ability of the contaminant), the location and extent of contamination, the hydrogeology of the site, and the goal of the remediation project, i.e., source remediation or plume control.

The cost of bioremediation project varies with the methodologies used and on the location, quantity, and characteristics of the wastes present. If groundwater is impacted, an average of $100 is spent for each gallon of gasoline spilled, according to a 1988 study by the American Petroleum Institute.

Bioremediation of contaminated soils is inexpensive compared to some of the non-biological methods, such as thermal treatment. Conventional land treatment costs range from $50 to $80/cubic yard.

Bioremediation of fuel contaminated groundwater is cost effective, but the process is often unable to achieve the low effluent concentrations achieved by other methods. An *in situ* bioremediation scheme can cost $8 to $15 per pound of contaminant $66 - $123/cubic yard of material treated. Since the cost of bioremediation is mostly site specific these reported values are an over simplification, and are only provided as a point of reference to compare other treatment technology with.

5.5.3.9. Additional Issues and Research Needs

In light of the potential of bioremediation process to eliminate contaminants without causing cross-media pollution and its over all cost effectiveness, there is a growing need for further and research in this area [13, 15, 17] There are several outstanding issues that need to be addressed and research needs to be supported, especially since an expanding requirement for bioremediation technologies can be seen in the future. Some of the issues and research needs have been listed below:

- Quantitative determination of *in situ* biodegradation rates
- Determining biodegradation kinetics for complex mixtures of chemicals, rather than for single chemicals
- Benefits of seeding contaminated sites with acclimated microorganisms
- Role of indigenous microorganisms in effective bioremediation and quantifying the impact of stimulating biological activity by addition of nutrients, water, oxygen, etc.
- Transport of contaminants and microorganisms in soil environments

A biotrench is a technology that involves developing a trench down gradient of the limits of the contaminant plume. The trench contains microbes fed by nutrients in the waste. As water flows through the trench microbes act on the contaminants and "clean water" flows out on the downgradient side (see Figure 5.5).

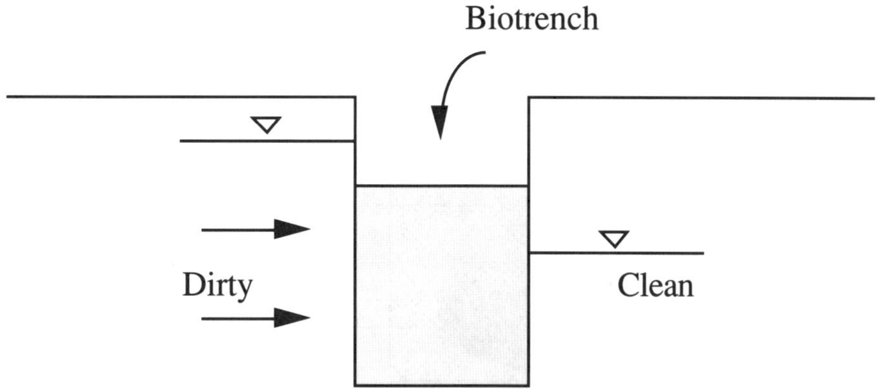

FIGURE 5.5
Biotrench
Biological Remediation

5.5.4. Bioventing

5.5.4.1. Feasibility and Conceptual Approach

Bioventing uses air movement to provide oxygen for aerobic degradation using either indigenous or introduced microorganisms. While some organics will be present in the exiting air streams, in-situ biodegradation is the main goal, and occurs mainly for less volatile materials, such as diesel fuel. For bioventing to work, sufficient water must be present in the unsaturated zone. In addition, nutrients, such as orthophosphate and ammonium nitrogen may need to be introduced to achieve optimal degradation rates [13, 15, 17].

There are three ways of achieving air flow in bioventing [14].

- Single well vapor extraction: Vacuum extraction is applied to one well causing ambient air to enter into the surrounding soil (see Figure 5.6 a)
- Two wells are used with air pumped into one well and extracted from the other well (see Figure 5.6.b.)
- Two wells are used with vacuum extraction applied to one well causing air to enter into the soil primarily through the other well (see Figure 5.6.c)

The level of vacuum that can be applied in the extraction well can vary from high vacuum (100 inches Hg) to low vacuum (20-30 inches Hg).

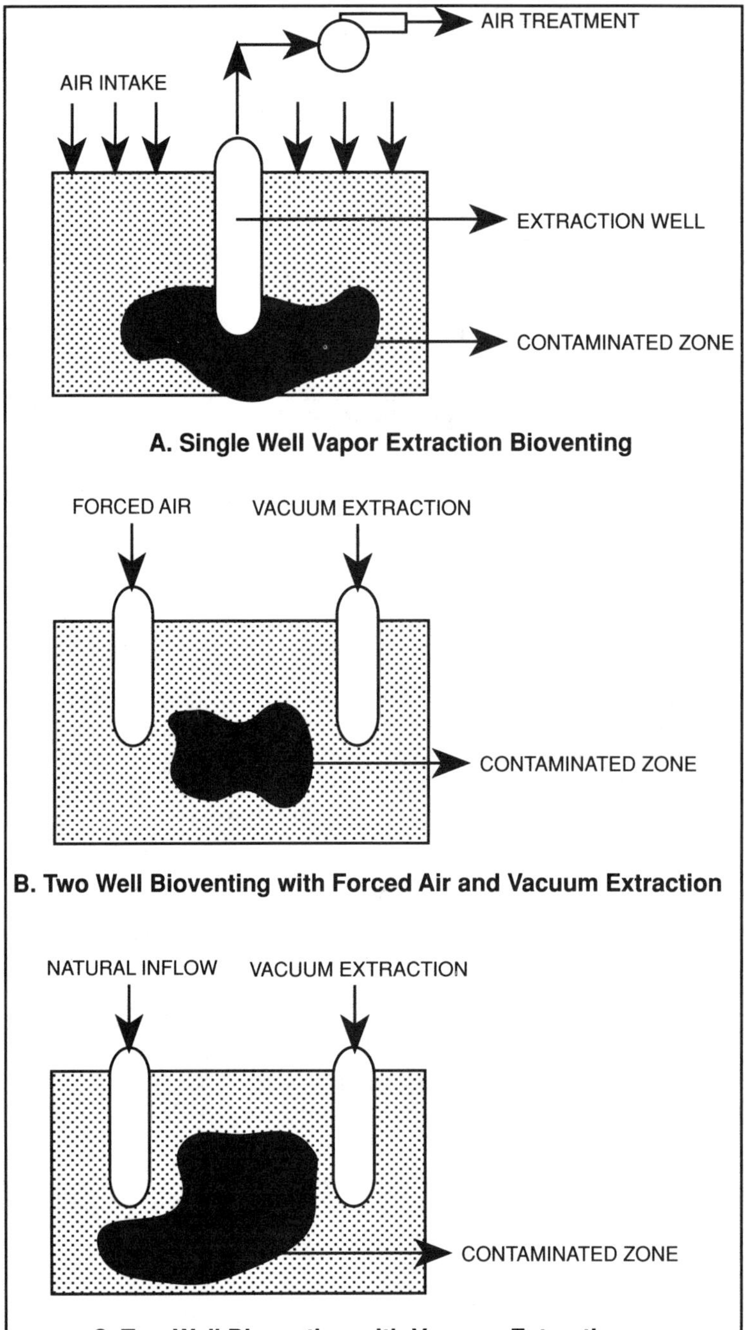

FIGURE 5.6
Methods of Achieving Air Flow in Bioventing
Biological Remediation

5.5.4.2. Factors Affecting Bioventing

Factors affecting bioventing can be classified into two categories [18, 19, 20, 21]:

- Substrate characteristics
- Environmental factors

Substrate limiting factors are as follows:

Chemical contaminant(s) must be volatile and biodegradable. Furthermore, since bioventing is based on air flow, a soluble contaminant will reside in the soil moisture to a greater extent than chemicals with low solubility. For maximum effectiveness of bioventing, compounds with high volatility (high vapor pressure) and high biodegradability are optimal. For many contaminants, volatility and biodegradability are synergistic since the compounds that are not volatile biodegrade easily. This is especially true for hydrocarbon fuels wherein light products are treated primarily by volatilization and heavy products by biodegradation.

The major problem is caused by high solubility, as in the case of Methyl ethyl ketone (MEK). Bioventing for MEK is not very efficient due to its high solubility in soil moisture, which slows the process of mass transfer from the adsorbed phase to the gas phase. For most effective bioventing, the mass transport has to occur form the adsorbed solid phase to the gas phase, which depends on the compound's vapor pressure. For soluble compounds, the mass transport occurs from the solid phase to the liquid phase and then to the gas phase, which then becomes dependent on the Henry's Law constant and solubility. The major factor that offsets the negative impact of solubility is that in bioventing, air is supplied which promotes the rate of biodegradation. Hence, the decrease in volatilization due to solubility is offset by an increase in biodegradation.

Environment limiting factors are as follows:

(1) Permeability of the soil is very critical for forced air circulation through the soil matrix. The flux of air flow through a soil matrix can be expressed using Darcy's Law:

$$F = \left(\frac{DBo}{v}\right)\frac{\partial P}{\partial x}$$

where:

F = flux

v = gas viscosity

Bo = soil permeability

D = gas density

$\partial P/\partial x$ = pressure gradient

Generally, significant decrease in flux occurs when the permeability is less than 10^{-4} cm/sec. At this point, the flux drops exponentially with decreasing permeability.

(2) The second major environmental factor in bioventing is the heterogeneity of the soil matrix. Generally, a soil matrix consists of high permeability sections interspersed between low permeability areas. Sometimes contamination can spread to low permeable areas through high permeability soil formations. The key is to access the high permeability formations by suitably locating trenches and vacuum wells.

(3) The third important factor is the degree of water saturation, since air flow rate is proportional to air filled soil porosity. Soil moisture contents approaching 100% saturation, reduce the amount of air filled porosity, which lowers the effectiveness of bioventing. When contamination occurs below the water table, either the water must be removed by dewatering, or air must be sparged into the water.

(4) Other environment limiting factors include temperature, pH, and availability of nutrients

5.5.4.3. Advantages/Disadvantages

The major advantages of bioventing are as follows [22, 23]:

(1) Allows direct extraction of organic compounds and enhances biodegradation. A bioventing system is capable of removing hundreds to thousands of pounds per day of highly volatile organic compounds. Table 5.3 gives calculated mass removal (lbs/day) rates for a compound with an average molecular weight of 100 and air flow of 100 SCFM.

Table 5.3
Estimate of mass removal rates of some volatile compounds

level of vacuum	Contaminant Removal Rates (lbs/day)		
	100% of vapor pressure	10% of vapor pressure	1% of vapor pressure
0-15mm Hg (Example Toluene)	0-400	0-20	0-4
15-30 mm Hg (Example Benzene)	400-2500	20-120	4-10
30-45 mm Hg	2500-5000	120-500	10-75
45-60 mm Hg	5000-7000	500-800	75-100
60-90 mm Hg (Example Acetone)	7000-10,000	800-1000	100-120

% of vapor pressure depends on level of vacuum created during bioventing using vacuum extraction.

Enhanced oxygenation due to air flow can also cause significant increases in biodegradation. Table 5.4 gives values for oxygen supply (lbs/day) depending on air flow rate and percent utilization of the oxygen supplied.

Table 5.4
Oxygen Supply Rates

Air flow rate (SCFM)	Percentage utilization		
	100 (lbs-O_2/day)	10 (lbs-O_2/day)	1 (lbs-O_2/day)
25	800	90	8
75	5000	500	60
100	7000	800	80
150	10,000	1000	100

As can be seen from the above table, even a small air flowrate of about 25 SCFM can provide significant amounts of oxygen, which even at low utilization rates can provide substantial enhancement of biodegradation activity.

(2) Bioventing is a relatively inexpensive process for in-situ bioremediation, provided, that no air vented treatment is required. Table 5.5 gives the cost of supplying oxygen ($/day) by vapor extraction in bioventing as compared to other methods such as peroxide addition and aeration [19, 24].

Table 5.5
Cost Estimates for Supplying Oxygen ($/day)

Type of Technology	Percent utilization 100	10	1
Peroxide	80	200	800
Aeration	100	200	300
Bioventing using vapor extraction	0.8	3.0	25

It is assumed that peroxide is added at 2000 ppm concentration at 10 GPM flowrate. Aeration is achieved by injection of aerated water (dissolved oxygen 10 ppm) at 20 GPM flow rate or by sparging in wells. It is assumed that in bioventing the air flowrate is 100 SCFM and hence no subsequent vapor treatment is needed.

(3) Minimal exposure of the public (or the contractors) to the affected zone. This is a basic advantage of all in-situ processes which results in minimal disruption of surface level activities (operation of airports, plants, etc.) once the process installation is completed. In-situ treatment, such as bioventing can proceed while normal site functions continue.

The major disadvantages of bioventing are as follows:

(1) A minimum permeability value of 10^{-5} cm/sec is needed for achieving effective bioventing. The limitations on permeability depends on the type of system being used for achieving air flow. The types of systems that are possible are as follows:

- High vacuum (20-30) inches Hg versus low vacuum (about 100 inches Hg)
- Single or multiple wells with forced pumping of air (Most venting applications end up with wells at 10-20 ft. spacing. Single or dual wells are only utilized for small contaminated areas.)

The main advantages of high vacuum are greater radius of influence, shorter remediation times, and ability to treat lower permeable formations. Low vacuum systems are lower in cost, have lower maintenance costs, and allows for more uniform treatment. The main disadvantages of high vacuum bioventing are: significant waste upwelling causing well fouling; increase a channeling of the air flow; and higher air velocities resulting in increased frictional losses.

In single well vapor extraction, for homogeneous soils the radius of influence is theoretically symmetrical around the well and hence a proper distribution of wells is required in the contaminated area.

Using two wells with air being forced into one well results in better air distribution, but limits the area of influence primarily to the area between the two wells and can force the contaminant away form the receiving well, if the injection well is located outside of the capture zone of the extraction well.

(2) Bioventing systems suffer from "aging" as the contaminant region decreases. This aging is primarily caused by diffusive flow rather than advective flow in the contaminated region. Usually this problem can be reduced by changing the configuration of the extraction wells thereby changing the air flow pattern. Changes in air distribution need to be optimized for effective bioventing throughout the treatment period.

(3) One potential disadvantage of an in-situ process, such as bioventing, is that contaminant materials are mobilized by the movement of air and water around the contaminated zone. This requires close monitoring of air and water around the contaminated zone to ensure that the contaminant does not escape the air and water capture systems. Flow models used to describe contaminate transport should be calibrated with actual site data so that the fate of the contaminant can be determined with a certain degree of confidence.

(4) Due to variations in site geology and contaminant distribution, the treatment time cannot be guaranteed, in spite of laboratory and pilot tests. Usually there are regions where the concentration remains high (hot spots) as compared to the treatment standard.

(5) Due to the fact that bioventing is a relatively new technology, public acceptance of the technology may require detailed explanations and meetings. Although bioventing has a history of full-scale success in cleaning sites, public acceptance is not guaranteed. The practitioner needs sufficient experience base to apply bioventing successfully.

5.5.4.4. Design Considerations of Bioventing

The design of the bioventing system requires proper specification of pumps and piping. The driving force for the creation of a vacuum in the soil is a positive displacement blower, a centrifugal blower, or a vacuum pump, depending on the type of system used as shown in Figure 5.6. Factors that determine blower design are level of vacuum, areal extent of contamination, air permeability of soil, and pressure losses through the pump, piping and collection system. Due to safety considerations, sites where explosive gases are present, explosion-proof housing for motors and liquid sealed vacuum pumps with minimal spark potential have to be used.

The piping used to connect the wells to the blower and/or vacuum pump is simple for single well systems but becomes increasingly complicated as more injection and extraction wells are used. Usually the piping is 6 inches in diameter and material of construction can be polyvinylchloride (PVC) (Schedule 40 or 80), polypropylene, high density polypropylene, or stainless steel. The piping can be above ground or buried underground for sites in active use. In colder climates the piping has to be insulated to prevent condensed water from freezing. Vapors exiting the extraction wells are connected to knock-out drums to separate fine silt particles and then cooled to condense moisture. If a high vacuum is used which can cause significant increases in the water table level, large quantities of water are removed and in cases were this water is contaminated, suitable water treatment systems have to be used. In addition is there are, problems with excess moisture in the vapors expensive condensation treatment may be required.

5.5.5. Site Characterization

The surface area and depth of the contaminated site is important for evaluating bioventing costs, in addition to time for achieving cleanup goals also critical. Capital cost are low, but operation and maintenance can be very high for extended operations. Traditionally, site areas have varied between 20-75,000 m^2 in the 3-10 m depth range. Sites where bioventing has been applied can be categorized into two classes: gas filling stations, where the surface areas typically varied between 400-1000 m^2, and large chemical industry (refinery) sites, where the areas of contamination may be as large as 20,000-75,000 m^2.

Bioventing is typically applied to sandy subsurface conditions, although some application sites have included occasional clay lens. Only in exceptional cases, has bioventing been applied at sites with extensive overburden clay or fractured bedrock. Permeability is an important factor in determining feasibility of bioventing, and for the projects reviewed, permeability varied between 10^{-2} to 10^{-8} m/s, with average value of 10^{-3} m/s. Generally, a value of 10^{-5} cm/s is regarded as the minimum permeability for successful application of bioventing. Soil pH also affects sorption of ionizable compounds in addition to limiting the types of microorganisms in the subsurface. Methanogens, which are responsible for mineralization of some aromatic compounds, are inhibited at pH values below 6. Biodegradation of many organic pollutants may be limited by insufficient concentrations of oxygen or unfavorable redox conditions. Temperature also influences microbial metabolism, and usually the temperature varies seasonally up to a depth of 10 m The impact of temperature is most pronounced up to the frost line (typically up to a depth of 2 m). Determination of iron and manganese concentrations is important because high concentrations of these metals can cause precipitation under aerobic conditions. Other heavy metals can be important, because at toxic levels they can inhibit the activity of microorganisms. Inorganic nutrients like nitrogen and phosphorous may be limiting when the carbon/nitrogen/phosphorous (C:N:P) ratio is unfavorable. Determination of ammonium, nitrate and nitrite levels is important before applying bioventing.

Site characterization involves determining the microbiological, hydrogeological, and chemical/physical parameters. Preliminary biotreatment studies at the laboratory scale are needed to determine the optimal stimulatory conditions for effective biodegradation at a given site.

5.5.6. Approximate Costs

Costs for bioventing depend on several factors:

- Geology and soil structure
- Type and concentrations of contamination
- Distribution of contamination in the soil
- Total surface and volume of contamination
- System characteristics of the bioventing process: type of air circulation, level of vacuum, number of wells etc.
- Time of application

Generally one-third of the total cost is for preliminary research and installation costs. The remaining two-thirds of the total costs are spent on operation and maintenance costs.

For a one acre site contaminated with gasoline with a treatment period of 1.5-2 years, the cost of bioventing is approximately \$250,000 without vapor treatment and \$350,000-\$400,000 with the vapor treatment. Enhancement of biodegradation by adding nutrients lowers the cost of vapor treatment. However, nutrient addition can add \$20-40,000 to the total cost.

6. HEALTH AND REGULATORY ISSUES

6.1. Health and Safety Issues

The two main public health and safety questions that determine the feasibility of bioremediation are:

- Can the hazard posed by the contaminant be removed in a timely manner, and can the hazard be contained during treatment. This may reflect in additional treatment cost which may be balanced against other more rapid treatment.
- Does the selected bioremediation process pose any hazards, such as toxic by-products, fire or explosion hazards? Other hazards include dust and vapors generated by land farming activities.

Information regarding health and safety issues can be obtained from risk assessments and the development of a health and safety plan for the project.

6.2. Regulatory Issues

Local, state, and federal regulations that are applicable need to be considered. For example, Land Disposal Restriction (LRD) regulations may prevent the use of certain bioremediation technologies, such as above ground biopiles, while in-situ treatment may be allowed. LDR's would not affect above-ground biopiles unless the waste was to be permanently disposed in the biopile. However, if contaminated soil was classified as hazardous under RCRA, one would need a RCRA treatment/storage permit for the waste pile, which can be a lengthy and expensive process requiring significant closure monitoring. Although, an in-situ treatment operation would not require a RCRA permit.

APPENDIX

1. BASIC MECHANISM OF BIODEGRADATION

As explained above all microbes require a carbon and energy source for their growth.

Types of carbon include:

- Organic (wide variety available)
- Inorganic (e.g.. dissolved carbon dioxide)

Types of energy include:

- Light energy such as sunlight
- Chemical reactions
  - — The amount of energy available from these sources depends not only on the energy source but also on the terminal electron acceptor used.

Most of the energy in biological reactions is obtained from electron pair transfers (oxidative phosphorylation) i.e., electrons are transferred from one compound to the other in the order of decreasing energy level. Each time these electrons are transferred a small amount of energy is generated which is either stored in the form of ATP - the energy currency of a cell, or used up for cell functions. These electrons are donated by the parent food molecule. They pass through a number of intermediate molecules releasing energy in every transaction they pass through and are finally accepted by a terminal electron acceptor (e.g.. oxygen is the terminal electron acceptor for aerobes) which gets reduced and does not transfer the electrons further down—thus stopping the reaction. Availability of this terminal electron acceptor is essential for the constant energy supply required by the cells.

Reference may be made here to the "electron tower", which is a table showing various electron acceptors in the order of decreasing energy levels (App. Table 1.1). For a biological reaction to be energetically favorable i.e. in order to have $\Delta G < 0$, the energy source must be positioned above the terminal electron acceptor in the energy tower. The farther apart the two are, the greater is the energy yield per molecule of substrate oxidized. The spontaneity of a biological reaction, in the way it is written, can be quantified by the Nerest Equation:

$$\Delta G = -nF\Delta E$$

Where:
ΔG = Free energy change of the reaction. If ΔG is <0 the reaction is spontaneous. If ΔG is >0 then energy is required for the reaction to proceed.
n = Moles of electrons transferred per mole of reactant/substrate consumed.
F = Faraday's Constant = 25,000 cal/volt
ΔE = Change in redox potential (volts)

Oxygen is placed very low in this tower and this is the reason that more energy is yielded using oxygen as the terminal electron acceptor than most other terminal electron acceptors. This also explains why facultative bacteria yield more energy aerobically than that in absence of air.

App.Table 1.1
Reduction Potentials of Some Redox Pairs of Biochemical and Microbiological Importance

Redox Pair	E'_o(V)
SO_4^{2-}/HSO_3^-	-0.52
CO_2/formate	-0.43
H^+/H_2	-0.41
Ferredoxin ox/red	-0.39
Flavodoxin ox/red*	-0.37
$S_2O_3^{2-}/HS^- + HSO_3^-$	-0.36
NAD/NADH	-0.32
Cytochrome c_3 ox/red	-0.29
CO_2/acetate$^-$	-0.29
S_o/HS^-	-0.25
SO_4^{2-}/HS^-	-0.25
CO_2/CH_4	-0.24
FAD/FADH	-0.22
$HSO_3^-/S_3O_6^{2-}$	-0.20
Acetaldehyde/ethanol	-0.20
Pyruvate$^-$/lactate$^-$	-0.19
FMN/FMNH	-0.19
Dihydroxyacetone phosphate/ glycerolphosphate	-0.19
Flavodoxin ox/red	-0.12
HSO3-/HS-	-0.11
Menaquinone ox/red*	-0.075
APS/AMP + HSO_3^-	-0.060
Rubredoxin ox/red	-0.057d
Acrylyl-CoA/propinyl-CoA	-0.015
Glycine/acetate$^-$ + NH_4^+	-0.010
$S_4O_6^{2-}/S_2O_3^{2-}$	+0.025
Furmate/succinate	+0.030
Cytochrome *b* ox/red	+0.030
Ubiquinone ox/red	+0.11
Cytochrome c_1 ox/red	+0.23
$S_4O_6^{2-}/S_2O_3^{2-} + HSO_3^-$	+0.23
NO_2^-/NO	+0.36
Cytochrome a_3 ox/red	+0.385
NO_3^-/NO_2^-	+0.43
Fe_3^+/Fe_2^+	+0.77
O_2/H_20	+0.82
NO/N_2O	+1.18
N_2O/N_2	+1.36

*Separate potentials are given for each electron transfer in this potentially two-electron transfer.

Cell yield - quantifies the amount of cells synthesized and is defined as the mass of microbial cells produced per unit mass of substrate consumed

Y = grams of cell mass/ grams of substrate consumed

2. AEROBIC BIODEGRADATION

2.1 Types of Microorganism

2.1.1 Methanotrophs and methylotrophs

Methanotrophs can oxidize methane and some other carbon compounds utilizing them as sole carbon and energy source. Organisms that can grow using only one - carbon compounds are generally called methylotrophs.

Presence of C1 metabolizers in the environment is of prime importance. As illustrated below, methanotrophs and methylotrophs play a key role in the degradation or initiation of degradation of a vast category of compounds. App. Table 2.1 lists some of the important species of these microorganisms and respective growth substrates.

App. Table 2.1
Classification of C_1 - Utilizing Microorganisms

Group	Examples	Growth substrates or C_1 compounds metabolized
BACTERIA		
Obigate methylotrophs (Type 1 methanotrophs)	*Methylobacter* *Methylococcus* *Methylomonas* *(Pseudomonas methanica)* *Methylocystis*	methane, methanol, carbon monoxide (oxidized only)
Obligate methylotrophs (Type II methanotrophs)	*Methylosinus* *(Methanomonas methanooxidans)*	
Facultative methanotroph (similar to Type II obligate methylotroph)	*Methylobaterium organophilum*	methane, methanol, variety of heterotrophic substrates
Obligate methylotrophs unable to utilize methane	Bacterium 4B6	methylamine, di- and tri-methylamine
	Methylomonas M15 *Methylophilus methylovora* *Methylophilus methylotrophus*	methanol
	Organism W1	methanol, methylamine

App. Table 2.1
Classification of C_1 - Utilizing Microorganisms (continued...)

Facultative methylotrophs (All grow on some heterophic substrates)	*Bacillus* PM6 Bacterium 5H2	methylamine, di- and tri-methylamine, tetramethyl-ammonium and trimethyl-sulfonium compounds
	Bacterium 5B1	methanol, formate, methylamine di- and tri-methylamine, tri-methylamine-N-oxide
	Hyphomicrobium sp.	methanol, formate, methylamine, tri-methylamine, formamide, urea
	Pseudomonas AM1	methanol, formate, methylamine, trimethylamine
	Pseudomonas SL4	formamide
	Pseudomonas aminovorans	formate, methylamine, di- and trimethylamine, tri-methylamine-N-oxide
	Pseudomonas MA	methylamine
	Pseudomonas MS	methylamine, di- and tri-methylamine, trimethyl-sulfonium compounds
	Pseudomonas M27	methanol, formate, methylamine
	Pseudomonas 3A2	methanol, formate, methylamine, di- and tri-methylamine, tri-methylamine-N-oxide
Autotrophs	*Vibrio oxalaticus*	formate
	Pseudomonas oxalaticus	
	Paracoccus denitrificans	methanol, methylamine, formate
	Thiobacillus novellus	
	Microcyclus aquaticus	
	Acromobacter sp.	methanol
	Pseudomonas sp.	
	Mycobacterium sp.	
	Hydrogenomonas carboxydovorans	
	Pseudomonas carboxydovorans	carbon monoxide
	Seliberia carboxydohydrogena	
Phototrophs	*Rhodopseudomonas palustris*	formate
	Rhodopseudomonas gelatinosa	methane, methanol
	Rhodopseudomonas acidophila	formate
	Rhodomicrobium vannielii	
	Rhodospirillium rubrum	methanol
	Rhodospirillium tenue	
	Chloropseudomonas sp.	formate
Nonphotosynthetic, nonmethanogenic anaerobes	*Rhodopseudomonas* sp.	
	Clostridium pasteurinum	
	Clostridium welchii	carbon monoxide
	Clostridium thermoaceticum	
	Clostridium formicoaceticum	
	Unidentified sulfate reducers	methane

App. Table 2.1
Classification of C_1 - Utilizing Microorganisms (continued...)

Methanogenic anarobes	*Methanobacterium* sp. *Methanospirillum* sp. *Methanococcus* sp. *Methanosarcina* sp.	methane, methanol, methyl-amines, formate dimethylsulfoxide, methane thiol
Miscellaneous bacteria	*Norcardia salmonicolor* *Bacillus* strain *Actinomycete* strain *Nocaridia* strain *Bacillus* pumilus	carbon monoxide
	Pseudomonas nonliquefaciens *Bacterium thiocyanoxidans*	thiocyanate, cyanide
	Thiobacillus thiocyanoxidans *Pseudomonas stutzeri* *Thiobacillus thioparus*	thiocyanate
	E. coli	dimethyl sulfoxide
Wide range of bacteria but especially ureolytic bacilli	*Bacillus pasteurii*	urea
	Bacillus sphaericus	N-monomehtyl urea
Fungi	*Trichoderma lignorum*	methanol, methylamine, formate
	Gliocladium deliquescens	methanol, formaldehyde
	Paecilomyces varioti *Aspergillus niger*	dimethyl sulfoxide
Yeasts	*Kloeckera* sp. *Torulopsis glabrata* *Candida methanolica* *Torulopsis methanolovescens* *Candida boidinii* *Hansenula* sp *Pichia pinus* *Pichia methanolica* *Pichia pastoris*	methanol
	Pink yeast strains	methane
	Many yeast species	methylamine
	Saccharomyces cerevisiae	dimethyl sulfoxide

Methanotrophs:

- Are primarily gram negative obligate aerobes
- Can grow only on methane
- May utilize methanol and dimethylether as a substrate
- Occur wherever methane is exposed to aerobic environments such as soil, sediments, natural waters, etc.
- May reach concentrations of 5E7/ml in sediments

The roles of methanogens, methylotrophs, and methanotrophs are very significant in the environment because they not only convert the C1 compounds to non toxic products, but also, in some cases, initialize the biodegradation of other organic compounds.
Methanotrophs have been found to:

- Degrade complex hydrophobic compounds which are not growth substrates
- Fix atmospheric nitrogen and oxidize ammonia to nitrite and nitrates

- Metabolize potent metabolic inhibitors like cyanide and carbon monoxide
- May remove pollutants, especially xenobiotic compounds

A schematic illustrating their importance in the environment is shown in Appendix Figure 2.1.

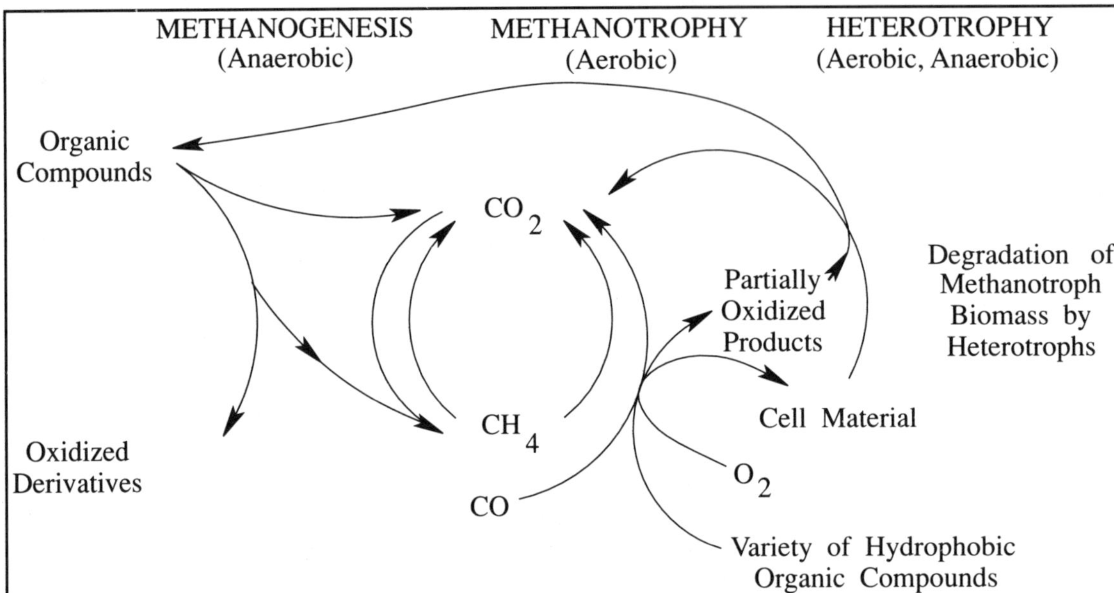

APPENDIX FIGURE 2.1
Carbon Recycling in the Biosphere Via Methane
Biological Remediation

2.1.2. Multi-carbon metabolizers

Many microorganisms may not be able to use aliphatic hydrocarbons as growth substrates; however, they may oxidize them in the presence of other growth substrates. This phenomena, known as co-metabolism or co-oxidation, causes oxidation of many aliphatic hydrocarbons. *Pseudomonas mechanica*, for example, could oxidize ethane, propane, and butane in the presence of methane, but none of these gases could serve as a growth substrate for the organism.

Soil microbial communities growing on light aliphatic hydrocarbons have also been found to be capable of co-metabolically degrading trichloroethylene and related compounds normally recalcitrant to biodegradation in aerobic surface environments This microbial process has high potential as a basis for methodologies for removal of chlorinated hydrocarbon solvents from polluted ground water, both *in situ* and in surface based systems.

A list of some of these microorganisms is provided in App. Table 2.2.
Populations of these organisms vary with the habitat. For example, they may be more prevalent in oil polluted environments. Enrichment of hydrocarbonoclastic microbes in areas containing petroleum, has been applied in prospecting for oil and gas fields. Geographical location factors in determining population density of these microbes include availability of nutrients, type of climate, and other physical conditions.

App. Table 2.2
Genera of Bacteria and Yeasts the Reportedly Contain Aliphatic Hydrocarbon-Oxidizing Species

Bacteria	Yeasts
Achromobacter	*Candida*
Acinetobacter	*Cryptococcus*
Actinomyces	*Debaryomyces*
Aeromonas	*Endomyces*
Alcaligenes	*Hansenula*
Arthrobacter	*Mycotorula*
Bacillus	*Pichia*
Beneckea	*Rhodotorula*
Brevibacterium	*Saccharomyces*
Corynebacterium	*Selenotila*
Flavobacterium	*Sporidiobolous*
Methylobacter[a]	*Sporobolomyces*
Methylbacterium[a]	*Torulopsis*
Methylococcus[a]	*Trichnosporon*
Methylocystis[a]	
Methylomonas[a]	
Methylosinus[a]	
Micromonospora	
Mycobacterium	
Nocadia	
Pseudomonas	
Spirillum	
Vibrio	

a Methane is the only hydrocarbon used as a sole carbon source. Aliphatic hydrocarbons larger than methane may be cooxidized.

2.2 Biodegradation Pathways for Typical Alkanes and Aromatics

Alkanes may occur as the source pollutant or be the product of prior multi-carbon substrates. Biodegradation pathways for the single carbon compounds are simpler and will aid in the understanding the pathways of the more complex compounds.
Methane is the most common C1 substrate involved in microbial metabolism since many compounds are converted into methane under anaerobic conditions.
The microorganisms involved include:

Methanogens - the class of microorganism that converts many organic compounds into methane under aerobic conditions

Methanotrophs - a class of microorganisms that converts methane into carbon dioxide in aerobic environments

Methylotrophs - more versatile nutritionally than Methanotrophs — cannot use methane as a carbon and energy source

Some of the other significant C1 compounds, which can also be utilized substrates, are listed in App. Table 2.3.

App. Table 2.3
C_1 Compounds of Major Environmental Importance

Compound	Formula	Comments
Methane	CH_4	Ubiquitous end product of anaerobic fermentation processes. Generated by free-living and rumen-inhabiting methanogens.
Methanol	CH_3OH	Generated during the breakdown of natural methyl-esters and ethers (e.g., pectin) and released by methanotrophs.
Formaldehyde	HCHO	Common combustion product, intermediate microbial oxidation of other C_1 compounds and methylated biochemicals (e.g., lignin)
Formate ion	$HCOO^-$	Present in plant and animal tissues. Common product of carbohydrate fermentation.
Formamide	$HCONH_2$	Natural product formed from plant cyanides.
Carbon dioxide	CO_2	Combustion, respiration, and fermentation end product. A major reservoir of carbon on Earth.
Carbon monoxide	CO	Combustion product, common pollutant. Product of plant, animal, and microbial respiration, highly toxic.
Cyanide ion	CN^-	Generated by plants, fungi, and bacteria. Industrial pollutant, highly toxic.
Dimethylether	CH_3OCH_3	Generated form methane by methanotrophs. Industrial pollutant.
Methylamine Dimethylamine Trimethylamine Trimethylamine-N- oxide	CH_3NH $(CH_3)_2NH$ $(CH_3)_3N$ $(CH_3)_3NO$	Wide spread occurrence in tissues of plants, animals, fishes. Nitrogenous degradation products, excretion products.
Tetramethylammonium salts	$(CH_3)_4N^+$	Halogen salts used as disinfectants.
Trimethylsulfonium salts	$(CH_3)_3S^+$	Industrial chemical.
Urea	$CO(NH_2)_2$	Important animal excretion product.
Carbamic acid	NH_2COOH	Hydrolysis product of urea, fertilizer, N- methyl- and N, N-dimethyl derivatives used as pesticides.
N- Monomethylurea	$CH_3NH.-CO.NH_2$	Nitrogenous degradation product.

Research has shown that the obligate methanotrophs convert methane to carbon dioxide by two electron oxidation steps, as shown in Figure 2.2

- Methane is converted to methanol in presence of oxygen.

  There is evidence to indicate that the oxygen in methanol is incorporated from molecular oxygen suggesting the involvement of an oxygenase. All methanotrophs have methane mono-oxygenase required for incorporating molecular oxygen in methane to yield methanol. The methane mono-oxygenases examined so far have extremely broad substrate specificities and can at least partially oxidize a wide variety of substances including carbon monoxide, ammonia, higher alkanes, alkenes and aromatic and alicyclic compounds. It can initiate biodegradation of many multi-carbon compounds which are otherwise non-growth substrates. In some cases oxidations are solely the result of mono-oxygenase activity as in the case of propane being converted to propene-oxide.

- Methanol is then subsequently converted to formaldehyde, formate, and, eventually, carbon dioxide by dehydrogenases

    Formation of formaldehyde is of prime importance since many microorganisms incorporate carbon at the formaldehyde level into biomass via various pathways like ribulose monophosphate cycle (App. Figure 2.3), serine pathway (App. Figure 2.4), or dissimilatory hexulose phosphate pathway (App. Figure 2.5). Formaldehyde thus acts as the introductory point for the organic carbon into the biomass (App. Figure 2.6).

    Many C1 compounds are converted to formaldehyde for incorporation into the biomass. Some of the formaldehyde formed is incorporated while the rest is converted to CO_2. In many species which use ribulose monophosphate pathway for formaldehyde fixation, a cyclic pathway for oxidation of formaldehyde, the dissimilatory hexulose phosphate pathway is operational.

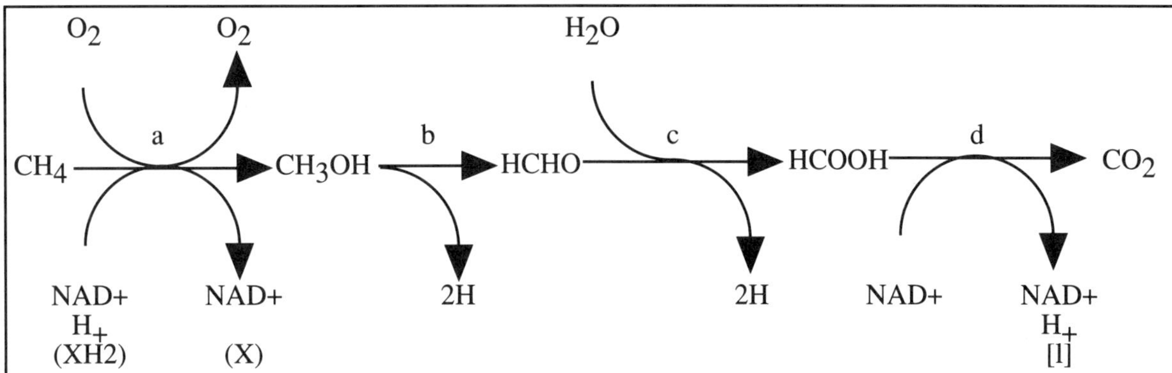

APPENDIX FIGURE 2.2
Oxidation of Methane by Methanotrophs:
(A) Methane Monooxygenase, (B) Methanol Dehydrogenase, (C) Formaldehyde Dehydrogenase, (D) Formate Dehydrogenase
Biological Remediation

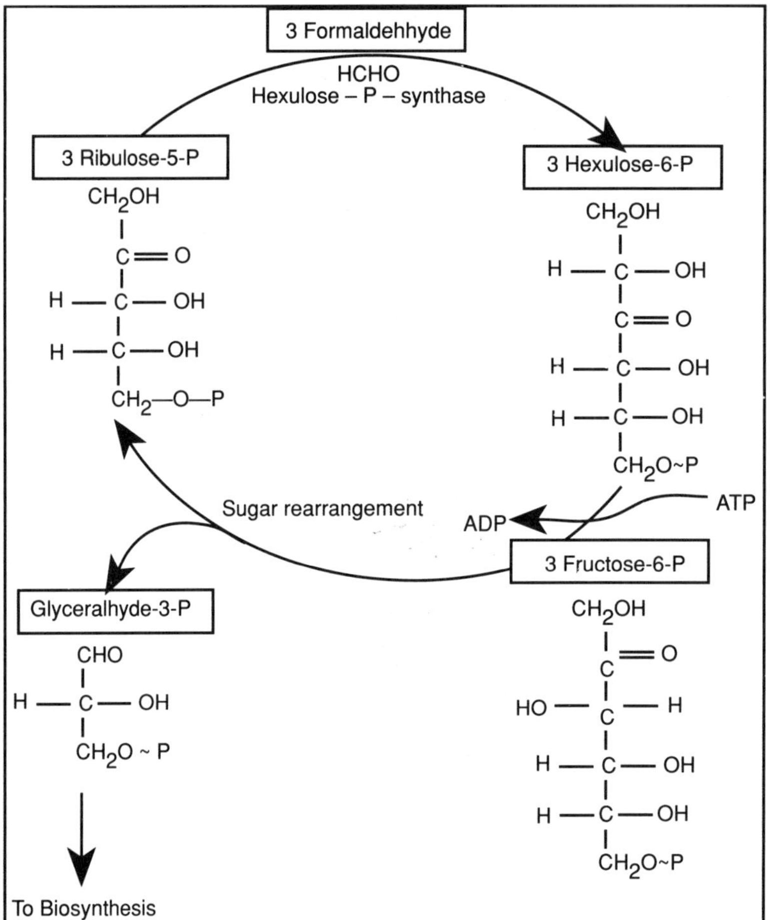

Overall: 3 formaldehyde + ATP → gylceraldehyde – 3 – P + ATP

APPENDIX FIGURE 2.3
Ribulose Monophosphate Pathway
Biological Remediation

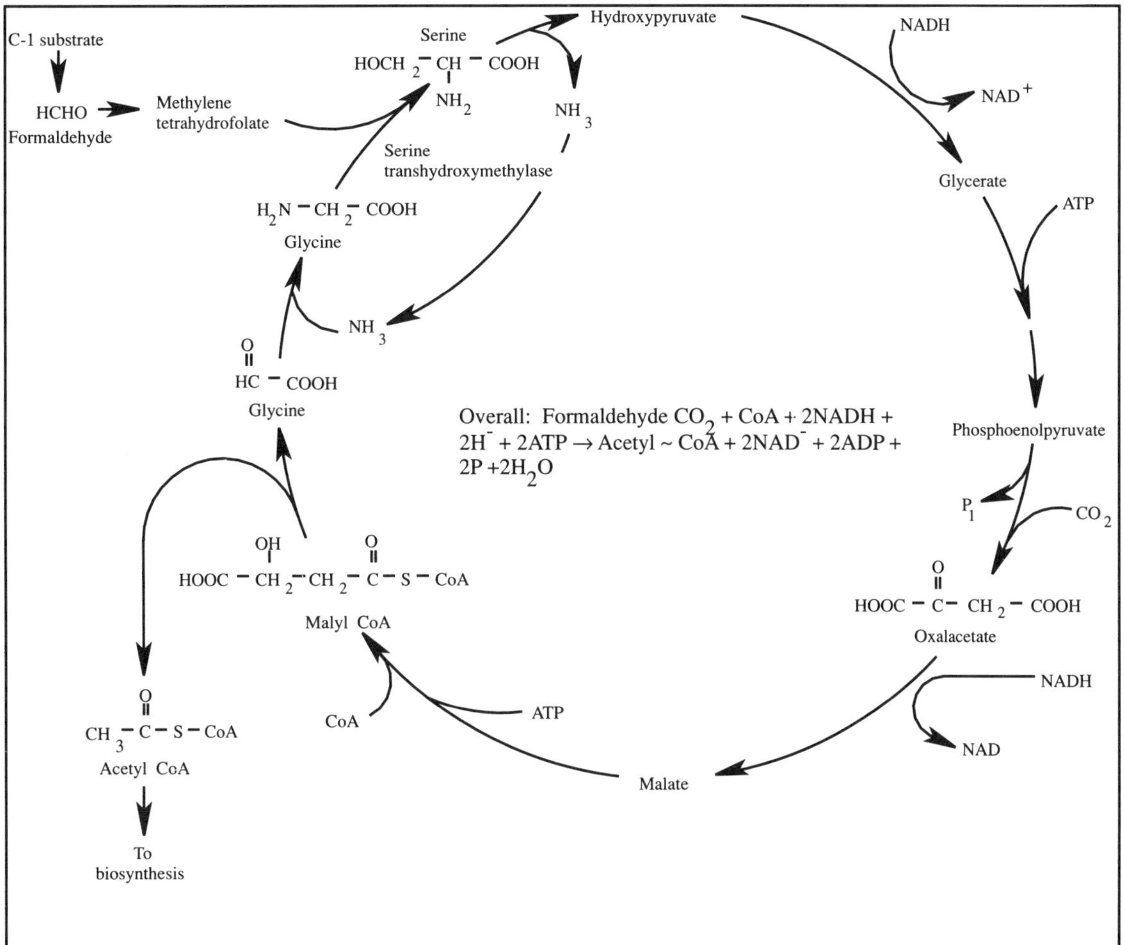

APPENDIX FIGURE 2.4
The Serine Pathway for the Assimilation of C-1 Units Into Cell Material
Biological Remediation

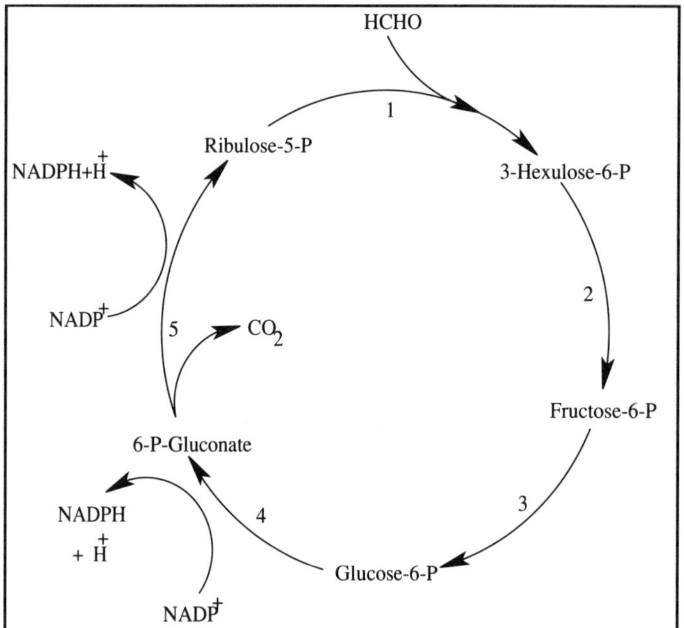

APPENDIX FIGURE 2.5
The Dissimilatory Hexulose Phosphatecycle
Biological Remediation

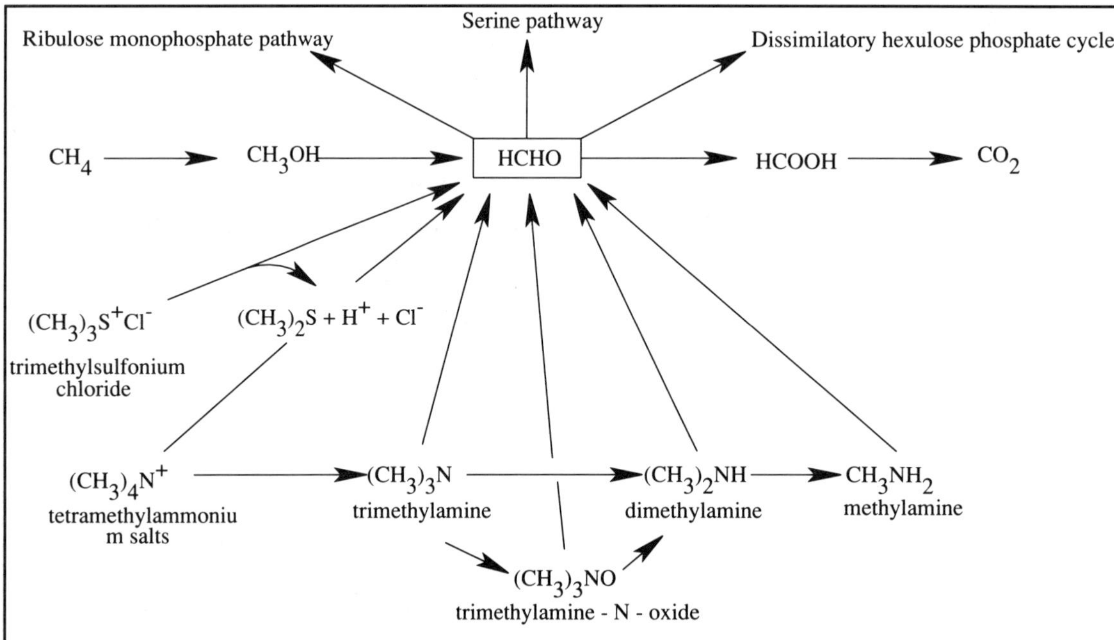

APPENDIX FIGURE 2.6
The Central Position of Formaldehyde in Microbial C1 Metabolism
Biological Remediation

In the case of aliphatic hydrocarbons, some generalizations can be made regarding their utilization for biodegradation:

- Aliphatic hydrocarbons assimilate more readily than aromatic hydrocarbons.
- Long chain hydrocarbons assimilate more readily than short chain hydrocarbons.

  n-Alkanes, shorter than n-nonane, may be oxidized but are not assimilated usually.

  For yeasts, there is an increase in the conversion of n-alkane to biomass and a decrease in the rate of oxidation with an increase in chain lengths after n-nonane. The reason for this behavior could be the toxicity of the short chain liquid hydrocarbons. This toxicity can be attributed to the greater solubility in the aqueous phase. The toxic levels in the aqueous phase could be reduced by partitioning the toxic hydrocarbon between aqueous phase and a second non-toxic hydrocarbon.

- Saturated aliphatic hydrocarbons degrade more easily than unsaturated ones
- Branched chain hydrocarbons degrade less readily than straight chain ones

Of the many mechanisms proposed for the oxidation of aliphatic hydrocarbons, all propagate oxidation of the terminal methyl group. The original proposal of terminal oxidation was based on the detection of homologous fatty acids in the culture media. Presence of esters, presumably formed by alcohols and homologous fatty acids, confirms oxidation at C1 position. This also indicates that the terminal methyl group is first oxidized to a primary alcohol followed by subsequent oxidation to fatty acid. Once fatty acid is formed, it is subjected to ß-oxidation yielding acetyl-CoA. Acetyl-CoA thus generated can be further oxidized via TCA cycle to yield energy or can be converted to hexose and other cell constituents via the glyoxylate cycle. Methyl group oxidation of n-alkanes is shown in Appendix Figure 2.7 as an example.

Some of the oxidative mechanisms are described by the mechanisms shown in Appendix Figure 2.8. There are problems associated with these mechanisms. For example, if the primary alcohols are formed by hydration of alkenes, as proposed by mechanism 2, then molecular oxygen should not be required for the oxidation. This definitely is not true since most of the hydrocarbon degradation processes have been known to utilize oxygen. If, on the other hand, primary alcohols are formed by way of oxygenation of alkenes then the oxidative products from n-alkanes and 1-alkenes should be the same, which, again, has been found to be incorrect.

The most widely accepted pathway for hydrocarbon degradation is mechanism 3. As indicated by this pathway, direct incorporation of oxygen is required for the formation of the primary alcohol. This requires presence of mixed-function oxidase or mono-oxygenase. Thus, organisms or systems having these enzymes can undergo aerobic oxidation easily.

Another mechanism, which is not so common or established is the sub-terminal oxidation of the methyl group. The phenomena was observed when propane and butane co-metabolized to acetone and 2-butanone, respectively. Subterminal oxidation is demonstrated by Pencillium species and strains of Pseudomonas aeruginosa. Existence of subterminal oxidation is also apparent when secondary alcohols and methyl-ketones are obtained from oxidation of n-alkanes by methane utilizing bacteria. A schematic showing the subterminal oxidation of n-alkane is shown in Appendix Figure 2.9. In this mechanism, studies conducted on 2-tridecanone oxygenase demonstrated that the oxygen atom inserted during conversion of the ketone to an ester is on the side of carbonyl group with the longest alkyl group.

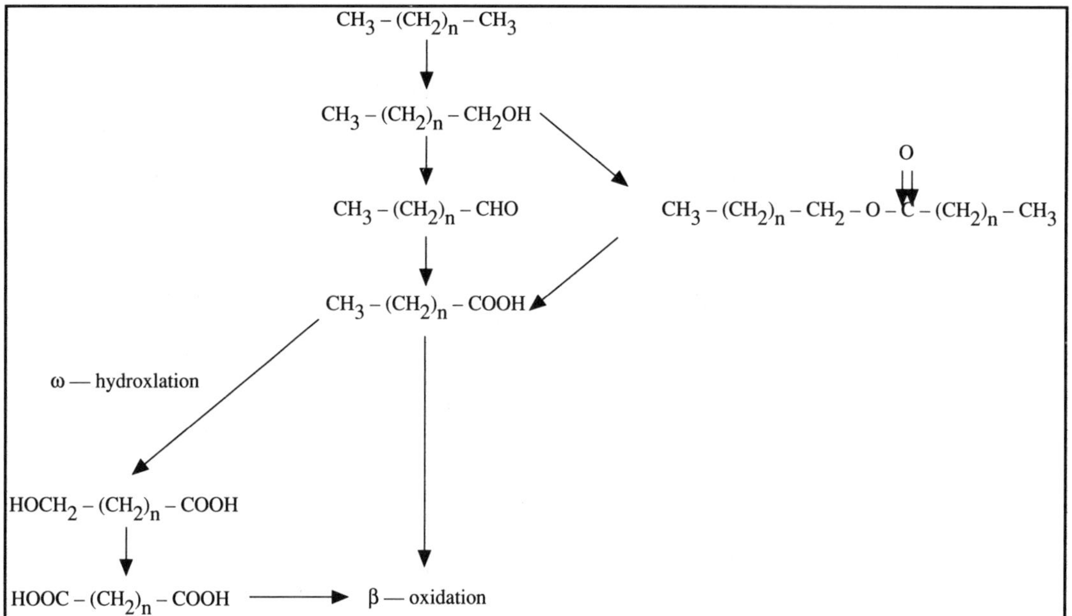

APPENDIX FIGURE 2.7
Methyl Group Oxidation of n - alkanes
Biological Remediation

$$R-CH_2-CH_3 \rightarrow R-CH_2-CH_2-OOH \xrightarrow{NAD(P)H} R-CH_2-OH + H_2O + NAD(P) \quad [1]$$

$$R-CH_2-CH_3 + NAD(P)^+ \rightarrow R-CH=CH_2 + NAD(P)H + H^+$$

$$R-CH=CH_2 + H_2O \rightarrow R-CH_2-CH_2OH$$

$$R-CH_2-CH_3 + O_2 + NAD(P)H + H^+ \rightarrow R-CH_2-CH_2OH + NAD(P)^+ + H_2O$$

$$R-CH_2-CH_2OOH \rightarrow R-CH_2-\overset{O}{\overset{\|}{C}} + OOH$$

$$\downarrow$$

$$R-CH_2-CHO \rightarrow R-CH_2-CH_2OH$$

$$\downarrow$$

$$R-CH_2-COOH$$

APPENDIX FIGURE 2.8
The Formation of Alkyl Hydroperoxides in Mechanism [1] Has Been Assumed From Indirect Evidence. Alkyl Hydroperoxides Were Oxidized in Alkane-Grown Acinetobacter
Biological Remediation

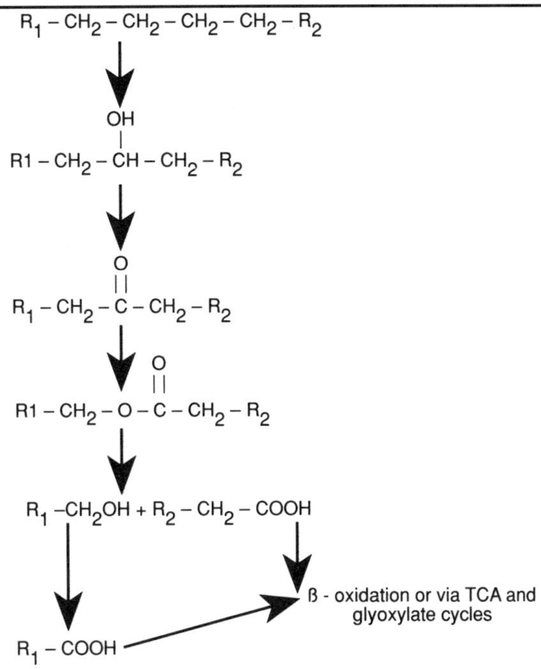

APPENDIX FIGURE 2.9
Subterminal Oxidation of Aliphatic Hydrocarbons; R = H or Alkyl Group
Biological Remediation

3. ANAEROBIC BIODEGRADATION

3.1 Types of Microorganism

Anaerobic environments, or environments of low reduction potential, include:

- Muds and other sediments of lakes
- Rivers and oceans
- Bogs and marshes
- Water logged sites
- Deep underground areas (such as oil pockets)
- Most ground water environments

Low reduction potential - is due to a lack of replacement of oxygen, necessary for the bacteria that use oxygen for respiration, and will cause the habitat to become anaerobic.

Anaerobiosis occurs in only two groups of microorganisms:

- Protozoa
- Bacteria

<u>Clostridia</u> - the best known obligately anaerobic bacteria, are a group of gram positive spore forming rods which are wide spread in soils and lake sediments.

Other obligately anaerobic bacteria are found among the methanogenic bacteria and the sulfate reducing bacteria.

Some substrates metabolized/oxidized by methylotrophic bacteria and their important characteristics are given in App. Table 3.1 and App. Table 3.2

Characteristics of sulfate and sulfur-reducing bacteria are provided in App. Table 3.3.

Sensitivity to oxygen greatly varies among obligate anaerobes. Some organisms can tolerate traces of oxygen while others cannot.

App. Table 3.1
Substrates Used by Methylotrophic Bacteria*

Substrates used for growth	Substrates oxidized, but not used for growth (co-metabolism)
Methane, CH_4	Ammonium, NH_4^+
Methanol, CH_3OH	Ethylene, $H_2C{=}{=}CH_2$
Methylamine, CH_3NH_2	Chloromethane, CH_3Cl
Dimethylamine, $(CH_3)_2NH$	Bromomethane, CH_3Br
Trimethylamine, $((CH_3)_3N$	Higher hydrocarbons (ethane, propane)
Tetramethylammonium, $(CH_3)_4N+$	
Trimethylamine N-oxide, $(CH_3)_3NO$	
Trimethylsulfonium, $(CH_3)_3S^+$	
Formate, $HCOO^-$	
Formamide, $HCONH_2$	
Carbon monoxide, CO	
Dimethyl ether, $(CH_3)_2O$	
Dimethyl carbonate, $CH_3OCOOCH_3$	

*A single isolate does not use all of the above, but at least one methylotrophic bacterium has been reported to oxidize each of the listed compounds.

App. Table 3.2
Some Characteristics of Methane-Oxidizing Bacteria

Organism	Morph-ology	Flagellation	Resting Stage	Internal Membranes*	Carbon Assimilation Pathway	DNA (Mole % GC)
Methylomonas	Rod	Polar	Cyst-like body	I	Ribulose monophosphate	50-54
Methylobacter	Rod	Polar	Thick-walled cyst	I	Ribulose monophosphate	50-54
Methylococcus	Coccus	None	Cystlike body	I	Ribulose monophosphate	62
Methylosinus	Rod or vibrioid	Polar tuft	Exospore	**I**	Serine	63-66
Methlocystis	Vibrioid	None	Poly-b-hydroxy-butyrate-rich cyst	**I**	Serine	—
Methylobacterium	Rod	Nonmotile	None	**I**	Serine	58-66

* Internal membranes: Type I, bundles of disk-shaped vesicles distributed throughout the organism; Type II, paired membranes running along the periphery of the cell.

App. Table 3.3
Characteristics of Sulfate- and Sulfur-Reducing Bacteria

Genus	Characteristics	DNA (mole % GC)
Group I sulfate reducers: Non-acetate oxidizers		
Desulfovibrio	Polarly flagellated, curved rods, no spores; Gram negative; contain desulfoviridin; seven species recognized; one thermophilic	46-61
Desulfomonas	Long, fat rods; nonmotile; no spores; Gram negative; contain desulfoviridin; habitat, intestinal tract; one species	66-67
Desuforomaculum	Straight or curved rods; motile by peritrichous or polar flagellation; Gram negative; desulfoviridin absent; produce endospores; four species, one thermiphilic, one species capable of utilizing acetate as energy source	37-46
Archaeoglobus	Achaebacterium; extreme thermophile, temperature optimum, 83°C; contains some unique coenzymes of methanogenic bacteria, makes small amount of methane during growth; H_2, formate glucose, lactate, and pyruyate serve as electron donors, SO_4^{2-}, $S_2O_3^{2-}$, or SO_3^{2-} as electron acceptor; one species	46
Desulfobullbus	Ovoid or lemon-shaped cells; no spores; Gram negative; desulfoviridin absect; if motile, by single polar flagellum; utilizes only propionate as electron donor with acetate + CO_2 as produces; one species	59

App. Table 3.3
Characteristics of Sulfate- and Sulfur-Reducing Bacteria (continued...)

Group II sulfate reducers: Acetate-oxidizers		
Desulfobacter	Rods; no spores, Gram nefatice; desulfoviridin absent; if motile, by sigle polar flagellum; utilizes only acetate as electron donor and oxidizes it to CO_2 via the citric acid cycle; one species	45
Desulfobacterium	Rods, some with gas vesicles, marine; capable of autotrophic growth via the acetyl-CoA pathway, three species	—
Desulfococcus	Spherical cells; nonmotile; Gram negative; desulfoviridin present, no spores; utilizes C_1-C_{14} fatty acids as electron donor with complete oxidation to CO_2; capable of autotrophic growth via the acetyl-CoA pathway; one species	57
Desulfonema	Large filamentous gliding bacteria; Gram positive; no spores desulfoviridin present or absent; utilizes C_2-C_{12} fatty acids as electron donor with complete oxidation to CO_2; capable of autotrophic growth via the acetyl-CoA pathway; one species	34-41
Desulfosatcina	Cells in packets (sarcina arrangement); Gram negative; no spores; desulfoviridin absent; utilizes C_2-C_{14} fatty acids as electron donor with complete oxidation to CO_2; capable of autotrophic growth via the acetyl-CoA pathway; one species	—
Dissimilatory sulfur reducers:		
Desulfuromonas	Straight rods, single lateral flagellum; no spores; Gram negative; does not reduce sulfate; acetate, succinate, ethanol, propanol used as electron donor; obligate anaerobe; three species	50-63
Campylobacter	Curved, vibrio-shaped rods; polar flagella; Gram positive; no spores; unable to reduce sulfate but can reduce sulfur, sulfite, thiosulfate, nitrate, or fumarate anaerobically with acetate or a variety of other carbon/electron donor sources; facultative aerobe	40-42

3.2 Biodegradation Pathways for Typical Alkanes and Aromatics

Hydrocarbon degradation is usually an aerobic process. Evidence including observations on petroleum hydrocarbon, indicates that oxygen is required for the biodegradation. Yet, reports on anaerobic degradation and consideration of anaerobic pathways persist in literature. Anaerobic degradation of aromatics, for example, is well established.

Anaerobic degradation of aromatics can occur by:

- Photodegradation
- Nitrate respiration
- Methanogenic fermentation

Benzene and several of its alkyl derivatives, all of which have been considered to be recalcitrant to anaerobic biodegradation earlier, have been observed to undergo relatively rapid and extensive biodegradation in anoxic aquifer materials.

Presence of anaerobic degradation of aliphatic hydrocarbons however, is not so well established, though there is evidence of its existence.

A strain of Pseudomonas, for example, was found to grow on n-alkanes under anaerobic environment when nitrate was present. This strain could not grow aerobically with other conditions being the same. An NAD dependent n-alkane dehydrogenase was isolated from the extracts of this specie. In purified form this enzyme could convert n-decane to 1-decene, which, as explained below, is the first step for anaerobic oxidation of alkanes.

In another study Desulfovibrio, a sulfate reducing obligate anaerobe, was found to degrade aliphatic hydrocarbon anaerobically. A study was made to assess the thermodynamic feasibility of utilizing hydrocarbons as energy source for sulfate reducers. It was found that none of the long chain hydrocarbons are adequate as a sole energy source for the microbes but can be cometabolized in the presence of other energy sources.

In a study on anaerobic oxidation of methane by sulfate reducers, it was found that the microbes could not oxidize methane to carbon dioxide unless acetate was present in the medium. It was also established that acetate, and not methane was assimilated by the microbes. It was concluded that in this case acetate was acting as the carbon source with methane as the electron donor (or source of energy) and sulfate as the terminal electron acceptor. It would appear from this study that if methane, which is considered biologically inert under anaerobic conditions, can be oxidized then other aliphatic hydrocarbons could be oxidized too. Studies on oxidation of hexadecane under similar conditions, however has contradicted this phenomena.

Anaerobic oxidation proceeds in a similar manner to aerobic oxidation.

- The alkane is first dehydrogenated to the corresponding alkene
- The alkene is then hydrated at the double bond to yield an alcohol

  This step is not well documented and there is no evidence that the oxygen in the alcohol is derived from the water molecule. There is evidence however that the conversion from alkene to alcohol can be established in the absence of molecular oxygen by a hydrase mechanism.

- The next step in the anaerobic oxidation is oxidation of the alcohol to the corresponding fatty acid
- The ß-oxidation yields acetyl-CoA
- A glyoxylate or TCA cycle, then, yields cell mass or energy respectively

The initial hydroxylation of alkanes and alkenes seems to be the rate limiting step. Once the alcohol is formed further oxidation is achieved easily as long as there is a balance between oxidation and reduction.

In a study done on cycling of natural and xenobiotic forms of carbon in anaerobic subsurface environment, biodegradation of halogenated aromatics was observed. Carbon and electron flow was primarily through methanogenesis and sulfate reduction. The anaerobic decomposition of halogenated aromatic substrates was preceded by the initial removal of aryl halides under methanogenic conditions. However, the anaerobic metabolism of alkylated aromatic compounds proceeded via methyl hydroxylation reactions.

REFERENCES

1. Alexander. M. "Biodegradation of Chemicals of Environmental Concern". *Science*. **211**: 1981. pp. 132-138.

2. Alexander. M. "Biodegradation of Organic Chemicals". *Environmental Science and Technology*. **18**: 1985. pp. 106-111.

3. Battersby N. S. "A Review of Biodegradation Kinetics in the Aquatic Environment". *Chemosphere*. **21**: 1990. pp. 1243-1284.

4. Howard R. E. "Determining "Real World" Biodegradation Rates". *Environmental Toxicology and Chem*. **4**: 1985. pp. 129-130.

5. Grady C. P. "Biodegradation: It's Measurements and Microbial Basis". *Biotechnology and Bioengineering*. **27**: 1985. pp. 660-674.

6. Tabak H. H. and R. Govind. Determination of Biodegradation Kinetics with the use of Respirometry for Development of Predictive Structure-Biodegradation Relationship Models., *Emerging Technologies in Hazardous Waste Treatment*. (ASCI&EC Symposium, Atlanta, GA, 1991)

7. Larson R. J. and R. L. Perry. "Use of the Electrolytic Respirometer to Measure Biodegradation in Natural Waters". *Water Research Bulletin*. **15**: 1981. pp. 697-702.

8. Boethling R. S. and M. Alexander. "Effects of Concentration of Organic Chemicals on their Biodegradation by Natural Microbial Communities". *Applied and Environmental Microbiology*. **53**: 1979. pp. 2617-2623.

9. Vogel T. M. and P. L. McCarthy. "Biotransformation of Tetrachloroethylene to Trichloroethylene, Dichloroethylene, Vinyl Chloride, and Carbon Dioxide under Methanogenic Conditions". *Applied and Environmental Microbiology*. 1985. pp. 1080-1083.

10. Michelic J. R. and R. G. Luthy. "Microbial Degradation of Acenapthene and Napthalene Under Denitrification Conditions in Soil-Water Systems". *Appli. Environ. Microbiol*. **54**: 1988. pp. 1188-1198.

11. Johnson, Eyeing, and Polissar. *The Kinetic Basis of Molecular Biology*. John Wiley & Sons, New York 1954.

12. A. W. &. M. Association. Bioremediation: The State of Practice in Hazardous Waste Remediation Operations, Live Satellite Seminar, 1992.

13. A. W. & M. Association. Bioventing and vapor Extraction: Uses and Applications in Remediation Operations, Live Satellite Seminar, 1992.

14. Brown R. A. The Advantages and Disadvantages of Bioventing and Vapor Extraction Technology, in Bioremediation, *The State of Practice in Hazardous Waste Remediation Operations*. A. W. a. M. Association. Live Satellite Seminar, 1992.

15. Dupont R. R., W. Doucette and R. E. Hinchee. *Assessment of In-Situ Bioremediation Potential and the Application of Bioventing at a Fuel-Contaminated Site.* (Butterworth Publishers, Stoneham, MA)

16. Wilson J. T. and C. H. Ward. "Opportunities for Bioremediation of Aquifers Contaminated with Petroleum Hydrocarbons". *J. Ind. Microbiol.* **27**: 1986. pp. 109-116.

17. Conner J. R. "Case Study of Soil Venting". *Poll. Eng.* **7**: 1988. pp. 74-78.

18. Hutzler N. J., B. E. Murphy and J. S. Cierke. *State of Technology Review: Soil Vapor Extraction Systems.* (Final Report to the U.S. Environmental Protection Agency Hazardous Waste Engineering Laboratory, OH, 1988).

19. Marley M. C., S. D. Richter, B. L. Cliff and P. E. Nangeroni. Design of Soil Vapor Extraction Systems - A Scientific Approach, *soil Vapor Extraction Technology Workshops on Soil Vacuum Extraction.* (U.S. Environmental Protection Agency, Edison, NJ, 1989)

20. Mutch R. D. J., A. N. Clarke and D. J. Wilson. In-Situ Vapor Stripping Research Project: A Progress Report, *Soil Vapor Extraction Workshop.* (U.S. EPA Risk Reduction Engineering Laboratory, Releases Control Branch, Edison, NJ, 1989)

21. Payne F. C. In-situ Removal of Purgeable Compounds from Vadose Zone Soils, *Purdue Industrial Waste Conference.* 1988.

22. U. S. EPA *State of Technology Review: Soil Vapor Extraction System Technology.* EPA/600/2-89/024. (Hazardous Waste Engineering Research Laboratory, Cincinnati, OH, 1990).

23. U. S. EPA *Engineering Bulletin: In-Situ Soil Vapor Extraction Treatment.* EPA/540/2-91/006. (Risk Reduction Engineering Laboratory, Cincinnati, OH, 1991).

24. Hinchee R. E. and D. C. Downey. The Role of Hydrogen Peroxide in Enhanced Bioreclamation, *Proceedings, Petroleum Hydrocarbons and Organic Chemicals in Groundwater: Prevention, Detection and Restoration.* (National Water Well Association, Dublin, OH, 1988), pp. 715-722.

K - Thermal Processes

THERMAL PROCESSES

Instructional Objectives:

- Understand the chemical and physical properties associated with thermal destruction.
- Describe heat and mass transport in the incinerator.
- Develop an understanding of how to apply basic concepts of chemical kinetics, equilibrium, stoichiometry and thermochemistry to the design of thermal processes.
- Develop a design approach for thermal processes design.
- Describe technologies used for control of air pollution.

1. GENERAL DESCRIPTION:

1.1. Introduction

Thermal destruction is the controlled destruction of compounds by thermal oxidation. Compared to alternative technologies, EPA's research data indicates that incineration has the highest degree of destruction and control for the broadest range of organic waste streams [1]. Thermal destruction usually converts contaminants to carbon dioxide, water and other combustion products. The specific products of thermal destruction vary depending on the type of waste, and operating conditions.

Rotary kiln incineration is the most common method of thermal destruction for contaminated materials. Rotary kiln incinerators can easily be designed to process large volumes of wastes. Feed systems can be altered to accommodate large particles and residence times can be increased to ensure complete oxidation of contaminants.

Most thermal destruction processes consist of a waste feed system, an air burner system, a combustion chamber, a combustion monitoring system, equipment for air pollution treatment and monitoring, and ash removal. The destruction process usually operates in two units. The organic contaminants are first desorbed from the material and combustion is partially initiated in the primary combustion chamber. Complete destruction of organic contaminants then occurs in the secondary combustion chamber or afterburner. Typical thermal destruction processes are illustrated in Figure 1.1 [2].

1.2. Compounds Destroyed by Thermal Destruction

Destruction and Removal Efficiencies (DRE's) of 99.99% have been demonstrated for halogenated, non-halogenated, aliphatic, aromatic and polynuclear organic compounds [1]. Additionally, thermal destruction is a proven technology for treating soils contaminated with organic compounds, usually accomplishing over 99.99% destruction and removal efficiencies.

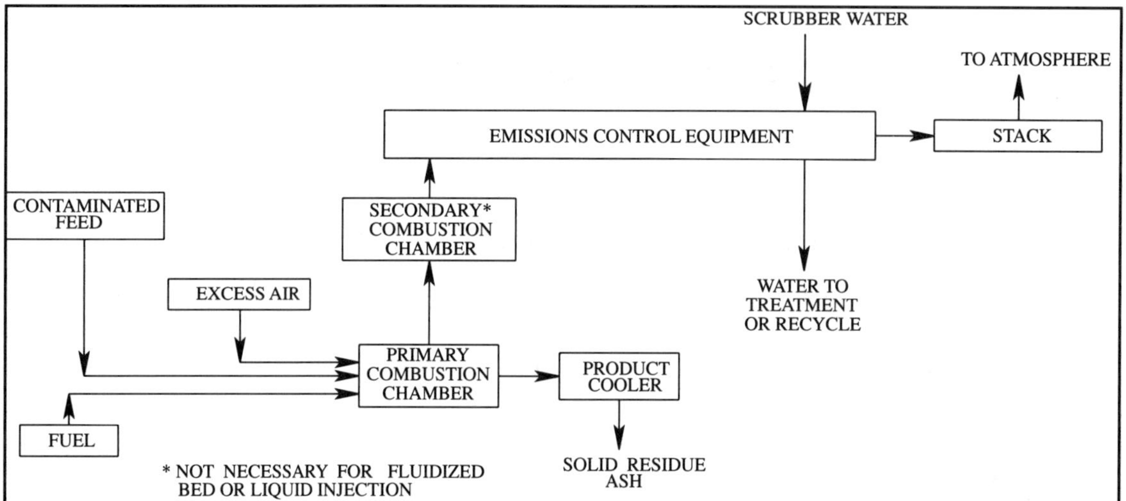

FIGURE 1.1
Thermal Destruction Flow Diagram
Thermal Processes

Under RCRA and TSCA, the EPA set standards for operating hazardous and toxic waste incinerators. In general, EPA requires that a DRE of 99.99% for each hazardous constituent and 99.9999% for toxic constituents must be achieved. DRE's include both destruction by combustion, and removal by air pollution control equipment. Also, combustion efficiency must be at least 99.9%. (See section 9, Regulations, of this module for a definition of DRE and combustion efficiency.)

2. FUNDAMENTALS

2.1. Introduction

Thermal destruction processes involve many complex interactions. The following disciplines govern the thermal destruction process:

- Thermodynamics
- Chemical kinetics
- Heat and mass transfer
- Fluid mechanics

This complex relationship is further complicated by shifts in waste composition, resulting in changes in heat release rate and combustion characteristics.

2.2. Waste Characterization

2.2.1. Introduction

Waste characterization is a major factor in assessing the feasibility of destroying a hazardous waste by thermal destruction processes. The design of the process and the air emissions control system are determined by the waste characteristics.

2.2.2. Proximate Analysis

Proximate analysis is an assay of the moisture, ash, volatile matter, and fixed carbon content in a waste. A proximate analysis is a relatively low cost laboratory test which is described in detail in ASTM D3172.

2.2.3. Ultimate Analysis

An ultimate analysis is an assay of the carbon, hydrogen, sulfur, nitrogen, oxygen and ash content of a waste. For thermal destruction of hazardous wastes, analysis for halogens, metals, and other elements that could influence the combustion process or would be critical to the design of air pollution controls is necessary. In addition to analysis for elemental components of a waste, analysis may also be performed for compounds that may be in the waste, such as PCB's, dioxins or other organics. ASTM D3176 describes the standard method of performing an ultimate analysis.

2.2.4. Heating Value

The heating value of a waste is a measure of the energy released when the waste is burned and is commonly expressed in units of British Thermal Units per pound (Btu/lb=2.326J/g). A minimum heating value of approximately 5000 Btu/lb (11,630 J/g) is required to sustain combustion. The heating value of a waste, or heat of reaction for the combustion reaction, is required for establishing an energy balance for the combustion chamber and in assessing the need for auxiliary fuel.

2.2.4.1. Calculation of Heating Value

The enthalpy change of a chemical reaction is the amount of heat that is evolved or absorbed during the course of the reaction [3]. The standard state enthalpy is defined as:

$$\Delta H^{\circ} = \left(\sum_i \nu_i \overline{H}_i^{\circ}\right)_{products} - \left(\sum_i \nu_i \overline{H}_i^{\circ}\right)_{reactants} \qquad (1)$$

where:

ΔH^{o} = change in standard state enthalpy
ν_i = stoichiometric coefficient
$\overline{H}_i^{\circ}$ = enthalpy of species i in kcal/mole at standard conditions of 25°C and 1 atm pressure

The heating value is equal to the enthalpy of combustion but of opposite sign. The heating value is dependent upon the phase of the water formed in the products, the phase of the fuel, and the combustion conditions (constant pressure or constant volume). When condensed water appears in the products of combustion, the heating value is commonly known as the higher heating value. When water vapor appears in the products of combustion, the heating value is known as the lower heating value. The difference between the lower and the higher heating value is the latent heat of vaporization of water (10,520 kcal/kg mol at 20°C). Heating value has been proposed as one of the criteria to determine the incinerability of hazardous wastes.

2.2.4.2. Effect of Temperature on Heat of Reaction

To determine the heat of reaction, H_r, at temperature T_2 if the heat of reaction at temperature T_1 is known, consider the following equilibrium statements:

$$(\text{heat change})_{\text{path A}} = (\text{heat change})_{\text{path B}}$$

(provided that the process is carried out at constant pressure or constant volume) (2)

$$\Delta H_{\text{reactants}} - \Delta H_{\text{products}} = \Delta H_{r,T_1} - \Delta H_{r,T_2} \tag{3}$$

where:

$$\Delta H_{\text{reactants}} = \sum_i \nu_i \int_{T_1}^{T_2} C_{p_{\text{reactants}}} dT \tag{4}$$

$$\Delta H_{\text{products}} = \sum_i \nu_i \int_{T_1}^{T_2} C_{p_{\text{products}}} dT \tag{5}$$

where:

C_p = molar heat capacity (constant-pressure specific heat) (cal/mol-K or Btu/lbmol-°F)

Figure 2.1 shows the relationship between heat of reaction at two different temperatures and two different reaction paths [4].

Combining equations (1) and (3) and integrating yields:

$$\Delta H_{r,T} = \sum_i \nu_i \left(\overline{H}_i^{\circ} + \int_{T_\circ}^{T} C_p dT \right)_{\text{products}} - \sum_i \nu_i \left(\overline{H}_i^{\circ} + \int_{T_\circ}^{T} C_p dT \right)_{\text{reactants}} \tag{6}$$

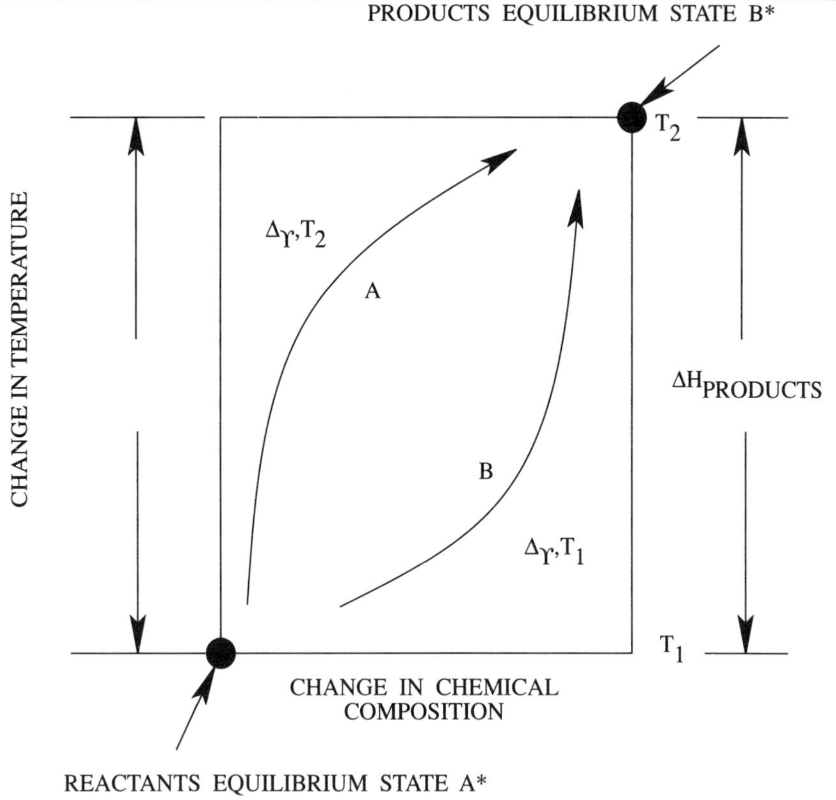

FIGURE 2.1
The Relationship Between Heats of Reaction at Two Different Temperatures and Two Different Reaction Paths
Thermal Processes

Molar heat capacity is a function of temperature only, and is usually represented by a power series in T:

$$C_p = \alpha + \beta T + \gamma T^2 + \ldots. \tag{7}$$

Tabulated values of α, β, and γ for various compounds are given in Table 2.1 [5].

Table 2.1
Molar Heat Capacities[a,b]

Compound	Formula	a	b x 10^3	g x 10^6
	Normal Paraffins (gases)			
Methane	CH_4	3.381	18.044	-4.300
Ethane	C_2H_6	2.247	38.201	-11.049
Propane	C_3H_8	2.410	57.195	-17,.533
n--Butane	C_4H_{10}	3.844	73.350	-22.655
n--Pentane	C_5H_{12}	4.895	90.113	-28.039
n--Hexane	C_6H_{14}	6.011	106.746	-33.363
n--Heptane	C_7H_{16}	7.094	123.447	-38.719
n--Octane	C_8H_{18}	8.163	140.2.17	-44.127
Increment per C atom above C_8	—	1.097	16.667	-5.338
	Normal Monoolefins (gases) (1-alkenes)			
Ethylene	C_2H_4	2.830	28.601	-8.726
Propylene	C_3H_6	3.253	45.116	-13.740
1-Butene	C_4H_8	3.909	62.848	-19.617
1-Pentene	C_5H_{10}	5.347	78.990	-24.733
1-Hexene	C_6H_{12}	6.399	95.752	-30.116
1-Heptene	C_7H_{14}	7.488	112.440	-35.462
1-Octene	C_8H_{16}	8.592	129.076	-40.775
Increment per C atom above C8	—	1.097	16.667	-5.338

Table 2.1
Molar Heat Capacities[a,b] (continued...)

Compound	Formula	a	b x 10^3	g x 10^6
	Miscellaneous Gases			
Acetaldehyde[c]	C_2H_4O	3.364	35.722	-12.236
Acetylene	C_2H_2	7.331	12.622`	-3.889
Ammonia	NH_3	6.086	8.812	-1.506
Benzene	C_6H_6	-0.409	77.621	-26.429
1.3-Butadiene	C_4H_6	5.432	53.224	-17.649
Carbon dioxide	CO_2	6.214	10.396	-3.545
Carbon monoxide	CO	6.420	1.665	-0.196
Chlorine	Cl_2	7.576	2.424	-0.965
Cyclohexane	C_6H_{12}	-7.701	125.675	-41.584
Ethyl alcohol	C_3H_6O	6.990	39.741	-11.926
Hydrogen	H_2	6.947	-0.200	0.481
Hydrogen chloride	HCl	6.732	0.433	0.370
Hydrogen sulfide	H_2S	6.662	5.134	-.0854
Methyl alcohol	CH_4O	4.394	24.274	-6.855
Nitric oxide	NO	7.020	-0.370	2.546
Nitrogen	N_2	6.524	1.250	-0.001
Oxygen	O_2	6.148	3.102	-0.923
Sulfur dioxide	SO_2	7.116	9.512	3.511
Sulfur trioxide	SO_3	6.077	23.537	-0.687
Toluene	C_7H_8	0.576	93.493	-31.227
Water	H_2O	7.256	2.298	0.283

Table 2.1
Molar Heat Capacities[a,b] (continued...)

Compound	Formula	a	b x 10^3	g x 10^6
	Chlorinated Organic Liquids			
Methyl chloride	CH_3Cl	8.994	10.280	-21.218
Dichloromethane	CH_2Cl_2	13.672	7.616	-33.272
Chloroform	$CHCl_3$	18.453	4.932	-37.923
Carbon tetrachloride	CCl_4	23.777	2.224	-39.923
Ethyl chloride	C_2H_5Cl	15.267	17.060	-48.798
1,1-Dichloroethane	$C_2H_4Cl_2$	20.014	14.315	-54.917
1,1,2,2-Tetrachloroethane	$C_2H_2Cl_4$	28.397	10.338	-66.739
n--Propyl chloride	C_3H_7Cl	20.703	24.070	-69.351
1,3-Dichloropropane	$C_3H_6Cl_2$	24.840	21.810	-68.214
n--Butyl chloride	C_4H_9Cl	26.258	31.208	-89.482
1-Chloropentane	$C_5H_{11}Cl$	31. 812	38.345	-109.612
1-Chloroethylene	$C_2H_6Cl_2$	14.058	10.939	-40.586
trans--1,2-Dichloroethylene	$C_2H_6Cl_2$	18.050	8.845	-42.981
Trichloroethylene	C_2HCl_3	22.410	6.394	-46.257
Tetrachloroethylene	C_2Cl_4	26.595	4.106	-46.208
3-Chloro-1-propene	C_3H_5Cl	18.728	18.514	-56.495
Chlorobenzene	C_6H_5Cl	28.558	25.786	-11.6.150
p--Dichlorobenzene	$C_6H_5Cl_2$	33.265	23.025	-116.848
Hexachlorobenzene	C_6Cl_6	50.759	13.286	-115.620

a Selected mainly from values given by H.M. Spencer, J. Justice, and G. Flanagan, *J. Am. Chem. Soc.,* **56**, 2311. 1934; **64**, 2511, 1942; **67**, 1859, 1945: *Ind. Eng. Chem.* **40**, 2152, 1948; also from personal notes: L. Theodore and J. Reynolds, 1986.

b Constants for the equation $C_p = a + bT + gT^2$, where T is in K and C_p is in Btu/lbmol-°F or cal/gmol-°C

c Applicable range 298-1000K

2.2.4.3. Dulong Approximation of Heating Value

A modified Dulong formula can sometimes be used to estimate the heating value from the elemental analysis:[6]

$$\frac{\text{Btu}}{\text{lb}} = 145.4C + 620\left(H - \frac{O}{8}\right) + 41S \qquad (8)$$

where:

C, H, O, S = weight percentages of carbon, hydrogen, oxygen and sulfur
(H - O/8) = represents hydrogen that is combined with oxygen as moisture as well as hydrogen and oxygen that are combined in some other form

2.3. Heat Transfer

2.3.1. General

The objectives of heat transfer in hazardous waste incineration operations should include:

- Maximization of heat transfer rates
- Utilization of heat exchange, whenever practical
- Minimization of heat loss by the use of insulation

There are three modes of heat transfer:

Conduction-the transfer of heat by means of a temperature gradient, without displacement of adjacent matter.

Convection-the transfer of heat by the mixing motion of one fluid with another or between a surface and an adjacent fluid.

Natural convection-result of motion caused by variations in density caused by temperature differentials under the influence of gravity.

Forced convection-motion induced by mechanical means such as pumps or fans.

Radiation-the transfer of energy through space by electromagnetic waves (no requirement for an intervening medium).

2.3.2. Conduction

The mechanism of heat transfer for thermal conduction through homogeneous solids can be expressed by Fourier's law. Fourier's law may be written as:

$$\frac{dq}{dA} = -k\frac{\partial T}{\partial n} \tag{9}$$

where:

q = rate of heat flow across surface in direction normal to surface
A = area of isothermal surface perpendicular to the flow of heat
k = thermal conductivity of homogeneous solid
T = temperature
n = distance measured normally to surface

Conduction under the conditions of time invariant temperature distribution is called steady-state conduction. For steady-state one-dimensional flow, Fourier's law may be expressed as:

$$\frac{q}{A} = -k\frac{dT}{dn} \tag{10}$$

2.3.2.1. Heat Flow Through a Cylinder

For thermal conduction through the walls of a hollow cylinder, Fourier's law can be applied and written in the form:

$$q = -k\frac{dT}{dr}2\pi rL \tag{11}$$

where:

q = rate of heat flow across surface in direction normal to surface
k = thermal conductivity
T = temperature
r = radius
L = length of cylinder
$2\pi rL$ = area perpendicular to the flow of heat

Rearranging and integrating gives:

$$\int_{r_i}^{r_o}\frac{dr}{r} = \frac{2\pi Lk}{q}\int_{T_o}^{T_i}dT \tag{12}$$

$$\ln r_o - \ln r_i = \frac{2\pi Lk}{q}(T_i - T_o) \tag{13}$$

$$q = \frac{k(2\pi L)(T_i - T_o)}{\ln\left(r_o/r_i\right)} \tag{14}$$

where:

r_o, T_o = outside radius and temperature
r_i, T_i = inside radius and temperature

2.3.3. Convection

Heat transfer by convection is dependent upon the properties of the convection medium and the geometry of the system. The general equation of the convection of heat is given by:

$$\frac{dq}{dA} = h_c\left(T_2 - T_1\right) \tag{15}$$

where:

q = heat transmission by convection
A = area of heat transfer surface
h_c = film coefficient of heat transfer, dependent upon system geometry and the temperature difference between the surface and ambient fluid (must be determined empirically, see reference [7])
T_1 = ambient temperature
T_2 = temperature at the interface

2.3.4. Radiation

When a particle is heated, radiant energy is emitted at a rate that is dependent upon the temperature, color, and texture of the particle. Heat transfer due to radiation can be expressed by the following equation:

$$q = \sigma A_s \varepsilon_w \left(T_w^4 - T^4\right) \tag{16}$$

where:

σ = Stefan-Boltzman constant
q = quantity of heat transferred
A_s = area of heat transfer surface
ε_w = emissivity of surface
T_w = absolute temperature of surface
T = absolute temperature of surroundings

2.3.5. Combined Heat Transfer

The total heat loss from a hot body includes losses by conduction-convection and radiation. Assuming black surroundings, the total heat loss can be expressed by the following equation:

$$\frac{q_T}{A} = \frac{q_c}{A} + \frac{q_r}{A} = h_c\left(T_w - T\right) + \sigma\varepsilon_w\left(T_w^4 - T^4\right) \tag{17}$$

where:

q_T/A = total heat flux
q_c/A = heat flux by conduction-convection
q_r/A = heat flux by radiation
h_c = convective heat transfer coefficient
ε_w = emissivity of surface
T_w = absolute temperature of surface
T = absolute ambient temperature

2.4. Combustion

2.4.1. General

Ideally, the principal products of combustion are carbon dioxide, water vapor and inert ash. Although combustion and incineration devices are designed to optimize complete combustion of the fuel-waste mixture, they never attain the ideal. Depending on the chemical composition of the fuel-waste mixture and the combustion conditions, small quantities of other products may be formed. These products and unreacted components of the waste become incinerator emissions, which must be controlled by air pollution control equipment.

Heat release by combustion is transferred by conduction, convection and radiation to the incinerator system components and to the incoming fuel-waste mixture. The heat transferred to the incoming fuel-waste mixture is required for ignition.

An understanding of the combustion process is essential to the design of incineration systems. The following sections will introduce the concept of the three "T's" of combustion, combustion reaction modes and discuss combustion emissions.

2.4.2. The Three "T's" of Combustion

Incinerators should be designed around the three "T's" of combustion, namely, time, temperature and turbulence in the presence of oxygen [6].

Time-accounted for in design through the space and volume of the combustion chamber(s) of the incineration system. Longer residence time generally improves destruction efficiency by increasing reaction time

Temperature-heat is the driving force to sustain combustion. Supplemental heat may be supplied by auxiliary burners to preheat the incinerator or to support combustion of materials having high moisture levels and low heating values [6].

Turbulence-promotes mixing between the waste compounds and oxygen. Greater turbulence increases combustion efficiency.

Time, temperature and turbulence are functions of the incinerator system design. Mathematically, the three "T's" are related to three characteristic times, given by the following equations:

For temperature,

$$\tau_c = \frac{1}{k} \tag{18}$$

For residence time,

$$\tau_r = \frac{V}{Q} = \frac{L}{u} \tag{19}$$

For turbulence,

$$\tau_m = \frac{L^2}{D_e} \tag{20}$$

where:

τ_c, τ_r, τ_m = chemical, residence and mixing times (s)
k = rate constant (s^{-1} or l/mol-s)
V = volume of reaction zone (m^3)
Q = volumetric flow rate (m^3/s)
L = length of reaction zone (m)
u = velocity (m/s)
D_e = effective (turbulent) diffusion coefficient (m^2/s), i.e. eddy diffusivity

2.4.3. Incineration Reaction Modes

In general, incineration of wastes involves three simultaneous chemical reaction modes:

- Oxidation
- Pyrolysis
- Radical attack

2.4.3.1. Oxidation

The stoichiometry of complete combustion (theoretical air combustion) of a typical waste, $C_zH_yO_xCl_wS_vN_u$ ($y \geq z$), can be expressed as [5]:

$$C_zH_yO_xCl_wS_vN_u + \left[z+\phi+v-\frac{1}{2}x\right]O_2 + \frac{79}{21}\left(z+\phi+v-\frac{1}{2}x\right)N_2 \rightarrow$$
$$zCO_2 + 2\phi H_2O + wHCl + vSO_2 + \left[\frac{1}{2}u+\frac{79}{21}\left(z+\phi+v-\frac{1}{2}x\right)\right]N_2 \tag{21}$$

where:

z, y, x, w, v, u = number of moles of C, H, O, Cl, S, N present in the waste-fuel mixture

$\phi = 1/4(y - w)$ when $y > w$

$\phi = 0$ when $y \leq w$

2.4.3.2. Pyrolysis

Pyrolysis is the thermal degradation of organic matter in the absence of oxygen. Although incineration requires excess air to ensure complete combustion of a waste, small fractions of the waste may not contact oxygen. These small waste fractions may undergo pyrolysis in the high temperature environment of the incinerator. For example, the pyrolysis of cellulose and PCB would be, respectively:

$$\text{cellulose: } C_6H_{10}O_5 \rightarrow 2CO + CH_4 + 3H_2O + 3C \tag{22}$$

$$\text{PCB: } C_{12}H_7Cl_3 \rightarrow 12C + 3HCl + 2H_2 \tag{23}$$

The thermal degradation of such compounds will generally produce compounds such as CO, CH_4, and H_2O, which will be in the gaseous phase, and carbon char (C), which could be in the solid or liquid phase [8]. Pyrolysis is sometimes intentional (see section 4.3.3, Fixed Hearth Incinerators).

2.4.3.3. Radical Attack

A radical is an atom or group of atoms with one or more unpaired electrons. Radicals are very reactive chemically as they tend to form bonds by pairing the spin of unpaired electrons.

Incineration flames are characterized by temperatures around 1000°C and a radical-rich gas flow. The radical-rich gas consists of atomic hydrogen (H•), atomic oxygen (O•), hydroxyl radicals (•OH), atomic chlorine (Cl•), methyl radicals (•CH_3), and chloroxy radicals (ClO•). The decomposition of a waste is facilitated by radical attack.

2.4.4. Combustion Emissions

2.4.4.1. Ideal Products of Combustion

For ideal complete combustion, the following reactions are generally assumed:

- All carbon reacts with oxygen to form carbon dioxide
- All hydrogen reacts with oxygen to form water vapor, unless otherwise noted
- All nitrogen from the waste, fuel, or air takes the form of the diatomic molecule, N_2
- All organic halogens convert to hydrogen halides (HCl)
- All inorganic halogens are assumed to be inert and exit the process as ash residual [9]
- All sulfur reacts with oxygen to form sulfur dioxide
- Alkali metals convert to hydroxides
- Non-alkali metals completely oxidize
- Phosphorous converts to phosphorous pentoxide

2.4.4.2. Combustion By-Products

CO_2, H_2O and N_2 emitted from combustors such as waste incinerators are termed combustion products. All other combustion emissions are termed by-products [10]. Formation of combustion by-products is dependent upon:

- Elemental composition of the waste-fuel mixture
- Incinerator operating parameters

Non-ideal incinerator behavior may decrease destruction efficiencies and cause generation and emission of products of incomplete combustion (PIC's). PIC's are defined as organic compounds which are present in the emissions from the incineration process which were not present in the feed.

There are several modes of failure that may explain non-ideal incinerator behavior. They include:

Volatile Puffs - the emanation of surges of volatiles when fresh waste is introduced to the primary combustion chamber too rapidly or in too high a concentration [11]. Sources include rapid pyrolysis of solids, rapid evaporation of volatile liquids, or rupture of delivery containers. Volatile puffs can cause failure by:

- flooding regions of the combustion chamber with more volatiles than can be burned by the available oxygen
- reducing ambient temperatures - the puff acts as a heat sink
- traveling to the secondary combustion chamber, where, if too large, it can cause failure of the incinerator system.

> Experimental work with a rotary kiln incinerator by the US EPA has shown that, for liquid wastes, puffs are easily generated, and that their magnitude and intensity increases with increasing kiln temperature and rotational speed [12].

Mixing Upsets - the diversion or interruption of waste or oxidant flow by constricted delivery tubes or blunt bodies (delivery cartons) within the combustion chamber, feed rates in excess of design specifications, or by volatiles puffs. Mixing upsets can result in failure by allowing volatiles to escape destruction or to convert to other potentially hazardous compounds.

Chemical Effects - free radical scavengers can lower the concentrations of •OH radicals which are important high temperature oxidants [11]. Free radical scavengers can also disrupt or extinguish flames by interrupting radical chain propagation reactions that maintain stable flame chemistry.

Temperature Deviations - low temperatures may be caused by:

- excess feed rates, or high moisture content or low heating value wastes
- volatiles puffs
- cold walls, particularly in small combustion chambers with high wall area to volume ratios

Time Deviations - reduced retention times can be caused by:

- flow disturbances

2.4.4.3. Carbon Monoxide

Several researchers have developed global models for the formation of carbon monoxide during the oxidation of a hydrocarbon [13].

> "A global model ignores many of the detailed steps that are required in mechanistic models, and ties the kinetics to the main stable reactants and products [13]."

The following two-step model is a simple global model for the oxidation of a hydrocarbon, C_xH_y [13]:

$$C_xH_y + \left(\frac{x}{2}+\frac{y}{4}\right)O_2 \rightarrow xCO + \left(\frac{y}{2}\right)H_2O \tag{24}$$

$$xCO + \frac{x}{2}O_2 \rightarrow xCO_2 \tag{25}$$

Oxidation reactions can be facilitated by atoms and radicals (especially in flames) and water vapor (which enhances oxidation), however, equations (24) and (25) have been successfully used to establish a kinetic model of VOC oxidation in vapor incinerators [13].

Assuming pseudo-first order kinetics, the following rate equations apply:

$$r_R = -k_1[R]\left[O_2\right] \tag{26}$$

$$r_{CO} = k_1[R]\left[O_2\right] - k_2[CO]\left[O_2\right] \tag{27}$$

$$r_{CO_2} = k_2[CO]\left[O_2\right] \tag{28}$$

where:
R = any hydrocarbon

In the presence of excess oxygen, the rate equations become:

$$r_R = -k_1[R] \tag{29}$$

$$r_{CO} = k_1[R] - k_2[CO] \tag{30}$$

$$r_{CO_2} = k_2[CO] \tag{31}$$

Solving the first-order series reaction,

$$R \xrightarrow{k_1} CO \xrightarrow{k_2} CO_2 \tag{32}$$

for CO yields:

$$CO = \frac{k_1(R)_o}{k_2 - k_1}\left(e^{-k_1 t} - e^{-k_2 t}\right) \tag{33}$$

The solution is discussed in references [14, 15].

2.4.4.4. Nitrogen Oxides

Several oxides of nitrogen are known to occur, of these, NO (nitric oxide) and NO_2 (nitrogen dioxide) are the two most important because they are emitted in large quantities [13] and are tied to ozone depletion.

> "About 95% of all NO_x from stationary combustion sources is emitted as NO [13]."

Oxides of nitrogen are formed by the thermal oxidation of atmospheric nitrogen in the air used for combustion (thermal NO_x) and by oxidation of nitrogen in the fuel-waste mixture (fuel NO_x). See references [13] and [16] for a discussion of NO_x formation chemistry and control. Emission of NO_x can be controlled by utilization of combustion chamber control or with downstream APCD's. In general, low combustion temperatures reduce NO_x emissions but usually increase PIC and POHC emissions

2.4.4.5. Sulfur Oxides

Sulfur oxides are formed from the oxidation of sulfur in the fuel-waste mixture. Most SO_x formed during the combustion process are sulfur dioxide (SO_2), and about 1 to 5% are sulfur trioxide (SO_3) [1]. Emissions of sulfur oxides can be controlled with downstream APCD's.

2.4.4.6. Hydrogen Halides

Combustion of halogenated compounds generally produces a mixture of the hydrogen halide and gaseous halogen. For chlorinated compounds, the predominant product is hydrochloric acid (HCl), which is favored over chlorine (Cl_2) at combustion temperatures because of chemical reaction equilibrium between HCl and Cl2. At combustion chamber temperatures, approximately 1% of Cl will appear as Cl2 in flue gasses [5]. In addition to being an air contaminant, HCl can create corrosion problems downstream of the combustion chamber. HCl can be controlled with downstream scrubbing systems.

2.4.4.7. Polychlorinated Dioxin/Furan

> "The terms dioxin and furan refer to families of 75 related chemical compounds known as polychlorinated dibenzo-p- dioxin's (PCDD's) and 135 related chemical compounds known as polychlorinated dibenzofurans (PCDF's), respectively [1]."

Since the first published report of PCDD/F emissions from a municipal incinerator, a great amount of attention has been given to this class of combustion by-product because dioxin is one of the most toxic compounds known to man and is highly persistent in the environment. For modern municipal waste incinerators, PCDD/F emissions are similar in magnitude to hazardous waste incinerators, however, older municipal waste incinerators are likely to emit higher levels [17]. Table 2.2 compares dioxin and furan emissions from waste incinerators [18].

Table 2.2
Dioxin and Furan Emissions from Waste Incinerators (ng/Nm^3)

Source	Polychlorinated dibenzo-p-dioxin	Polychlorinated dibenzofuran
Medical	117 - 450	52 - 30,300
Municipal	1 - 10,700	2 - 37,500
Hazardous	ND - 16	ND - 56

ND - not detectable

Several mechanisms have been proposed to explain the presence of PCDD/F's in incinerator emissions, they include:

- PCDD/F was originally present in the waste and has passed through the incinerator - this is unlikely since most wastes contain insufficient quantities of PCDD/F to account for levels observed in emissions [19].
- PCDD/F's are produced from related chlorinated compounds such as PCB's, chlorinated phenols, or chlorinated benzene's.
- PCDD/F's are formed via de novo syntheses, i.e., they are formed by reactions between simpler organic compounds in the presence of chlorine donors.

Formation reactions can be both homogeneous and heterogeneous in nature. Figure 2.2 shows potential PCDD/PCDF formation mechanisms [19].

PCDD/F emissions can be controlled by:

- Reduction of precursor compounds during the combustion process
- Use of downstream air pollution control devices such as scrubber-electrostatic precipitation or scrubber-baghouse combinations.

2.4.4.8. Metals

The emission of metals from waste combustion devices is of concern because of the presence of metals in many hazardous wastes and because of the potential threat to human health. Combustion will not destroy metals but will partition the metal between emission streams. Metals may exit the combustion system by any one of several pathways, including:

- Siftings or bottom ash from the primary combustion chamber
- Fly ash from heat recovery and particulate control systems
- Scrubber wastewaters
- Stack emissions

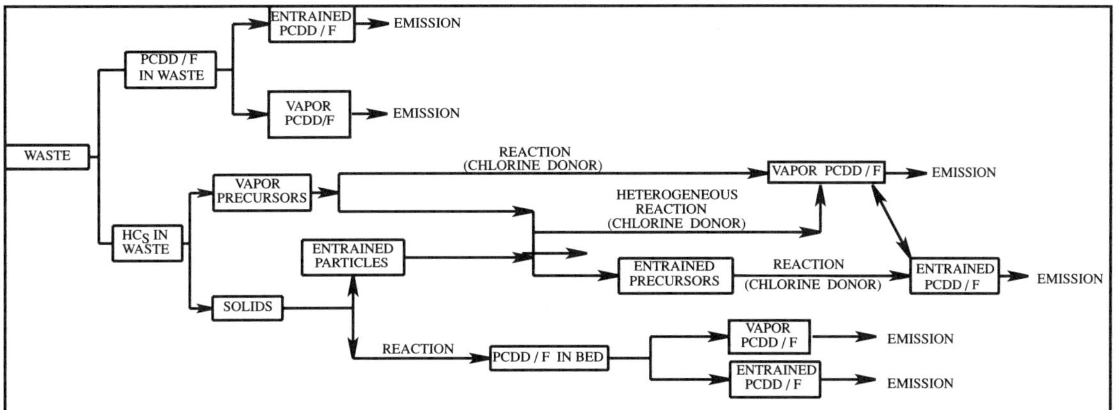

FIGURE 2.2
Potential PCDD/F Formation Mechanisms
Thermal Processes

An understanding of the fate or partitioning of metals within an incineration system is important in order to examine the impact of metals upon emissions. Figure 2.3 shows the mechanisms thought to control metals behavior during the combustion process [20]. Most metals are present in a waste as inorganic compounds. They pass through the incinerator unchanged by the combustion environment and end up in bottom ash. Volatile metals will vaporize and diffuse into the exhaust gas stream. As the exhaust gas cools, the vapors will condense on ash particles. The partitioning of metals within the incineration system depends upon:

- Combustion chamber temperature
- Waste chlorine content
- Flue gas cleaning system operation

2.4.4.8.1. Combustion Chamber Temperature

Combustion chamber temperature has the strongest effect on the partitioning of metals within the incinerator system. Figure 2.4 shows the effect of temperature on vapor pressure [20]. As temperature increases, vapor pressure increases, resulting in increased metals emissions.

2.4.4.8.2. Waste Chlorine Content

Metals may react with chlorine in the combustion chamber to form compounds that are generally more volatile than the original species [20]. Figure 2.5 illustrates the effect of chlorine content on vapor pressure of various metals [20]. Increased chlorine content also tends to promote formation of hexavalent chromium, which is more toxic than other forms of chromium [17].

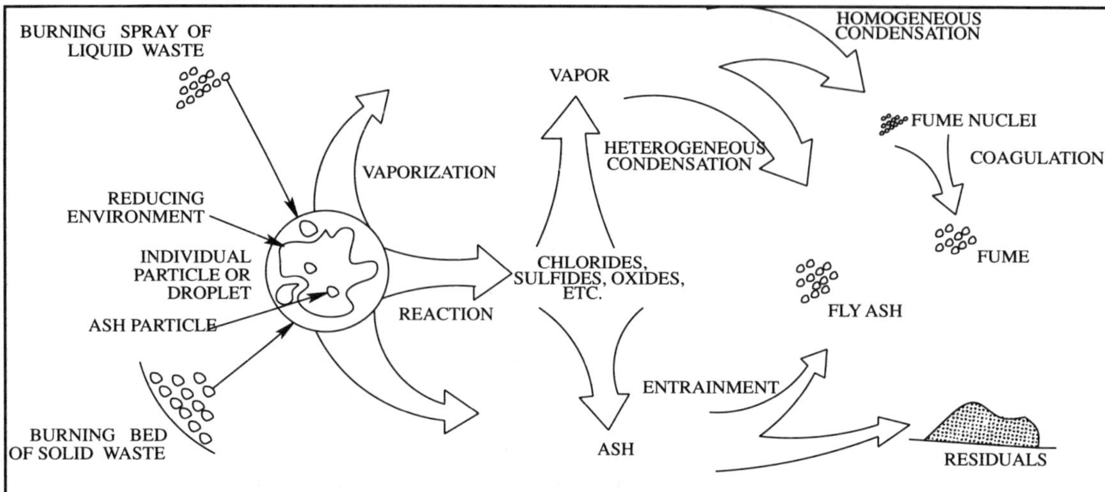

FIGURE 2.3
Metals Behavior in the Combustor
Thermal Processes

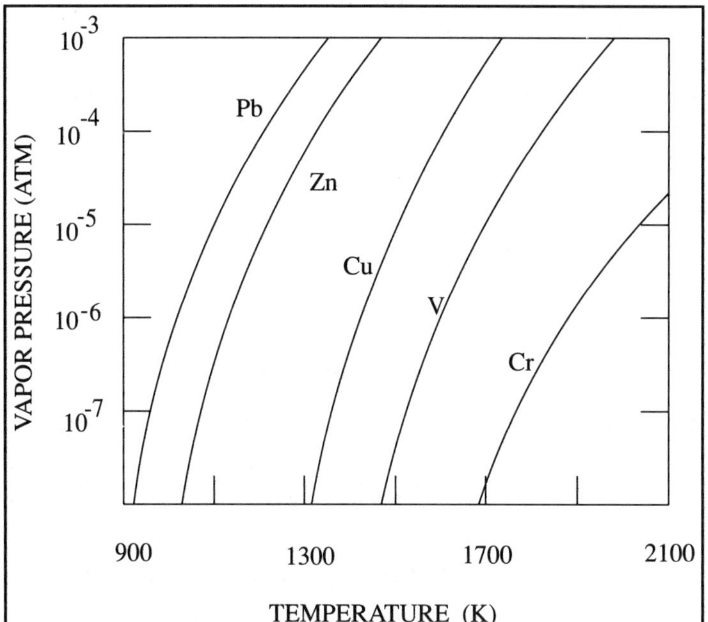

FIGURE 2.4
Effect of Temperature on Vapor Pressure
Thermal Processes

2.4.4.8.3. Flue Gas Cleaning System Operation for metals

The efficiency of downstream APCD's is a function of operating parameters, size distribution of particulate's, and type of APCD. Homogeneous condensation forms particles less than 1µm in

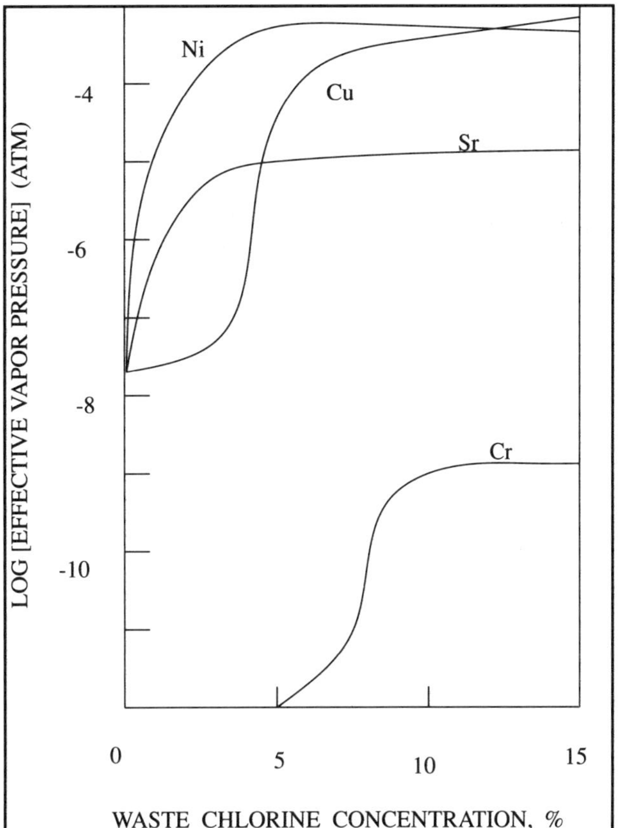

FIGURE 2.5
Effect of Chlorine Content on Vapor Pressure of Various Metals
Thermal Processes

diameter while heterogeneous condensation generally occurs on larger particles, that can be captured by APCD's. Therefore, stack emissions increase as the ratio of homogeneous to heterogeneous condensation increases.

2.4.4.9. Ash/Residuals

Ash and residuals emissions from hazardous, municipal and infectious waste incineration systems are dependent upon the quantity of noncombustibles in the incoming waste feed. Non-volatile noncombustibles generally ends up in bottom ash while volatile noncombustibles end up in fly ash, which can be captured by downstream APCD's. Waste incineration residuals can include:

- Scrubber blowdown wastewaters
- Lime/limestone-laden fly ash from APCD's
- Spent activated carbon removed from APCD service
- Bottom ash

The Resource Conservation and Recovery Act (RCRA) defines ashes and residues from hazardous waste incineration as hazardous waste, therefore, these materials must be disposed of as a hazardous waste unless a waiver is obtained. Such waivers are generally very difficult to obtain [17].

2.4.4.10. Unburned POHC's

Combustion by-products can include unburned principal organic hazardous constituents (POHC's). RCRA incineration emission standards require a destruction and removal efficiency (DRE) of 99.99% for each POHC in the waste feed and TSCA (the Toxic Substances Control Act) requires a DRE of 99.9999% for PCB's (see section 9. Regulations). Emissions of unburned POHC's are therefore very low. Higher DRE's could theoretically be accomplished by increasing residence times, turbulence, or operating temperatures or by increased APCD efficiencies.

3. APPLICATIONS

3.1. General Applicability

Thermal destruction of hazardous wastes can generally be employed :

- When heat content of waste is greater the 2500 BTU/lb [8]
- When waste contains less than 1000 ppm halogens [8]

Thermal destruction technology is applicable to numerous waste streams. including:

- Hazardous wastes
- Municipal wastes
- .Toxic substances
- Laboratory wastes
- Infectious (medical) wastes
- Spent pesticides
- Wastewater sludges
- Contaminated soils
- Liquids/wastewaters

3.2. Advantages and Limitations of Thermal Destruction

3.2.1. Advantages

Thermal destruction offers the following advantages:

- Hazardous compounds destroyed
- Volume and weight reduction
- Short residence times

- Control of air discharge
- Heat recovery
- Possibility of on-site operation
- Applicable to numerous waste types
- General regulatory acceptance

3.2.2. Limitations

Thermal destruction of hazardous wastes is limited because:

- Some materials are not incinerable
- Some organics generate toxic PIC's
- High capital and operating costs compared to alternative technologies
- Skilled operators are required (ASME certified)
- Supplemental fuel is often required
- Public opposition

4. PROCESS DESCRIPTION

4.1. Introduction

A typical incineration system consists of the following major components:

- Waste storage, preparation and feeding systems
- Combustion chamber(s)
- Air pollution control
- Residue and ash handling
- Process monitoring systems

Different incineration systems have been developed to handle the various types of hazardous wastes. The four most common types of incinerator designs are:

- Liquid injection
- Rotary kiln
- Fixed hearth
- Fluidized bed

The selection of the incinerator system is dependent upon the physical and chemical properties of the wastes to be incinerated. Figure 4.1 shows the key components and their configuration in a typical incineration system [1].

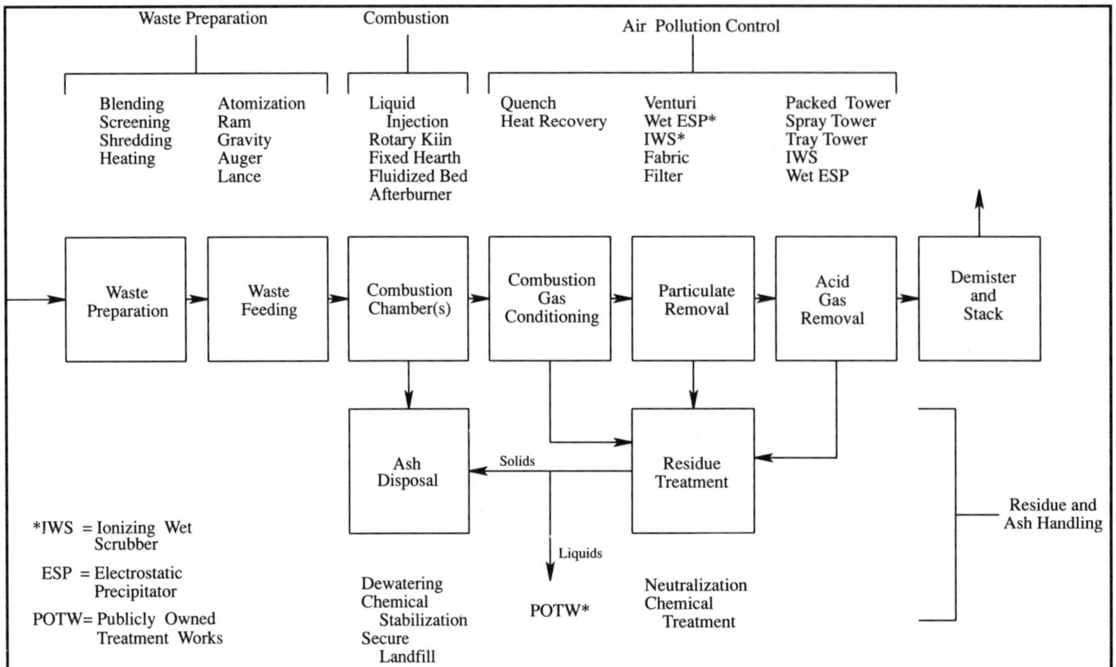

FIGURE 4.1
General Orientation of Incineration Subsystems and Typical Process Component Options
Thermal Processes

4.2. Waste Storage, Preparation and Feeding

The waste storage, preparation and feed systems are an integral part of the incineration facility design and operation. The selection of the waste storage, preparation and feed systems is dependent upon the following factors:

- Physical form: solid, liquid, gas, or mixture
- Solids: size, form, and quantity
- Liquids: viscosity as a function of temperature, specific gravity, suspended solids content
- Sludges: density, viscosity, percent solids
- Gases: density
- Special characteristics: cohesiveness, abrasiveness, explosiveness, corrosiveness, toxicity
- Disposal rate: peak, average, minimum

For the storage and preparation of liquid or gaseous wastes, main storage tanks, temporary holding tanks and batching vessels are required. Tanks or vessels may also be required to store fuel oil or gas for incinerator ignition and auxiliary burners. Liquid wastes are typically blended then pumped into the combustion chamber through nozzles or atomizing burners. Liquid wastes with suspended solids may have to be screened to avoid clogging of the nozzles and atomizing burners. Blending is required to provide a uniform feed to the combustion chamber.

Bulk solid wastes are typically stored in enclosed bins or silos, lined pits, or stockpiles. Solids may require shredding to control particle size and/or drying to reduce moisture content. Solid wastes may be fed to the combustion chamber by:

- Rams
- Gravity feed
- Air lock feeders
- Vibratory of screw feeders
- Belt feeders

Sludges can be fed to the combustion chamber using progressive cavity pumps or water cooled lances.

4.3. Combustion Chambers

The most important component of any incineration system is the combustion chamber. Combustion chamber selection is dependent upon the physical and chemical characteristics of the waste and its ash content. The name of the incineration system is derived from the type of primary combustion chamber used.

The secondary combustion chamber (afterburner) is essentially an air pollution control device designed to increase destruction efficiencies.

Table 4.1 provides selection considerations for the four major types of combustion chambers as a function of waste type (source: reference [1]). Table 4.2 provides operating parameters for incineration systems [1, 6].

4.3.1. Liquid Injection Incinerators

Liquid injection incinerators are applicable to hazardous liquids, sludges, and slurries that can be pumped and atomized after heating to some reasonable temperature. Generally, a viscosity of 10,000 SSU (Saybolt seconds) or less is required [5]. These incinerators are usually simple refractory-lined vertical or horizontal cylinders equipped with one or more waste burners. Liquid wastes are injected through the burners, atomized to fine droplets with mean particle diameters as low as 1 µm, and burned in suspension.

Temperature, residence time, and turbulence must be optimized to achieve high destruction efficiencies. Typical combustion chamber temperature and residence times are 1500 - 2200°F and 0.3 - 2 seconds, respectively.

Horizontal liquid injection incinerators are preferred for wastes with low inorganic content while vertically aligned liquid injection incinerators are used with high inorganic and ash content wastes. Figure 4.2 shows a typical horizontal liquid injection combustion chamber [1]. Typical liquid injection incinerators have a capacity of 10 x 10^6 Btu/hr and can be as high as 70 to 100 x 10^6 Btu/hr [21].

Table 4.1
Incinerator/ Waste Selection Chart

Waste Type	Rotary Kiln	Liquid Injection	Fluidized Bed[a]	Fixed Hearth (controlled air)
Solids:				
Granular, homogeneous	X		X	X
Irregular, bulky (barrels, etc.)	X			X[b]
High melting point (tars, etc.)	X	X[c]	X	X
Organic compounds with fusible ash constituents	X			X
Unprepared, large, bulky material	X			
Gases:				
Organic vapor-laden	X	X	X	X
Liquids:				
High organic-strength aqueous wastes often toxic	X[d]	X	X	
Organic liquids	X[d]	X	X	
Solids/liquids:				
Waste contains halogenated aromatic compounds	X	X[e]		
Aqueous organic sludge	X[f]		X	

[a] Suitable for pyrolysis operation
[b] Handles large material on a limited basis
[c] If material can be melted and pumped
[d] If equipped with auxiliary liquid injection nozzles
[e] If liquid
[f] Provided waste does not become sticky upon drying

Table 4.2
Operating Parameters for Incineration Processes

Process	Temperature Range	Residence Time
Liquid Injection	1500 - 2200 °F	0.3 - 2.0 seconds
Rotary Kiln	1200 - 2300 °F	Liquid/gas - seconds Solids - hours
Fixed Hearth	1200 - 1800 °F	Liquid/gas - seconds Solids - 0.25 - 1.5 hours
Fluidized Bed	1400 - 2000 °F	1.0 - 5.0 seconds

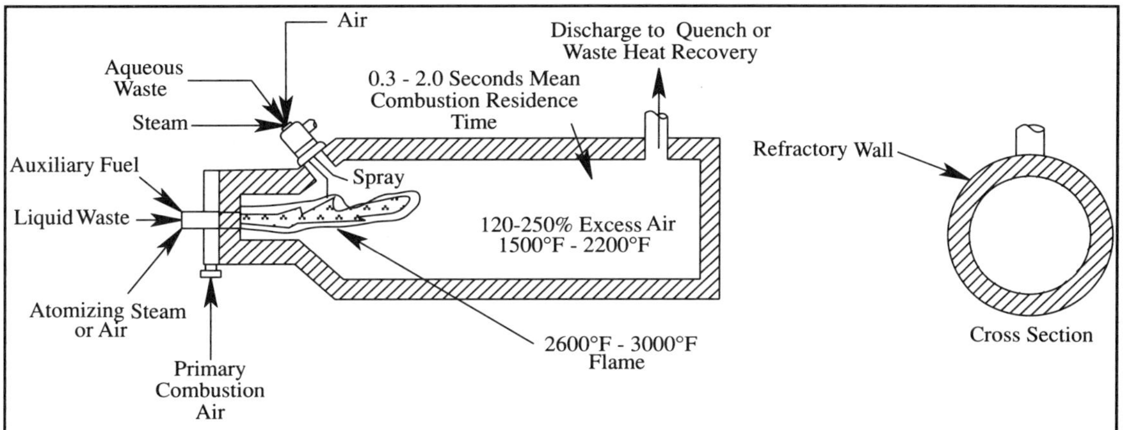

FIGURE 4.2
Typical Liquid Injection Combustion Chamber
Thermal Processes

4.3.1.1. Atomizers

The critical aspect of achieving a high destruction efficiency in a liquid injection incinerator is good atomization. Burners or nozzles may be oriented either axially, radially or tangentially to the combustion chamber. Maximization of turbulence can be attained using vortex, swirl, or tangentially aligned burners. Burners should be located so that the flame does not impinge on refractory walls, which can result in flame quenching and incomplete combustion.

Liquid waste atomization can be achieved with any of the following types of atomizers:

- Rotary cup
- Single-fluid pressure
- Two-fluid, low pressure air
- Two-fluid, high pressure air or steam

Table 4.3 lists typical kinematic viscosity and solids-handling limitations for the various types of atomizers [5].

4.3.1.2. Advantages of Liquid Injection Incineration

Liquid injection incineration offers the following advantages:

- Capable of incinerating a wide range of liquid wastes
- No continuous ash removal system is required other than for air pollution control equipment
- Fast temperature response to changes in the waste/fuel flow
- Virtually no moving parts
- Low maintenance costs
- No secondary combustion chamber needed if primary chamber has high residence time (generally 2 seconds)

Table 4.3
Kinematic Viscosity and Solids Handling Limitations of Various Atomization Techniques[a]

Atomization Type	Maximum Kinematic Viscosity (SSU)	Maximum Solids Mesh Size	Maximum Solids Concentration (%)
Rotary cup	130 - 175	35-100	20
Single-fluid pressure	150		0
Internal low pressure air[b]	100		0
External low pressure air[b]	200-1500	200[d]	30[d]
External high pressure air[c]	150 - 5000	100 - 200	70
External high pressure steam[c]	150 - 5000	100 - 200	70

[a] From T. Bonner et al., Hazardous Waste Incineration Engineering, Noyes Data Corp., Park Ridge, NJ, 1981.
[b] Lower than 30 psi.
[c] Greater than 30 psi.
[d] Depends on nozzle i. d.

4.3.1.3. Limitations of Liquid Injection Incineration

Liquid injection incineration has the following limitations:

- Only wastes that can be atomized through a burner nozzle can be incinerated
- Heat content of waste must maintain adequate ignition temperatures or supplemental fuel must be burned
- Burners susceptible to plugging

4.3.2. Rotary Kiln Incinerators

Rotary kiln incinerators are applicable to the destruction of solid, liquid, gaseous, sludge and containerized wastes. The rotary kiln is a refractory-lined cylinder mounted at a slight angle (rake). As the kiln rotates, the waste moves through the kiln and is mixed by tumbling action. The primary function of the rotary kiln is to convert solid wastes to gas via volatilization and partial combustion reactions. Complete combustion then occurs in an afterburner connected to the discharge end of the kiln. Inert ash is removed from the lower end of the kiln.

Residence time of solid waste is typically between 0.5 and 1.5 hours. Rotational speed (typically 1-5 rpm) and rake control the residence time within the combustion chamber. Rotary kiln incinerators usually have a length-to-diameter ratio (L/D) between 2 and 10. In general, large L/D ratios and slower rotational speed are used when the waste requires long residence time to ensure complete combustion. Gas residence time is controlled by the gas velocity in the incineration system.

The design of a rotary kiln incinerator incorporates high temperature seals between the stationary end plates and the rotating cylinder. These seals are difficult to maintain and present a problem in that there is a potential for release of unburned wastes from the incineration system. Operation at a negative pressure helps to eliminate this problem.

The kiln is supported by at least two trunnion supports that are adjustable in the vertical direction. The angle to the horizontal, or rake, will normally vary between 2% and 4% (1/4 in to 1/2 in per foot of length).

Kiln temperatures are typically between 1200 and 1800°F but can be as high as 2550°F. At temperatures between 2000 and 2200°F, ash will begin to deform and fluidize, depending on the chemical composition of the waste and the presence of oxygen in the kiln, and the kiln will begin to operate in a slagging mode. This is undesirable in kilns that are designed to operate in a non-slagging mode and temperatures must be kept below slagging temperatures. Temperatures in the range of 2600°F are required to maintain slagging.

Feed rate to the kiln is generally limited so that the amount of waste being processed is limited to between 5 to 20% of the kiln volume, so that mixing of the waste and heat transfer to the waste is maximized. Typical rotary kiln incinerators have capacities between 25 and 60 x 10^6 Btu/hr and can be as high as 90 x 10^6 Btu/hr [1, 21]. Figure 4.3 shows a typical rotary kiln combustion chamber with a vertical afterburner [1].

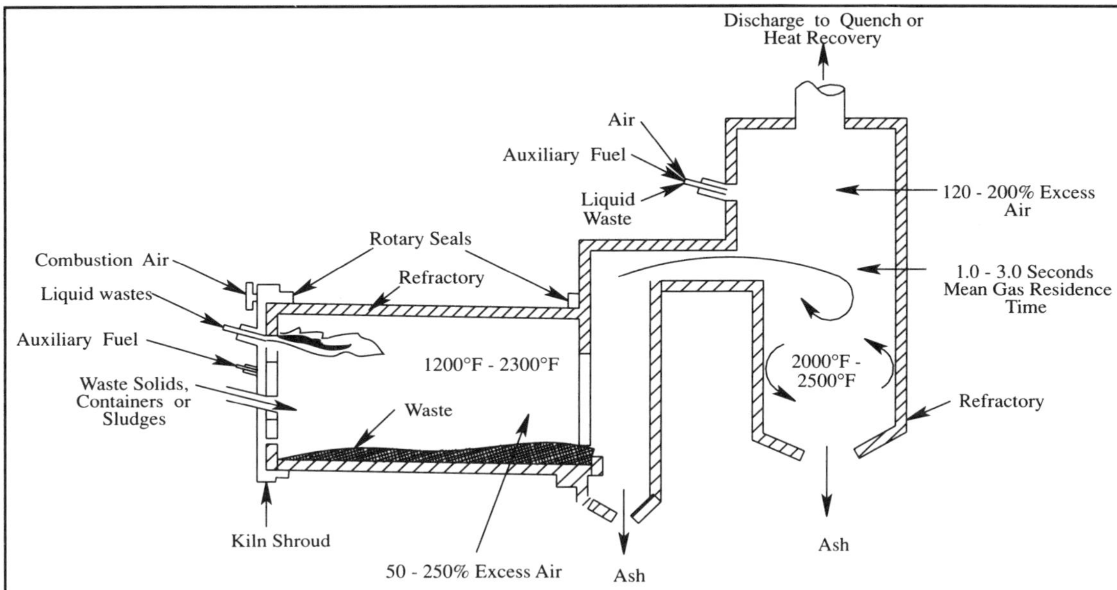

FIGURE 4.3
Typical Rotary Kiln/Afterburner Combustion Chamber
Thermal Processes

4.3.2.1. Advantages of Rotary Kiln Incineration

The advantages offered by rotary kiln incineration are:

- Can maintain high residence time and mixing of solid wastes
- Can handle a variety of wastes including drums and bulk containers
- Ash removal is continuous and does not interfere with operations

4.3.2.2. Limitations of Rotary Kiln Incineration

Rotary kiln incineration has the following limitations:

- High capital costs for installation
- Spherical or cylindrical items may roll through kiln before being combusted
- High particulate carryover to downstream systems
- Problem maintaining seals at either end of the kiln
- Aqueous sludges may dry and form rings on refractory walls

4.3.3. Fixed Hearth Incinerators

Fixed hearth incinerators (also called controlled air, starved air, or pyrolytic incinerators) employ a two-stage combustion process similar to rotary kilns. Waste is usually ram fed to the first stage, or primary combustion chamber, where it is burned in an oxygen deficient atmosphere (50 to 80% of stoichiometric air requirements). Most of the volatile fraction of the waste is pyrolytically destroyed in the primary combustion chamber. The starved air conditions of the primary chamber minimize particulate carryover by maintaining low turbulent velocities. Complete combustion of the resulting pyrolytic products occurs in the secondary chamber where 100 to 200% excess air (beyond stoichiometric requirements) is supplied to ensure complete combustion.

Fixed hearth incinerators are usually used to treat solid wastes, although liquid or gaseous wastes can be injected into the secondary chamber. Feed rates are limited because of physical limitations of ram-feeding and moving waste material through the primary chamber. Fixed hearth units are smaller than rotary kilns, making them attractive for smaller on-site installations. They are most often used for destruction of medical (infectious) wastes. Figure 4.4 shows a typical fixed hearth combustion chamber [1].

4.3.3.1. Advantages of Fixed Hearth Incineration

Fixed hearth incineration offers the following advantages:

- Small space requirement for system as secondary chamber is usually located above primary chamber
- Low particulate emissions
- Low capital costs relative to rotary kiln

4.3.3.2. Limitations of Fixed Hearth Incineration

Fixed hearth incineration has the following limitations:

- May have poor burnout of carbonaceous material
- Operation is limited to batch mode
- Ash transport through primary chamber is difficult

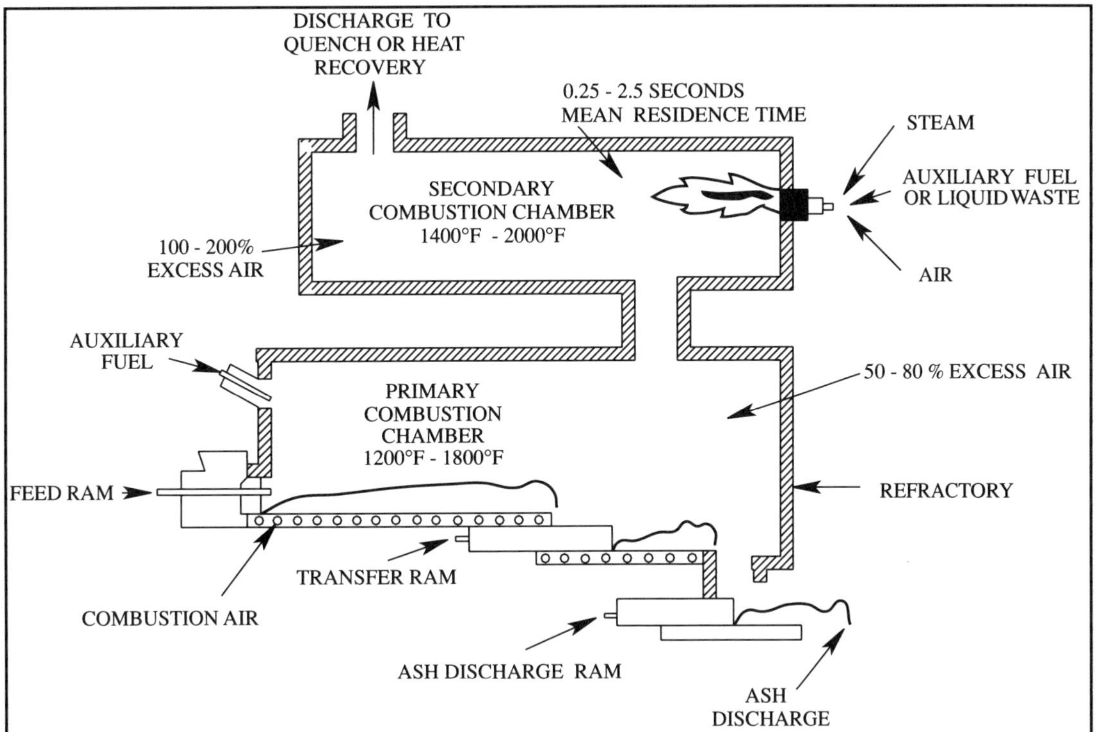

FIGURE 4.4
Typical Fixed Hearth Combustion Chamber
Thermal Processes

4.3.4. Fluidized Bed Incinerators

The fluidized bed combustion chamber consists of a refractory lined vessel partially filled with inert granular material such as sand or alumina. Combustion air is blown through a distributor plate at the base of the chamber at a rate that is sufficient to cause the bed to expand and fluidize (bubbling bed) or become entrained (circulating bed). In the circulating bed design, entrained material is blown overhead, separated in a cyclone and returned to the combustion chamber. In the bubbling bed design, the bed expands 40% to 60% in volume and bed depths usually range between 1.5 and 10 feet when fluidized. Figure 4.5 illustrates a typical bubbling bed type combustion chamber [1].

Fluidized bed combustion chambers are used primarily for incineration of sludges or shredded solid material. Materials are usually limited to 2 inches in diameter to allow for good distribution within the bed.

Waste material enters the combustion chamber through nozzles located above or within the bed depending on the moisture content of the waste. Generally, wastes with relatively high moisture contents are introduced above the bed, which allows for drying before the waste reaches the bed thus the bed can be sized smaller. Residence time is controlled by bed depth, which should be maintained at a minimum value consistent with required residence time for complete combustion.

The versatility of the fluidized bed incineration process is due to its ability to control residence time, excellent gas-solids contacting and stable temperature control.

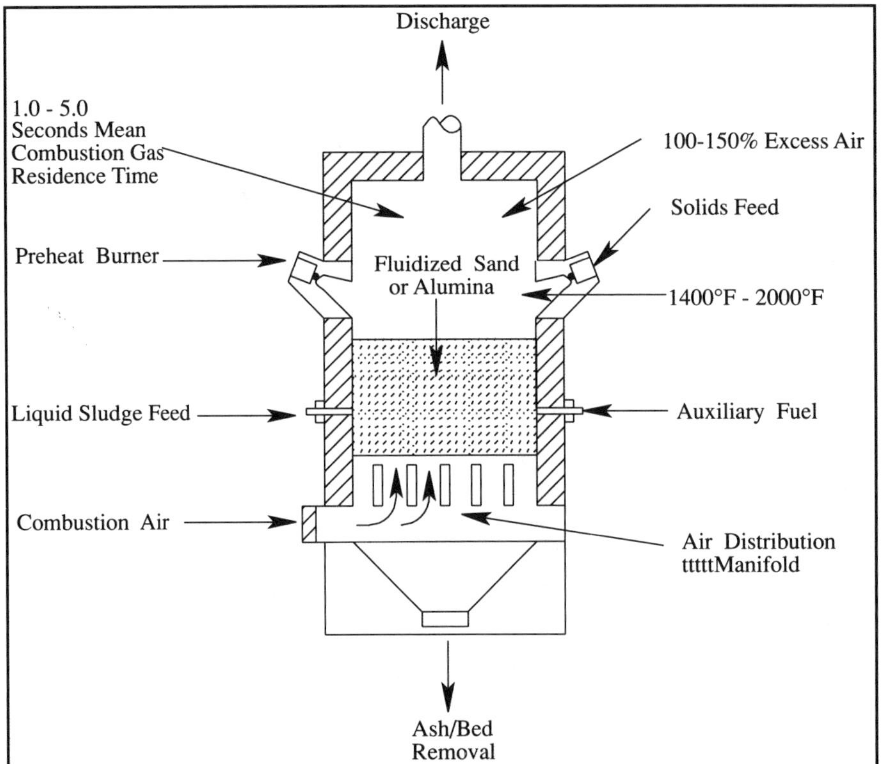

FIGURE 4.5
Fluidized Combustion Chamber
Thermal Processes

4.3.4.1. Advantages of Fluidized Bed Incineration

Fluidized bed incinerators offer the following advantages:

- Design simplicity, few moving parts and low maintenance costs
- High combustion efficiency
- The bed is thermally secure, after shutdown the bed will lose as little as 10°F per hour [9]; start-up after shutdown will require little preheat time
- Afterburners are generally not required due to high residence time in freeboard

4.3.4.2. Limitations of Fluidized Bed Incineration

Fluidized bed incineration has the following limitations:

- High operation costs
- Ash removal is difficult
- Certain wastes may cause the bed to agglomerate
- High particulate emissions compared to other systems

- Waste material must be sized prior to injection into the combustion chamber
- Applicable waste types are limited

4.3.5. Secondary Combustion Chambers/Afterburners

Secondary combustion chambers, or afterburners, are typically used to complete gas-phase combustion reactions of emissions from primary combustion chambers. The afterburner functions as an air pollution control device and is essential to some incinerator systems, such as the rotary kiln or fixed hearth, while others are capable of achieving high destruction efficiencies without one, such as liquid injection and fluidized bed incineration systems. The afterburner operates and is designed using the same principles of combustion that govern other combustion processes. The afterburner is illustrated in Figures 4.3 and 4.4. Liquid wastes can be fired in the afterburner, thereby increasing the utility of the rotary kiln or fixed hearth incineration systems.

4.3.6. Oxygen-Enriched Incineration

Oxygen-enriched incineration is a somewhat innovative thermal technology. The use of oxygen-enriched air can improve overall performance and efficiency of the incineration process. However, the use of oxygen for combustion at high-enrichment levels can result in local heating and high thermal NO_X formation [22].

> The rate of formation of thermal NO_X is extremely temperature sensitive, and becomes rapid only at flame temperatures (3000-3600°F) [13]. Increased flame temperatures associated with oxygen-enriched incineration will increase the rate of NO_X formation.

Increasing the percentage of oxygen in the combustion chamber (oxygen-enrichment) by mixing pure oxygen with air reduces the percentage of nitrogen in the combustion chamber. The percentages of oxygen and nitrogen entering the combustion chamber can be calculated by the following equations:

$$X + 0.21(100 - X) = \%O_2 \tag{34}$$

$$0.79(100 - X) = \%N_2 \tag{35}$$

where: X = percent oxygen enrichment.

Therefore, 40% oxygen enrichment will increase oxygen from 21% to 53% in the combustion gas and nitrogen will decrease from 79% to 47%. Oxygen-enriched incineration systems are capable of providing 100% oxygen to the combustion chamber. However, for rotary kiln incinerators operating at a slight vacuum, this would be difficult to achieve due to leaks at the combustion chamber seals [23].

Typical incineration processes require about 150% excess air, which contains 79% nitrogen. Nitrogen does not contribute to the combustion reaction and its presence in the combustion chamber is detrimental to the incineration process. Nitrogen effects the combustion process by:

- Acting as a heat sink - extra energy is required to raise the nitrogen to combustion temperatures

- Increasing volumetric gas flow rate through the combustion chamber and downstream air pollution control devices, which decreases gas residence time within the combustion chamber and increases capacity requirements of the APCD's

Oxygen-enhanced combustion systems can be retrofitted to existing incinerators. The EPA's Office of Research and Development has tested two oxygen-enhanced systems retrofitted to mobile rotary kiln systems :

- Union Carbide Corporation - Linde Oxygen Combustion System (OCS)
- American Combustion Inc. - PYRETRON Thermal Destruction System (TDS)

Both systems have displayed a number of advantages over traditional incineration and a few disadvantages. In summary:

Technology Advantages [23]:

- Waste throughput can be significantly increased

  Throughput is normally limited by air blower capacity, reducing the volume of air required for combustion will increase system throughput.

- Fuel consumption is lowered due to the reduced sensible heat loss to the flue
- DRE should be improved due to the higher oxygen concentration in the fuel-oxidant mixture and longer residence times
- Pollution control of the exhaust gas is less costly and more effective due to reduced flue gas volumes

Technology Disadvantages [23]:

- Higher flame temperatures can result in higher NO_x formation and local overheating

  Because of less diluents, such as N_2, in the combustion off-gas system, oxygen-enriched incineration, in general, has higher flame temperatures compared to conventional air-only incineration.

- Safety considerations - many materials (i.e. sealing materials and lubricants) that do not ignite in air can ignite in an oxygen-enriched environment, materials to be used should be evaluated for oxygen service [22]
- Increased operating costs due to need for oxygen

Table 4.4 lists adiabatic flame temperatures of selected fuels at stoichiometric ratios in air and oxygen [22]. Table 4.5 summarizes the EPA's test results with the Linde OCS [23]. Figure 4.6 highlights the relationship between % O_2 in the oxidant and flue gas composition [22].

Table 4.4
Adiabatic Flame Temperatures* of Selected Fuels at Stoichiometric Ratios

Fuel	Air °C	Air °F	Oxygen °C	Oxygen °F
Acetylene	2330	4220	3140	5680
Carbon Monoxide	2130	3860	2950	5340
Heptane	2020	3660	2830	5120
Hydrogen	2130	3860	2800	5080
Methane	1940	3520	2750	4990

*Adiabatic flame temperature is the maximum temperature attainable under ideal conditions, actual flame temperatures will be lower due to radiative heat loss and turbulent mixing with colder gases.

Table 4.5
EPA Mobile Incineration System: Oxygen Demonstration Result Summary

	Air Case*	Oxygen Case	Percent Change
Contaminated Soil (lb/hr)	1478	4000	+171%
Firing Rate (MMBtu/ton soil)			
Kiln	4.9	3.9	
Afterburner	3.9	5.4	
Total	8.8	9.3	+6%
Specific Fuel Use (MMBtu/ton soil)	11.9	4.7	-61%
Pure Oxygen Input (lb/hr)			
Kiln**		574	
Afterburner		0	
Total		574	
Specific Oxygen Consumption (ton O_2/ton soil)		0.14	
Fuel Savings (MMBtu/ton O_2)		50	
Kiln Superficial Velocity (ft/sec)	8.1	3.3	-59%
Afterburner residence time† (sec)	2.6	2.6	0
Quenched gas flow (DSCFM)	3250	2500	-23%

* Average of four runs. Maximum throughput was 2000 lb/hr.
** 53% oxygen entering the kiln (equivalent to 40% O_2 enhancement).
† Calculated data.

4.4. Air Pollution Control

Combustion emissions may require treatment in an air pollution control system to reduce acid gas, nitrogen oxide, particulate or combustion by-product emissions. The most common air pollution control systems used at hazardous waste facilities utilize the following air pollution control devices (APCD's) in series [1]:

- Quench - for gas cooling, heat recovery and conditioning
- Venturi Scrubber - for particulate removal
- Packed Tower Absorber - for acid gas removal
- Demister - for elimination of a visible vapor plume.

Innovative air pollution control is available for NO_X and Hg control, however, these devices are expensive and generally found only on commercial large scale systems. State and local permits may require on-site systems to meet stringent emissions standards requiring use of innovative technology. In California, air quality standards limit the feasibility of on-site thermal destruction, because use of the required technology for air pollution control makes the process cost inefficient [17].

The following sections will briefly discuss various air pollution control devices used in incineration systems. For further details about the APCD's and design, refer to reference [13].

4.4.1. Quench Systems

The quenching system reduces combustion gas temperatures and depending upon the cooling method, the volumetric flowrate of the exhaust gas, thus reducing the size of downstream APCD's and equipment (ductwork, fans, etc.). Heat can also be recovered during the quenching process. The three major types of quench systems are:

- Air dilution - increases downstream volumetric flowrate
- Water injection - may add entrained vapor to flue gas
- Heat exchange - offers the most advantages of the three types of quench systems (decreases volumetric flowrate and energy recovery)

4.4.2. Scrubbers

Scrubbers are devices used to remove fumes, mists and suspended solids from the exhaust gas. The scrubber collects particles by contact with a liquid, usually water. There are a number of scrubber designs in which the dominant mechanism for particulate removal is inertial impact with liquid droplets. These include spray-chamber, cyclone, orifice and venturi-jet scrubbers.

4.4.3. Absorption Towers

Absorption towers are used to control acid gas emissions from combustion sources. Gas absorption involves mass transfer from the gas to the liquid phase and is generally limited by resistance at the gas/liquid interface. Several types of absorption towers are used including packed bed, plate towers and spray towers.

4.4.4. Electrostatic Precipitators

Electrostatic precipitators are used to remove particles, either dry dusts or liquid droplets, from combustion exhaust gasses. The process involves:

- Ionization of gas flowing between electrodes
- Charging, migration, and collection of particles on oppositely charged plates
- Removal of particles from the plates

The collection forces of the electrostatic precipitator act only on the particles and not the gas stream, resulting in a high removal efficiency with low air pressure drop.

4.4.5. Baghouses

Baghouses are used to remove dust and fumes from exhaust gas and typically consist of circular tubes (5-6" inside diameter, 5-10' long and spaced 2" apart) made of cotton, wool, glass fibers or synthetic material. The baghouse removes particles from the gas stream by any of the following mechanisms:

- Filtration through the fibers
- Interception
- Impingement
- Gravitational settling
- Diffusion
- Electrostatic attraction.

There are three common types of baghouses, distinguished by the mechanism used for cleaning the filters. They are the reverse air, shaker, and pulse-jet type baghouses.

4.5. Ash and Residue Handling

Inorganic components of hazardous waste exits the incineration system as bottom ash from the combustion chamber, as contaminants in air pollution control residues and wastewaters, and in stack emissions. RCRA defines ashes and residues from hazardous waste incineration as hazardous waste, therefore, ashes and residues must be handled and disposed of properly.

Ash treatment and disposal generally consists of:

- Air cooling or quenching with water
- Accumulation and storage in a secure storage lagoon or in drums
- Dewatering and/or chemical stabilization/solidification
- Disposal in a permitted hazardous waste land disposal facility

Air pollution control residues are usually aqueous waste streams that contain particulates, absorbed acid gas and dissolved organic contaminants. Residue treatment and disposal generally consists of:

- Collection in sumps and tanks
- Particulate and dissolved contaminant removal
- Acid neutralization and recycle to air pollution control processes

4.6. Combustion Process Monitoring

Accurate measurement of combustion emissions is critical to the thermal destruction process. Monitoring equipment and measurement methods have been developed to insure compliance with RCRA standards. Performance monitoring may have any of the following purposes:

- Regulatory compliance
- Process performance monitoring and direct process control
- Performance measurements for research and equipment development purposes

Figure 4.7 shows locations of potential sampling points for assessing incinerator performance [1].

5. FEASIBILITY

5.1. Introduction

The feasibility of destroying hazardous wastes by incineration processes is based upon waste characteristics.

5.2. Waste Characteristics

Factors to be considered in evaluating a hazardous waste for destruction by incineration include:

- Dioxin/furan and PCB content
- Halogen content
- Moisture content
- Inert (ash) content
- Heating value
- Potential health and environmental effects of the effluent
- Metals content
- Corrosiveness
- Thermal or shock instability

5.2.1. Dioxin/Furan and PCB Content

Thermal destruction of PCDD/F or PCB containing wastes is regulated by the Toxic Substances Control Act (TSCA). The EPA, under authority of the TSCA, requires a destruction and removal efficiency (DRE) of 99.9999% for PCDD/F and PCB wastes. This high DRE is difficult to achieve unless the combustion system and APCD's have been specifically designed for this purpose [17].

5.2.2. Halogen Content

The presence of halogenated organic compounds in the waste can result in the formation of hydrogen halides in the combustion emissions. Suitable air pollution control devices must be utilized to reduce these emissions. Generally, alternative treatment methods should be investigated for wastes containing greater than 1000 ppm halogens because the presence of hydrogen halides in the emissions will increase the cost of the incineration process (due to their corrosive characteristics) and the associated air pollution control devices [8]. Also, large amounts of halogens may act as a flame retardant [24].

5.2.3. Moisture Content

For rotary kiln and fixed hearth incineration, it is undesirable to incinerate high moisture content wastes since the moisture acts like a heat sink and decreases combustion chamber temperatures. If wastes with high moisture content must be incinerated it may be necessary to first dry the waste before introduction to the combustion chamber. Waste heat recovered by the quenching operation can be used for drying.

5.2.4. Inert (Ash) Content

The ash content of a waste must be determined in order to evaluate waste handling and feed systems, combustion chamber design, potential for slagging and particulate emissions. The ash content of a waste is very important in determining the type of combustion chamber used for the incineration of the waste.

5.2.5. Heating Value

The heating value of a waste must be considered in assessing the need for auxiliary fuel. Generally, a minimum heating value of 5000 Btu/lb is required to sustain combustion [5].

5.2.6. Potential Health and Environmental Effects of the Combustion Emissions

Potential health and environmental effects of the emissions (gaseous, liquid and solid) must be assessed in order to design and evaluate alternative treatment and disposal methods. Also, pollutants expected to be in the emissions from the combustion process will determine treatment and disposal methods (APCD's and ash/blowdown disposal).

5.2.7. Metals Content

Wastes containing large quantities of volatile metals may be poor candidates for incineration. Volatile metals will vaporize and diffuse into the exhaust stream, where removal by downstream APCD's is difficult to achieve. Incineration of metals-containing wastes will also produce metals-containing ash and residuals that must be treated accordingly before disposal.

5.2.8. Corrosiveness

Wastes that can cause severe corrosion within the incineration system (high halogen or phosphorous containing wastes) can influence system operation and performance.

5.2.9. Thermal or Shock Instability

Wastes that exhibit the properties or thermal or shock instability pose an explosion hazard. Explosions within rotary kiln units have resulted in worker injuries and fatalities [17].

5.3. Burnability Flowchart

Figure 5.1 schematically illustrates a conceptual strategy for determining the burnability of a waste (adapted from [8]).

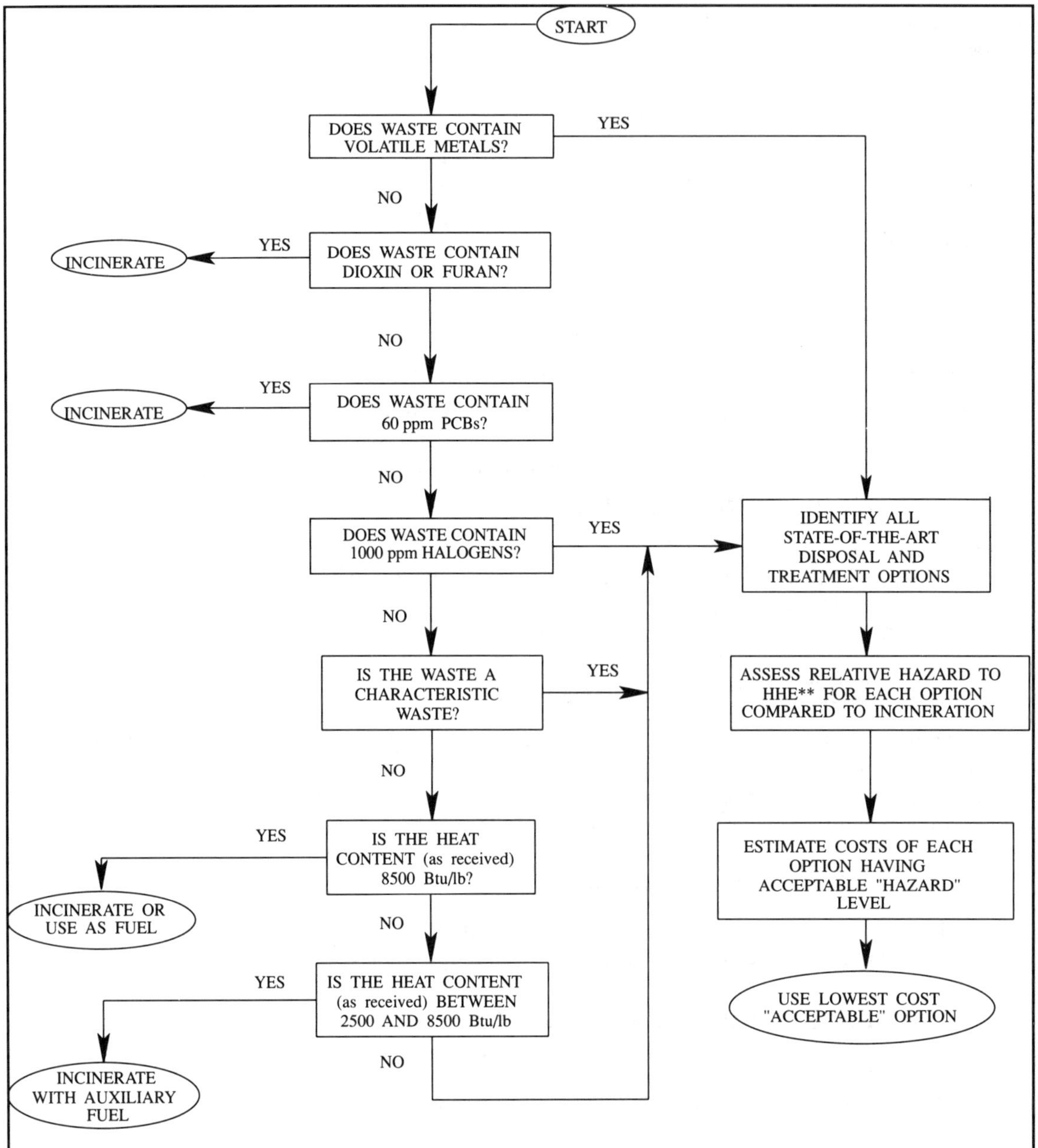

Characteristic waste means ignitable, corrosive, reactive, EP (extraction procedure) toxicity waste

**HHE - Human Health and Environment

FIGURE 5.1
Conceptual Strategy for Determining Waste Burnability
Thermal Processes

5.4. Regulatory and Practical Feasibility [17]

- While techniques are well-demonstrated from agency point-of-view, they are often subject to great community concerns. In particular, with CERCLA (Superfund) cleanups requiring public participation, this could substantially delay cleanup implementation.
- Equipment availability:
  - Fixed incinerators may not have adequate current capacity for some sites due to increased loads caused by land disposal regulations.
  - Mobile incinerators (and other mobile thermal treatment units) have a limited number of treatment units available. Treatment unit availability within the required time frames should therefore be considered in technology evaluation and selection.

6. DESIGN

6.1. Introduction

Incineration system vendors frequently maintain files that are consulted for similar applications when designing new facilities [5]. Six major steps are generally required when designing a hazardous waste thermal treatment facility. They are [5]:

1. Conceptualize and define the process based upon knowledge of waste characteristics, expected design loads and remedial time frame.
2. With the aid of a process flow sheet, define a solution to the problem.
3. Design the process equipment.
4. Perform an economic analysis.
5. Evaluate alternative designs.
6. Compile a design report.

These six steps are applicable to the design of any engineered system and thermal processes design is not an exception. The following discussion of process equipment design will define a general design protocol.

6.2. Process Equipment Design

An understanding of the combustion process is essential to the design of a thermal destruction system. The following areas must be evaluated to understand combustion processes and to design system components:

- Mass and energy balance
- Thermodynamic and kinetic analysis
- Heat transfer
- Residence time
- Temperature
- Turbulent Mixing

6.2.1. Mass and Energy Balance

Mass and energy balance calculations are based on the concepts of mass and energy conservation (the first law of thermodynamics).

Based on an ultimate analysis of the fuel/waste mixture and assuming the stoichiometry of complete combustion (see section 2.4.3.1.), a mass balance can be performed on the combustion chamber. The mass balance will give an estimate of the amount of products formed by the reactants. The products can be expressed as either a mass or volumetric flow assuming that the combustion gasses act as an ideal gas. Downstream APCD's and ash handling systems can then be designed to handle the expected loads.

The energy balance determines the energy transferred within the combustion system and is based upon changes in enthalpy and molar heat capacity of products of combustion reactions (see [4, 5], or a chemical thermodynamics textbook).

Conservation of mass and energy and thermochemical principles (heat of reaction, etc.) are used to calculate fuel requirements, operating temperatures, and combustion air requirements for the combustion process.

6.2.2. Thermodynamic and Kinetic Analysis

Thermodynamic and kinetic analysis will provide answers to two questions regarding chemical reactions, respectively:

1. How far will the reactions proceed?
2. How fast will the reactions proceed?

Generally, two reaction zones exist within the combustion chamber: the flame and post-flame zones. Flame zone destruction is facilitated by radical attack. Post-flame destruction, or thermal destruction, rates are much slower than flame zone rates and are characterized by strong oxidation and weak pyrolysis reactions. Post-flame zone reaction kinetics are the basis for determining reaction velocity constants. Most of the combustion reaction takes place in the post-flame zone. Rate constants for specific compounds and waste samples are available through the EPA [25].

6.2.3. Heat Transfer

Heat transfer to incoming waste and through reactor walls must be considered when designing an incineration system. However, modeling of heat transfer within the combustion environment is a complicated and often times neglected aspect of hazardous waste incineration system design. Heat transfer models have been developed and verified using pilot-scale rotary kiln combustion chambers (see [26]).

6.2.4. Residence Time

Residence time of gas phase reactants is based upon combustion chamber volume and volumetric flowrates obtained from mass and energy balance calculations.

For rotary kiln combustion chambers, average solids residence time can be estimated using the following experimentally derived empirical relationship [5]:

$$\tau = \frac{0.19\ L}{NDS} \tag{36}$$

where:

τ = retention time (min)
L = kiln length (ft)
N = kiln rotational velocity (rpm)
D = kiln diameter (ft)
S = kiln slope, rake (ft/ft)

6.2.5. Temperature

Sufficient combustion chamber temperatures must be maintained to ensure complete combustion of waste compounds. Energy balance calculations can be used to estimate adiabatic flame temperatures, the temperature achieved by the flue gas assuming no heat transfer to the surroundings. Auxiliary fuel required to maintain a given temperature is also found by performing energy balance calculations.

6.2.6. Turbulent Mixing

The combustion system must be designed to provide adequate mixing between waste compounds/fuel and oxygen to ensure complete combustion.

7. WET AIR OXIDATION

7.1. Introduction

Wet air oxidation (WAO), an established thermal treatment technology, developed and patented by Zimpro Inc. (Rothschild, Wisconsin), has been proven to be effective in destroying a wide variety of hazardous organics dissolved or suspended in wastewaters. The process refers to the aqueous phase oxidation of organics at elevated temperatures and pressures.

The process can effectively treat dilute aqueous waste streams that are either too dilute for incineration (low heating values) or too toxic for biological treatment. The process is applicable to the destruction of the following waste types:

- Inorganic and organic cyanide compounds
- Aliphatic and chlorinated aliphatic compounds
- Aromatic hydrocarbons
- Aromatic and halogenated aromatic compounds containing non-halogen functional groups

Halogenated aromatic compounds without other non-halogen functional groups, such as chlorobenzene and PCB's, are resistant to treatment by WAO [27, 28].

7.2. Theoretical Principles of Wet Air Oxidation

Any substance capable of being burned can be oxidized in an aqueous phase at temperatures between 250°F and 700°F [29]. Pressures of 300 to 3000 psig are maintained to prevent excess evaporation of the liquid phase.

Liquid water serves two functions in the WAO process:

- Catalyzes oxidation - reactions proceed at relatively lower temperatures than required by flame combustion (incineration).
- Heat transfer medium - removes excess heat by evaporation and enables the process to be self-sustaining with relatively low organic feed rates.

The enhanced solubility of oxygen in aqueous solution at higher temperatures provides the driving force for oxidation [27].

The products of the WAO process are generally CO_2 in the gas phase, and low molecular weight organics in the aqueous phase (i.e. formic or acetic acids). One of the advantages of WAO is that the products of WAO stay in the aqueous phase, where they can usually be biodegraded. Generally, compounds of sulfur and nitrogen, chlorinated hydrocarbons, and metals are oxidized to sulfate, ammonia, hydrochloric acid, and dissolved or suspended salts of their highest oxidation states, respectively.

The enthalpy difference between the influent and effluent streams is the only energy required for the WAO process. In comparison, incineration requires that: 1) the sensible heat and heat of vaporization of the liquid, and 2) heat for elevating water vapor, combustion products and excess air to combustion temperatures be supplied to the process [27].

Chemical Oxygen Demand (COD) is a convenient measure of the fuel value of an aqueous organic solution. With wastewaters, a COD of 1g/l is equal to a heat value of 50 to 56 Btu/gal [30]. Generally, feed COD concentrations of approximately 15,000 mg/l are required to sustain autogenous oxidation (self-sustaining, with no auxiliary fuel) [27]. Incineration generally requires 300,000 to 400,000 mg/l to sustain autogenous oxidation.

Temperature has the greatest effect on WAO reaction rates. Experimentation with bench and pilot scale units by Zimpro indicate that, in most cases [31]:

- 150°C (300°F) - the lower limit for appreciable reaction
- 250°C (482°F) - temperature required for 80% reduction in COD
- 300°C (572°F) - minimum temperature for 95% reduction in COD

Bench and pilot scale studies also show that initial reaction rates over the first 30 minutes are relatively fast. After 60 minutes, reaction rates slow down considerably, resulting in very little increase in percent oxidation.

Results from performance tests conducted by Zimpro, under partial support from the EPA, on a number of waste classes is shown in Table 7.1 (adapted from [31]).

Table 7.1
Wet Air Oxidation Performance Test Results

	Operating Parameters			Removal (%Percent)			
Waste Class	Temperature (°C/°F)	Pressure (psig)	Residence Time (min)	Target Compound	COD	DOC	BOD_5
Pesticides	280/537	1600	182	98.0-99.8[a]	95.3	96.1	93.8
Dinoseb							
Methoxychlor							
Carboryl							
Malathion							
Solvent still bottoms	268/514	1550	118	—	72.3	47.4	73.6
Phenolic wastes	268/514	1610	113	99.77	89.3	—	—
Organic sulfur wastes	268/514	1610	113	94.0	89.3	—	—
General organic wastewater	277/431	1550	120	—	90.7	96.7	—
Cyanide wastewaters	257/495	1220	80	99.7	88.8	88.4	—

[a] Removal range of the four pesticide wastes

7.3. Process Description

Figure 7.1 shows a general schematic of a continuous WAO process [29]. The process may be either continuous or batch. The WAO system consists of the following components:

- High pressure liquid feed pump
- Oxygen source (either an air compressor or a liquid-oxygen vaporizer)
- Reactor
- Heat exchangers
- Process regulators

The waste stream is pumped to the system using a positive displacement, high energy pump. Air or oxygen is introduced at either the discharge end of the pump or directly into the reactor. Before entering the reactor, the waste stream is preheated by heat exchange with the hot effluent.

Four basic reactor configurations for application to hazardous waste treatment exist, each characterized by either reactor design or operating parameters:

- Aboveground vertical-column reactor

  Designed by Zimpro Inc. - consists of a co-current bubble column which may or may not be baffled depending on the required degree of mixing.

- Stirred-tank reactor

  Commonly referred to as the Wetox System (owned by Zimpro, Inc.) - consists of a series of completely mixed flow reactors.

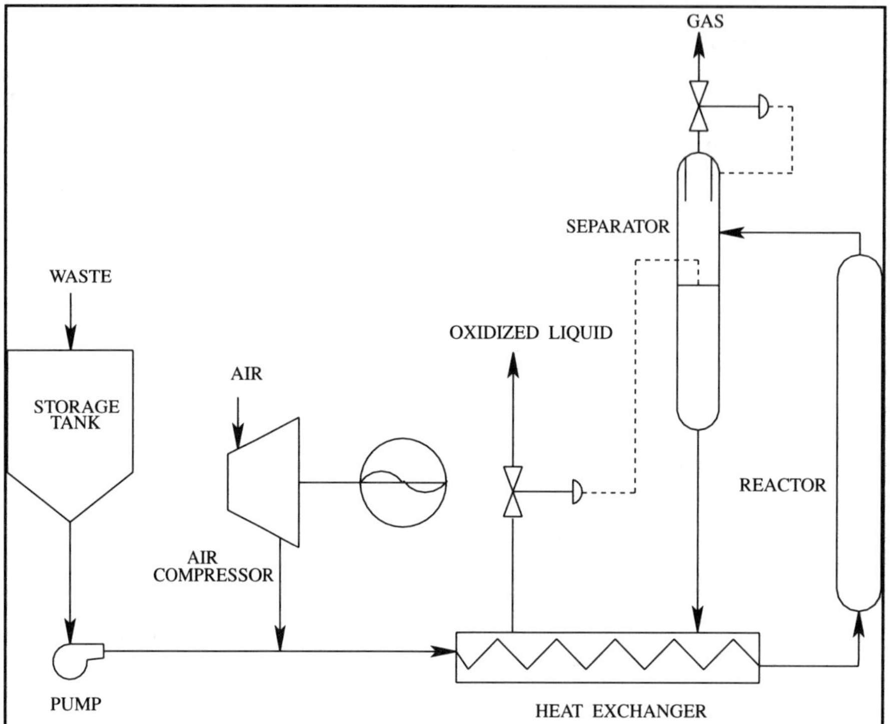

FIGURE 7.1
Flowsheet of Wet Air Oxidation Process
Thermal Processes

- Supercritical-water

  The supercritical-water wet-oxidation technology was developed by MODAR Inc., of Natick, Mass. The reactor is identical to the aboveground vertical-column owned by Zimpro but the operating temperature is above the critical temperature of water (374°C, 705°F) and the pressures are above 22,000 kPa (32,000 psi). Supercritical water oxidation offers the advantage of increased oxygen and organic solubility and decreased inorganic solubility at temperatures above critical.

- Underground vertical-tube (Deep-shaft WAO)

  Utilizes oil-field drilling technology to provide a drilled, cased, and cemented well as a reactor. Wastewater flows down the center of two concentric tubes and returns in the annular space. The reactor extends to a depth of approximately 1 mile. Pressure is developed naturally by hydrostatic head and heat results from the exothermic oxidation reaction. Operating conditions can be either subcritical or supercritical.

The temperature in the reactor is raised by the heat of oxidation and is regulated by system pressure. Steam may be injected into the reactor to supplement heat of oxidation.

The effluent is cooled by heat exchange with the influent waste stream. A separator drum separates the liquid and condensable gas from the non-condensable gas. Treatment of the effluent streams may consist of:

- Non-condensable gas - scrubbing or secondary combustion may be necessary depending upon the products formed during oxidation.
- Liquids and condensable gasses - may be further treated by biological methods [32] or if the oxidation products are recyclable they may be separated and re-used.

7.4. Wet Air Oxidation Design

The design protocol and principles outlined in Section 2 of this module, are applicable to the design of WAO systems.

The following parameters must be determined prior to design of a WAO system:

- Waste characteristics
- Temperature and residence times required for desired degree of destruction
- Air requirements of oxidation reaction stoichiometry

A general procedure for WAO design is as follows:

1. Characterize the waste and determine if WAO is a viable treatment method. Waste types which are applicable to WAO include but are not limited to:
   - Inorganic and organic cyanide compounds
   - Aliphatic and chlorinated aliphatic compounds
   - Aromatic hydrocarbons
   - Aromatic and halogenated aromatic compounds containing non-halogen functional groups

   Halogenated aromatic compounds without other non-halogen functional groups, such as chlorobenzene and PCB's, are resistant to treatment by WAO [27, 28]. Results from numerous bench and pilot scale studies performed by Zimpro may be consulted to determine applicability of waste to WAO.

2. Determine the temperature and residence time required to achieve a given degree of destruction by performing bench or pilot scale tests. Several tests should be conducted to determine degree of destruction as a function of both temperature and residence time. Zimpro maintains several pilot scale systems, capable of operating over a range of parameters. Table 7.2 summarizes the pilot plant design capabilities and operating parameters.

Table 7.2
Pilot Plant Design Capabilities and Operating Parameters

Construction Material	Maximum Operating Pressure (psia)	Maximum Operating Temperature (°F)	Reactor Residence Time (min)	Maximum Flow Rate (gal/hr)
Stainless Steel	4,000	650	10 - 180	100
Nickel	3,000	625	10 - 60	100
Titanium	3,000	625	30 - 60	15 - 60
Titanium	1,600	550	15 - 60	10

3. Using the following oxidation reaction stoichiometry, determine the air required for complete oxidation:

$$C_xH_y + zO_2 \rightarrow xCO_2 + \frac{y}{2}H_2O \tag{37}$$

where:
z = x + y/4

4. Choose the system components capable of operating at the parameters obtained from the preliminary studies. The system components should be compatible with the waste (i.e. capable of resisting chemical effects due to the wastes characteristics).

8. LOW TEMPERATURE THERMAL DESORPTION

8.1. Introduction

Low temperature thermal desorption can be described as ex-situ soil vapor extraction. This process uses ambient air, heat, or mechanical agitation to increase the rate of mass transfer of contaminants into the gas phase. Once in the gas phase, the contaminants can be further treated by either thermal or physical methods. The low temperature thermal desorption process is used especially for contaminated soils.

The process can effectively remove volatile organic compounds from the soil matrix. Low temperature thermal desorption is applicable to the following waste types:

- Halogenated aromatic and aliphatic compounds
- Volatile non-halogenated organic compounds
- Semi-volatile non-halogenated organics (to a limited extent)

Specific waste contaminants shown to be effectively treated by a low temperature thermal desorption process are listed in Table 8.1 [33]. Some polynuclear aromatic compounds may be effectively treated under optimum operating parameters. Low temperature thermal desorption is not as effective as thermal destruction in treating PNA's since removal efficiencies average about 65% [2].

Figure 8.1 shows a mobile low temperature thermal desorption system process schematic [34].

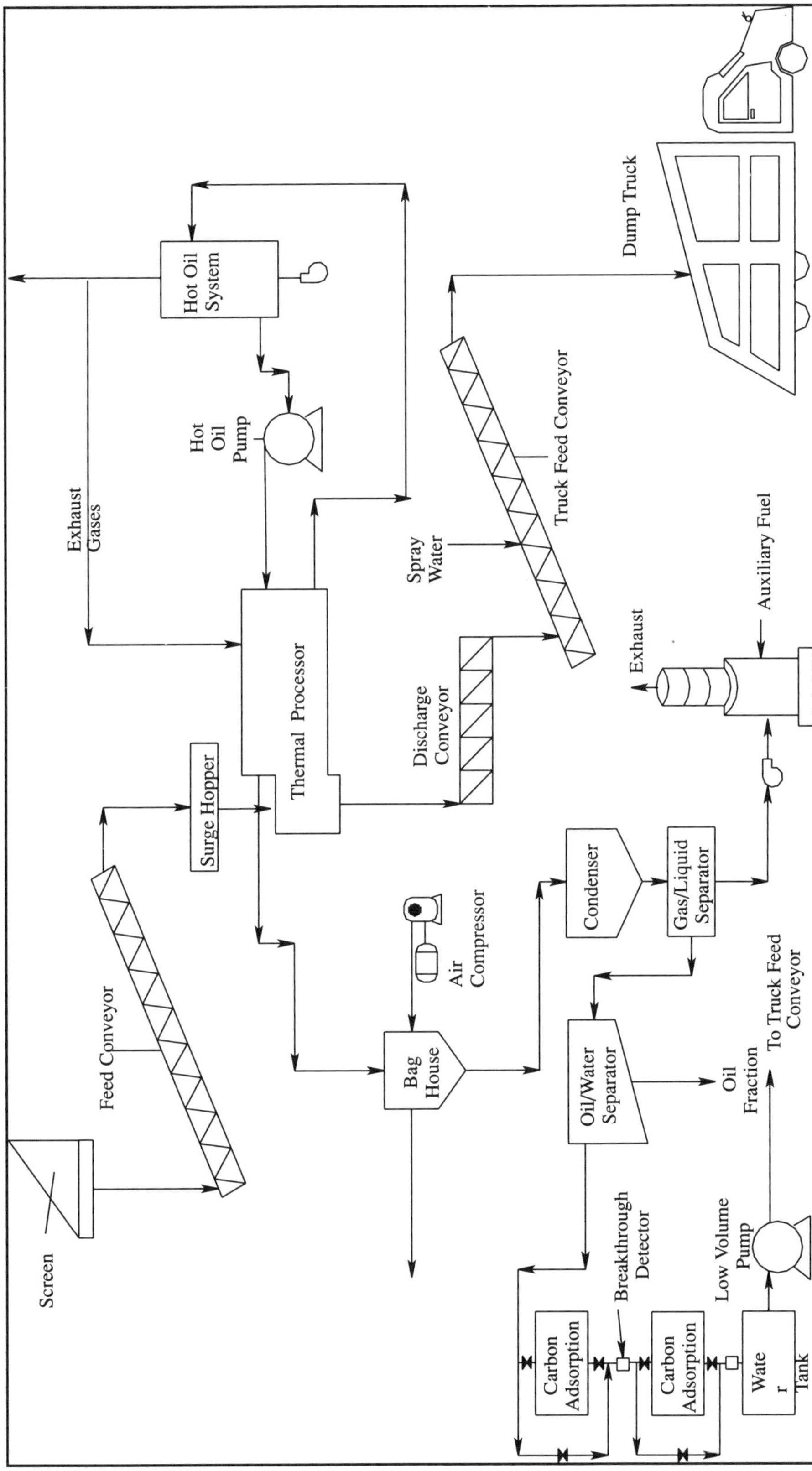

FIGURE 8.1
Process Schematic Mobile Lo-Temperature Thermal Treatment System
Thermal Processes

8.2. Theoretical Principles of Low Temperature Thermal Desorption

The basic principle of the low temperature thermal desorption process is the use of ambient air, heat, or mechanical agitation to increase the rate of mass transfer of contaminants into the gas phase.

The design of the low temperature thermal desorption process should ensure that contaminant transport is not diffusion limited by providing sufficient pre-processing of the soil. Mass transfer theory is discussed in Section 2, "Fundamentals," of the Soil Vapor Extraction Module and in various textbooks.

The low temperature thermal desorption process can be either continuous or batch.

8.3. Process Description

Figure 8.1 shows a process schematic of a mobile low temperature thermal desorption system [34]. The process generally consists of the following components:

- Solids handling equipment
- Aeration or solids mixing unit
- Gas stream handling equipment
- Liquid handling equipment
- Monitoring systems

Contaminated soil may contain debris and particles of various size. The first step in the process is to remove debris and oversized material (51 mm or 2 inch diameter) from the feed. Sorted material is then conveyed to the aeration unit.

Table 8.1
Specific Waste Contaminants

Volatiles:		
Acetone	1,2-Dichloroethene	1,1,1-Trichloroethane
Benzene	1,1,1-Trichloroethene	1,1,2-Trichloroethane
Carbon tetrachloride	*trans*-1,2-Dichloroethene	Tetrachloroethene
Chloroform	2-Hexanone	Toluene
Chloromethane	Methylene chloride	Trichloroethane
1,1-Dichloroethene	1,1,2,2-Tetrachloroethane	Vinyl chloride
Semivolatiles:		
Ethyl benzene	Naphthalene	Xylene

The low temperature thermal desorption process can be either batch or continuous depending upon the type of aeration system used.

Batch processes generally involve:

- Placement of the contaminated soil into the unit at a fixed bed height

- Circulation of air through the unit (air enters the unit either above the bed or through a perforated header below the bed)
- Mechanically agitated beds to enhance mass transfer

Several continuous low temperature thermal desorption processes have been developed. They typically operate at temperatures between 121°C (250°F) and 232°C (450°F) and can process up to 15000 lb/hr. Continuous feed systems usually utilize a rotary screw indirect heat exchanger as a thermal processor. This type of processor has been used for over thirty years in the mineral industry to dry, heat, and cool a variety of materials [33]. The soil is indirectly heated by circulating hot oil through the hollow flights of the screw. Heat transfer can be controlled by adjusting the oil temperature and flowrate. The oil is heated using conventional oil heaters. The following systems utilize the rotary screw heat exchanger:

- Roy F. Weston, Inc., of West Chester, Pa., has patented a mobile continuous-feed low temperature thermal desorption process known as LT^3 (Low-Temperature Thermal Treatment). The key component of the LT^3 system is the thermal processor, which consists of heated conveying screws.

- A processor known as the LPR (Large Particle Recycle) System has been developed which utilizes a hollow screw indirect heat exchanger for treatment or waste sludges containing volatiles. The waste sludge is prevented from caking on the screw as it dries by the scouring action of hard, multi-faceted limestone particles (1/4" to 1/2" dia.). The particles are added at the feed end and removed by screening at the discharge end of the processor.

- Canonie Environmental Service Corporation of Porter, Indiana, has patented a mobile continuous thermal desorption process known as LTTA (Low Temperature Thermal Aeration). The key component of the LTTA system is the materials dryer, which consists of a rotating drum (4 ft. in diameter and 20 ft. long), where hot air flows concurrently with the soils, collecting vaporized organics [17].

The solids effluent is cooled and tested (TCLP) to determine a method of disposal. The solids can be used for backfill if shown to meet the standards set by the appropriate environmental agency.

The vaporized contaminants are purged from the aeration unit by a continuous, low-flow rate gas purge. The purge gas must have a low oxygen content (less than the lower explosion limit, about 8%) to prevent combustion inside the thermal processor. Nitrogen or a mixture of air and exhaust gas from the hot oil system can be used as purge gas [34, 35]. The effluent gas is further treated by either thermal or physical methods.

Liquids which are separated from the effluent gas contain water and condensed volatile organics. This stream must be treated before disposal. Treatment may include oil/water separation, and carbon adsorption for the liquid water fraction.

8.4. Low Temperature Thermal Desorption Design

The design protocol and principles outlined in Section 2 of this module, are applicable to the design of low temperature thermal desorption systems.

The following parameters must be determined prior to the design of a low temperature thermal desorption system:

- Waste characteristics
- Soil characteristics
- Temperature and residence times required for desired degree of mass transfer and contaminant removal
- Required degree of mixing

A general procedure for low temperature thermal desorption design is as follows:

1. Characterize the waste and quantify the extent of contamination to determine if low temperature thermal desorption is feasible. Table 8.1 lists specific waste contaminants that have been shown to be effectively treated by low temperature thermal desorption [33]. As the volume of contaminated soil increases, the unit cost of the process decreases. Weston estimates that to offset the fixed cost of mobilization, a project should have at least 1000 yd^3 (765 m^3) of contaminated soil [34].
2. Soil characteristics can influence thermal desorption of organic compounds [35]. Characteristics that can influence desorption include:
   - Particle size - effects the surface area where volatilization occurs and diffusion pathways between soil particles.
   - Moisture content - high moisture content reduces heating value and increases the overall energy required by the process. Weston suggests an upper limit of 20%.
   - Organic matter - effects the sorption of organic contaminants. Soils with high organic content provide high sorption potential for organic contaminants, reducing the feasibility of low temperature thermal desorption.
3. Bench and pilot scale treatment studies must be utilized to obtain temperatures and residence times required for effective treatment.
4. The process should be designed to promote mixing of the soil matrix to maximize mass transfer.

9. REGULATIONS

9.1. Introduction

Incineration has been accepted as a proven technology to dispose of hazardous, municipal, infectious and toxic wastes. Regulations governing the operation of incinerator systems are defined by the Resource Conservation and Recovery Act (RCRA) and the Toxic Substances Control Act (TSCA).

9.2. TSCA

Under TSCA, the EPA set standards for operating PCB and dioxin incinerators. The general requirements of the TSCA standards are:

- Liquid wastes - must be maintained at 1200°C +/- 100°C for a 2-second dwell time and 3% excess oxygen in the stack gas, or

- Maintained at 1600°C +/- 100°C for a 1.5-second dwell time and 2% excess oxygen in the stack gas.
- Combustion efficiency must be at least 99.9%, where:

$$\text{Combustion efficiency} = \left[\frac{C_{CO_2}}{C_{CO_2} + C_{CO}} \right] \text{x } 100 \tag{38}$$

where:

C_{CO_2} = concentration of carbon dioxide

C_{CO} = concentration of carbon monoxide

- Non-liquid wastes - mass of emissions from the incinerator must be less than 0.001 g POHC/kg of the POHC introduced into the incinerator (this is how DRE is defined)
- Scrubbers must be used to control HCl emissions
- A trial burn is required to demonstrate that the incinerator meets the standards

The standards set by EPA under TSCA require a destruction and removal efficiency (DRE) of 99.9999% for PCB and dioxin containing wastes. Low temperature thermal desorption is considered to be an alternative to incineration, and must be able to treat to below 2 ppm PCB's, a value defined to represent treatment equivalence to an incinerator [17].

9.3. RCRA

RCRA divides solid wastes into three major categories:

- Hazardous waste under Subtitle C
- Municipal waste under Subtitle D
- Medical (infectious) waste under Subtitle J

In general, RCRA covers all hazardous wastes except for PCB's and dioxins.

9.3.1. Hazardous Waste

The general requirements of the RCRA standards for hazardous waste incineration are:

- An incinerator must achieve a DRE of 99.99% for each POHC (Principal Organic Hazardous Constituent) designated for each waste feed, where:

$$DRE = \left[\frac{W_{in} - W_{out}}{W_{in}}\right] x\ 100 \tag{39}$$

where:

W_{in} = mass feed rate of POHC

W_{out} = mass emission rate of POHC to the atmosphere

- An incinerator burning hazardous waste may not emit more than 1.8 kg/hr of HCl and must control HCl emissions, such that:

$$HCl_{emissions} \leq 1.8 \text{ kg/hr} \tag{40}$$

$$\text{or } HCl_{emissions} \leq 1\% \text{ of the HCl in the exhaust gas} \tag{41}$$

- Particulate emissions from incinerators burning hazardous waste must be less than 180 mg per dry standard cubic meter corrected to 7% O_2 in the stack gas, where:

$$P_c = P_m \left(\frac{14}{21 - Y}\right) \tag{42}$$

where:

P_c = corrected concentration of particulate matter

P_m = measured concentration of particulate matter

Y = measured concentration of oxygen in the stack gas, using the Orsat method for oxygen analysis for dry flue gas

- A trial burn is required to demonstrate that the hazardous waste incinerator meets RCRA standards
- Other local permits may be required, especially by state agencies and if the waste is not regulated under RCRA or TSCA. For state cleanups, air permits are generally required. In addition, other non-incineration-specific permits and/or variances could be required, depending on the location (stream encroachment, general hazardous activities variances due to locations in protected areas, etc.) [17].

10. COSTS

10.1. Introduction

The costs of using thermal processes include equipment and operations and maintenance (O&M) costs. Remediation costs are highly variable and depend upon:

- Waste composition and characteristics
- Fuel and labor costs
- Construction costs
- Permitting

Generally, costs for fuel, labor, construction and permitting are regionally dependent. Waste composition and characteristics are highly variable and affect thermal treatment costs significantly. Waste composition and characteristics effect thermal treatment costs by determining:

- The type of thermal process equipment to be used
- The amount of auxiliary fuel required
- Waste storage, feed, and handling methods

Major equipment costs for thermal treatment include:

- Waste storage and loading facilities
- Thermal process unit (i.e. combustion chamber, desorption unit)
- Air pollution control equipment
- Energy recovery equipment

Ancillary equipment costs include:

- Piping
- Storage and batching tanks and vessels
- Pumps, fans and stacks
- Monitoring and instrumentation equipment

The primary costs in operating a thermal treatment process are for labor and energy. Energy costs are dependent upon:

- Heat input rate
- Heat application method
- Waste characteristics
- The heat recovery system

Additional O&M costs include:

- Ash and residue treatment and disposal

- Air pollution control equipment operations
- Monitoring and analysis costs
- Insurance costs

10.2. Incineration

Figures 10.1, 10.2, and 10.3 show freight on board (FOB) equipment costs as a function of heat input for rotary kiln, liquid injection, and fixed hearth incineration systems [36]. The cost estimates are accurate to +/- 20% of actual costs [36]. Freight on board costs include:

- Waste storage and handling equipment
- Energy recovery equipment
- Air pollution control systems

Table 10.1 details a remediation cost estimate for an on-site cleanup of 197,200 tons of contaminated soil and other materials [37]. The total cleanup costs amount to approximately $550/ton.

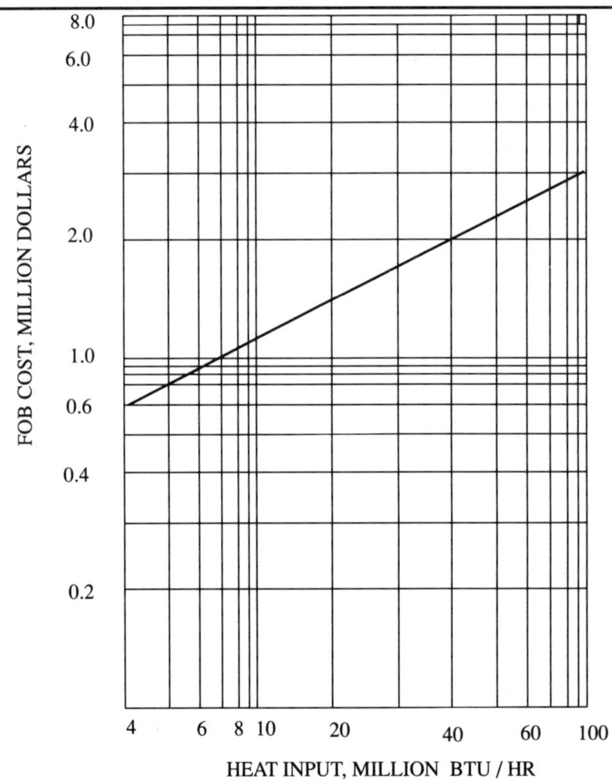

NOTE: FOB = FREIGHT ON BOARD

FIGURE 10.1
Rotary Kiln Incinerator Facility Cost as a Function of Heat Input
Thermal Processes

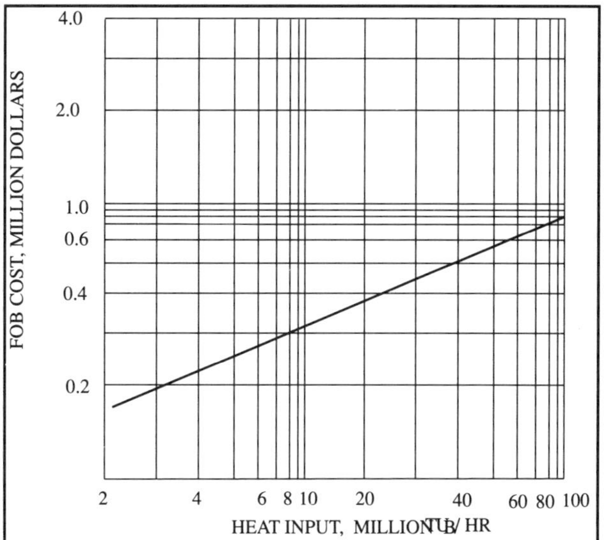

NOTE: FOB = FREIGHT ON BOARD

FIGURE 10.2
Liquid Injection Incinerator Facility Cost as a Function of Heat Input
Thermal Processes

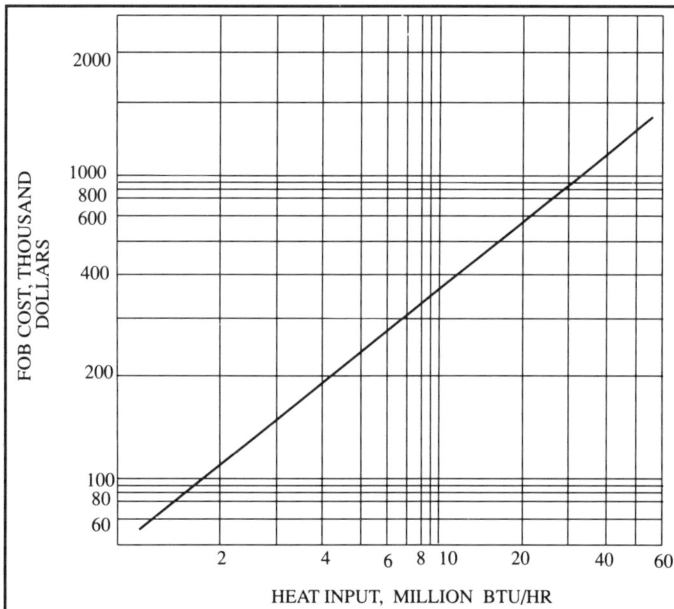

NOTE: FOB = FREIGHT ON BOARD

FIGURE 10.3
Hearth Incineratory Facility Cost as a Function of Heat Input
Thermal Processes

Table 10.1
Remediation Cost Estimate

Activity	Cost, $\$x10^3$	$/ton
1. Mobilization/demobilization	1,602	8
2. Site preparation	8	-
3. Excavation	2,713	14
4. Excavated mat'l transport	533	3
5. Excavated mat'l storage bldg.	432	2
6. Air emissions tent	1,145	6
7. Emissions control foam	1,984	10
8. General operations	990	5
9. Material handling	733	4
10. Lime basins	879	-
11. Incineration	24,663	125
a. Fuel *	3,885	20
b. Lime*	1,664	8
12. Water	60	-
13. Power equipment	480	2
14. Closure	81	-
SUBTOTAL	36,303	183
15. Fill	2,287	12
16. Residuals transport	6,390	32
17. Residuals disposal	30,004	152
TOTAL CONSTRUCTION COST	74,984	379
Owner's Cost		
18. Legal, administration, permitting (7%)	5,250	27
19. Engineering (5%)	3,750	19
20. Services during construction (8%)	6,000	30
SUBTOTAL	89,984	456
21. Contingency (20%)	18,000	91
TOTAL REMEDIATION COST	107,984	548

* Included in Incineration Cost

10.3. Wet Air Oxidation

Figures 10.4 and 10.5 show installed plant costs as a function of plant capacity and operating costs as a function of unit capacity, respectively [31].

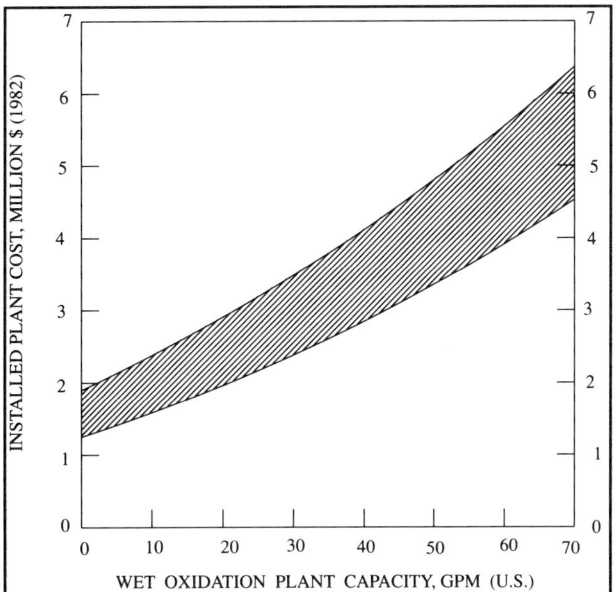

FIGURE 10.4
Installed Plant Cost vs. Capacity for Wet Air Oxidation Units (Based on High Pressure Units, Low Pressure Units Will Be Substantially Less)
Thermal Processes

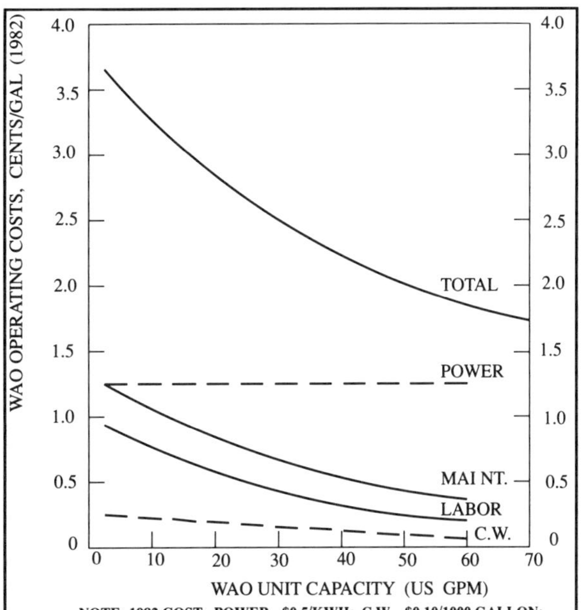

NOTE: 1982 COST: POWER - $0.5/KWH; C.W. - $0.10/1000 GALLON; MAINTENANCE - 1% CAPITAL COST; LABOR - $25,000/YR/OPERATOR

FIGURE 10.5
Wet Air Oxidation Unit Operating Costs as a Function of Flow Rate
Thermal Processes

10.4. Low Temperature Thermal Desorption

The costs of the low temperature thermal desorption process is most significantly affected by the volume of material to be processed [34]. Costs are dependent upon:

- Soil type
- Contaminant concentrations
- Moisture content of the soil

Weston, Inc. of West Chester, Pa., estimates that to offset the fixed cost of mobilization, and to keep unit costs competitive with alternative treatment technologies, a project should have at least 1000 yd^3 (765 m^3) of contaminated soil [34].

10.4.1. Low Temperature Thermal Desorption Case Study

Canonie's LTTA system was recently used at an operating facility in Newark, New Jersey to treat approximately 7,000 tons of soils containing volatile organic compounds (VOC's) including toluene, ethyl benzene, and xylenes as well as base/nuetral compounds (BN's) including benzo-(a)-pyrene and chrysene. The project took approximately 8 months to complete, at an overall cost per ton (including mobilization, demobilization, excavation, site restoration, sampling and analysis, and permitting) of approximately \$350 [17].

REFERENCES

1. Oppelt, T. E. Incineration of Hazardous Waste, Acritical Review, *JAPCA*. **37**: 5. 1987, pp. 558-586.

2. Camp Dresser and M. *Summary of Treatment Technologies Effectiveness for Contmatinated Soil*.(EPA-1540/2-89/053 Cincinatti, June, 1990).

3. Snoeyink, V. L. and D. Jenkins. *Water Chemistry*. 2nd. (John Wiley and Sons, New York, New York, 1980).

4. Kuo, K. K. *Principles of Combustion*. (Wiley-Interscience, New York, New York, 1986).

5. Theodore, L. and J. Reynolds. *Introduction to Hazardous Waste Incineration*. (Wiley-Interscience, New York, New York, 1987).

6. Cheremisinoff, N. P. *Handbook of Heat and Mass Transfer, Volume One: Heat Transfer Operations*. (Gulf Publishing Co., Houston, Texas, 1986).

7. Guyer, E. C. *Handbook of Applied Thermal Design*. (McGraw Hill, New York, New York, 1989).

8. Lee, C. C. and G. L. Huffman. Incineration of Solid Waste, *Environmental Progress*. **8**: 3. 1989, pp. 143-151.

9. Brunner, C. R. *Site Cleanup by Incineration*. (HMCRI, Silver Spring, MD, 1988).

10. Lee, C. C. and G. L. Huffman. *Regulatory Framework for Combustion By-Products from Incineration Sources*. (U.S. Environmental Protection Agency, Cincinnati, August, 1990).

11. Peters, W. A. Solids Pyrolysis and Volatiles Secondary Reactions in Hazardous Waste Incineration: Implications for Toxicnnts Destruction and PIC's Generations, *Hazardous Waste and Hazardous Materials*. **7**: 1. 1990, pp. 89-102.

12. Wendt, J. O. L., W. P. Linak and P. M. Lemieux. Predition of Transient Behavior During Batch Incineration of Liquid Waste in Rotary Kilns, *Hazardous Waste and Hazardous Materials*. **7**: 1. 1990, pp. 41-54.

13. Cooper, C. D. and F. C. Alley. *Air Pollution Control: A Design Approach*. (P. W. S. Engineering, Boston, 1986).

14. Fogler, H. S. *Elements of Chemical Reaction Engineering*. 2nd. (Prentice Hall, Englewood Cliffs, New Jersey, 1992).

15. Levenspiel, O. *Chemical Reaction Engineering*. 2nd. (John Wiley and Sons, New York, New York, 1972).

16. Linack, W. P. and et. al. Applications of Staged Combustion and Reburning to the Co-firing of Nitrogen Containing Waste, *Hazardous Waste and Hazardous Materials*. **8**: 1. 1991, pp. 1-15.

17. Washburn, S. Environ Corporation. 1992.

18. Steverson, E. M. Provoking a Firestorm: Waste Incineration, *ES&T*. **25**: 11. 1991, pp. 1808-1814.

19. Barton, R. G. and et. al. Dioxin Emissions from Waste Incinerators, *Chemosphere*. **2**: 1990, pp. 1825-1839.

20. Barton, R. G. and et. al. Fate of Metals in Waste Combustion Systems, *Combustion Science and Technology*. **74**: 1990, pp. 327-342.

21. Lee, C. C. and et. al. Hazardous/Toxic Waste Incineration, *Hazardous Waste Management*. **36**: 8. 1986, pp. 922-931.

22. Ding, M. G. The Use of Oxygen for Hazardous Waste Incineration, *Thermal Processes*. H. M. Freeman. (Technomic, 1990).

23. Lee, C. C. and G. L. Huffman. Update on Innovative Thermal Destruction Technologies, *Thermal Processes*. H. M. Freeman. (Technomic, 1990).

24. Seeker, W. R. Waste Combustion, *Twenty-Third Symposium (International) on Combustion*. (The Combustion Institute, Pittsburgh, PA, 1990), pp. 867-885.

25. Ruby, W.A. *Design Considerations for a Thermal Decomposition Analytical System*. EPA-600/2-80/098. Cincinnati, OH, 1980.

26. Owens, W. D. and et. al. Thermal Analysis of Rotary Kiln Incineration: Comparison of Theory and Experiment, *Combustion and Flame*. **86**: 1991, pp. 101-114.

27. Dietrich, T. L. and et. al. Wet Air Oxidation of Hazardous Organics in Wastewater, *Environmental Press*. **4**: 3. 1985, pp. 171-177.

28. Copa, W. R. Wet Oxidation, *Standard Handbook of Hazardous Waste Treatment and Disposal*. H. M. Freeman. (McGraw Hill, New York, 1989).

29. Sittig, M. *Incineration of Industrial Hazardous Wastes and Sludges*. (Noyes Data Corporation, Park Ridge, NJ, 1979).

30. Teletzke, G. H. Wet Air Oxidation, *Chemical Engineering Progress*. **60**: 1. 1964, pp. 33-38.

31. Freeman, H. M. *Innovative Thermal Hazardous Organic Waste Treatment Processes*. (Noyes Publications, 1985).

32. Wihelmi, A. R. and R.B. Ely. A Two-Step Process for Toxic Wastewaters, *Chemical Engineering*. 1976, pp. 105-109.

33. Velazquez, L. A. and R. Nielson. Low-Temperature Thermal Treatment for Removal of Organic Contaminats from Soil, *Thermal Processes*. H. M. Freeman. (Technomic, 1990).

34. Nielson, R. K. and M. G. Cosmos. Low Temperature Thermal Treatment (LT3) of Volatile Organic Compounds From Soil: A Technology Demonstrated, *Environmental Progress*. **8**: 2. 1989, pp. 139-142.

35. Helsel, R., E. Alperin and A. Groen. *Engineered-Scale Demonstration of Thermal Desorption Technology for Manufactured Gas Plant Site Soils*. 1989. (prepared for Illinois Haz. Waste Research and Information Center and Gas Research Institute)

36. O'Brien & Gere Engineers, I.*Hazardous Waste Site Remediation*.(Van Nostrand Reinhold, New York, 1988).

37. Brunner, C. R. On-Site Incineration, *Civil Engineering*. April. 1989), pp. 68-71.

L - Risk Assessment

HUMAN HEALTH RISK ASSESSMENT

1. INTRODUCTION

Risk assessments must be conducted because it is generally not possible to obtain a direct measure of risk to people under actual conditions of exposure. Risks need to be understood in advance of exposure to insure proper safety precautions.

Risk assessment was initiated by the Food and Drug Administration in 1973 to address carcinogens. When the low-dose risk was quantified, the insignificant risk dose, or "safe" dose, could be selected. Uses for risk assessment in regulations gradually expanded from 1975 to the present, and, in 1983, the National Academy of Science established a framework for risk assessments.

Driving forces for risk assessment in the early 1970's were:

- The identification of more carcinogens that are more widespread in environment
- Methodology in literature
- The institution of many new laws
- Requirements for explicitness

1.1. Definition of Risk, Risk Assessment, Risk Management

Risk - The probability of occurrence of an unwanted adverse effect on human life, human health, or the environment.

Risk Assessment - The process of estimating the probability of adverse health effects of human exposure to chemicals or environmental hazards based upon information on the nature and severity of the chemical's toxicity, and the magnitude of human exposure to the chemical. A human health risk assessment includes:

- — Description of the potential adverse health effects associated with a given chemical, based on an evaluation of results of epidemiological, clinical, toxicological, and environmental research
- — Extrapolation from those results to predict health effects in humans under different conditions of exposure
- — Judgments as to the number and characteristics of persons exposed and the intensity and duration of exposure
- — Quantification of overall magnitude of the risk problem
- — Characterization of the uncertainties inherent in the process of inferring risk.

Risk Management - The process of evaluating and selecting alternative regulatory actions.

1.2. Uses of a Risk Assessment

- To identify the risk associated with various levels of exposure.
- To identify limits on exposure so that a specific level of risk will not be exceeded.

- To provide risk managers with a basis for decisions.

Human health risk assessments can be used for the following specific applications:

- Pre-marketing safety evaluation of new chemicals such as pesticides, food additives, drugs, and products of biotechnology
- Post-marketing safety evaluation of existing chemicals
- Evaluation of health impacts of emissions from industrial facilities and other sources
- Evaluation of health impacts of chemicals migrating from hazardous waste sites
- Selection among several remedial alternatives for hazardous waste sites
- Determination of workplace health risks
- Evaluation of potential risks of new facilities
- Evaluation of product liability and toxic tort claims

2. METHODOLOGY FOR A HUMAN HEALTH RISK ASSESSMENT

In 1983, a special National Academy of Science committee reported that most human and environmental health hazards can be evaluated by dissecting the analysis into four parts. The four parts are hazard identification, dose-response evaluation, exposure assessment, and risk characterization. This methodology, described by the committee's report "Risk Assessment in the Federal Government", has been widely adopted.

2.1. Hazard Identification

Hazard identification involves gathering and evaluating toxicity data on the types of health injury or disease that may be produced by a chemical and on the conditions of exposure under which injury or disease is produced. It also may involve characterization of the behavior of a chemical within the body and the interactions it undergoes with organs, cells, or even parts of cells. These data may be of value in determining whether the forms of toxicity known to be produced by a chemical agent in one species or population group are also likely to be produced in the human population group of interest. Hazard identification is conducted to determine the scientific validity of inferring toxic effects observed in one setting, based on effects observed in other settings (e.g., whether chemicals found to be carcinogenic or teratogenic in experimental animals also are likely to be so in humans).

2.2. Dose-Response Evaluation

Dose-response assessment involves describing the quantitative relationship between the amount of exposure to a chemical and the extent of toxic injury or disease. Data are derived from animal studies or, less frequently, from studies in exposed human populations. There may be many different dose-response relationships for a chemical agent depending on the conditions of exposure (e.g., single versus repeated exposures) and the response (e.g., cancer, birth defects) being considered.

2.3. Exposure Assessment

Exposure assessment involves describing the nature and size of the various populations exposed to a chemical agent and the routes, magnitude, and duration of their exposures. The evaluation could be of past exposures, current exposures, or exposures anticipated in the future.

2.4. Risk Characterization

Risk characterization involves the integration of the data and analyses in the other three steps of risk assessment to quantify the likelihood that the human population of interest will experience any of the various forms of toxicity associated with a chemical under specified conditions of exposure.

3. HAZARD IDENTIFICATION

The first step of a human health risk assessment involves a review of published literature to obtain toxicity data on the chemicals of concern and identify the types of toxicity and their relevance to humans.

3.1. Sources of Toxicity Information

Toxicity information may be derived from human or animal data collected through means such as:

- Epidemiology studies in groups of individuals (typically those exposed at relatively high levels)
- Controlled experiments in laboratory animals
- "Test-tube" experiments

3.1.1. Human Data

Information on adverse health effects in human populations potentially attributable to environmental chemicals is obtained from four sources:

- Summaries of self-reported symptoms in exposed persons
- Case reports prepared by medical personnel
- Descriptive epidemiology studies
- Analytic epidemiology studies

The first two sources are generally considered to represent only anecdotal data that, when considered alone, are of limited use in determining causes of disease.

3.1.1.1. Descriptive Epidemiological Studies

Descriptive epidemiologic studies involve the identification of the amount and distribution of disease in populations. Such studies often employ the correlational approach in which differences in disease rates in human populations are associated with differences in environmental exposures among population subgroups. Thee studies are useful in drawing attention to previously unsuspected problems and in generating hypotheses that can be further tested. However, because the unit under observation is a population, rather than individuals, they are generally not useful in identifying specific cause-and-effect relationships.

3.1.1.2. Analytic Epidemiological Studies

Analytic epidemiologic studies are the primary means for determining causal relationships between specific environmental exposures and human health effects. Analytic epidemiology studies test specific hypotheses using "observational" study designs. Observational study designs are of two major types, cohort and case control. Cohort studies involve the evaluation of the incidence of disease among individuals exposed by choice or accident to a particular environmental agent, as compared with disease incidence in similar but nonexposed individuals. Case-control studies involve comparisons of individuals with a particular disease with similar individuals without the disease for differences in past exposure to suspected agents.

The determination of causality of a disease through controlled laboratory experiments is fairly straightforward; however, identifying causal relations in epidemiological studies may be more difficult. Prior to accepting any association as causal, all alternative explanations should be considered and confounding variables should be addressed. Positive criteria that can aid in the judgment concerning causality include:

- Strength of the association
- Biologic credibility of the hypothesis
- Consistency of the findings
- Temporal sequence
- Presence of a dose-response relationship

3.1.1.3. Case Studies

Case studies, when available, provide information on acute toxic human effects of the chemical based on specific exposure circumstances.

3.1.2. Animal Data

Animal toxicity testing is used to identify the effects produced by the chemical in laboratory animals and to predict the possible hazard in humans based on the exposure of experimental animals to toxic agents in high doses. These animal tests are designed to characterize the toxic effects of a chemical. Tests for acute toxicity, subchronic toxicity, and chronic toxicity will be discussed below. Other tests which explore more specific effects are teratogenicity (effects on offspring), reproduction, mutagenicity (genetic mutations), chemical interactions, and behavioral effects.

3.1.2.1. Acute Toxicity

Acute toxicity tests provide:

- Quantify the acute toxicity (median lethal dose or LD_{50}) of a chemical for comparison to other substances
  - Two species (ideally one rodent and one nonrodent) and two routes of administration are used in lethality tests
- Identify the acute toxic effects and clinical manifestations of a chemical
- Provide dose-ranging guidance for other studies

Table 3.1: Acute Animal LD_{50}s of Selected Chemical Agents (will kill approximately 50% of the population)

Agent	LD50 (mg/kg)
Ethyl alcohol	10,000
Sodium chloride	4,000
Ferrous sulfate	1,500
Morphine sulfate	900
Phenobarbital sodium	150
DDT	100
Picrotoxin	5
Strychnine sulfate	2
Nicotine	1
d-Tubocurarine	0.5
Hemicholinium-3	0.2
Tetrodotoxin	0.10
Dioxin (TCDD)	0.001
Botulinus toxin	0.00001

3.1.2.2. Subchronic Toxicity

Subchronic studies are conducted to characterize the dose-response relationship following repeated exposures to a chemical and also provide reasonable predictions of appropriate doses for chronic exposure studies.

The duration of a subchronic study is usually 90 days and is usually conducted in two species (rat and dog) using the route of intended exposure. At least three doses are generally used: a high dose that produces toxicity but does not cause more than 10 percent fatalities; a low dose that produces no apparent toxic effects; and an intermediate dose.

3.1.2.3. Chronic Toxicity

Chronic exposure studies are often used to determine the carcinogenic potential of chemicals. It is performed similar to subchronic studies except the exposure period is usually for the lifetime of the animal (i.e., 2 years for rodent species).

Table 3.2: Effect of Experimental Group Size on Minimum Incidence of a Toxic Effect

Number of animals in each group	Minimum percentage incidence in experimental groups that can be regarded as significant
10	40.0
50	8.0
100	4.0

3.2. Pharmacokinetics

Pharmacokinetics and/or toxicokinetics is the quantification of the time course of chemical absorption, distribution, biotransformation within the body and excretion of the chemical.

Absorption may occur through:

- Gastrointestinal tract
  - Stomach
  - Small intestine
- Lungs
- Dermal
- Injected

Absorption rates vary widely between species and depend on the substance being absorbed.

Distribution of a chemical depends on:

- Characteristics of blood flow, mass of tissue, affinity of tissue for chemical
- Storage depots
  - Plasma proteins
  - Fat
  - Bone
- Barriers
  - Blood/brain
  - Placental
  - Testicular
- Redistribution

The metabolism of a chemical affects its toxicity as it is converted to a less lipid soluble, which is usually, but not always, less toxic, and always easier to excrete. Metabolism is broken down into two phases: oxidation and synthesis. Metabolism occurs primarily in the liver, which may be the site of action in the case of an active metabolite.

Excretion of the chemical occurs through all the normal passageways of the body: urine (pH effect), bile, exhaled air, milk, saliva, sweat, tears, etc.

Mathematical models are used to describe and predict the body burden, duration of the chemical in the body, rate of absorption, distribution of the chemical in the body, rate and process of metabolism and biotransformation in the body, and the rate of elimination from the body.

See Figure 3.1 for a schematic of toxicokinetics.

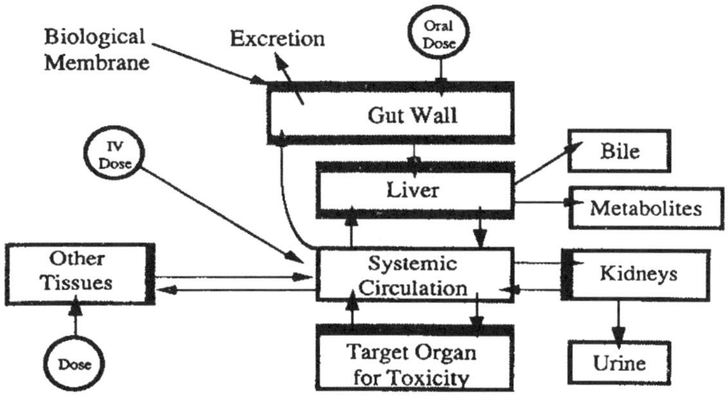

FIGURE 3.1
Toxicokinetics
Risk Assessment

4. DOSE-RESPONSE EVALUATION

The second step in conducting a human health risk assessment involves an evaluation of the relationship between exposure levels (and resulting doses) and the incidence of adverse effects and/or the severity of adverse effects and the time of onset of these effects.

Information on identified hazardous substances is evaluated and applied in the dose response analysis. The following table indicates the level of toxicity for a lethal dose of a hazardous substance.

Table 4.1: Toxicity Rating Chart

	Probable Oral Lethal Dose for Humans	
Toxicity rating or class	**Dose**	**For average adult**
1. Practically nontoxic	> 15 g/kg	More than 1 quart
2. Slightly toxic	5 - 15 g/kg	Between pint and quart
3. Moderately toxic	0.5 - 5 g/kg	Between ounces and pint
4. Very toxic	50 - 500 mg/kg	Between teaspoon and ounces
5. Extremely toxic	5 - 50 mg/kg	Between 7 drops and teaspoon
6. Super toxic	< 5 mg/kg	A taste (less than 7 drops)

4.1. Dose Response

The relationship between degree of exposure to a substance (dose) and magnitude of the substance-induced effects (response) is described by a dose-response curve. By definition, no response is seen in the absence of the substances. As the amount that is taken into the bodies of individuals in a population increases, the response becomes apparent and its frequency increases. When expressed as frequency of occurrence (or probability) within a population with increasing dose, the resulting graph is the familiar bell-shaped curve. This normal frequency distribution represents differences in susceptibility among individuals to chemicals which is known as biologic variation. The left and right tails to the frequency distribution represent hypersusceptible and resistant individuals in a population, respectively.

Hypothetically, dose-response curves fall into two groups; those in which no response is observed until some minimum (threshold) dose is present and those in which no threshold is apparent (nonthreshold). Most drugs and environmental chemical substances follow a threshold pattern. This dose-response curve indicates that certain individuals within the population are more or less susceptible to the chemical substance's toxicity than the majority. Nonthreshold dose-response curves are thought theoretically to be the case with some types of chemical carcinogens, i.e., any amount may produce a response, e.g., tumors.

Figure 4.1 shows graphic representations of three different dose-response relationships.

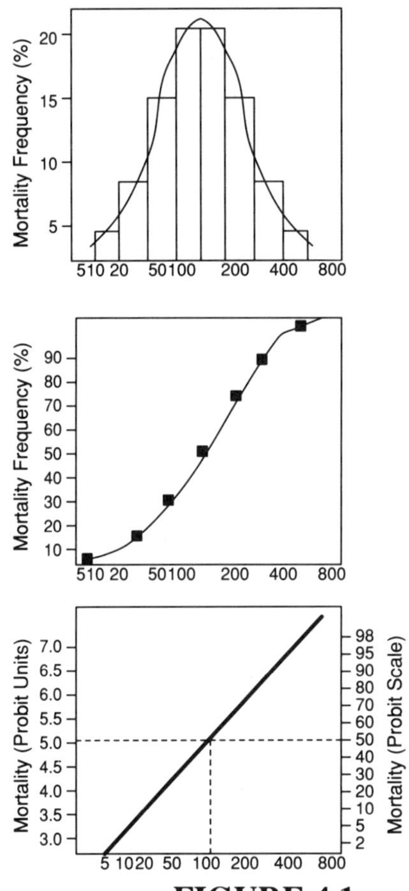

FIGURE 4.1
Dose-Response Relationships
Risk Assessment

4.2. Dose Extrapolation from High to Low Doses

For any given chemical and route of exposure, the severity and frequency of the effect (i.e., the response) generally increase with increasing doses. The type of adverse effect may also change with dose. For example, there may be no detectable effect of anesthetic gases at low doses, lethargy or headaches with increasing dose levels, sleep induction at higher doses leading to death at very high doses. An understanding of such a dose-response relation is critical to an assessment of risk. To the extent that the dose-response relation is not well defined, uncertainty about the estimated risk will exist.

Even for carcinogens which have been extensively examined for toxicity, adjustments must frequently be made for differences between laboratory conditions and those of actual human exposure. One such adjustment involves extrapolation of effects seen at the high doses used in animal experiments to the lower doses typical of human exposure. Higher doses are used in laboratory experiments, in part, to compensate for the small number of experimental animals used and to maximize the chance of observing a response.

Extrapolation of the dose-response curve, often referred to as risk extrapolation, is the major approach for regulating carcinogens and noncarcinogens in the environment. Risk extrapolation is the process of evaluating risks associated with low-dose exposure to a certain chemical substance, based upon the response observed at high doses.

Figure 4.2 shows an example of virtually safe dose estimated through low dose extrapolation.

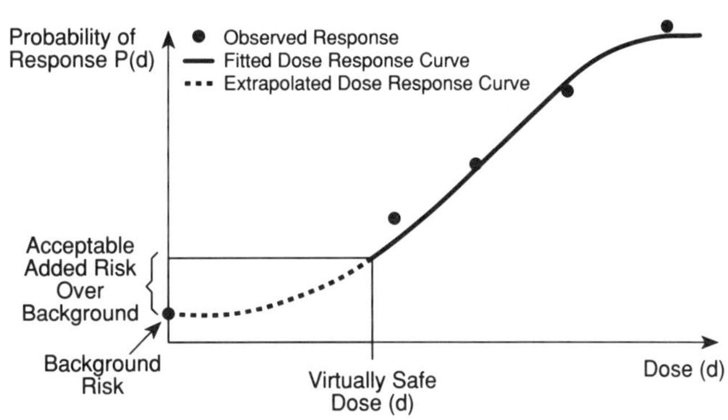

FIGURE 4.2
ESTIMATION OF A VIRTUALLY SAFE DOSE
BY LOW DOSE EXTRAPOLATION
Risk Assessment

4.3. Models

Scientists have developed several mathematical models to estimate low-dose carcinogenic risks from observed high-dose risks. Such models describe the expected quantitative relationship between risk and dose and are used to estimate a value for the risk at the dose of interest. The accuracy of the projected risk at the dose of interest is a function of how accurately the mathematical model describes the true, but unmeasurable, relationship between dose and risk at the low-dose levels. While most of the available models fit the experimental, high-dose data, the predicted risks at low doses may vary significantly among the various models.

None of the models is chemical-specific; that is, each is based on general theories of carcinogenesis or on certain statistical principles rather than on data for a specific chemical. The four types of dose-response extrapolation models are distribution, mechanistic, pharmacokinetic and time-to-tumor.

4.3.1. Distribution Models

The distribution models are based on the assumption that every member of a population has a critical dosage below which there will be no response to the exposure in question (i.e., a threshold). It is presumed that the critical dosage varies among individuals and that this variability can be described in terms of a probability distribution.

Log-probit model - assumes a normal logarithmic dose-response distribution and is an extrapolation of the line obtained by plotting the experimental data on a probit versus $\log_{10}$ dosage scale to an "acceptable" level of risk such as 10 (1/10,000 percent).

Mantel-Bryan model - is similar to the log-probit model; however, rather than extending the slope of the observed log dose-probit response, a fixed slope of one is used (based on empirical knowledge of experimental carcinogenesis data that the slope is usually much greater than one) starting from the upper confidence limit of the observed proportion of animals with tumors at the experimental exposure level.

4.3.2. Mechanistic Models

Mechanistic models are based on one of the presumed mechanisms of carcinogenesis. It is assumed that a tumor originates from a single cell.

One-hit model - is based on an assumption regarding the biologic mechanism of action for carcinogens, specifically that a tumor can be induced by the exposure of DNA to a single molecule of a carcinogen. This model assumes a linear dose-response relationship in the low dose region. The one-hit model will almost always yield the highest estimate of low-dose risk.

Gamma multihit model - is an extension of the one-hit model, which assumes that more than one hit is required at the cellular level to initiate the process of carcinogenesis.

Multistage (Armitage-Doll) model - is based on the biological justification that cancer is believed to be a multistage process that can be approximated by a series of multiplicative linear functions. The multistage model, which yields risk estimates either equal to or less than the one-hit model, is based on the multistage theory of the carcinogenic process that a cell must progress through several stages of development to form a cancer; and the rate of progression between some of these stages is influenced by a carcinogen.

Linearized multistage model - is most often used in risk assessments. This model assumes a linear dose-response relationship in the low dose region, which results in estimates of potential risk that are similar to those of the one-hit model. The linearized multistage model is used in EPA assessments for extrapolation to low doses of carcinogens. At the higher doses where experimental data exist, this model behaves like the multistage model and accommodates the nonlinearity often observed. At the lower doses from which the data are extrapolated, this model constrains the results to be linear and, thus, resembles the more conservative one-hit model. Other plausible models usually predict lower risks than the linearized multistage model.

Figure 4.3 shows dose-response curve extrapolations calculated by the above mathmatical models.

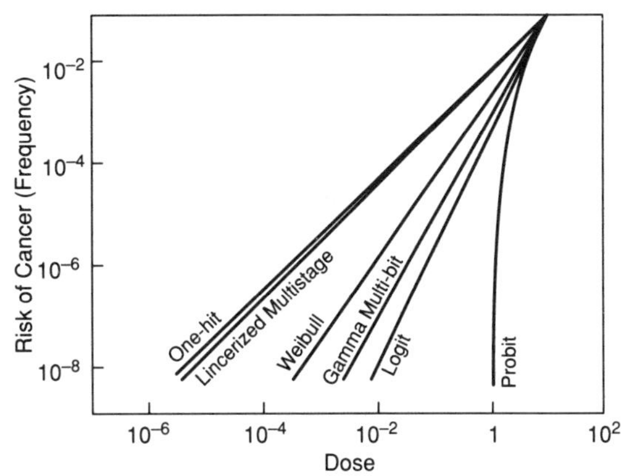

FIGURE 4.3
DOSE-RESPONSE CURVE EXTRAPOLATIONS
BY SIX MATHEMATICAL MODELS
Risk Assessment

4.3.3. Pharmacokinetic Models

Pharmacokinetic models estimate the amount of reactive metabolites formed and bound to DNA ("effective") dose and uses this value as the dosage rather than the administered dose.

4.3.4. Time-To-Tumor Model

The time-to-tumor model uses the time-to-observance in addition to the proportion of animals possessing tumors. The time-to-tumor is the time at which a tumor is detected or observed by palpation or by gross or microscopic examination of an animal at the time of death.

4.4. Reference Doses (RfDs) and Reference Concentrations (RfCs)

4.4.1. Development of RfDs/RfCs

In the past, chronic Acceptable Daily Intake (ADI) was used as an estimate of a daily exposure level for the human population, including sensitive subpopulations, that is likely to be without appreciable risk for adverse health effects during a lifetime. The inhalation RfC considers toxic effects for both the respiratory system (portal-of-entry) and for effects peripheral to the respiratory system (extrarespiratory effects). In general, the RfC is an estimate of a daily inhalation exposure of the human population (including sensitive subgroups) that is likely to be without an appreciable risk of deleterious effects during a lifetime. RfDs and RfCs have replaced the term ADI and specific RfDs and RfCs that have been verified by the USEPA are represented in Integrated Risk Information System (IRIS), and are used in evaluating noncarcinogenic effects.

4.4.1.1. Identification of No-Observed-Effect Levels for Noncarcinogenic Effects

Noncarcinogenic effects are thought to have a threshold dose below which the effect will not occur. The dose-response evaluation involves estimation of the threshold dose by determination of the no-observed-effect level (NOEL) from observations in exposed people or experimental animals. In some cases, all examined exposures produce an effect and only a lowest-observed-effect level (LOEL) is available. Not all effects of exposure to a chemical are adverse to health. Thus, the distinction is often made between a NOEL and a no-observed-adverse-effect level (NOAEL) and between a LOEL and a lowest-observed-adverse-effect level (LOAEL). The NOAEL is the exposure level at which there are no statistically or biologically significant increases in the frequency or severity of adverse effects between the exposed population and its appropriate control. Some effects may be produced at the NOAEL but are not considered to be adverse or precursors to specific adverse effects. In experiments with more than one NOAEL, the regulatory focus is primarily on the highest NOAEL, the highest exposure level without adverse effects. NOEL is the exposure level at which there are no statistically or biologically significant increases in the frequency or severity of any effect between the exposed population and its appropriate control. The NOAEL or NOEL is based on the highest dose that produces no effects in the most sensitive species/strain/sex.

4.4.1.2. Uncertainty Factors

NOELs and NOAELs are divided by uncertainty factors to obtain an RfDs/RfCs, which is used as a basis for risk characterization. Uncertainty factors generally are multiples of 10, with each factor representing a specific area of uncertainty inherent in the extrapolation of available data. An uncertainty factor of 10 is used to account for intrahuman variation and is intended to protect sensitive human subpopulations (e.g., elderly, children). An uncertainty factor of 10 is used to account for the extrapolation from animals to humans and is intended to account for the variability between humans and other mammals. In the absence of a chronic study, a chronic RfD/RfC can be derived from a subchronic study, in which case an additional uncertainty factor of 10 would be applied to account for the uncertainty of deriving a chronic RfD/RFC from subchronic data. In the absence of a NOAEL a LOAEL can be used to derive the RfD/RfC, in which case an uncertainty factor of 10 would be applied to account for the uncertainty associated with extrapolating from LOAELs to NOAELs. An additional uncertainty factor of 10 can be applied if the database for the chemical is incomplete (e.g. no reproductive studies).

Modifying factors, which can range from 1 to 10, are also applied to reflect a qualitative professional assessment of additional uncertainties. The use of safety factors or uncertainty factors was introduced in the early 1950s for substances such as food additives and pesticide residues in food. The safety factor approach was originated in 1954 by Lehman and Fitzhugh of the FDA, who indicated that variability in biological response across several species was usually in the range of two- or three-fold and did not appear to exceed ten. They also indicated that the variability among extensively outbred individuals and individuals of all ages and degrees of susceptibility (e.g., persons in the general population) appeared to be less than one order of magnitude.

4.4.2.Carcinogenic Effects - Cancer Slope Factors (CSFs)

For carcinogenic effects, USEPA uses a two-part evaluation in which the chemical is first assigned a weight-of-evidence classification, and then a slope factor is calculated. The available data on the chemical is evaluated to determine the likelihood that it is a carcinogen. First, the evidence is characterized separately for human studies and animal studies; then, the characterizations of the two types of data are combined. Based on the extent to which the chemical has been shown to be a carcinogen in experimental animals and/or humans, the chemical is given a provisional weight-of-evidence classification. USEPA scientists then adjust the provisional classification upward or downward, based on other supporting evidence of carcinogenicity.

The weight-of-evidence classifications are broken down into five categories:

Group A - Human carcinogen (sufficient human evidence)

Group B - Probable human carcinogen (B1, limited human evidence, or B2, sufficient animal evidence with inadequate human evidence)

Group C - Possible human carcinogen (limited animal evidence)

Group D - Not classifiable (no adequate human or animal evidence)

Group E - Evidence of noncarcinogenicity (no evidence for carcinogenicity in two animal species, or both human and animal studies)

The Upper-Bound Estimate is:

- Plausible upper limit to risk consistent with some proposed mechanism of carcinogenesis
- Not likely to be lower than the true risk, but likely to exceed the true risk (which could be zero)

A toxicity value is calculated in the second part of the evaluation. Based on the evaluation that a chemical is a known or probable human carcinogen, a toxicity value that defines quantitatively the dose-response relationship (i.e., the cancer slope factor) is calculated. Cancer slope factors (CSFs) are typically calculated for potential carcinogens classified as A, Bl and B2 based on the USEPA's categorization of weight-of-evidence of carcinogenicity. In deriving the CSF, the appropriate data set is identified. In the absence of human data of high quality, animal data is used. The preferred animal data are species that respond most similarly to humans (metabolism, physiology and pharmacokinetics); however, when no clear choice is possible, the most sensitive species is given the greatest emphasis. When no single study is available, and several studies collectively support the estimate, the geometric mean of estimates from all studies may be adopted as the slope.

The development of the CSF entails applying a model to the available data set and using the model to extrapolate from the relatively high doses administered to experimental animals (or the exposures noted in epidemiologic studies) to the lower exposure levels expected for human contact in the environment. These mathematical models are discussed in Section 4.3.

CSFs are used in risk assessments to estimate an upper-bound lifetime probability of an individual developing cancer as a result of exposure to a particular level of a potential carcinogen. The CSF is the result of application of a low-dose extrapolation procedure. Generally, CSFs are a plausible upper-bound estimate of the probability of a response per unit intake of a chemical over a lifetime and is presented as a risk per (mg/kg)/day.

Table 4.2: Slope Factors for Selected Chemicals

Chemical	Oral Route		Inhalation Route	
	Class	SF $(mg/kg/d)^{-2}$	Class	SF $(\mu g/m^3/d)^{-1}$
Benzene	A	$2.9 X 10^{-23}$	A	$8.3 X 10^{-6}$
Benzo(a)pyrene	B2	11.5	B2	$1.7 X 10^{-3}$
Chloroform	B2	$6.1 X 10^{-3}$	B2	$2.3 X 10^{-5}$
Polychlorinated biphenyls	B2	7.7	B2	ND
Trichloroethylene	B2	$1.1 X 10^{-3}$	B2	$1.7 X 10^{-5}$
Vinyl chloride	A	1.9	A	8.4

5. EXPOSURE ASSESSMENT

Exposure assessments are designed to determine the degree of contact an individual has with a chemical. Several factors will affect the extent of exposure, including the duration and frequency of the exposure, route of exposure, amount of chemical absorbed into the body by each route, and sensitivity of the population receiving the exposure.

The first step in the exposure assessment typically is to list all of the possible relevant sources of exposure. Generally, the list identifies the points at which the chemical enters the environmental medium along with estimated rates of entry. Exposure assessors also must identify the specific locations of the sources and the ways in which the substances involved are used.

The duration of the exposure usually designates the period of time over which the individual is exposed. An acute exposure generally involves one contact with the chemical, usually for less than a day. A chronic exposure involves exposure to a chemical for a substantial portion of the individual's lifetime (i.e., inhalation of pollutant emissions, ingestion of drinking water, ingestion of local produce, dairy, or meat). Exposures of intermediate duration are referred to as subchronic exposures.

As a result of exposure, a dose of an agent may occur when the agent enters the body and is absorbed into the bloodstream. The received (or systemic) dose is ordinarily measured as total weight of the agent absorbed divided by the body weight of the individual. If the degree of absorption is unknown, dose also may be expressed as total weight of agent administered divided by body weight. The units of dose are usually expressed as milligrams of agent/kilogram of body weight (mg/kg bw). It is also necessary to specify the time over which a dose is received, e.g., mg/kg bw/day.

For carcinogens, the standard practice of regulatory agencies is to average exposure over the entire lifetime. For noncarcinogenic compounds, the maximum daily dose (MDD) is often used for both acute and chronic exposures. This procedure is equivalent to assuming that maximum exposure occurs every day, and thus, will overestimate risks from intermittent exposures. Alternatively, exposure to noncarcinogenic compounds may be averaged over the total period of exposure (but not over a lifetime). The different phases of the exposure assessment are discussed below.

5.1. Sources of Chemical Releases/ Exposures

Chemical releases to the environment may occur during transportation, storage, and disposal of chemicals both before and after processing. Storage and transportation can entail "normal" levels of release to the environmental media and occasionally higher levels as a result of accidents. Manufacturing accidents may also cause a major release to the environment.

The major source for many chemical releases, however, is simply use of a chemical or of chemical-containing products; industrial, commercial, household and other consumer uses; deliberate applications in the environment (e.g., pesticides); and uses causing direct exposures, such as in food, drugs, or cosmetics.

Some common sources of contamination include:

- Agricultural runoff
- Pesticide runoff
- Accidental spills
- Surface spills
- Septic tanks
- Waste pits

Common sources of contaminant emissions include:

- Hazardous waste landfill
- Construction and excavation
- Accidental spill
- Stack and automobile emissions
- Toxic emissions
- Wind-blown dust
- Burning wood

5.2. Estimation of Source Characteristics and Exposure Concentrations

While it is often fairly easy to describe the source information needed to perform an exposure assessment, it is not always easy to obtain that information. It is difficult to predict the approach that will yield the most useful results with the least effort in a specific instance. Selection of the best strategy requires sound professional judgment in the absence of scientific certainty.

Site characteristics that may influence environmental fate include:

- Ambient moisture
- Ambient humidity levels
- Ambient temperatures
- Ambient wind velocity and direction
- Geologic characteristics
- Hydrologic characteristics
- Soil characteristics
- Topographic features of site and surrounding area

- Vegetative cover of site and surrounding area
- Watershed/land use characteristics

(See Figure 5.1)

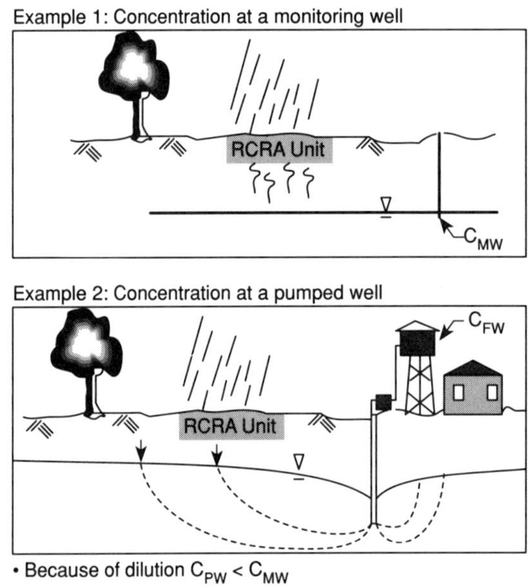

FIGURE 5.1
CONCENTRATION AT EXPOSURE
Risk Assessment

5.2.1. Direct Measurement

Information on sources and their strengths can be derived from direct physical measurements. Direct measurement is the preferred way of estimating source strength.

5.2.2. Indirect Measurement

A second measuring technique, mass balance, depends partly on direct measurement and partly on inference. The materials balance approach, a chemical engineering standard, can be applied in many situations. Losses to the environment are determined from the measured mass of a chemical in the finished product minus the mass of the chemical in raw materials; this loss is partitioned throughout the various environmental media.

5.2.3. Models

Engineering calculations and mathematical models may also be used for estimating chemical releases. It is possible to predict the emission of volatile chemicals from a landfill using models based on the physical-chemical properties of the chemicals. If enough data exist, it may be possible to predict a chemical's path through the environment with an environmental fate model.

5.2.4. Regulatory Limits

Regulatory limits are sometimes used to predict the source strength of chemicals released to the environment. For example, federal law stipulates that hazardous waste incinerators achieve a 99.9999% destruction and removal efficiency for PCBs. Models based both on this stipulation and on data about the components of the waste stream entering the incinerator are often used to predict releases to the environment although such an approach can lead to significant over- or underestimates.

The exposure concentration of chemical(s), which may be contacted over the exposure period, may also be determined by actual measurements or by mathematical modeling.

Mathematical models are often important in predicting chemical concentrations in the environment for future chemical releases when direct sampling of potentially affected media is not possible. Mathematical modeling incorporate site specific assumptions to predict the movement and concentration of the chemical through various environmental medium.

Emission models - are used to predict the rate at which chemicals may be released into the air from a source. Dispersion models predict associated concentrations in air at potential receptor locations.

Surface and ground water models - incorporate physical/chemical data on each chemical, along with hydrology data, to estimate water concentrations.

Fate and transport modeling - incorporate physical/chemical parameters to estimate exposure concentrations through the food chain and various biological media (e.g., food pathways, human milk). Factors such as biodegradation and bioaccumulation are considered in fate and transport modeling (see Figure 5.2).

Estimates of exposure concentrations often rely on sampling data obtained from the preliminary site characterization. Statistical issues which arise when the estimates of exposure concentrations are based on preliminary sampling data include the use of geometric versus arithmetic means and consideration of sampling data points, which are "outliers" or may represent "hot spots." Limitations and uncertainty of using such sampling data are discussed in Section 7.4.

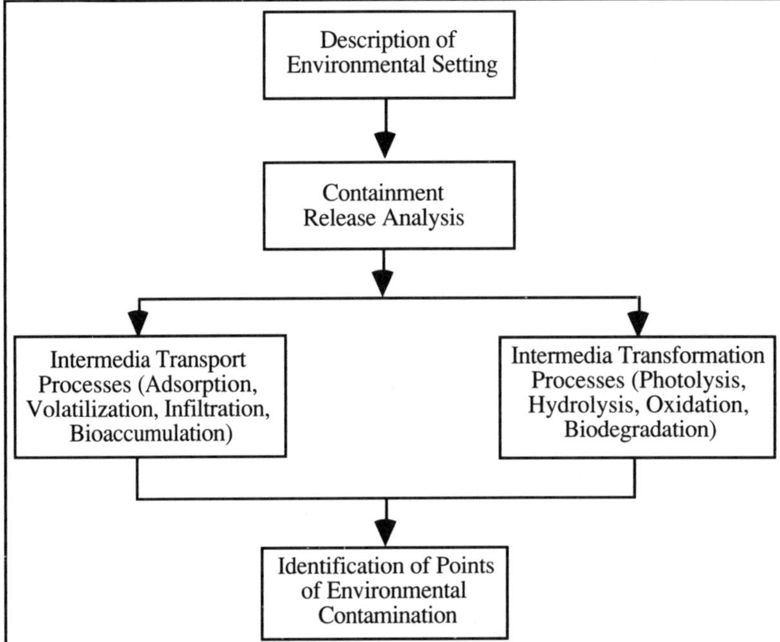

FIGURE 5.2
Environmental Fate and Transport
Risk Assessment

5.3. Identification of Population(s) at Risk

The populations at potential risk of exposure to chemical(s) are identified based on the data obtained from the sources of chemical releases/exposures and site characterization.

- Residential
- Worker

The exposed population must be defined and divided into classifications of receptors. For example, the difference in the definitions of Maximum Exposed Individual and Average Exposed Individual:

Maximum Exposed Individual:

- Lives at peak off-site receptor location
- Has a backyard vegetable garden
- Has a diary cow for personal milk production
- Consumes fish on a regular basis
- Breathes air entire life of project at peak receptor

Average Exposed Individual:

- Breathes air at an average exposure location
- Food chain exposure based on average in area
- More realistic exposure scenario

5.4. Identification of Exposure Pathways

Exposure pathways describe the course a chemical or physical agent takes from the source to the exposed population. The exposure pathway analysis determines the significant pathways of human exposure by linking the sources of exposure, location and types of environmental releases with population location and activity patterns. Figure 5.3 provides a flowchart that maps routes of exposure through environmental pathways for humans.

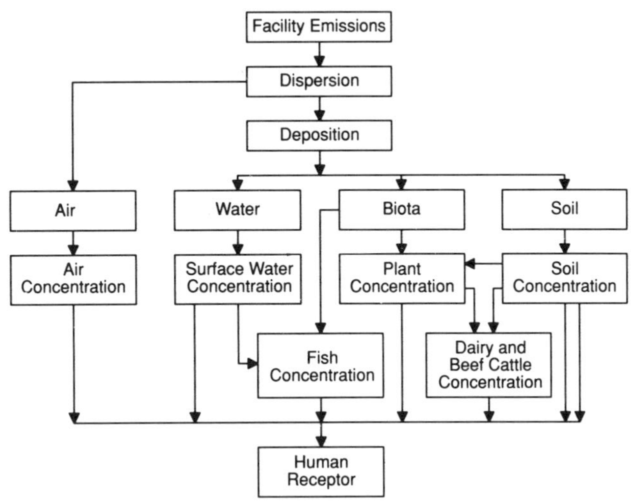

FIGURE 5.3
ENVIRONMENTAL PATHWAYS AND ROUTES OF EXPOSURE
Risk Assessment

5.4.1. Environmental Fate of the Chemical

An evaluation of the environmental fate and transport of chemicals is necessary to predict future exposures and to help link sources with currently contaminated media. This qualitative analysis is conducted to identify media that are receiving or may receive exposure to the target chemicals and evaluate the retention or accumulation within a single medium.

5.4.2. Demographics and Activities of the Population(s) at Risk

Demographics and activity of the exposed populations must be considered in an exposure assessment. The composition of the population will determine certain physiological parameters that affect exposure. For example, body weights differ among adult males, adult females, children and infants, as do their breathing rates, consumption of food and liquid, and body surface areas. Other characteristics of the population, such as the presence of hypersensitive individuals, also may need to be considered.

5.5. Routes of Exposure

The major routes of exposure for humans are ingestion, inhalation, and dermal absorption. Ingested contaminants may be absorbed in the gastrointestinal tract; inhaled contaminants may be absorbed in the lungs; and contaminants which contact the skin may be absorbed dermally. The degree of absorption among these routes differs depending in part on physical/chemical characteristics (e.g., particle size, lipid solubility) of a chemical.

The frequency and duration of exposure to a substance may influence the resultant toxicity. Some substances are readily eliminated from the body so that additional exposure doses at a later date will not accumulate to levels that would be sufficient to produce adverse effects; however, others may accumulate at varying degrees and thus chronic exposure may result in toxicity. Another factor that influences toxicity is the body's natural capability to repair insults at the molecular (DNA) level. These biological mechanisms to repair toxic insults at the DNA level vary from tissue to tissue and the capacity to repair toxic insults may be rate-limiting. Low levels of toxic insults may be handled by the body's repair mechanisms; however, higher levels of toxic insults may be overwhelming and thus result in toxicity.

5.6. Human Exposure Assumptions

The following is a list of parameters used in assessing human exposure to chemicals:

<u>Route-specific contact rate</u> - the amount of chemical contacted per unit time or event.

<u>Body weight</u> - the average body weight over the exposure period.

<u>Averaging time</u> - specific period of time over which exposure is averaged (pathway specific for noncarcinogenic effects and 70 year lifetime for carcinogenic effects.)

<u>Exposure time, duration, and frequency</u> - describes how often and for how long exposure will occur.

<u>Exposure concentration</u> - describes the average chemical concentration contacted over the exposure period. The total amount of chemical that enters the body per unit of time depends upon: the concentration of the chemical in the medium; amount of medium contacting or entering the body per unit of time; absorption of the chemical through the lungs, skin, gastrointestinal tract; the number of different media to which a person is exposed.

5.7. Calculations for Chemical Intake via Various Routes of Exposure

Once chemical concentrations in media, degree of contact with media, and absorption rates have been determined for a chemical, a dose is calculated. The two different measures of dose often used in a risk assessment are Maximum Daily Dose (MDD) and the Lifetime Average Daily Dose (LADD).

The MDD is used in toxicity assessments and relative safety assessments. In toxicity assessments, the NO(A)EL dose is divided by uncertainty factors to derive an ADI or RfD. The MDD to ADI ratio is used to assess whether the toxic effect experienced by animals is likely to be experienced by individuals receiving a specified dose. Relative safety assessments simply compare the MDD to the NO(A)EL without first incorporating the uncertainty factor. The resultant MDD to NO(A)EL ratio is usually termed the margin of safety.

For carcinogenic risk assessments, the LADD is used as a measure of "cumulative lifetime dose." For cancer induced by chemicals, risk is assumed to be a function of cumulative dose. Thus the risk from receiving one unit of a carcinogen for 1,000 days is assumed equal to that from receiving 10 units for 100 days.

The LADD is easily estimated by multiplying the average daily dose (ADD) by the fraction of the total lifetime that an individual is exposed (exposure days divided by days per lifetime).

6. RISK CHARACTERIZATION

6.1. Quantification of Hazard Indices

To assess risks of noncarcinogenic effects, the RfD/RfC is usually compared with the maximum daily dose (MDD) that a person might daily sustain for extended periods of time. The ratio of MDD to RfD/RfC is referred to as the Hazard Index (HI). Generally, if the MDD is less than or equal to the RfD/RfC (an HI < 1), no risk is assumed to exist for all members of the population. The MDD may even exceed the RfD/RfC without a significant risk arising, although, as the MDD approaches the experimental NOEL, the risk of toxicity may be significant. There is no clear demarcation between "safe" and "unsafe" exposure levels.

6.2. Quantification of Carcinogenic Risk

Carcinogens are considered by regulatory agencies to pose a finite risk at finite dose levels and the probability of developing cancer increases with increasing dose. While this suggests that any exposure to a carcinogen poses some risk of cancer, the probability may be extraordinarily small so that, for all practical purposes, no risk exists. The potency of the chemical, as measured by its slope factor, is multiplied by the exposure (LADD) to obtain the upper-bound probability of developing cancer. Regulatory agencies and other groups involved in risk management decision making must determine the tolerable level of risk based on risk management considerations that are not purely scientific.

Th quantitative expression for an Upper Bound on a lifetime risk of cancer may read:

$$\frac{\textbf{Risk}}{\textbf{Unit Dose}} \times \textbf{Dose} = \textbf{Risk}$$

Other information necessary to estimate risks for carcinogens:

- Weight-of-evidence of carcinogenicity
- Plausibility of linear no-threshold model
- Nature of exposure analysis (e.g., worst-case, typical)
- Sensitivity of results to alternative assumptions

  Example: Chloroform byproducts in chlorinated drinking water pose excess lifetime cancer risk of 1 per 100,000 to 1 per 10,000, assuming: Chloroform is a human carcinogen, based on observations in animals; humans are as sensitive as the most sensitive animal species, strain, and sex; risk is a linear function of dose; and no-threshold in dose response.

6.3. Uncertainty Analysis

Uncertainty analysis, also known as sensitivity analysis, is useful for defining a range of risks associated with a particular exposure. It is based on an examination of the effect of different assumptions on the outcome of a risk assessment. Conservative assumptions set the upper bounds on potential risks, while more moderate assumptions indicate the statistically most likely estimates. Another type of uncertainty analysis, computerized Monte Carlo analyses, involves the use of probability distributions for each exposure variable rather than a single point estimate. The computer performs thousands of iterative risk calculations, varying the exposure variables according to the probability distribution, to generate a range of potential exposures and risks. Monte Carlo analysis can also provide some statistical probability of occurrence of a given risk level, for example, identifying the 90th percentile risk level.

Components that contribute to uncertainty include:

- Shape of curve at low doses
- Qualitative effects at low doses
- Metabolic pathways
- Toxicokinetics: time course of a chemical and its metabolites in the body
- Differences among sexes, strains and species
- Animal to human dose equivalence
- Difference in susceptibility among humans

7. LIMITATIONS OF A HUMAN HEALTH RISK ASSESSMENT

Data limitations and uncertainties are inherent in the risk assessment process.

7.1. Characterization Or Toxicity Values

7.1.1. Dose Extrapolation

Toxicity values are derived by extrapolating the occurrence of adverse health effects under certain conditions of exposure, based on knowledge of adverse effects that occur under a different set of conditions (e.g., at different dose levels, in different species, and possibly by different routes of exposure). Carcinogenic risks at low doses are extrapolated from effects observed at high doses, using various mathematical models that produce widely varying estimates of carcinogenic potency.

7.1.2. Evaluation of Exposure to Mixtures of Chemicals

Exposure to two or more chemicals concurrently can produce toxicological effects that are independent, synergistic, antagonistic, or additive. Data are quite limited regarding interactions among constituents of a mixture of chemicals. Since there is not yet an approved quantitative methodology for the evaluation of exposure to a mixture of chemicals, most human health risk assessments assume an additive effect of multiple chemical exposures. Therefore, the risks for individual chemicals are summed to present a total cancer risk. Similarly, individual hazard indices may be summed for those chemicals that produce similar toxic endpoints.

7.2. Human Exposure Assumptions

In the absence of actual data, human health risk assessments often rely heavily on exposure assumptions, which are critical to the outcome of the assessment.

7.3. Mathematical Models

Virtually all risk assessments rely to some extent on mathematical modeling to predict or estimate the nature and magnitude of human exposures. These models vary considerably in complexity; some involve only a single, simple equation, whereas others are extremely sophisticated and attempt to relate numerous variables through the use of multiple mathematical relationships.

7.4. Analytical Methodology

Estimation of exposure concentrations are often based on laboratory results from samples obtained during the initial site characterization. The specificity of the analytical methodology is critical in accurately identifying the chemicals of concern and estimating exposure concentrations used in the human health risk assessment. When appropriate, specialized testing (speciation) should be considered to identify the individual contaminants.

Statistical manipulations that may be required on the sampling data used in the human health risk assessment include comparisons with background concentrations, estimates of upper confidence limits on means, and determinations of the probability of identifying hot spots. It is imperative that the sampling strategies must be appropriate for use in a quantitative risk assessment; because if inappropriate, even the strictest QA/QC procedures associated with the sampling strategy will not ensure the useability of sample results. Therefore, risk assessors should be involved in discussions concerning sampling strategy for conducting the site characterization and preliminary field investigations.

8. RISK MANAGEMENT

8.1. Use of Human Health Risk Assessment Information in Decision-Making

The results of a human health risk assessment are but one consideration in most environmental decision making. Risk management encompasses any of those activities necessary to reach decisions about whether an assessed risk requires reduction and the degree of reduction necessary. The factors considered by regulatory risk managers consider and heavily weight scientific evidence in favor of the highest risk estimate to humans, but also takes into account the requirements of applicable statutes and the policy precedents established within the responsible agencies. In addition to health risk considerations, economic impact, technical feasibility, public opinion and allocation of limited resources may be factored into risk management decisions.

9. RISK COMMUNICATION AND RISK PERCEPTION

Risk communication is an interactive process of exchange of information or opinion: often involves multiple messages about the nature of risk or expressing concerns, opinions, or reactions to risk messages or to legal and institutional arrangements for risk management (National Research Council, 1989).

Risk perception is the study of human beliefs regarding the magnitude and acceptability of risks. People's tolerance for different hazards is not simply, or even largely a function of the expert's judgment regarding their probability of occurring. People generally consider many other characteristics.

Tolerated risks:

- Voluntarily assumed
- Personal benefit high
- Scientists agree
- Not catastrophic
- Natural in origin
- Hazard not fearsome
- Common event

Nontolerated risks:

- Imposed by others
- No perceived personal benefit
- Scientists disagree
- Catastrophic
- Man-made
- Highly dreaded hazard
- Rare event

Risk communication is an important and integral aspect of risk management. The risk communication process can be considered successful if it can improve or increase the base of accurate information that decision makers use and if it can satisfy those involved such that they are adequately informed within the limits of available knowledge. However, it must be understood that since people do not all share common interests and values, the responses may not always lead to consensus about controversial issues or to uniform personal behavior; that is, a better understanding will not necessarily lead people to the same conclusion.

Key elements in improving risk communication include:

- Setting realistic goals
  — Establish practical goals that explicitly accommodate the political/ legal mandates and constraints bounding the process and the roles of the potential recipients of the risk messages; and clearly show the contribution of improved understanding of issues and actions
- Openness
  — Ensure effective dialogue with potentially affected outsiders
- Safeguarding balance
  — Ensure that risk messages are not, and do not appear to be, distorted
- Competence
  — Incorporate expertise on the risk subject matter and on risk communication

The contents of risk messages should consider:

- Audience orientation
  — To reflect the perspectives, technical capacity and concerns of the target audience.
- Uncertainty
  — To convey the level of confidence of estimates and the significance of scientific uncertainty; that is, the existence of uncertainty should addressed.
- Risk comparisons
  — To present useful risk comparisons that help convey the magnitude of a particular risk estimate that occurs in the same decision context and that has a similar outcome

(i.e. Comparison of similar risks, risks with benefit, alternatives, and same risk at different times with a regulatory standard).

- Completeness
  - To provide a complete information base that contains information on the nature of the risk, the benefits that might be changed if risk were reduced, the available alternatives, the uncertainty in knowledge about risks and benefits, and the management issues.

Seven cardinal rules of risk communication (Center for Risk Communication, Columbia University):

- Accept and involve the public as a legitimate partner
- Plan carefully and evaluate your efforts
- Listen to your audience
- Be honest, frank, and open
- Coordinate and collaborate with other credible sources
- Meet the needs of the media
- Speak clearly and with compassion

The following tables provide some examples of risk comparisons.

Table 8.1: Comparative Risks of Death

	Number of deaths per year	Lifetime risks
Motor vehicle accident	46,000	1/65
Home accident	25,000	1/130
Lung cancer deaths in smokers	80,000	1/12

Table 8.2: Risk Estimated to Increase Chance of Death in any Year by 0.000001

Activity	Cause of death
Smoking 1.4 cigarettes	Cancer, heart disease
Spending 1 hour in a coal mine	Black lung disease
Living 2 days in New York or Boston	Air pollution
Living 2 months in Denver on vacation from New York	Cancer caused by cosmic radiation
One chest X-ray taken in a good hospital	Cancer caused by radiation
Eating 40 tbs. of peanut butter	Liver cancer caused by Aflatoxin B
Drinking 30 12-oz. cans of diet soda	Cancer caused by saccharin
Living 150 yrs. within 20 miles of a nuclear power plant	Cancer caused by radiation

Table 8.3: Annual Fatality Rates per 100,000 Persons at Risk

Activity	Rate
Motorcycling	2000
Living (all ages)	1000
Aerial acrobatics (planes)	500
Smoking (all causes)	300
Smoking (cancer)	120
Fire fighting	80
Coal mining	63
Farming	36
Driving (motor vehicles)	24
Police work (nonclerical)	22
Rodeo performer	3
Fires	2.8
Floods	0.06
Lighting	0.05
Meteorite	0.000006

Table 8.4: Calculated Lifetime Risks of Cancer to an Individual Exposed Over 70 Years

Chances per Million	
5,000	Average household exposure to radon
4,640	Pesticide residue on fresh foods
720	Consumption of 1/2-pound of fish per week from Lake Ontario
250	Peanut butter consumption by children, due to potential Aflatoxin
180	Acceptable benzene levels
20	Average diagnostic medical X-ray
4	Frequent flyer radiation exposure 20,000 miles per year
2.4	Drinking and showering with chlorinated water

- Project Summary -

UTILIZING A RULE-BASED EXPERT SYSTEM FOR MINIMIZING INDUSTRIAL HAZARDOUS WASTE: A DEMONSTRATION PROJECT

Anthony G. Collins
Department of Civil and Environmental Engineering
Clarkson University

Joseph V. DePinto
Department of Civil Engineering
State University of New York at Buffalo

ABSTRACT

Industries in New York State and across the United States need to implement waste minimization programs. The highest priority for these management programs is the elimination and reduction of wastes, then, any waste which cannot be eliminated or reduced should be evaluated to determine if any recycling opportunities exist either on- or off-site. Due to the complex interaction of production, process, and maintenance decisions at a typical manufacturing plant, a myriad of alternatives exist for effective waste reduction. The **WA**ste **M**inimization **M**anagement **A**dvisory **S**ystem (**WAMMAS**) is a computer program developed to assist the waste minimization efforts of industrial concerns. **WAMMAS** accomplishes this objective through a combination of tutorial activities and reports of successful waste minimization techniques. The program also presents an innovative approach to the industrial manufacturing decision making process through an expert system management aid to help account for the environmental impacts of an industrial process. Five components make up the **WAMMAS** system. The **Main** component is an introduction to waste minimization which reviews the concepts and terms involved with reducing the waste generated by an industrial process. The **Waste Audit Tutor** is an interactive tutorial that educates the program user on recommended waste audit procedures. Reduction strategies for various processes used by New York State industries are given in the **Waste Reduction Information** component and the **Case Studies** component documents successful waste minimization projects. The last component, the **Expert System Examples**, demonstrates the feasibility of the application of expert system technology to manage the hazardous waste generated by an industrial process. The advisory system has been developed to be used by a range of individuals, from owner/operators of small industrial concerns to the environmental engineering specialists employed by major industries. The program can be obtained from:
New York State Center for Hazardous Waste Management
State University of New York at Buffalo
Jarvis Hall 207
Buffalo, New York 14260
(716) 636-3446

INTRODUCTION

The highest priority for a hazardous waste management program is the elimination of wastes. Any waste which cannot be eliminated should be reduced or then investigated to determine if any recycling opportunities exist either on- or off-site. Due to the complex interaction of production, process, and maintenance decisions at a typical manufacturing plant, a myriad of alternatives exist for effective waste reduction.

While many segments of the community (*e.g.* industrial, legislative, and academic) agree that currently suitable techniques exist to control, monitor, and reduce the production and disposal of industrial hazardous materials, little has been done to introduce these known techniques or identify areas requiring innovative approaches. One of the primary reasons for this situation is the lack of a real understanding as to how a waste minimization audit can identify waste minimization alternatives. A less obvious reason is the reluctance of many industries to allow external assessment of their production processes by knowledgeable waste reduction experts in what are often viewed as sensitive areas. This attitude slows the transfer of existing technology and the development of new approaches.

Rule-based expert systems are one of the new techniques available that have possible application to waste minimization. Potentially these systems can provide a rational input to production decisions regarding the control and minimization of hazardous wastes. Expert systems could be constructed from a knowledge base consisting of a waste minimization audit, applicable regulations, and production alternatives. The system output will provide management with production decision guidance regarding hazardous waste issues.

To meet the goals of improving technology transfer and evaluating the applicability of expert system technology, the **WA**ste **M**inimization **M**anagement **A**dvisory **S**ystem (**WAMMAS**) was developed under the sponsorship of the New York State Center for Hazardous Waste Management. This computer program is intended to focus industrial attention on reducing waste at its source rather than through the traditional, end-of-pipe techniques (treatment, storage, and disposal). A combination of tutorial activities and reports of successful waste minimization techniques meets the technology transfer objective. The program also presents an innovative approach to the manufacturing decision making process through expert system management aid examples that help account for the environmental impacts of an industrial process.

OBJECTIVES AND SCOPE OF THE STUDY

The development of **WAMMAS** is best understood when the chain of events influencing its creation are described. The original project proposal submitted to the New York State Center for Hazardous Waste Management called for 1) implementation of a waste minimization audit at ALCOA's Massena facility, accumulating information regarding relevant legislation and the identification of process alternatives to minimize waste , 2) development of a rule-based expert system using the information gained under the first objective as the database, 3) evaluation of process alternatives that would minimize hazardous waste generation through the use of the rule-based expert system, and 4) evaluation and documentation of the methodology of the approach.

At the request of the NYS Center for Hazardous Waste Management, a significant change to the original project proposal was incorporated. In addition to demonstrating the feasibility of the expert system approach to minimizing hazardous waste at an industrial site (ALCOA), a second component, a tutorial, was to be incorporated to assist with the dissemination of hazardous waste management principles to all New York State industrial generators of hazardous waste.

BACKGROUND

Many economic, regulatory, and legal incentives exist to motivate an organization to undertake a waste minimization program. Raw material, waste disposal, and energy costs continue to rise. Present state and federal regulations require a waste minimization program to be in place. This policy was announced by the federal government in the Hazardous and Solid Waste Amendments (HSWA) to the Resource Conservation and Recovery Act (RCRA) in 1984.

The New York State Department of Environmental Conservation has a set of similar policies via the 1987 Organization and Delegation Memorandum #87-13. These policies have lead to regulations calling for a waste minimization program to be in place to reduce the volume and toxicity of wastes generated. Other regulations restrict the dumping of industrial by-products in landfills. The list of compounds requiring permits is increasing as more information becomes available on these substances. There are also laws in place which make any party who handles hazardous substances liable for its ultimate disposal. The incentives to develop a comprehensive waste minimization program are, therefore, numerous and obvious.

How does a waste minimization program work? Traditionally industries treated and disposed of wastes generated by their process almost as an afterthought to their operation. Wastes were collected as they arose from the industrial process and were often treated *en masse*. These end-of-pipe wastes were generated by industries without regard to minimization and were typically high volume, dilute, and complex which made them difficult and costly to treat. The goal of hazardous waste management is to encourage management and staff to think in terms of a waste management hierarchy. The priorities are ranked as follows.

1. Waste Elimination
2. Waste Reduction
3. Recovery and Recycling
4. Treatment
5. Storage and Disposal

The first two priorities are referred to as source reduction. Together with the third priority, recycling, they define waste minimization. Only as a last resort should treatment, storage, and disposal techniques be utilized.

Source reduction is any activity that reduces or eliminates hazardous waste where it is generated, usually within an industrial process. This elimination or reduction may be accomplished in a number of ways:

- input material changes,
- process technology alterations,
- product substitution/modification, and
- good operating practices.

Recycling includes any recovery and reuse of materials either on- or off-site prior to treatment, storage, or disposal. Source reduction and recycling are effective tools for saving money, reducing liability, complying with regulatory limits, and protecting human health and the environment.

METHODOLOGY

WAMMAS has been developed to be used by a range of individuals, from owner/operators of small industrial concerns to the environmental engineering specialists employed by major industries. The advisory system demonstrates how a materials balance, which attempts to account for all inputs (raw and recycled materials and water) and outputs (product, processing losses, waste streams, treatment discharges, treatment residuals, etc.) for a given production system, is at the heart of a waste audit. A waste audit in turn should identify, amongst other things, the sources, quantities, and types of waste being generated. After reviewing this material, the advisory system user is then in a position to construct a materials balance for their own process, obtain information from the materials balance regarding their most important areas of concern and receive advice from the system not only to initiate a waste minimization strategy, but to also obtain constructive examples and ideas about waste minimization opportunities.

To provide a logical flow to this information, **WAMMAS** was divided into three sections specifically designed to tutor the user about waste minimization. The first section walks the user through a waste audit example, stopping periodically to interact with the user or highlighting key aspects of the audit procedure. Helping the user develop their own waste reduction ideas is the goal of the second section. The user is exposed to several waste generating processes and the applicable waste reduction techniques for those processes. The third section contains abstracts of several successfully completed waste minimization projects. After reviewing this portion of the program, the user would be familiar with waste audit procedures, waste reduction alternatives, and case studies of proven waste minimization applications.

To achieve the project objective of exploring expert system technology, a waste minimization audit was undertaken at the ALCOA Massena plant in a manner consistent with the currently accepted procedures for such audits. The materials balance model developed from the audit together with relevant legislation and process options formed input data for the decision making process of the expert system. After formulation of the relevant data into production rules, the expert system to assist production scheduling to minimize one waste generating source was completed following verification and validation. The component of the study was completed when the impact of using an expert system was evaluated as an aid to minimize hazardous waste generation at an industrial site.

The PCB sampling and analysis program at ALCOA took longer than initially estimated and the opportunity was taken to approach other industries with a view to evaluate their potential for expert system applications. These industries included Occidental Chemical Company in Niagara Falls, NY, Dresser-Rand, Olean, NY, a manufacturer of compressors, Frito-Lay, Kirkwood, NY, a food processing company, and Coyne Textile, Syracuse, NY, an industrial laundry.

Initial investigations of these industries revealed that while, in general, that the individual unit process operations remained constant within each industry, the waste stream concentrations emenating from these processes would fluctuate. The question then arose, "what causes fluctuations in waste streams at these facilities if the specific processes do not change their operating parameters?" The answer lay in the fact that there were situations where either concurrent operation of intermittent processes operating together or the properties of the raw materials used by the processes varied, both causing fluctuations in pollutant loadings to a wastewater treatment stream or facility potentially violating a discharge limit. Therefore, if the environmental impact of each individual process could be determined, then it would also be possible through the use of an expert system to take into consideration these environmental impacts along with other manufacturing decisions and schedule the processes to avoid these violations or peaks. This essentially became a scheduling problem, much like the ALCOA situation.

There became two necessary criteria for identifying potentially useful expert system applications of this type. The first is that the industry must have a waste stream that fluctuates in strength. Secondly, good environmental data must be available to develop correlations between the process operation and environmental consequences. The ALCOA scenario had both the criterion. Also, Coyne Textile Services had a situation which could potentially be developed. These two industries were selected to be the expert system demonstration examples.

HARDWARE/SOFTWARE ENVIRONMENT

The hardware requirements for **WAMMAS** are an IBM PC-AT or compatible computer with a harddrive, 3.5 inch disk drive and a VGA screen. The program requires approximately 3 MB of disk space during installation and approximately 2 MB once installed. **WAMMAS** is written in C and the expert system inferencing engine **CLIPS** (the **C** **L**anguage **I**ntegrated **P**roduction **S**ystem, a NASA product) is embedded within the C program. C allows a level of flexibility not available with other programming languages and also caters well to graphics. Currently a large number of commercial software packages are programmed in C for these reasons. A choice was made to steer clear of commercial software and program the graphics portion of the program to avoid royalty expenses and to maintain control over program development. Not only does this provide flexibility to design the system as desired, but will also allow easy alteration and updating of the system in years to come.

RESULTS

Five components make up the **WAMMAS** system. The **Main** component is an introduction to waste minimization which reviews the concepts and terms involved with reducing the waste generated by an industrial process. The **Waste Audit Tutor** is an interactive tutorial that educates the user on the waste audit procedure. Reduction strategies for various processes used by New York State industries are given in the **Waste Reduction Information** component and the **Case Studies** component documents successful waste minimization projects. The last component, the **Expert System Examples**, demonstrates the feasibility of the expert system approach to manage the hazardous waste generated by an industrial process. Each component of **WAMMAS** is discussed further.

MAIN COMPONENT

The **Main** component of **WAMMAS** serves a number of purposes. A waste minimization novice is exposed to the terms and concepts involved with the process of waste reduction. Due to the emerging state of waste minimization technology, the use of terminology is mixed. Therefore, the introduction of **WAMMAS** acquaints a user, even one accustomed to waste minimization, with the terms and concepts used in the **WAMMAS** system. Once familiar with the material, this section can be skipped during subsequent tutorial sessions. All other components of **WAMMAS** are accessed from this component.

The concept of a waste management hierarchy is introduced or reaffirmed in this component. The hierarchy emphasizes that the elimination of waste from an industrial process should be evaluated as the highest priority; waste should then be reduced to the maximum extent feasible. When the alternatives for eliminating and reducing industrial wastes are exhausted, recycling options should be considered. **WAMMAS** defines recycling as any on- or off-site use, reuse, or reclamation of a material. One option for industries is to market their waste as a raw material for another industry. This process of finding alternative markets for waste is attractive considering that an expense (the cost of waste disposal) is turned into a profit (the market value of the waste). Not only is it economically practical, but it is environmentally appealing even if no profit is made. The liability associated with waste disposal will be eliminated. As a last resort, the hierarchy stresses proper treatment and disposal practices. Treatment and disposal options are not covered by **WAMMAS**.

WASTE AUDIT TUTOR COMPONENT

A waste audit is the necessary initial undertaking to provide a detailed analysis of a production process that eventually guides the selection of waste minimization options. **WAMMAS** outlines four steps for undertaking a waste audit. The first step is to understand the process from an environmental standpoint. Information on raw materials and finished products is usually easily obtained. However, all streams entering or leaving the industrial process must be accounted for. With this in mind, the next step is to gather input information. It follows that the output information will then be collected. Once all of the information is assumed to have been accumulated, a mass balance is performed on the process. The mass balance ensures the auditor that all waste streams have been tracked through the industrial process. When the four steps described above are completed, the waste audit should:

- define the sources, quantities, and types of waste being generated,
- bring together information on unit processes, products, raw materials, water usage, and waste generation,
- highlight process efficiencies and inefficiencies,
- identify areas of wastage and loss problems,
- help set targets for waste reduction,
- permit development of cost-effective waste management strategies, and
- increase employee knowledge of the production process and concern for waste minimization.

While the procedures for conducting a waste minimization audit have been specified for generic production process situations, site-specific auditing requires considerable interpretation for effective implementation.

The **Waste Audit Tutor** component is designed to aid the user in tailoring the approach to their own situations. The user is guided, step by step, through an example of an industrial waste audit on a steel pickling process. This component interacts with the user in several ways. Questions with multiple choice answers are asked. If an incorrect response is given, a helpful hint is displayed and the user is prompted to try again. After the input and output information is gathered for the steel pickling example, the user is asked to solve a mass balance. The solution is a simple mathematical one. It is not the intent of **WAMMAS** to carry out an extensive waste audit, but to show the user that the mass balance is the heart of the waste audit by accounting for all inputs and outputs associated with an industrial process.

The information acquired during the waste audit is collected from many in-plant sources. The **Waste Audit Tutor** guides the user to a number of these sources. Use of a consultant to develop a data collection strategy for compiling input and output data may be unnecessary because most of the information needed for the mass balance can be acquired with a careful search of purchasing records, flow charts, hazardous waste manifests, and employees, just to name a few sources. The waste audit should greatly improve the understanding of an industrial process and serves as a basis for developing waste minimization alternatives.

WASTE REDUCTION INFORMATION COMPONENT

The **Waste Reduction Information** component was developed to transfer information about proven waste reduction techniques. This component of **WAMMAS** has a data base consisting of waste reduction options for a number of industrial operations. Reduction alternatives are provided for emissions of all phases (gas, liquid, or solid) if the technology exists for the operation.

The user selects the unit operation he/she is interested in from a list of processes contained in the database. The program then asks the user for specific information on their industrial process. This information is cross-referenced with the process information contained in the database. The user is then asked to compare his/her process to the process in **WAMMAS**' database. The program checks for inconsistencies between the user's process and the process contained within the database. Waste reduction suggestions are then given for streams which the user has matched, then reduction alternatives are given for streams which the user has not identified. For example, the program may say, "You have indicated that your bath process has only two output streams, are you aware that drag-out of residual solution is considered an output even though it is not measured? Drag-out can be reduced by optimizing the dip time or by decreasing the solution viscosity."

A summary of waste reduction options is displayed at the end of the interactive session. The user also has the option of computing a simple mass balance on the unit operation. The importance of the mass balance is again heavily stressed in this component of **WAMMAS**.

CASE STUDIES COMPONENT

The purpose of the **Case Studies** component is to transfer information on successful waste minimization projects. Each case study has the following information:

- company name,
- type of industry,
- material(s) subject to reduction,
- economic information (payback time, total cost),
- abstract of how the waste minimization technique was applied, and
- reference number indicating where more information can be obtained.

There is an emphasis on the industrial application of the five waste minimization techniques (input material substitution, technology modification, product alteration, good housekeeping, and recycling). When the case study is accessed, the appropriate waste minimization technique is highlighted at the bottom of the screen.

Each case study can be accessed by either the waste minimization technique used or by the material subject to reduction. Once the pathway is selected, a list of appropriate industries is displayed. The user can then choose the industry of interest.
The user also has the option of viewing the definition of any of the five waste minimization techniques. The definitions of the techniques used by **WAMMAS** are given to avoid any confusion and are included in this section to emphasize their application to industry.

EXPERT SYSTEM EXAMPLES COMPONENT

Good operating practices or the efficient management of an industrial process involve procedural, administrative, and organizational activities are one of five waste minimization techniques. Expert systems can contribute to this area of waste minimization. The expert system technology evaluation was originally a unique project, however, incorporation into **WAMMAS** expands the scope of educational benefits provided by the program. Anyone reviewing **WAMMAS** now has the option of learning about expert system technology as it can potentially be applied to waste minimization, thus providing one more alternative for the reduction of waste generation.

The expert system development could have taken one of two directions. A "feedback" type of system which would have diagnosed a problem with production wastewater and then decided what corrections could be made to eliminate the problem. Or a "feedforward" system which would decide on the best operational scenario given a set of operating conditions. The "feedforward" system was chosen because it conforms with the waste minimization theme of preventing pollution at the source.

As indicated previously, the development of the last component, the **Expert System Examples**, involved two industries, ALCOA-Massena and Coyne Textile Services. Both the expert system examples in **WAMMAS** are designed to help control fluctuations in the mass flow rate of contaminants from multiple wastewater sources to a discharge point. If treatment occurs before discharge, then a consistent mass flow rate will maximize the treatment efficiency and minimize the potential for discharge violations as would be the case for situations where treatment does not exist. For example, if two processes which both generate excessively high concentrations of a contaminant are running simultaneously, then a possible solution may be to simply schedule the processes to run at different times. The expert systems use this approach to check for violations in waste generation, investigate alternatives for correcting the problem, and disclose their conclusions to the user.

CONCLUSIONS

Reduction of industrial waste generation through a well planned waste minimization program is possible if industrial attention is focused on source reduction and recycling instead of the traditional methods of treatment, storage, and disposal. The knowledge and skills of industrial plant operators, engineers, and managers need to be continually developed and expanded to operate processes in an environmentally efficient way. Industries require educational innovation as well as technical innovation to develop alternative waste minimization strartergies.

A system such as **WAMMAS** has potential as an improved training method. **WAMMAS** provides a learning environment that is interactive and nonthreatening as well as having the capability for continual updating. This environment also allows the user to receive helpful, friendly advice. The **Expert System Examples** component demonstrates the potential of rule-based expert systems as aids in the management of waste minimization strategies. The focus of the two demonstration systems has been on scheduling. Two criteria were identified as being important to a successful implementation. The industry must have a waste stream that fluctuates in strength and environmental data must be available to develop correlations between process operation and environmental consequences.

SIGNIFICANCE OF THE RESEARCH

There are two significant aspects to this research project. First, a computer tutor has been produced that will facilitate the transfer of knowledge about waste minimization programs in general, environmental audits in particular, and a number of successful specific waste minimization projects. Second, it has been demonstrated that rule-based expert systems provide a useful and innovative technology to the management of complex industrial operations generating waste from multiple raw material sources and processes.

ACKNOWLEDGEMENTS

This work could not have been accomplished without the dedicated efforts of Donald E. Smith, Jr. and the interest and cooperation of ALCOA-Massena and Coyne Textile Services.

BIBLIOGRAPHY

The following presentations and publications have resulted from this project to date:

DePinto, Joseph V. and Collins, Anthony G., *WAMMAS - Waste Minimization Management*

Advisory System, New York State's Third Annual Hazardous Waste Reduction Conference, Albany, New York, June 1990.

Collins, Anthony G., DePinto, Joseph V., Smith, Jr., Doneld E. and Bristol, Lynnette, *Waste Minimization Advisory System*, 63rd Annual Conference Water Pollution Control Federation, Washington, DC, October 1990.

Collins, Anthony G., Smith, Jr., Donald E. and DePinto, Joseph V., *WAste Minimization Management Advisory System*, New York State's Fourth Annual Waste Reduction Conference, Albany, New York, June 1991.

Smith, Jr., Donald E., Collins, Anthony G. and DePinto, Joseph V., *WAste Minimization Management Advisory System*, *Proceedings*, 23rd Mid-Atlantic Industrial Waste Conference, University of Pittsburgh, Pittsburg, Pennsylvania, 295-305, June 1991.

Collins, Anthony G., Smith, Jr., Donald E. and DePinto, Joseph V., *A Computer Tutor for Waste Minimization*, Proceeding of Symposium: Educating Engineers for Waste Minimization, AIChE Summer National Meeting, Minniapolis, Minnesotta, August 1992.

Appendix

Instructor's Course Guides

Linearized Multistage Model

$$P_d = 1 - e^{-q_0 + q_1 d + q_2 d^2 + \ldots 1_k d^k}$$

- P_d is the risk at dose d
- e is the base of natural logarithms (2.71828...)
- $q_0, q_1, \ldots q_k$ are non-negative parameters estimated by the model based on the experimental data
- k is the number of stages in the model, generally one fewer than the number of experimental dose groups
- The Cancer Potency Factor, or Slope Factor, is the upper 95% statistical confidence limit on the linear term (q_1) of the best fitting multistage model (hence the term linearized multistage model)

EPA Weight of Evidence Categories

- Group A — Human carcinogen (sufficient human evidence)
- Group B — Probable human carcinogen
(B1 - limited human evidence, or
B2 - sufficient animal evidence with inadiquate human evidence)
- Group C — Possible human carcinogen
(limited animal evidence)
- Group D — Not classifiable
(no adequate human or animal evidence)
- Group E — Evidence of Non-Carcinogenicity
(no evidence for carcinogenicity in two animal species, or both human and animal studies)

EPA Weight of Evidence Categories

- Upper-Bound Estimate
  - Plausible Upper Limit to Risk Consistent with Some Proposed Mechanism of Carcinogenesis
  - Not Likely to Be Lower than the True Risk, but Likely to Exceed the True Risk (which Could Be Zero)

Issues in EPA's Derivation of Cancer Potency Factors

- Selection of data set
- Choice of low-dose extrapolation model
- Choice of interspecies scaling procedure

Step 2
Dose-Response Evaluation

- Data from human and animal studies reveal relationship between *exposure levels* (and resulting *doses*) and,
  - *Incidence* of adverse effects and/or
  - *Severity* of adverse effects
  - *Time of onset* of adverse effects

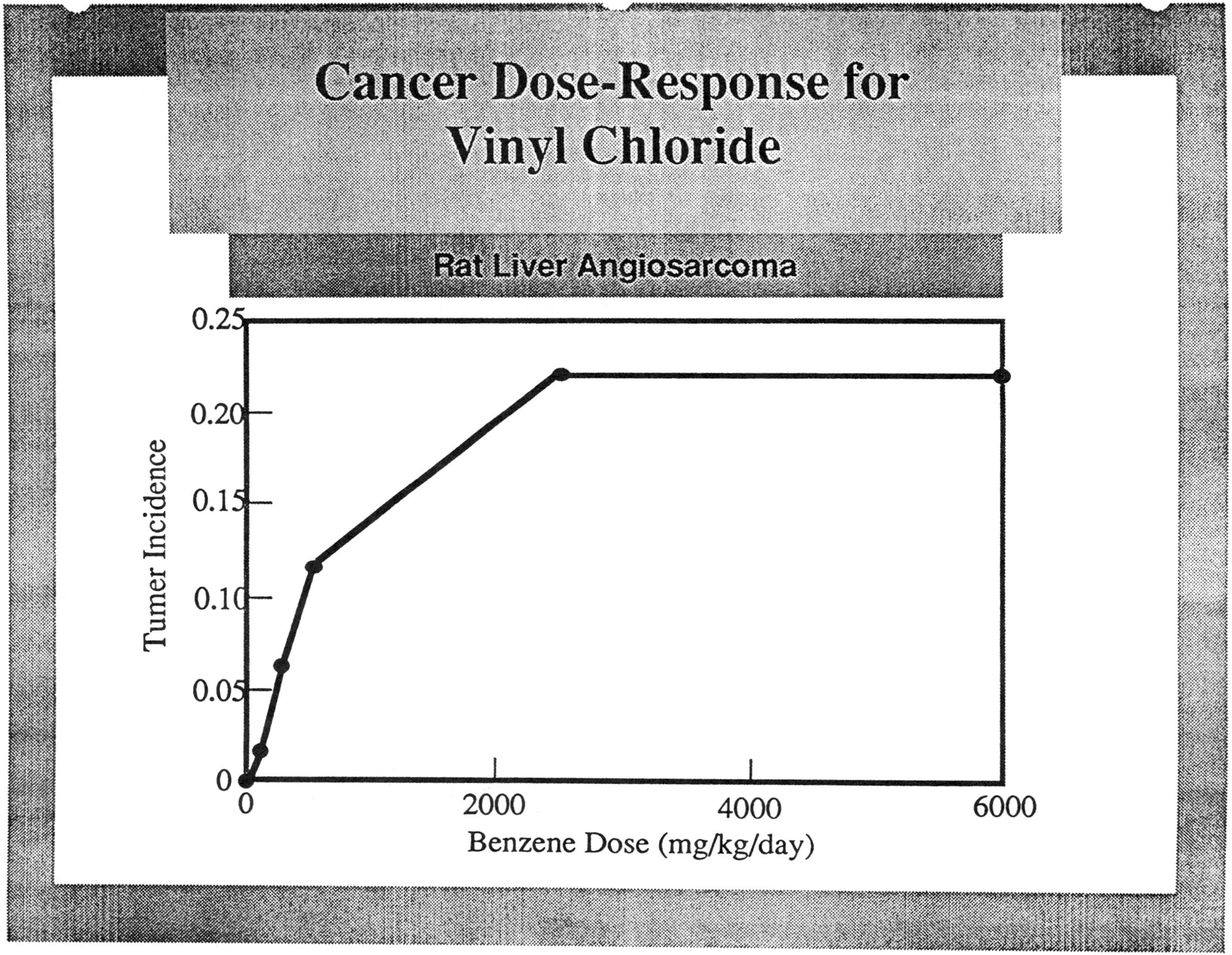
Cancer Dose-Response for
Vinyl Chloride
Rat Liver Angiosarcoma
Tumer Incidence
0.25
0.20
0.15
0.10
0.05
0
0
2000
4000
6000
Benzene Dose (mg/kg/day)

Dose-Response Curve Extrapolations by Six Mathematical Models

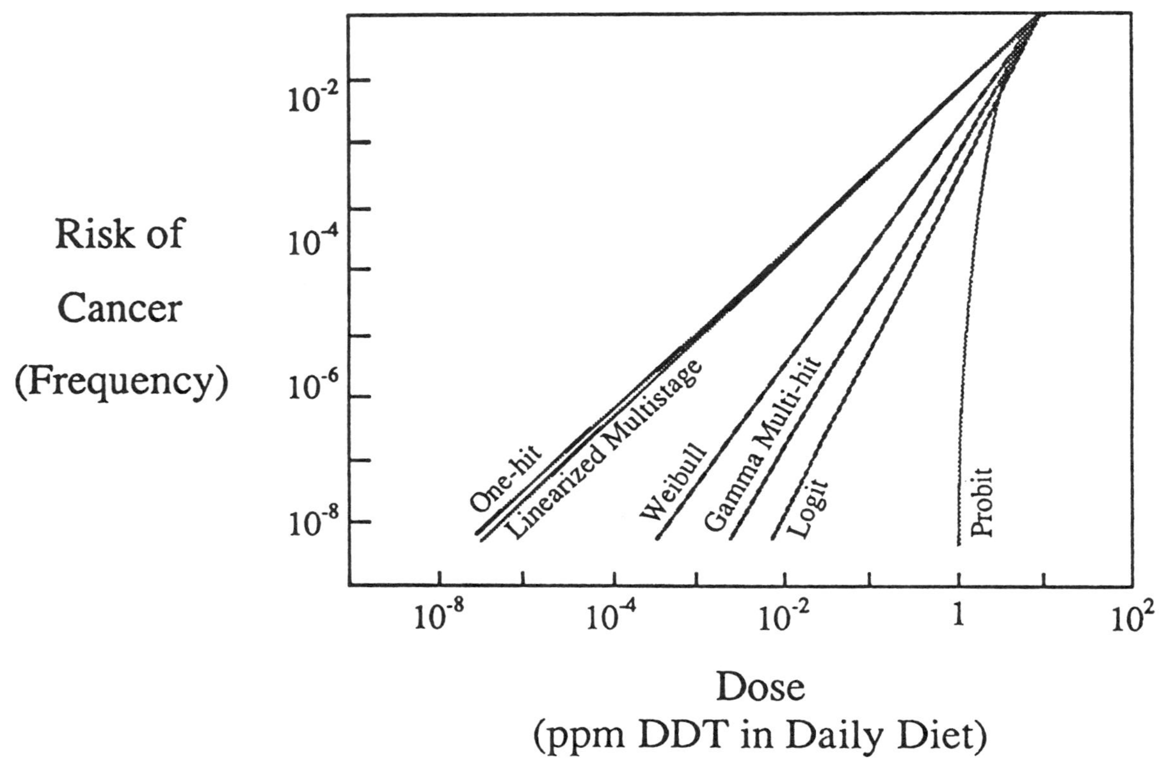

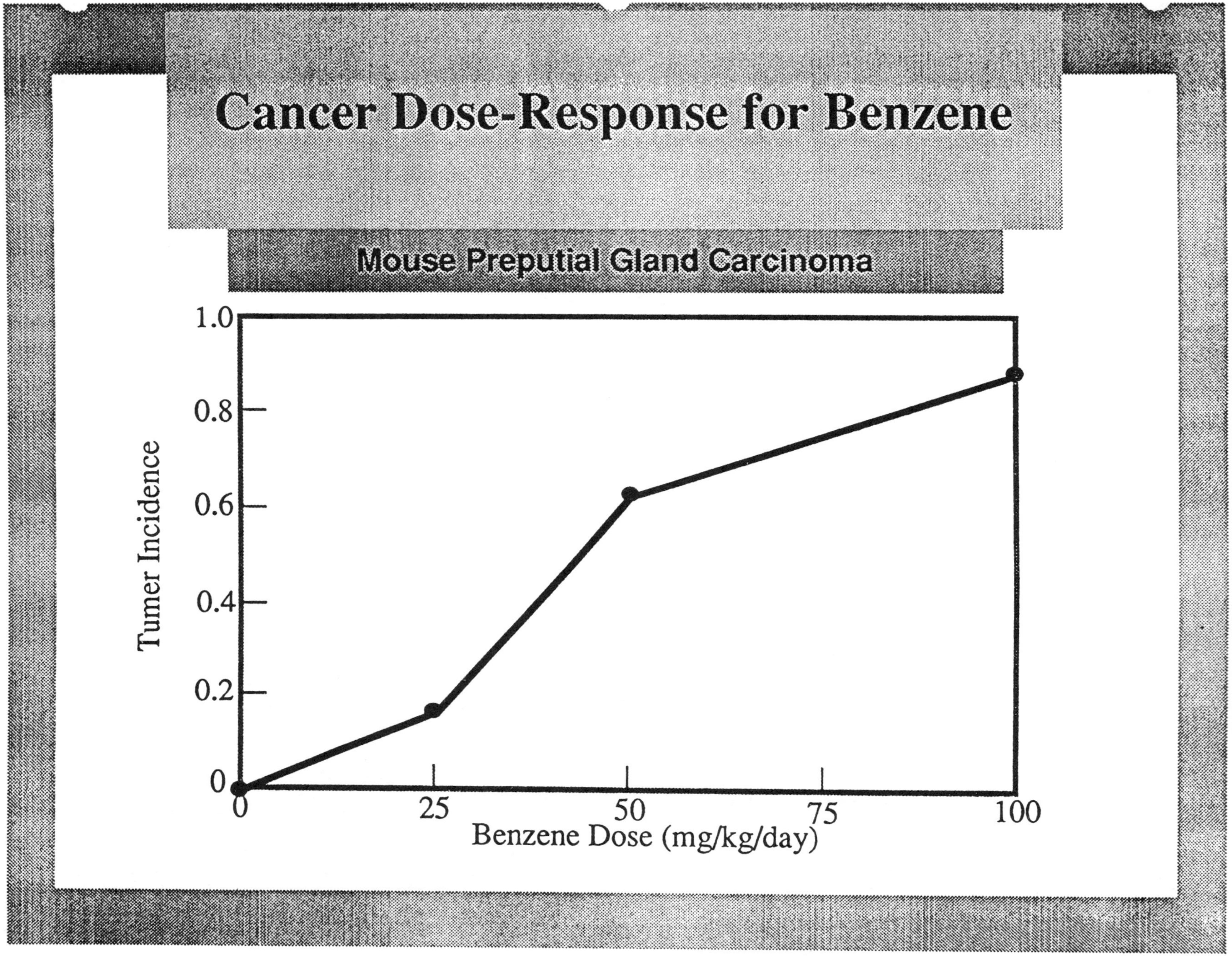

Cancer Dose-Response for Benzene
Mouse Preputial Gland Carcinoma
Tumer Incidence
1.0
0.8
0.6
0.4
0.2
0
0
25
50
75
100
Benzene Dose (mg/kg/day)

Selection of Data Set

- Human data preferred over animal data
- Absent good human data, use data from most representative species
- Absent data to contrary, use data from most sensitive species

Choice of Low-Dose Extrapolation Model

- A simple linear model is generally used with human data
- The default is the linearized multistage model
- EPA may use other models if clearly supported by specific data

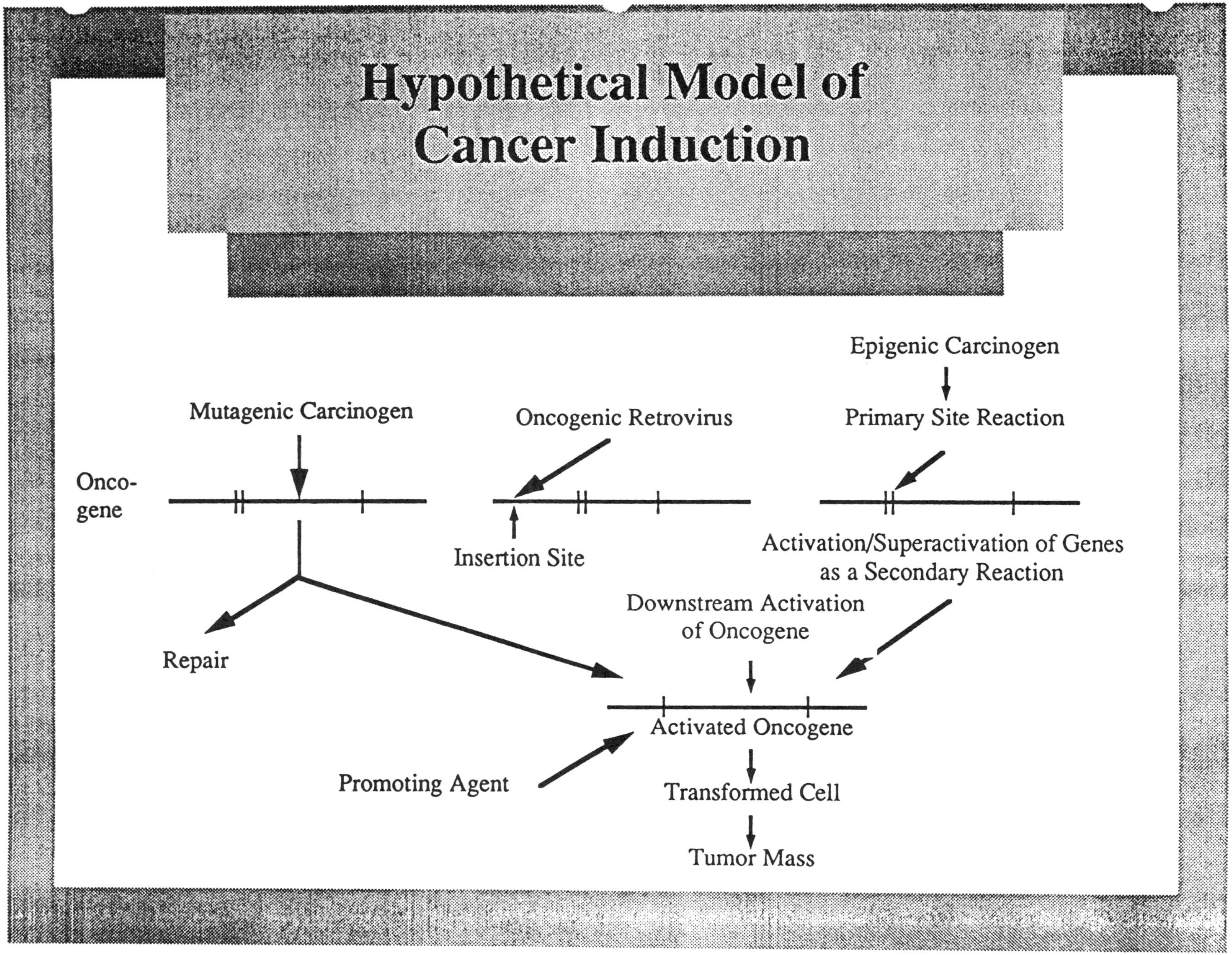
Hypothetical Model of
Cancer Induction
Onco-
gene
Mutagenic Carcinogen
Repair
Oncogenic Retrovirus
Insertion Site
Epigenic Carcinogen
Primary Site Reaction
Activation/Superactivation of Genes
as a Secondary Reaction
Downstream Activation
of Oncogene
Activated Oncogene
Promoting Agent
Transformed Cell
Tumor Mass

Toxic Effects Exhibiting Thresholds

Systemic Effects

- Determine no-observed effect level
  - NOEL is maximum dose producing no significant effect in exposed organisms
  - experimental doses should span both sides of the NOEL and LOEL
- Apply safety factors to determine acceptable daily intake (ADI)

Step 3: Exposure Assessment

- What dose(s) do or could people receive?
- How frequently?
- Through what environmental media?

Exposure Assessment

What is Included?

- Source(s)
- Pathway(s)
- Receptor(s)
- Dose(s)

Definition of Receptors

Maximum Exposed Individual:

- Lives at peak off-site receptor location
- Has a backyard vegetable garden
- Has a dairy cow for personal milk production
- Consumes fish on a regular basis
- Breathes air for entire life of project at peak receptor

Average Exposed Individual:

- Breathes air at an average exposure location
- Food chain exposure based on average in the area
- More realistic exposure scenario

Sources of Contaminant Emissions

- Hazardous Waste Landfill
- Construction and Excavation
- Accidental Spill
- Stack and Automobile Emissions
- Toxic Emissions
- Wind-Blown Dust
- Burning Wood

Common Sources of Contamination

- Agricultural Runoff
- Pesticide Runoff
- Accidental Spills
- Surface Spills
- Septic Tanks
- Waste Pits

Concentration at Exposure

Example 1: Concentration at a monitoring well

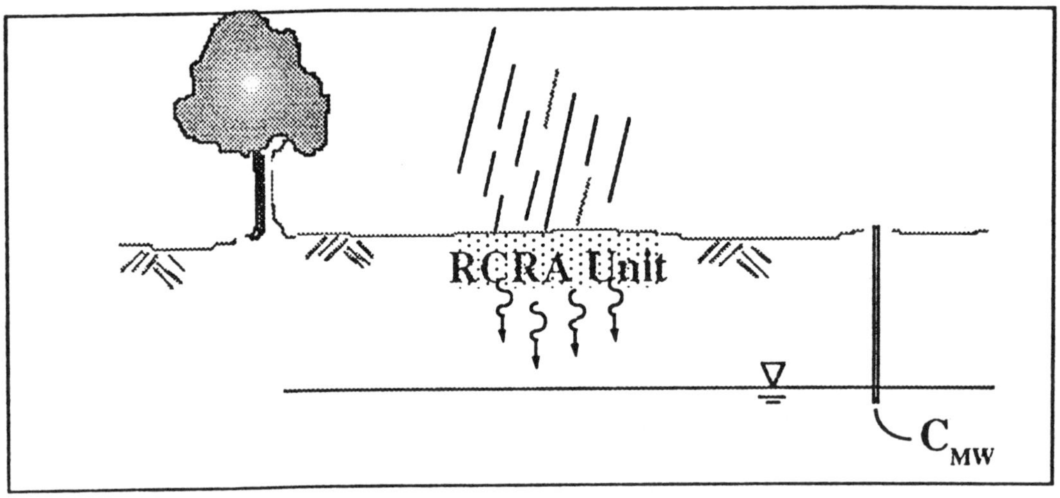

Example 2: Concentration at a pumped well

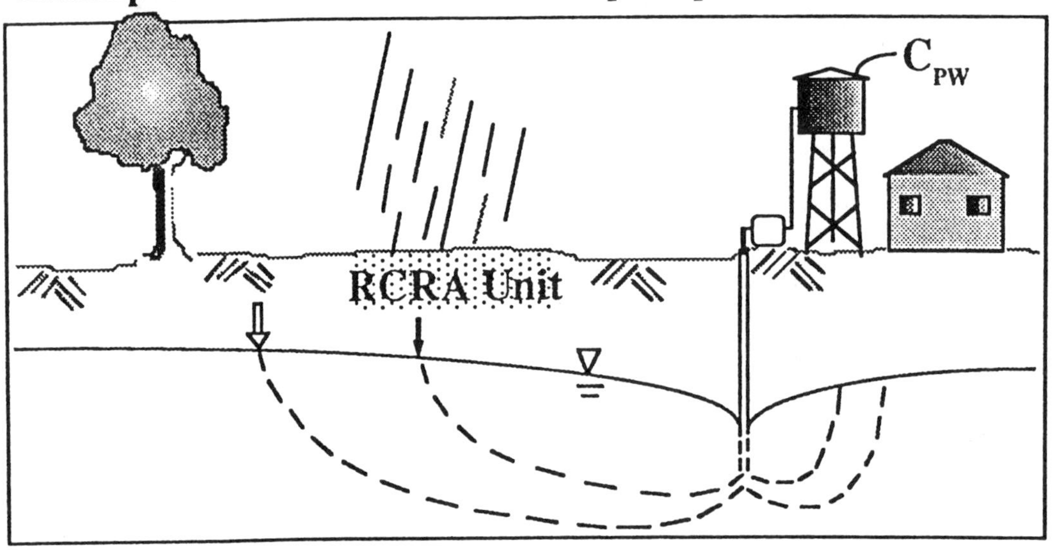

- Because of dilution $C_{PW} < C_{MW}$

Environmental Pathways and Routes of Exposure

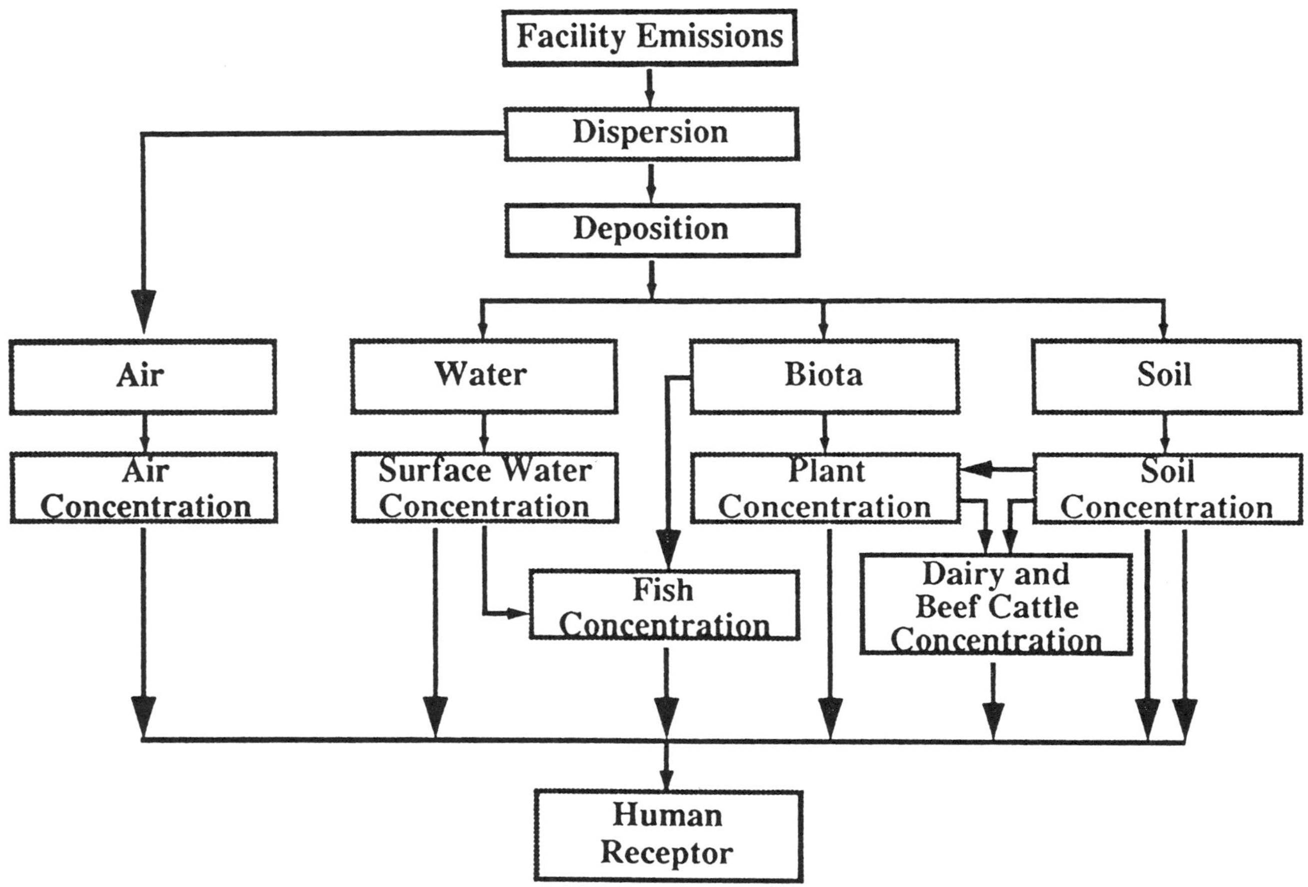

Exposure Pathways Anticipated For Evaluation of Chronic Intake

- Inhalation of Pollutant Emissions
- Ingestion of Outdoor Soil by Children
- Ingestion of Indoor Dust by Children
- Ingestion of Drinking Water
- Consumption of Vegetables from Typical Home Garden
- Consumption of Beef Products from Neighboring Farm
- Ingestion of Milk and Dairy Products from Neighboring Farm
- Consumption of Fish From Recreational Farm Pond

Environmental Fate and Transport

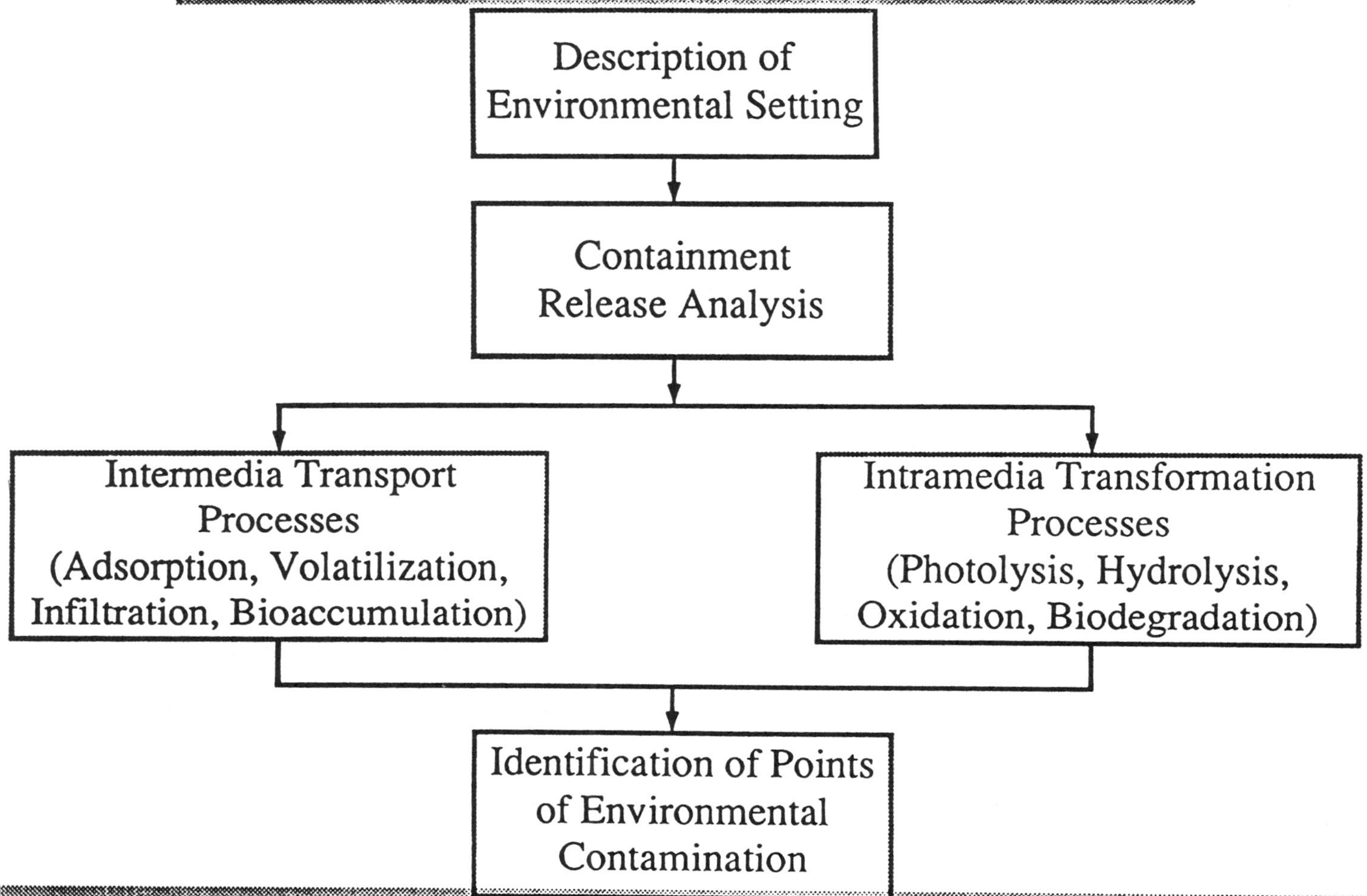

Dose-Response

- *All toxic effects other than cancer* are assumed not to appear until a *threshold dose* is exceeded.
  - "No-observes-effect level" (NOEL) is the maximum dose (in the most sensitive species) at which no excess toxicity (relative to controls) is observed.
  - The approximate toxicity threshold for the human population is set at some fraction of the experimental NOEL.
  - NOEL/Safety Factor = approximate human threshold for toxicity.

Dose-Response Relationships

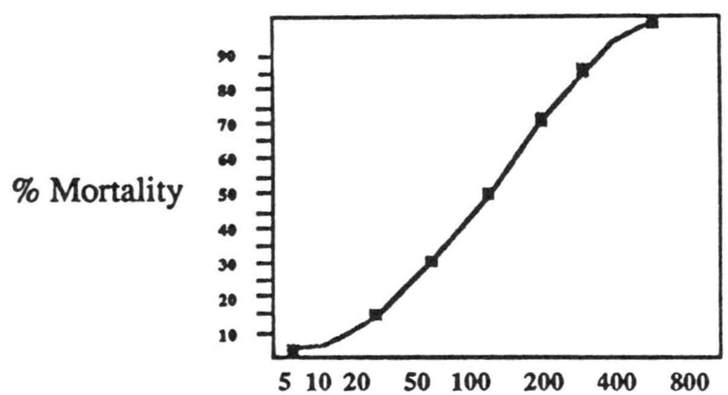

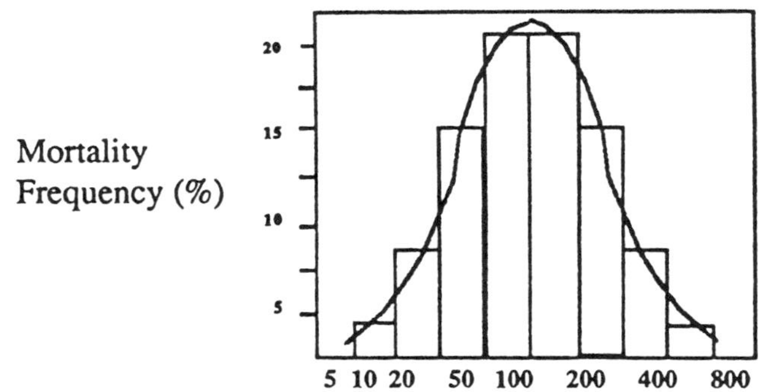

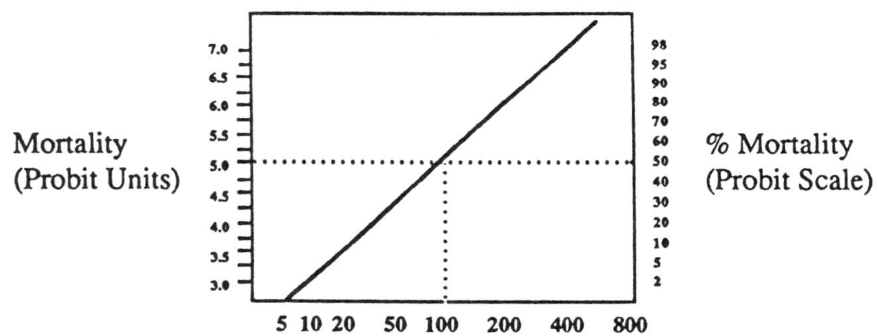

Diagram of Dose-Response Relationship

Dosage is most often expressed as mg/kg and plotted on a Log Scale.

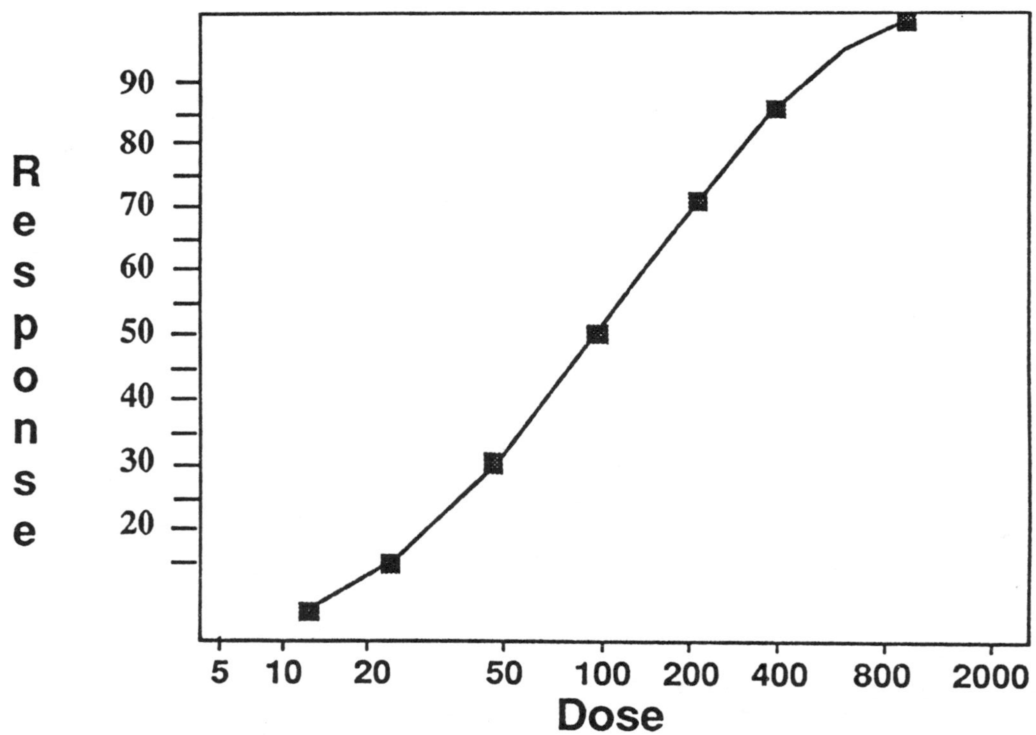

Dose-Response Curves for Chemicals Administered to a Uniform Population

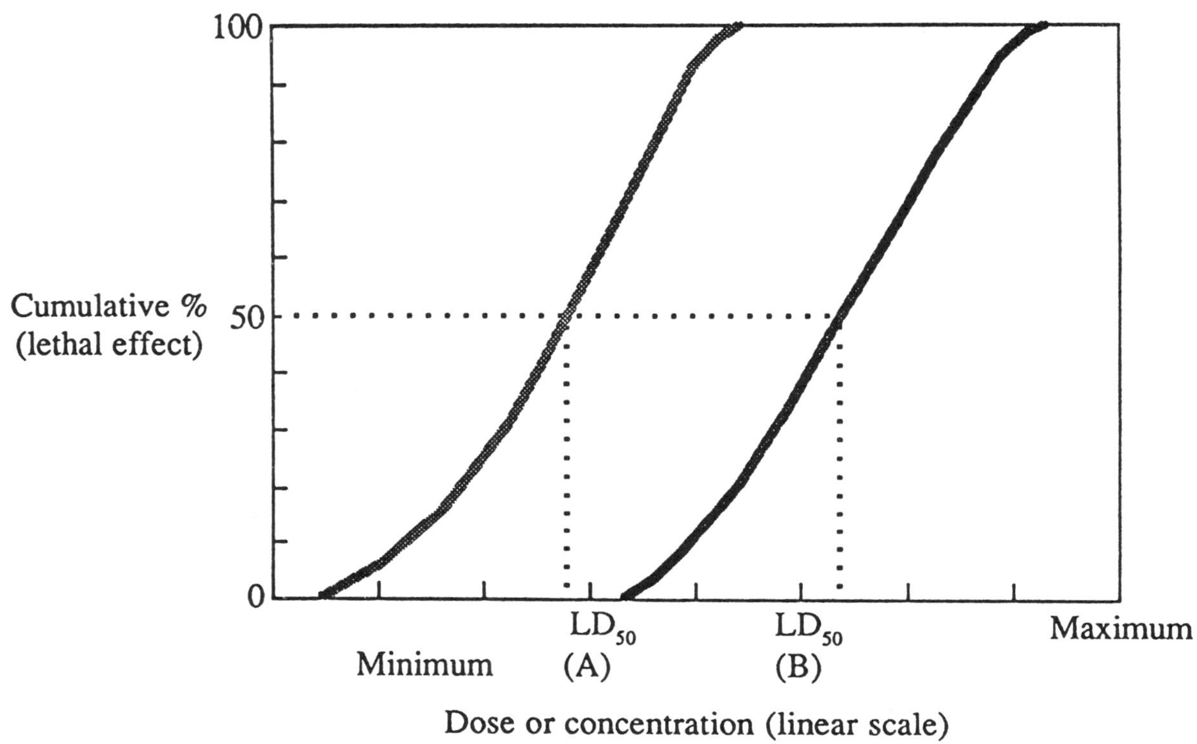

Dose-Response Curves for Chemicals Administered to a Uniform Population

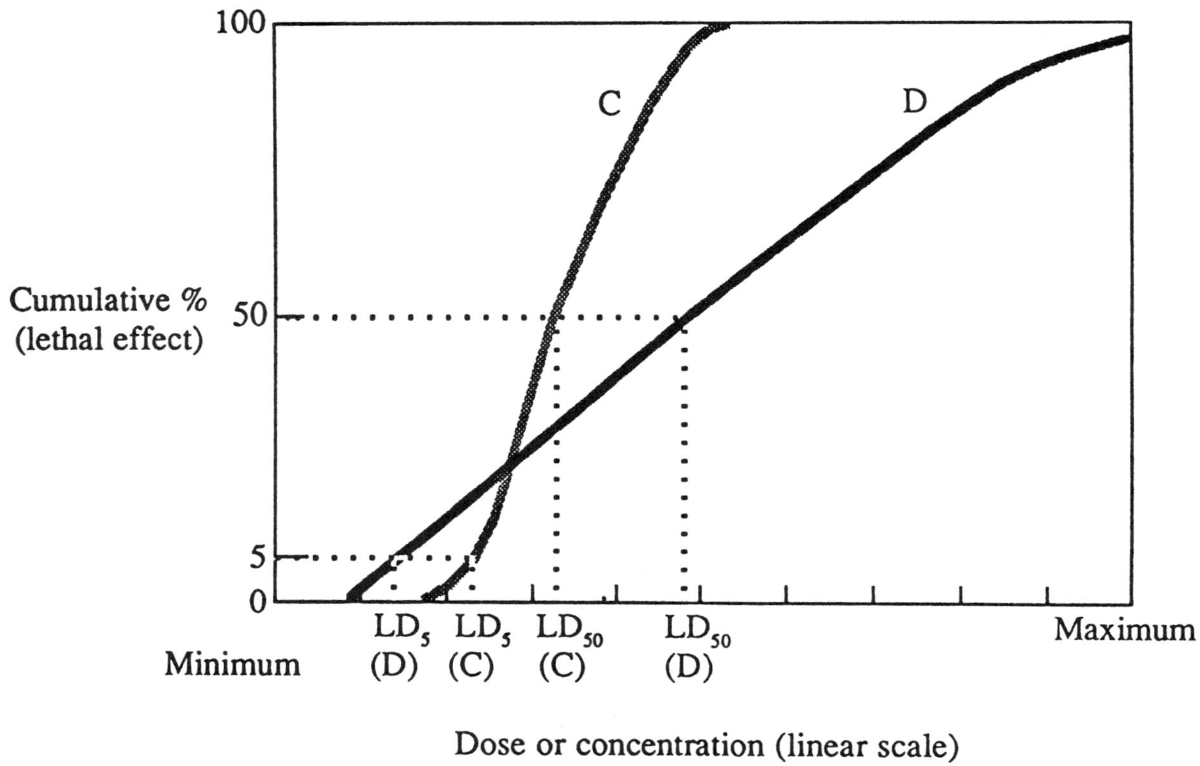

Estimation of a Virtually Safe Dose by Low Dose Extrapolation

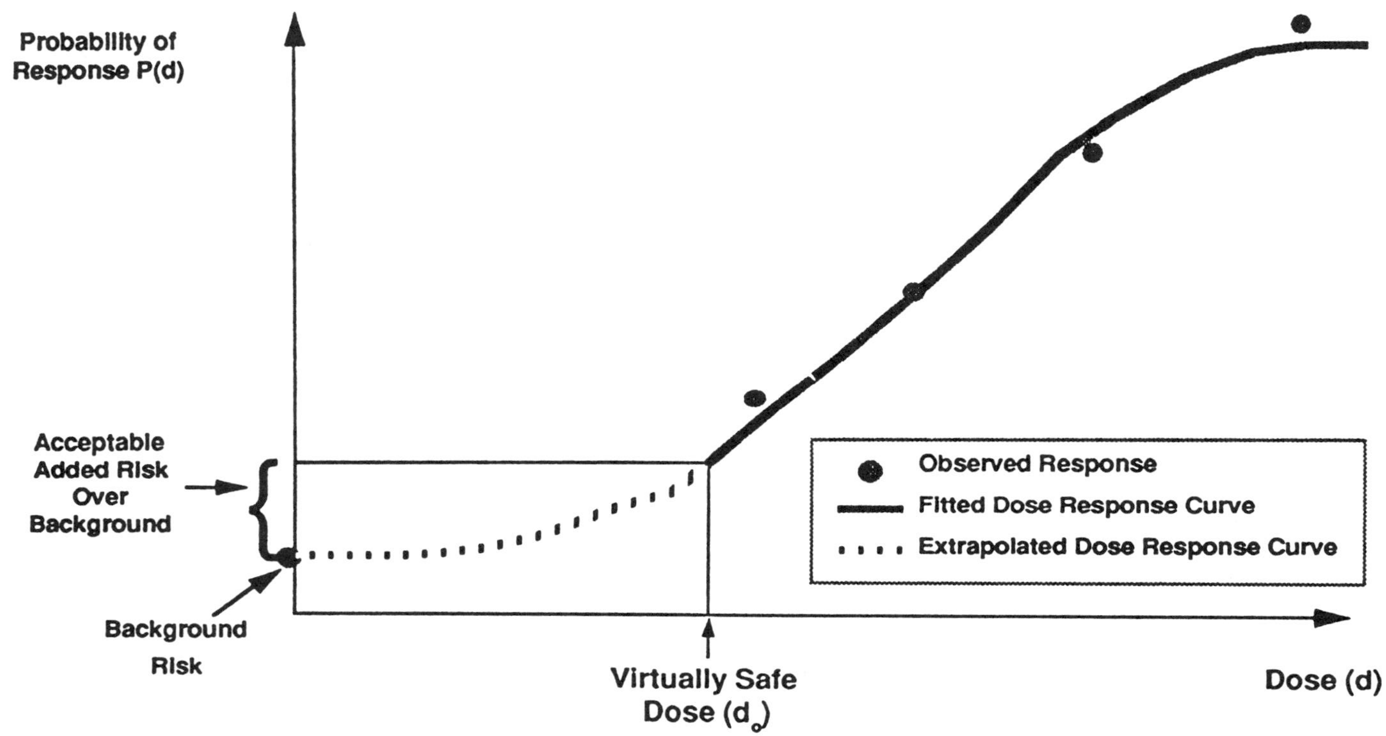

Slope Factors for Selected Chemicals

Chemical	Oral Route		Inhalation Route	
	Class	SF $(mg/kg/d)^{-2}$	Class	SF $(\mu g/m^3/d)^{-1}$
Benzene	A	2.9×10^{-23}	A	8.3×10^{-6}
Benzo(a)pyrene	B2	11.5	B2	1.7×10^{-3}
Chloroform	B2	6.1×10^{-3}	B2	2.3×10^{-5}
Polychlorinated biphenyls	B2	7.7	B2	ND
Trichloroethylene	B2	1.1×10^{-3}	B2	1.7×10^{-6}
Vinyl chloride	A	1.9	A	8.4×10^{-5}

Site Characteristics That May Influence Evironmental Fate

- Ambient Moisture
- Ambient Humidity Levels
- Ambient Temperatures
- Ambient Wind Velocity and Direction
- Geologic Characteristics

Site Characteristics That May Influence Evironmental Fate (continued)

- Hydrologic Characteristics
- Soil Characteristics
- Topographic Features of Site and Surrounding Area
- Vegetative Cover of Site and Surrounding Area
- Watershed/Land Use Characteristics

A Conceptual Diagram of Pathways of Containment Transport Throughout the Aquatic/Wetlands Ecosystem

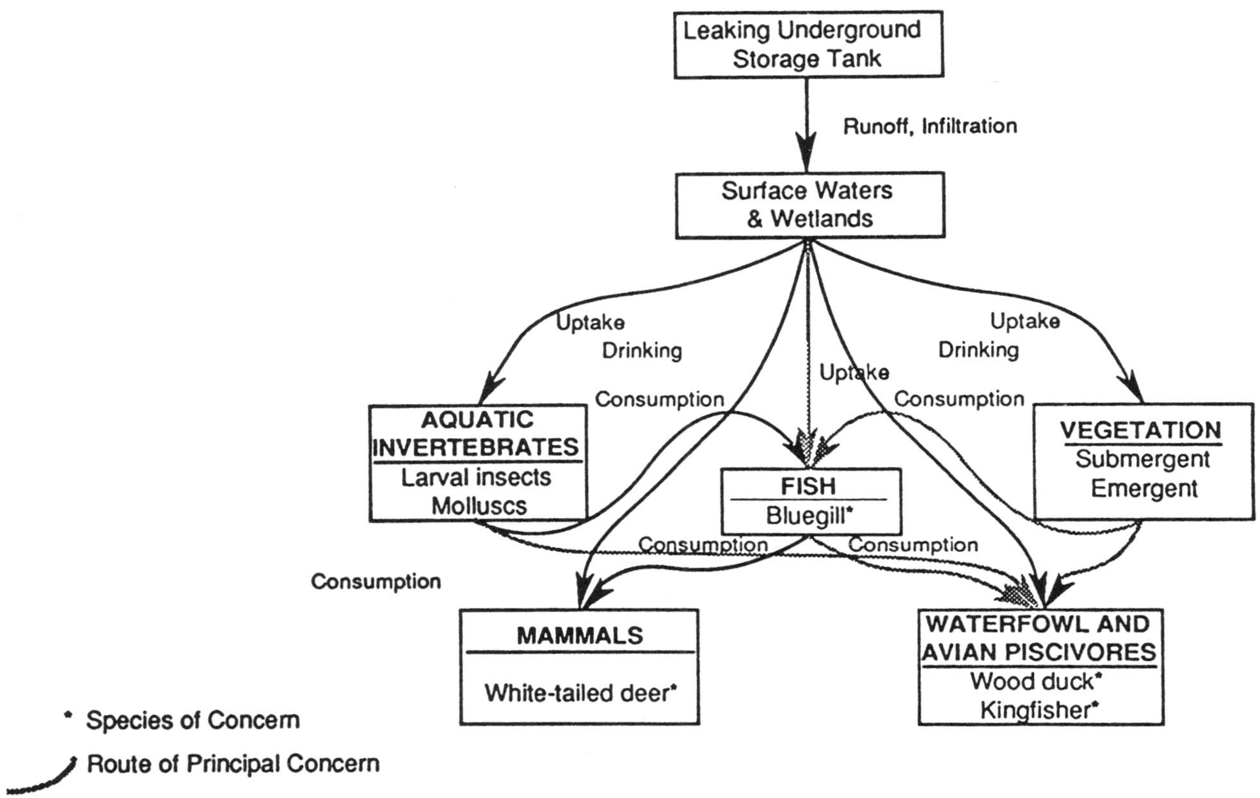

A Conceptual Diagram of Pathways of Containment Transport Throughout the Terrestrial Ecosystem

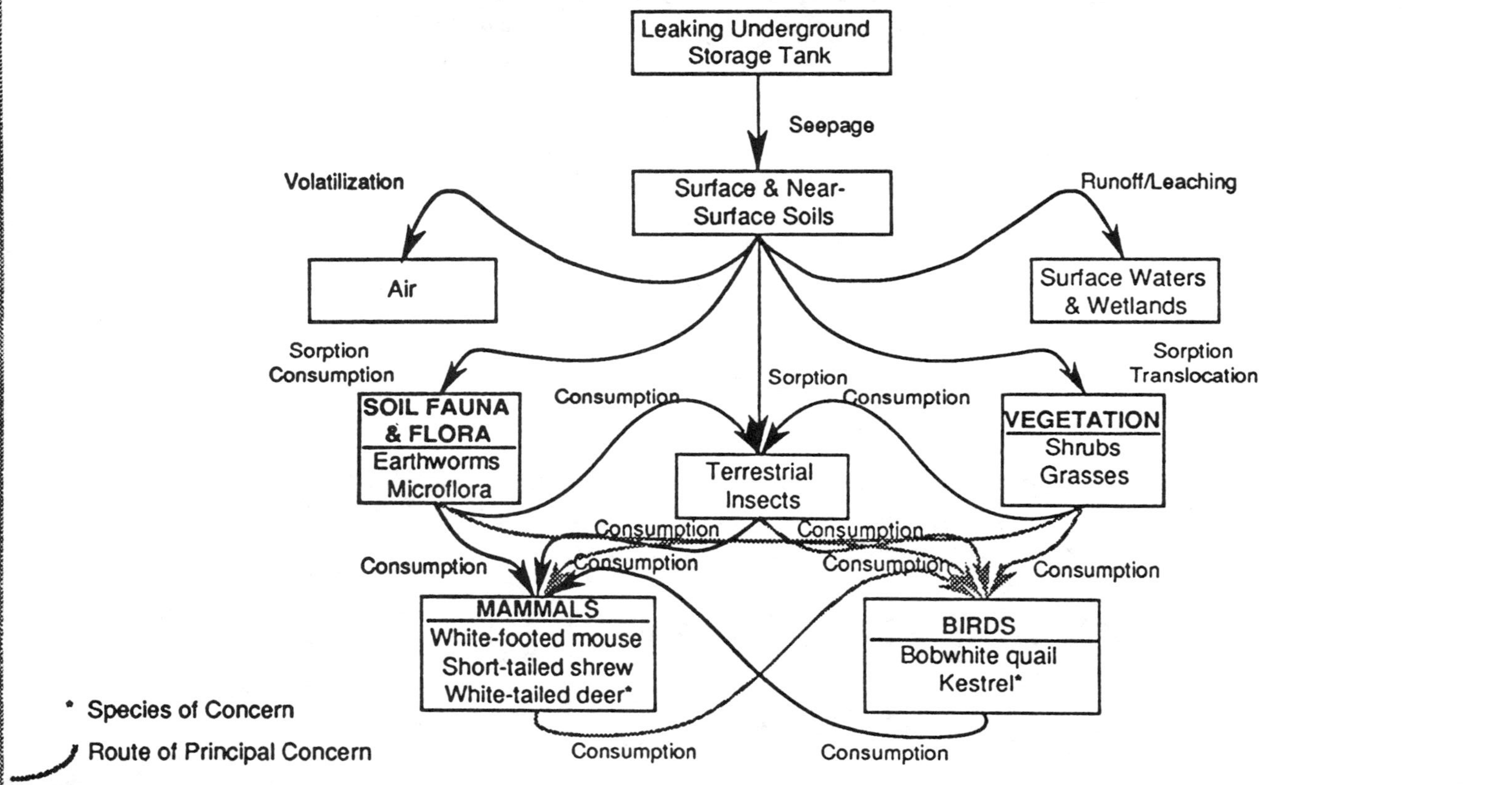

Exposure Results from Contact with a Medium Carrying the Chemical

The Media Provide the Pathways of Exposure

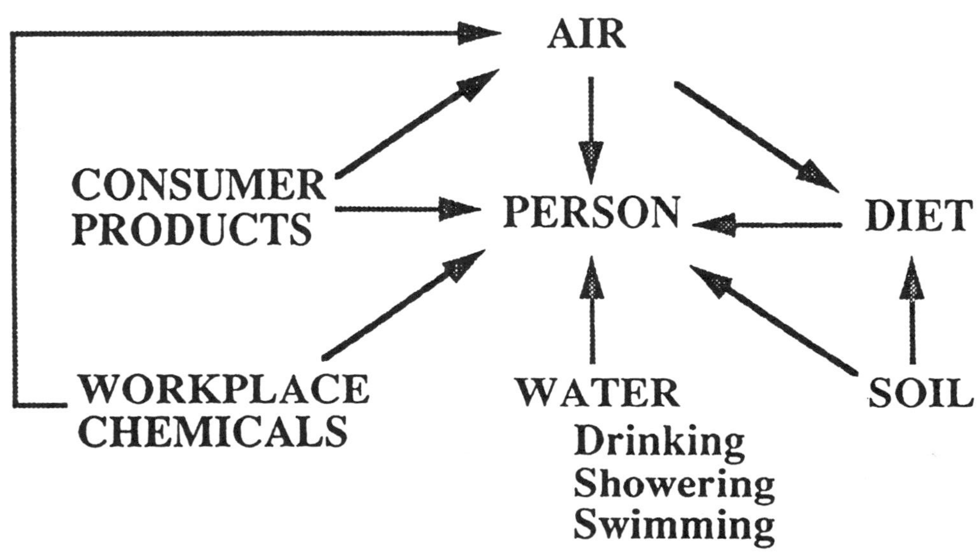

- Chemicals reach these media by intentional or inadvertant addition

The Total Amount of Chemical that Enters the Body Per Unit of Time Depends upon the ...

- Concentration of the chemical in the medium
- Amount of medium contacting or entering the body per unit of time
- Absorption of the chemical through the lungs, skin, gastrointestinal tract
- The number of different media to which a person is exposed

Routes of Exposure

Inhalation:
Air: $20 m^3$ / day

Ingestion:
Food: 2000 g / day
Water: 1.4 ℓ / day

Skin Contact:
Body Surface Area
Males: 19,400 cm^2
Females: 16,900 cm^2

Exposure Assessment

Matching Exposure and Dose

- Lifetime Average Daily Dose
- Maximum Daily Dose
- Concentration in:
  - air
  - food
  - water
  - soil

Risk Characterization

Carcinogens

- Quantitative expression:
  - Upper Bound[a] on lifetime risk of cancer

$$\frac{\textbf{Risk}}{\textbf{Unit Dose}} \times \textbf{Dose} = \textbf{Risk}$$

[a] Actual risk not likely to exceed, may be lower, could be zero

Risk Characterization

Carcinogens (continued)

- Chloroform byproducts in chlorinated drinking water pose excess lifetime cancer risk of 1 per 100,000 to 1 per 10,000

- Assuming:
  - Chloroform is a human carcinogen, based on observations in animals.
  - Humans are as sensitive as the most sensitive animal species, strain, and sex
  - Risk is a linear function of dose
  - No-threahold in dose-response

Risk Characterization

Carcinogens (continued)

- Other Information
  - Weight-of-evidence of carcinogenicity
  - Plausibility of linear no-threshold model
  - Nature of exposure analysis (e.g., worst-case, typical?)
  - Sensitivity of results to alternative assumptions

Components of Uncertainty

- Shape of Curve at Low Doses
- Qualitative Effects at Low Doses
- Metabolic Pathways
- Toxicokinetics: Time Course of Chemical and Its Metabolites in Body
- Differences Among Sexes, Strains and Species
- Animal to Human Dose Equivalence
- Difference in Susceptibility Among Humans

EPA Approach to Calculating a Reference Dose (RfD)

$$\text{RfD} = \frac{\text{NOAEL (No-Observed-Adverse-Effect-Level)}}{\text{Uncertainty Factor(s)} \times \text{Modifying Factor}}$$

- Uncertainty Factors (10):
  - Sensitive Population
  - Animal Data to Humans
  - Subchronic to Chronic
  - LOAEL to NOAEL
  - Incomplete Data Base
- Modifying Factor (1 - 10):
  - Additional Scientific Uncertainty

Uncertainty

- Emissions
  - Permitted Versus Estimated vs. Measured
- Exposure
  - Standardized Assumptions
  - Usually Conservative
- Toxicological
  - Cancer Potency Values: high to low dose extrapolation
  - Extrapolation of animal data to humans

Risk Management

- Level of Risk Reduction Sought Depends upon Law
  - Risk only?
  - Balancing?
  - Technical feasibility?

Risk Management (continued)

- General Population Precedents — Carcinogens
  - 1 per 1,000,000 and lower generally considered neglible
  - 1 per 10,000 down to one per 1,000,000 found tolerable in many cases
  - 1 per 10,000 and above generally found unacceptable
  - zero risk (no addition) is applied to direct food additives (Delaney)

- Occupational Population Precedents
  - OSHA has generally not pushed risks from occupational carcinogens below about 1 per 1,000 (technical feasibility)
  - OSHA Benzene ruling:
    Language regarding level of risk that a "reasonable person" might take steps to reduce

Science vs. Policy in Risk Assessment

- Risk assessments depend upon data, models, and assumptions.
- For many steps in risk assessment, several sets of data, or models, or assumptions may be available.
- Unless a specific set of data can be selected on scientific grounds, a policy choice will be made.
- Regulatory policy favors sets of data, models, and assumptions that yield the highest estimate of human risk.

Risk Communication

- Source
- Media
- Receiver

Risk Communication

- People's tolerance for different hazards is not simply (or even largely) a function of the expert's judgment regarding their probability of occurring!
- People generally consider many other characteristics:

Tolerated Risks	Non-Tolerated Risks
Voluntarily assumed	Imposed by others
Personal benefit high	No perceived personal benefit
Scientists agree	Scientists disagree
Not catastrophic	Catastrophic
Natural in origin	Man-made
Hazard not fearsome	Highly dreaded hazard
Common event	Rare event

- This and other aspects of risk perception need to be considered in any risk communication effort

Seven Cardinal Rules of Risk Communication

- Accept & Involve the Public as a Legitimate Partner
- Plan Carefully & Evaluate Your Efforts
- Listen to Your Audience
- Be Honest, Frank and Open
- Coordinate & Collaborate with Other Credible Sources
- Meet the Needs of the Media
- Speak Clearly and with Compassion

(Center for Risk Communication, Columbia University)

Alternative Types of Comparison

- Comparisons of Similar Risks
- Comparisons of Risks with Benefit
- Comparisons of Alternatives
- Comparisons of the Same Risk at Different Times
- Comparisons with Background Level
- Comparisons with a Regulatory Standard

(Center for Risk Communication, Columbia University)

Comparative Risks of Death

	Number of Deaths/Year	Lifetime Risks
• Motor vehicle accident	46,000	1/65
• Home accidents	25,000	1/130
• Lung cancer deaths in smokers	80,000	1/12

Annual Fatality Rates per 100,000 Persons at Risk

Activity	Rate
Motorcycling	2000
Living (all ages)	1000
Aerial acrobatics (planes)	500
Smoking (all causes)	300
Smoking (cancer)	120
Fire fighting	80
Coal mining	63
Farming	36
Driving (motor vehicles)	24
Police work (nonclerical)	22
Rodeo performer	3
Fires	2.8
Floods	0.06
Lightning	0.05
Meteorite	0.000006

Risk Estimated to Increase Chance of Death in any Year by 0.000001[a]

Activity	Cause of Death
Smoking 1.4 cigarettes	Cancer, heart disease
Spending 1hr. in a coal mine	Black lung disease
Living 2 days in New York or Boston	Air pollution
Living 2 months in Denver on vacation from New York	Cancer caused by cosmic radiation
One chest X-ray taken in a good hospital	Cancer caused by radiation
Eating 40 tbs. of peanut butter	Liver cancer caused by Aflatoxin B
Drinking 30 12-oz. cans of diet soda	Cancer caused by saccharin
Living 150 yrs. within 20 miles of a nuclear power plant	Cancer caused by radiation

Calculated Lifetime Risks of Cancer to an Individual Exposed Over 70 Years

Chances Per Million	
5,000	Average Household Exposure to Radon
4,640	Pesticide Residue on Fresh Foods
720	Consumption of 1/2-Pound of Fish Per Week from Lake Ontario
250	Peanut Butter Consumption by Children, Due to Potential Aflatoxin
180	Acceptable Benzene Levels
20	Average Diagnostic Medical X-Ray
4	Frequent Flyer Radiation Exposure 20,000 Miles/Year
2.4	Drinking and Showering with Chlorinated Water

Putting Risk in Perspective

Risk of Death	Occupation	Lifestyle	Accidents	Environmental Risks
1 in 100 10^{-2}	Stuntman	Smoking (1 pack a day)	Skydiving, Rock Climbing	
1 in 1,000 10^{-3}	Race Car Driver, Fireman, Miner, Farmer,Policeman	Heavy Drinking	Canoeing, Driving, All Home Accidents, Frequent Air Travel	
1 in 10,000 10^{-4}	Truck Driver, Banker, Engineer		Skiing	
1 in 100,000 10^{-5}	Insurance Agent	Using Contraceptive Pills, Light Drinking, Diagnostic X-rays	Home Fires, Fishing	Drinking Water Substances, Natural Background Radiation
1 in 1,000,000 10^{-6}		Smallpox Vaccination (per occassion)	Poisoning, Occassional Air Travel (one flight a year)	
1 in 10,000,000 10^{-7}		Eating Charcoal-Broiled Steak (once a week)		Living at the Boundary of a Nuclear Power Plant, Hurricane, Tornado, Lightning, Animal Bite or Insect Sting

Public Perception

- Voluntariness
- Control
- Fairness
- Process
- Morality
- Familiarity
- Memorability
- Diffusion

Hazard Identification

Toxicity is usually characterized by the body site/system adversely affected or by the disease produced

Systems
Nervous
Immune
Reproductive
Endocrine
Cardiovascular

Organs
Lung
Liver
Kidney
Skin
Eye

Diseases
Cancer
(Carcinogens)
Birth Defects
(Teratogens)

Definitions

- **Risk:** The probability of occurrence of adverse health effects in humans.
- **Risk Assessment:** The process of characterizing the nature and probability of adverse health effects of human exposure to environmental hazards.
- **Risk Management:** The process of evaluating and selecting among alternative regulatory actions.

Why is Risk Assessment Necessary?

- It is generally not possible to obtain a direct measure of risk to people under actual conditions of exposure.
- Risks need to be understood in advance of exposure.

Uses of Risk Assessment

- To identify the risk associated with various levels of exposure
- To identify limits on exposure so that a specific level of risk will not be exceeded

Specific Uses of Risk Assessment

- Pre-marketing safety evaluation of new chemicals
  - pesticides, food additives, drugs
  - products of biotechnology
- Post-marketing safety evaluation of "old" chemicals
- Evaluation of health impacts of emissions from industrial facilities and other sources
- Evaluation of health impacts of chemicals migrating from hazardous waste sites
- Selection among several remedial alternatives at hazardous waste sites
- Determination of workplace health risks
- Evaluation of potential risks of new facilities
- Evaluation of product liability, toxic tort claims

History of Risk Assessment in Regulation of Chemicals

- Introduced by FDA in 1973 to address carcinogens
- Low-dose risk quantified; insignificant risk dose selected ("safe" dose)
- Gradually expanding uses in regulation from 1975 to present
- National Academy of Science establishes framework in 1983

Driving Forces for Risk Assessment

Early 1970's

- More carcinogens
- More widespread in environment
- Methodology in literature
- Many new laws
- Requirement for explicitness

Steps in a Risk Assessment

National Academy of Science (NAS), 1983

- Step 1: Hazard Identification

  - What type of toxicity?
  - Relevance to humans?

Steps in a Risk Assessment (continued)

National Academy of Science (NAS), 1983

- Step 2: Dose-Response Evaluation
  - What is the quantitative relationship between dose and risk of toxicity in region of human exposure?

Steps in a Risk Assessment (continued)

National Academy of Science (NAS), 1983

- Step 3: Human Exposure Evaluation
  - What are the relevant routes of exposure?
  - What dose do people receive?

Steps in a Risk Assessment (continued)

National Academy of Science (NAS), 1983

- Step 4: Risk Characterization
  - What risk of toxicity?
  - How precise is the risk estimate?

Sources of Toxicity Information

- Epiemiology studies in groups of individuals (typically those exposed at relatively high levels)
- Controlled experiments in laboratory animals
- "Test-tube" experiments

Hazard Identification

Toxicity is usually characterized by the body site/system adversely affected or by the disease produced

Systems
Nervous
Immune
Reproductive
Endocrine
Cardiovascular

Organs
Lung
Liver
Kidney
Skin
Eye

Diseases
Cancer
(Carcinogens)
Birth Defects
(Teratogens)

Hazard Identification

- Chemicals vary greatly in the types of toxicity they produce.
- No single measure of toxicity is adequate to characterize a chemical's hazardous properties.
- All chemicals, natural and synthetic, produce toxicity under some conditions of exposure.

Hazard Identification

Weight-of-Evidence

- Toxicity data base varies greatly among chemicals.
- Strength of human and animal evidence also varies greatly.

Goal:

How certain are we that chemical X can produce certain forms of toxicity in humans ?

Regulatory Context:

Threshold for certainty fairly low... Any convincing animal data...

Personal Injury Context:

Legal vs. scientific "burden of proof". Threshold for certainty may be higher ?

Hazard Identification

Toxic effects are ...

Exposures
Acute
Subchronic
Chronic

Severity
Mild, Reversible
Seriuos, Irrversible
Life-Threatening

Responses
Immediate
Delayed

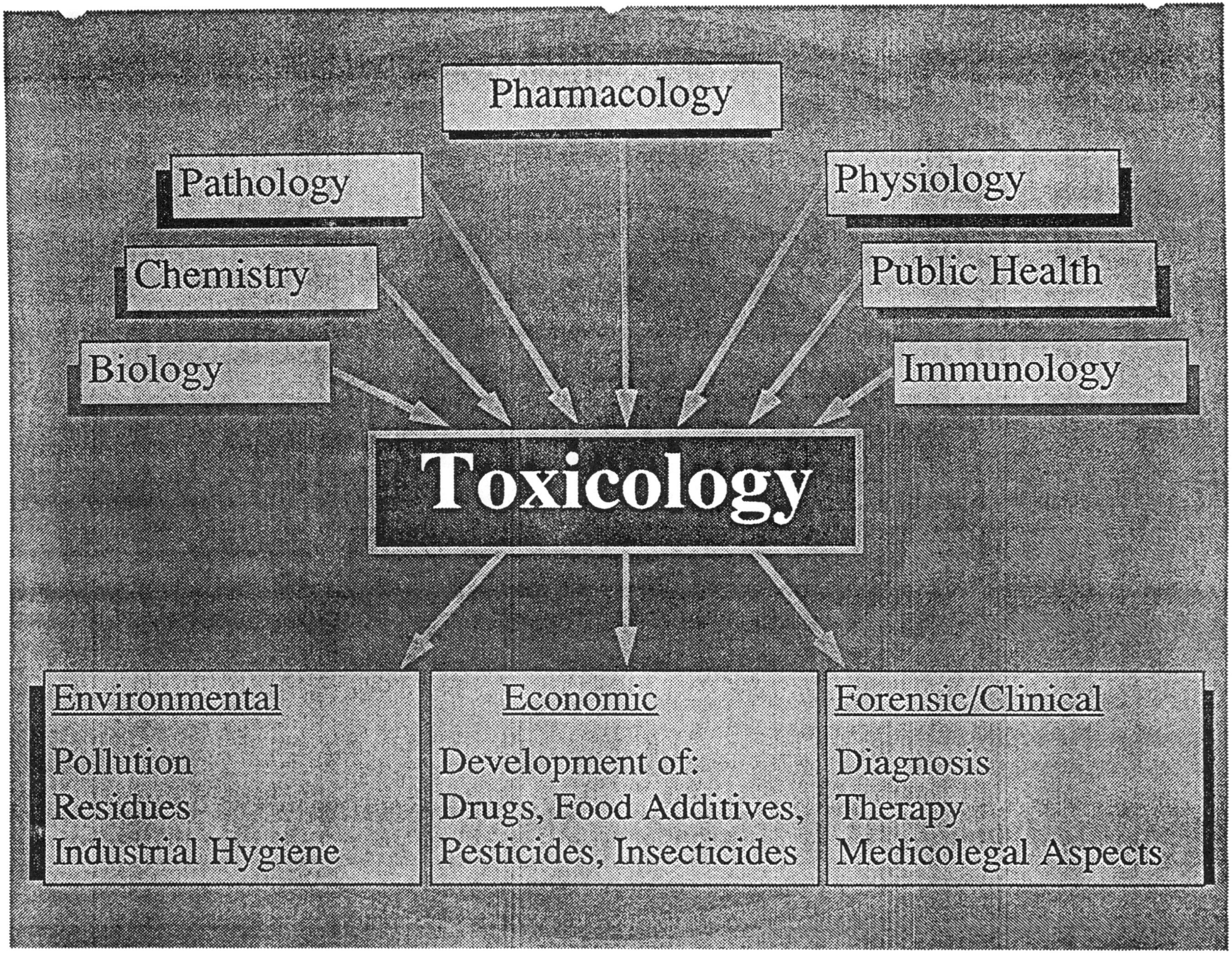
Pharmacology
Pathology
Chemistry
Biology
Physiology
Public Health
Immunology
Toxicology
Environmental
Pollution
Residues
Industrial Hygiene
Economic
Development of:
Drugs, Food Additives,
Pesticides, Insecticides
Forensic/Clinical
Diagnosis
Therapy
Medicolegal Aspects

Toxicity Versus Risk

Risk:
The probability that a substance will produce harm under specified conditions.

Safety:
The probability that harm will not occur under specified conditions.

Risk = Inherent Toxicity × Exposure

Various Disciplines of Toxicology

- Environmental:
  As applied, the study of chemicals in the environment and their effects on man, domestic animals, and wildlife.

- Forensic:
  A hybrid of analytical chemistry and fundamental toxicologic principles, primarily concerned with the medicolegal aspects of the harmful effects of chemicals on man and animals.

- Clinical:
  An area concerned with disease caused by, or uniquely associated with, toxic substances. Efforts are directed at treating patients poisoned with drugs or other chemicals, and at development of new techniques to treat these intoxications.

Toxicity Rating Chart

Toxicity Rating or Class	Probable Oral Lethal Dose For Humans	
	Dose	For Average Adult
1. Practically nontoxic	> 15 g/kg	More than 1 quart
2. Slightly toxic	5 - 15 g/kg	Between pint and quart
3. Moderately toxic	0.5 - 5 g/kg	Between ounces and pint
4. Very toxic	50 - 500 mg/kg	Between teaspoonful and ounces
5. Extremely toxic	5 - 50 mg/kg	Between 7 drops and teaspoonful
6. Super toxic	< 5 mg/kg	A taste (less than 7 drops)

Spectrum of Undesired Toxic Effects

- Side effects or undesirable effects

- Adverse, deleterious, or toxic effects
  - immediate vs. delayed
  - reversible vs. irreversible
  - local vs. systemic
  - idiosyncratic: abnormal reactivity that is genetically determined, but qualitatively similar to case-determined effects
  - allergic or sensitization reactions (not hypersensitization)

Epidemiologic Approach

- **Epidemiology**
  - The study of how disease rates, incidence rates, and death rates vary among different populations and within the same population over time
- **Advantages**
  - Based on observations in humans
- **Disadvantages**
  - Generally can not be performed until after human exposure has occurred
  - Identifying exposed and control groups is difficult
  - Confounding effects difficult to control

Epidemiologic Methods

- **Descriptive Epidemiology**
  Identifying patterns of disease in populations and correlating disease patterns with patterns of environmental exposures

- **Analytical Epidemiology**
  - **Case Control Studies**
    Identify the populations with and without a disease and compare past exposures
  - **Cohort Studies**
    Identify a population known to be exposed to a suspect agent and compare health status over time with a unexposed control group

Causalty - Weighing the Epidemiology Evidence

Factors Influencing a Determination That Exposure to a Chemical Is a Causative Factor

- Same type of association in more than one study, among different populations
- Exposure precedes outcome
- Biological plausibility
- Absence of confounders
- Increasing risk with increasing exposure
- Strength of the observed association (magnitudes of RRs, SMRs, or ORs), and statistical significance

Main Principles of Descriptive Animal Toxicity Tests

- Effects produced by a compound in laboratory animals, when properly qualified, are applicable to man.
- Exposure of experimental animals to toxic agents in high doses is a necessary and valid method of discovering possible hazards in man.

Types of Toxicity Tests

- Acute
- Subchronic
- Chronic
- Others:
  Tetratogenicity, Reproductive, Mutagenicity, Chemical Interactions, Behaviorial Effects

Acute Toxicity Tests

Single Dose

1. Lethality (i.e., LD_{50})
   - Two species (ideally one rodent and one non-rodent)
   - Two routes of administration

2. Topical effects (skin and eye)

Acute Animal LD_{50}s of Selected Chemical Agents

Will kill approximately 50% of the population

Agent	LD_{50} (mg/kg)
Ethyl alcohol	10,000
Sodium chloride	4,000
Ferrous sulfate	1,500
Morphine sulfate	900
Phenobarbital sodium	150
DDT	100
Picrotoxin	5
Strychnine sulfate	2
Nicotine	1
d- Tubocurarine	0.5
Hemicholinium-3	0.2
Tetrodotoxin	0.10
Dioxin (TCDD)	0.001
Botulinus toxin	0.00001

Subchronic Toxicity Tests

Daily Doses

1. Duration (three months)
2. Two species
   (usually 100 animals per dose level per sex, plus a control set)
3. Three dose levels (to generate dose-response curve)
4. Routes of expected exposure
5. Evaluate health weekly
6. Complete necropsy

Chronic Toxicity Tests

Daily Doses

1. Duration (2 - 7 years)
2. Species (similar to man)
3. Minimum 2 doses
4. Route of expected exposure
5. Evaluate health weekly
6. Complete necropsy

Special Toxicity Tests

- Teratogenicity (effects on offspring)
- Reproduction
- Mutagenicity (genetic mutations)
- Chemical Interactions
- Behavioral Effects

Effect of Experimental Group Size on Minimum Incidence of a Toxic Effect

Considered Significant at a Probability Level of 0.025 when the Effect is Absent in Controls and when Equal Numbers of Experimental and Controls Animals are Used

Number of Animals in each Group	Minimum Percentage Incidence in Experimental Groups that can be Regarded as Significant
10	40.0
50	8.0
100	4.0

Toxicokinetics

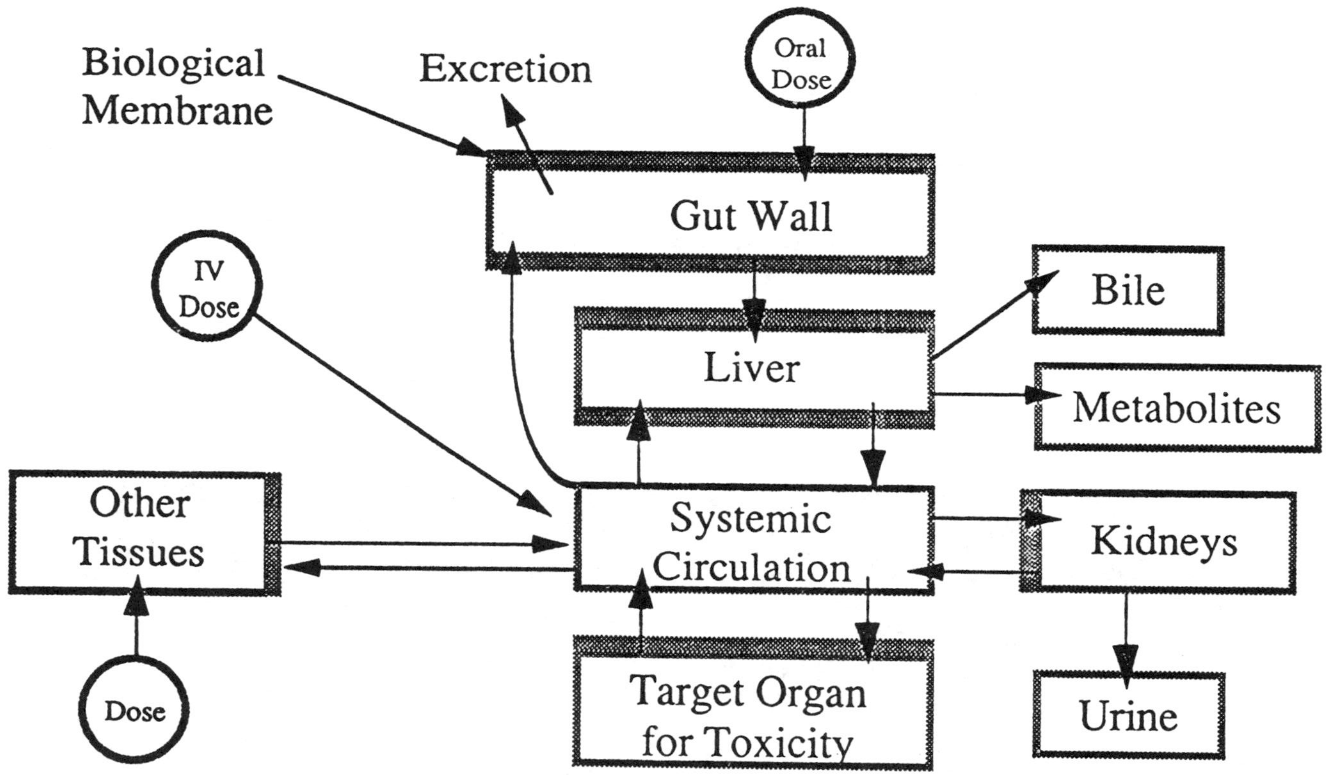

Pharmaco/Toxico-Kinetics

Rates of ADME

- Extrapolate between species
- High-low dose extrapolations
- Predict biological effects
- Extrapolate between routes

Chemical Disposition

- Absorption
- Distribution
- Metabolism
- Excretion

Absorption

- Gastrointestinal
  - Stomach
  - Small intestine
- Lungs
- Dermal
- Injected

Distribution

- Dependent on:
  blood flow, mass of tissue,
  affinity of tissue for chemical

- Storage depots
  - plasma proteins
  - fat
  - bone

- Barriers
  - blood/brain
  - placental
  - testicular

- Redistribution

Metabolism

- Converts compound to less lipid soluble
- Usually, but not always less toxic
- Always easier to excrete
- Phase I: Oxidation
- Phase II: Synthesis
- Primarily liver
- May be at site of action — active metabolite

Excretion

- Urine
  - pH effect
- Bile
- Exhaled Air
- Milk
- Saliva, sweat, tears, etc.

Cross-Biological Membranes

- Simple Diffusion
  - Lipid solubility
  - pH
- Special Transport
  - Active transport
  - Phagocytosis

Antagonistic Chemical Interactions

- Functional
  e.g., caffeine, alcohol

- Chemical
  two chemicals combined form a third non-toxic chemicale.g., chelates, milk and tetracycline

- Dispositional
  enhanced metabolism or elimination

- Receptor
  direct antagonist

Species Differences in *In Vivo* Absorption

	(% Dose absorbed)					
	Rat	Rabbit	Guinea pig	Pig	Squirrel monkey	Human
Haloprogin	95.8	113.0		19.7		11.0
Acetylcysteine	3.5	2.0		6.0		2.4
Cortisone	24.7	30.3		4.1		3.4
Caffeine	53.1	69.2		32.4		47.6
Butter yellow	48.2	100.0	34.9	41.9		21.6
Testosterone	47.4	69.6		29.4		13.2
Chlorophenothane	46.3			43.4	1.5	10.4
Lindane	51.2			37.6	16.0	9.3
Parathion	97.5			14.5	30.3	9.7
Malathion	64.6			15.5	19.3	8.2

Data on Mechanisms of Toxic Action

Influence on Risk Assessment

- Mechanism concerns
  - the path the chemical follows in the body
  - the nature of interactions in cells that produce changes leading to toxic response
- Such data can give clues regarding:
  - extrapolation from animals to humans
  - extrapolation from high to low-dose
- Much is understood about some chemicals at the molecular level
- Not yet a routine component of regulatory risk assessment
- Strong support among scientists

Relationships between Effects of Toxic Chemical Concentrations and Time after Oral Ingestion

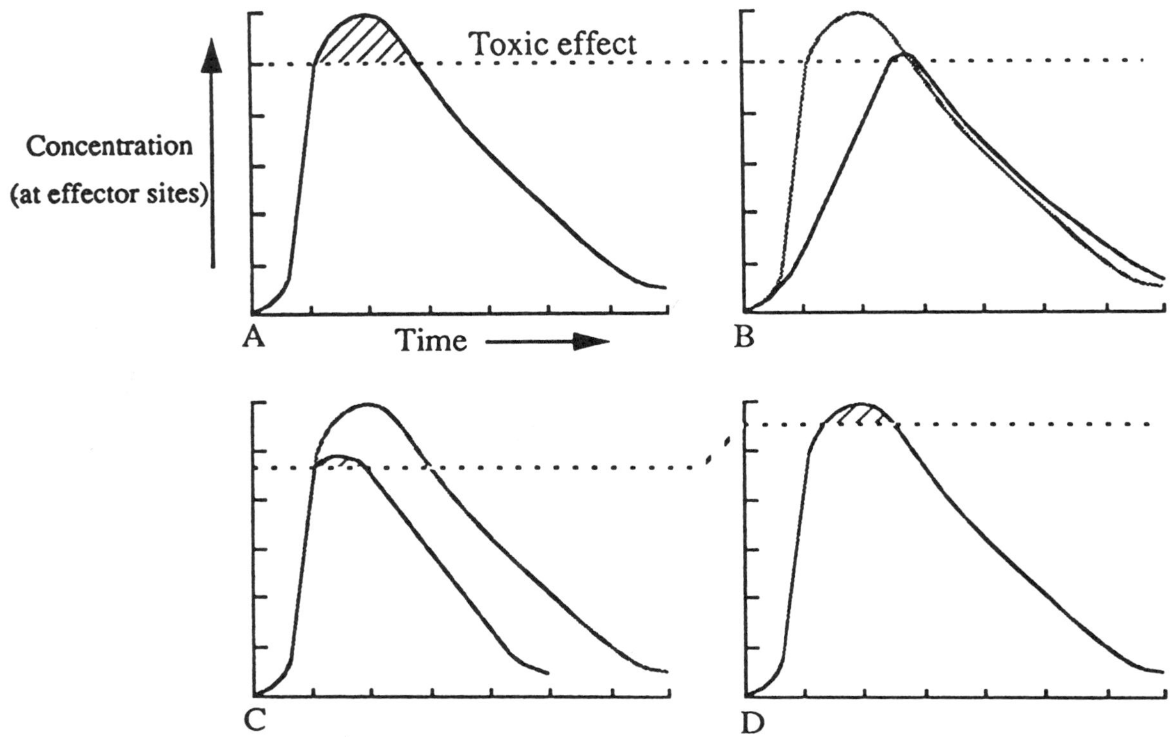

Chemical Interactions

Additive[a]	$2 + 3 = 5$
Synergistic	$2 + 3 = 20$
Potentiation	$0 + 2 = 10$
Antagonistic	$4 + 6 = 8$ $4 + 0 = 0$

[a] EPA assumes carcinogenic effects to be additive.